Lecture Notes in Computer Science 1041

Edited by G. Goos, J. Hartmanis and J. van Leeuwen

Advisory Board: W. Brauer   D. Gries   J. Stoer

Lecture Notes in Computer Science
Edited by G. Goos, J. Hartmanis and J. van Leeuwen

Advisory Board: W. Brauer D. Gries J. Stoer

Springer
Berlin
Heidelberg
New York
Barcelona
Budapest
Hong Kong
London
Milan
Paris
Santa Clara
Singapore
Tokyo

Jack Dongarra   Kaj Madsen
Jerzy Waśniewski (Eds.)

Applied Parallel Computing

Computations in Physics, Chemistry
and Engineering Science

Second International Workshop, PARA '95
Lyngby, Denmark, August 21-24, 1995
Proceedings

Springer

Series Editors

Gerhard Goos, Karlsruhe University, Germany

Juris Hartmanis, Cornell University, NY, USA

Jan van Leeuwen, Utrecht University, The Netherlands

Volume Editors

Jack Dongarra
University of Tennessee
107 Ayres Hall, Knoxville, Tennessee 37996-1301, USA

Kaj Madsen
Jerzy Waśniewski
Technical University of Denmark
DK 2800 Lyngby, Denmark

Cataloging-in-Publication data applied for

Die Deutsche Bibliothek - CIP-Einheitsaufnahme

Applied parallel computing : computations in physics,
chemistry and engineering science ; second international
workshop ; proceedings / PARA '95, Lyngby, Denmark, August
21 - 24, 1995. Jack Dongarra ... (ed.). - Berlin ; Heidelberg ;
New York ; Barcelona ; Budapest ; Hong Kong ; London ,
Milan ; Paris ; Santa Clara ; Singapore ; Tokyo : Springer, 1996
  (Lecture notes in computer science ; Vol. 1041)
  ISBN 3-540-60902-4
NE: Dongarra, Jack [Hrsg.]; PARA <2, 1995, Lyngby-Tårbæk>; GT

CR Subject Classification (1991): G.1-2, G.4, J.2, J.6, F1.2, D.1.3, I.6.8

ISBN 3-540-60902-4 Springer-Verlag Berlin Heidelberg New York

© Springer-Verlag Berlin Heidelberg 1996
Printed in Germany

Typesetting: Camera-ready by author
SPIN 10512562   06/3142 – 5 4 3 2 1 0   Printed on acid-free paper

Preface

The Second International Workshop on Applied Parallel Computing in Physics, Chemistry and Engineering Science (PARA95), and the Tutorial on ScaLAPACK and Parallel NAG Library on IBM SP2 and SGI Power Challenge were held in Lyngby, Denmark, August 21-24, 1995. The conference was organized and sponsored by the Danish Computing Centre for Research and Education (UNI•C), the Institute of Mathematical Modelling (IMM) of the Technical University of Denmark (DTU) and the Danish Natural Science Research Council through a grant for the EPOS project (Efficient Parallel Algorithms for Optimization and Simulation). Support was also received from the IBM and SGI computing organizations.

The purpose of the Workshop was to bring together scientists working with large computational problems in physics, chemistry, and engineering science, and specialists in the field of parallel methods and efficient exploitation of modern high-speed computers. Some classes of methods appear again and again in the numerical treatment of problems from different fields of science and engineering. The aim of this workshop was to select some of these numerical methods and plan further experiments on several types of parallel computers. The key lectures reviewed the most important numerical algorithms and scientific applications on parallel computers. The invited speakers were made up of physicists, chemists, and engineers, as well as numerical analysts and computer experts.

The workshop was preceded by a one-day tutorial on ScaLAPACK and the Parallel NAG Library. The ScaLAPACK (Linear Algebra Package for Parallel Distributed Memory Computers) and the Parallel NAG (Numerical Algorithms Group) Library were presented and discussed. During the tutorial, time was allocated for practical exercises on the IBM SP2 and SGI Power Challenge computers. More than 60 people attended the tutorial.

The workshop attracted more than 90 participants from around the world. Authors from over 17 countries submitted 60 papers, of which 12 were invited and 48 were contributed. The Third International Workshop (PARA96) will be held in Lyngby, August 18-21, 1996.

Jack Dongarra

December 1995   Kaj Madsen

Jerzy Waśniewski

Table of Contents*

---

* Italic style indicates the speaker.

  Bold italic style indicates an invited speaker.

A High Performance Matrix Multiplication Algorithm for MPPs

Ramesh C. Agarwal   Fred G. Gustavson   Susanne M. Balle
Mahesh Joshi   Prasad Palkar

IBM T. J. Watson Research Center,
Yorktown Heights, NY 10598

Abstract. A 3-dimensional (3-D) matrix multiplication algorithm for massively parallel processing systems is presented. Performing the product of two matrices $C = \beta C + \alpha A B$ is viewed as solving a 2-dimensional problem in the 3-dimensional computational space. The three dimensions correspond to the matrices dimensions m, k, and n: $A \in \mathbf{R}^{m \times k}$, $B \in \mathbf{R}^{k \times n}$, and $C \in \mathbf{R}^{m \times n}$. The p processors are configured as a "virtual" processing cube with dimensions p_1, p_2, and p_3. The cube's dimensions are proportional to the matrices' dimensions—m, n, and k. Each processor performs a local matrix multiplication of size $m/p_1 \times n/p_2 \times k/p_3$, on one of the sub-cubes in the computational space. Before the local computation can be carried out, each sub-cube needs to receive sub-matrices corresponding to the planes where A and B reside. After the single matrix multiplication has completed, the sub-matrices of C have to be reassigned to their respective processors. The 3-D parallel matrix multiplication approach has, to the best of our knowledge, the least amount of communication among all known parallel algorithms for matrix multiplication. Furthermore, the single resulting sub-matrix computation gives the best possible performance from the uni-processor matrix multiply routine. The 3-D approach achieves high performance for even relatively small matrices and/or a large number of processors (massively parallel). This algorithm has been implemented on IBM Powerparallel SP-2 systems (up to 216 nodes) and have yielded close to the peak performance of the machine. For large matrices, the algorithm can be combined with Winograd's variant of Strassen's algorithm to achieve "super-linear" speed-up. When the Winograd approach is used, the performance achieved per processor exceeds the theoretical peak of the system.

1 Introduction

A parallel high performance matrix multiplication algorithm based on a 3-dimensional approach is presented. We are interested in performing $C = \beta C + \alpha A B$ types of computation for double precision and double precision complex IEEE format, as well as all the cases where the matrices' transposes and their conjugates are involved. The algorithm described has been implemented in both the double precision and the complex double precision IEEE format, as well as for all combinations of matrix products involving matrices in normal form, their transposed, and their conjugates. For all formats, carrying out the implementations was done with equal ease and the performances obtained for the different types of computations were alike. The main idea in this algorithm is to map the matrices A, B, and C onto a "virtual" processing cube. Each processor then posseses a sub-matrix of A, B, and C, thereby carrying out a local sub-cube matrix multiply. The communication part is divided into very regular communication patterns. We refer readers interested in parallel matrix multiply algorithms to the descriptions given by Demmel, Heath, and van der Vorst [5], by Agarwal, Gustavson, and Zubair [2], and to look at the references given in the paper by Gupta and Kumar [9] as well as [14, 16]. A study of parallel matrix multiplication algorithms on distributed-memory concurrent computers, resulting in the ScaLAPACK's PUMMA library [3, 4], was carried out by Choi, Dongarra, and Walker.

In Section 2, we describe the parallel 3-D algorithm and its specific 3-D mapping scheme. Further details are presented in [1]. Section 3 illustrates that this approach yields very high performance on massively parallel processing systems—such as the IBM Powerparallel SP-2 systems—for both small and large problems (relative to the number of nodes and the memory size.) For large matrices, the algorithm can be combined in a natural way with Winograd's variant of Strassen's algorithm to achieve "super-linear" speed-up.

2   A 3-D Parallel PDGEMM Algorithm

The 3-D parallel matrix multiplication algorithm can be thought of as solving a 2-dimensional problem using a 3-dimensional structure in the computational space. This work is related to some research done by Johnsson and Ho [14] on algorithms for multiplying matrices of arbitrary shape on boolean cubes. The three matrices' dimensions are mapped onto a "virtual" processor cube, with its dimensions p_1, p_2, and p_3 proportional to the matrices dimensions m, n, and k. We define $A \in \mathbf{R}^{m \times k}$, $B \in \mathbf{R}^{k \times n}$, and $C \in \mathbf{R}^{m \times n}$. Parallel algorithms can be divided into inter-processor communication and local computation. This algorithm carries out the local matrix multiplication computation using the high performance uni-processor matrix matrix multiplication subroutine DGEMM available to us via the IBM ESSL [12, p.482-488]. The communication part of the algorithm is done using efficient MPI collective communication primitives: *all-gather* and *all-to-all* [6, 8]. In fact, multiple calls to the collective communication primitives are performed concurrently. A very attractive feature of this algorithm is that communication and computation are perfectly load-balanced.

We assume that the matrices A, B, and C are mapped onto the 3-D processor grid showed in Figure 1. We choose to have the matrix A distributed across the p_1-p_3 plane with p_2 being the orthogonal dimension. The matrix B is similarly laid out in the p_2-p_3 plane having p_1 as orthogonal dimension. The p_1-p_2 plane holds the output matrix C making p_3 the orthogonal dimension. In the upcoming description, we consider square matrices of dimension n and a total of p processors. Each processor initially contains n^2/p elements of A, of B, and C.

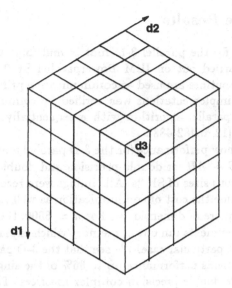

Fig. 1. 3-dimensional processor grid of dimension (p_1, p_2, p_3).

The computational part is carried out having the processing nodes configured as a virtually $p^{1/3} \times p^{1/3} \times p^{1/3}$ cube. Each processor in the cube performs the multiplication of a $n/p^{1/3} \times n/p^{1/3}$ block of A with a $n/p^{1/3} \times n/p^{1/3}$ block of B. The amount of data required in the local computation is $n^2/p^{2/3}$, performing $2 \cdot n^3/p$ arithmetic operations.

Before the local computation is carried out, blocks of A and B need to be broadcasted to assigned processors. We assume that the broadcast is performed using the 2-phase algorithm [11]. Broadcasts are also performed to assign the scattered block of C to their respective processors after the computation has completed. From [11], we know that the total nodal communication time T_{comm} for the three broadcasts and thereby in the algorithm is:

$$T_{comm} = 3 \cdot t_s + 9 \cdot t_w \cdot n^2/p^{2/3}. \tag{1}$$

The total nodal computational time T_{comp} is:

$$T_{comp} = 2 \cdot n^3/p.$$

Comparing Equation 1 with the communication costs for the parallel algorithms described by A. Gupta and V. Kumar [9] shows that the 3-dimensional matrix multiplication algorithm has, to the best of our knowledge, the least amount of communication of all known parallel matrix product algorithms. We also see that it moves a factor $p^{1/6}$ less data than the conventional 2-D algorithms—$n^2/p^{1/2}$. The 3-D algorithm can be combined with the asymptotically fast matrix multiply scheme developed by Strassen in a straightforward manner, thereby allowing to take fully advantage of the former's high efficiency.

3 Performance Results

Performance results for the parallel 3-D matrix multiply are presented. These experiments were carried out on IBM Powerparallel SP-2 systems [13]. MPI message passing subroutines are used as communication primitives [6, 7, 8]. The functionality of the implementations was verified by comparing the computed matrix C from the parallel algorithm with a sequentially computed $\tilde{C}$ using LAPACK DGEMM [12, p.482-488].

Figures 2 and 3 show performance for the 3-D parallel matrix multiply implementation of $C = C + AB$ for double precision and double precision complex IEEE format on various sizes of SP-2s. All timings were recorded using the wall clock and hence include the cost of communication as well as computation. The input matrices are square of dimension n. For $n = 5000$, the 27 node SP-2 considered for these experiments ran out of memory which explains why no MFlops rate is shown for that particular case. We see that the 3-D parallel matrix multiply implementation yields performance up to 86% of the single processor's peak performance for large double precision complex matrices. The double precision matrix multiply achieve a maximum performance of approximatively 82% of the machine's peak performance. When considering the MFlops rates for n larger than 2000 performance we see that they are approximatively constant (relative to the problem size for the different SP-2s considered) which indicates that our implementation is scalable for at least up to 216 nodes. Figures 2 and 3 illustrate that even for relatively small matrices and/or a large number of processors this approach yields very high performance.

Table 1 shows the Mflops rates per processor for the cases $C = C + AB$, $C = C + A^T B$, $C = C + AB^T$, and $C = C + A^T B^T$ when applied to the double precision IEEE format. From this table we see that the performance is the same for all cases.

n	# nodes	$C = C + AB$	$C = C + A^T B$	$C = C + AB^T$	$C = C + A^T B^T$
500	8	156	155	152	153
1000	8	196	197	193	194
2000	8	220	219	217	218
500	16	146	145	143	143
1000	16	186	186	184	185
2000	16	213	212	211	212
5000	64	245	245	243	244
5000	128	244	244	242	242

Table 1. Mflops rate per processor for the four cases for the double precision IEEE format

We generalized our approach to Winograd's variant of the Strassen algorithm [12]. Agarwal and Gustavson showed in the early eighties that the Strassen-Winograd approach was a practical serial algorithm [12]. Based on the Strassen approach [15], which trades the conventional 8 matrix multiplications and 4 matrix additions with 7 matrix multiplications and 18 matrix additions, Winograd

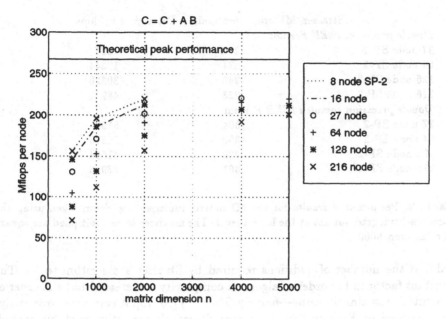

Fig. 2. Performance results for the 3-D parallel double precision matrix multiply implementation of $C = C + AB$ on various sizes of SP-2s.

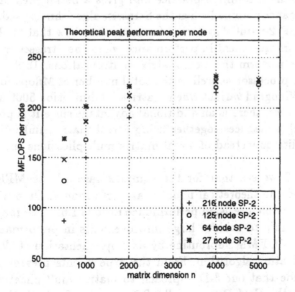

Fig. 3. Performance results for the 3-D parallel double precision complex matrix multiply implementation of $C = C + AB$ on various sizes of SP-2s.

	Strassen MFlops per node	Total # of MFlops
Double precision IEEE Format		
27 node SP-2	272	7337
64 node SP-2	271	17344
125 node SP-2	243	30320
216 node SP-2	223	48907
Double precision complex IEEE Format		
27 node SP-2	304	8198
64 node SP-2	369	23616
125 node SP-2	329	41094
216 node SP-2	361	77976

Table 2. Performance results for the 3D matrix multiply algorithm when using the Strassen-Winograd variant at the lowest level. The matrices to be multiplied are square of dimension 5000

reduced the number of additions required by Strassen's algorithm to 15. The constant factor in the order of algebraic complexity is increased but the order of magnitude remains the same—namely $O(n^{2.807})$ [10, 15]. A complete error analysis presented by Higham [10] shows that Strassen's algorithm is stable enough to be used in many applications.

The Mflops rates presented in Table 2 for the Winograd-Strassen algorithm are based on $2n^3$ arithmetic operations. Using the same number of operations for the two algorithms is common practice and gives a better idea of the improvement in performance obtained when the Strassen algorithm is used. Experiments carried out on SP-2 parallel systems of various size show that for large matrices there is a large improvement in performance when the Strassen-Winograd algorithm is used to perform the local matrix multiplications. Table 2 presents the Mflops rate per processor as well as the total number of Mflops for our generalization to the Winograd variant when matrices of dimension 5000 are multiplied. In the complex case, there is an additional advantage since it is possible to multiply two complex matrices together using 3 real matrix multiplications and 5 real matrix additions instead of 4 real matrix multiplications and 2 real matrix additions [12].

From table 2 we see that for the complex case all the MFlops rates per processor exceed the theoretical nodal peak performance. On a 216 node SP-2, the complex to double precision performance ratio is 1.62. The major conclusion is that is is possible to achieve large improvements in performance by replacing the genuine DGEMM subroutine by an implementation of Winograd's fast matrix multiplication algorithm. From the experiments presented in this section, we conclude that our 3-D approach to matrix multiplication yields high performance on the IBM Powerparallel SP-2 system for both the real and the complex IEEE format. These results demonstrate the possibility of using the sequential Winograd approach local at each node in parallel matrix multiplication implementations. An additional generalization of the 3-D matrix multiplication algorithm to the Winograd approach is investigated in [1]. This second approach consists in performing a sequential Winograd method on the global matrices down to a level where the resulting matrices fit into the local memory of the node. This algorithm is needed if the matrices to be multiplied are too big to allow the required expansion from the 3-D approach.

4 Conclusion

The 3-D approach to parallel matrix multiplication presented in this paper yields very high performance on massively parallel processing systems such as the IBM Powerparallel SP-2 system. The algorithm is designed such that it is perfectly load-balanced for both communication and computation. An important conclusion is that the fast matrix multiplication algorithm—the Winograd-Strassen algorithm—can be incorporated to this algorithm in a straightforward manner, thereby enhancing its efficiency without introducing complex bookkeeping.

Acknowledgements

We thank Prof. V. Kumar from the University of Minnesota and M. Zubair for their initial ideas regarding the analysis of communication in the 3-D algorithm.

References

1. R. C. Agarwal, F. G. Gustavson, S. M. Balle, M. Joshi, and P. Palkar. A 3-dimensional approach to parallel matrix multiplication. Technical report, IBM T. J. Watson Reasearch Center, Yorktown Heights, 1995. Under preparation.
2. R. C. Agarwal, F. G. Gustavson, and M. Zubair. A high performance matrix multiplication algorithm on distributed-memory parallel computer, using overlapped communication. *IBM Journal of Research and Development*, pages 673–681, 1994.
3. J. Choi, J. J. Dongarra, R. Pozo, and D. W. Walter. ScaLAPACK: A scalable linear algebra library for distributed memory concurrent computers. Technical report, UT, 1992.
4. J. Choi, J. J. Dongarra, and D. W. Walter. PUMMA: parallel universal matrix multiplication algorithms on distributed memory concurrent computers. Technical report, UT, 1994.
5. J. W. Demmel, M. T. Heath, and H. A. van der Vorst. Parallel numerical linear algebra. In *Acta Numerica 1993*, pages 111–197. Cambridge University press, 1993.
6. Message Passing Interface Forum. *MPI: A Message-Passing Interface Standard*, May 1995.
7. H. Franke, C. E. Wu, M. Riviere, P. Pattnaik, and M. Snir. MPI programming environment for IBM SP1/SP2. Technical report, IBM T. J. Watson Research Center, 1995.
8. W. Gropp, E. Lusk, and A. Skjellum. *Using MPI: Portable Parallel Pragramming with the message passing interface*. MIT Press, 1994.
9. A. Gupta and V. Kumar. Scalability of parallel algorithms for matrix multiplication. Technical report, Department of Computer Science, University of Minnesota, 1991. Revised April 1994.

10. N. J. Higham. Exploiting fast matrix multiplication within the level 3 BLAS. *ACM Trans. Math. Software*, 16:352–368, 1990.
11. A. Ho. Personal communications. IBM Almaden, 1995.
12. IBM. *Engineering and Scientific Subroutine Library, Guide and reference: SC23-0526-01*. IBM, 1994.
13. IBM. Scalable parallel computing. *IBM Systems Journal*, 34, No 2, 1995.
14. S. L. Johnsson and C-T. Ho. Algorithms for multiplying matrices of arbitrary shapes using shared memory primitives on boolean cubes. Technical Report TR-569, Yale University, 1987.
15. V. Strassen. Gaussian elimination is not optimal. *Numer. Math.*, 13:354–356, 1969.
16. R. van de Geijn and J. Watts. SUMMA: Scalable Universal Matrix Multiplication Algorithm. Technical report, Department of Computer Science, University of Texas at Austin, 1995.

Iterative Moment Method for Electromagnetic Transients in Grounding Systems on CRAY T3D

G. Ala*, E. Francomano* ° and A. Tortorici* °

* Dipartimento di Ingegneria Elettrica, Università di Palermo, Viale delle Scienze - 90128 Palermo, Italy

° CE.R.E., Centro Studi sulle Reti di Elaboratori, C.N.R., Viale delle Scienze - 90128 Palermo, Italy

Abstract In this paper the parallel aspects of an electromagnetic model for transients in grounding systems based on an iterative scheme are investigated in a multiprocessor environment. A coarse and fine grain parallel solutions have been developed on the CRAY T3D, housed at CINECA, equipped with 64 processors working in space sharing modality. The performances of the two parallel approaches implemented according to the work sharing parallel paradigm have been evaluated for different problem sizes employing variable number of processors.

1 Introduction

The simulation of transients electromagnetic phenomena in grounding systems can be carried out by employing a model based on the discretization of the electric field integral equation in frequency domain, solved by the Method of Moment. The time behaviour of the interesting electromagnetic quantities can be then obtained by using a Fast Fourier Transform algorithm. Such an approach, applied to study the performances of typical earthing structures in power systems, determines a very expensive computational task which can be advantageously handled with parallel processing techniques.

In a previous paper [2] the Authors investigated the electromagnetic problem by using the direct Moment Method in frequency domain via thin-wire approximation and point-matching procedure and carried out some large size examples by employing non conventional architectures such as IBM 3090 with vector facility and CRAY Y-MP with four processors.

By using the direct Moment Method the integral equation describing the physical problem is reported to a matrix equation which poses an heavy store demanding when the number of the unknowns increases. In order to overcome this limit the Authors in [3] employed an alternative computational strategy offered by the Conjugate Gradient method directly applied to the integral operator opportunely discretized. Furthermore, the close partitioning required in the discretized model in order to obtain the necessary reduction of the analytical error, makes the iterative method suitable to run on parallel environments.

In the present paper, parallel processing techniques in the computation of the electromagnetic transients in grounding systems approached in an iterative way are proposed.

Questions regarding computational load balancing and communication overhead are discussed referring to the distributed machine CRAY T3D with 64 processors.

In order to assess the performances of the parallel computational strategies developed, a practical example has been examined.

2 Iterative Formulation

The frequency domain technique applied in this study leads to an integral equation which can be generally expressed in the following form [6]:

$$Y(r) = \int_S G(r,r')X(r')dr',$$

where X and Y are the unknown and known quantities respectively, whilst G is the kernel function of the integral equation.

In this section the Conjugate Gradient method, which is the most commonly used in electromagnetics among the iterative approaches, is outlined.

By fixing an estimate of the unknown $X^{(0)}$, the Conjugate Gradient method computes at the initial step k=0 the residual vector:

$$T^{(0)}(r) = Y(r) - \int_S G(r,r')X^{(0)}(r')dr'.$$

At each iterative step, k=1,2,..., by integrating the transpose and complex conjugate of the kernel function G(r, r') applied to the residual vector just computed, the gradient vector is obtained:

$$S^{(k-1)}(r) = \int_S G^*(r',r)\, T^{(k-1)}(r')dr'.$$

The gradient vector and its square norm

$$a^{(k)} = \left\| S^{(k-1)}(r) \right\|^2$$

are involved in the computation of the variational vector :

$$H^{(k)}(r) = S^{(k-1)}(r) + \frac{a^{(k)}}{a^{(k-1)}}\; H^{(k-1)}(r)$$

which assumes the value of $S^{(0)}(r)$ when k=1.

Finally, at the end of each step, by computing the variational deviation:

$$F^{(k)}(r) = \int_S G(r,r')\, H^{(k)}(r')dr',$$

the square norm:

$$b^{(k)} = \left\| F^{(k)}(r) \right\|^2,$$

and the variational parameter :

$$\eta^{(k)} = a^{(k)}/b^{(k)},$$

the residual vector and the unknown vector are so modified:

$$T^{(k)}(r) = T^{(k-1)}(r) - \eta^{(k)}F^{(k)}(r)$$

$$X^{(k)}(r) = X^{(k-1)}(r) + \eta^{(k)}H^{(k)}(r).$$

The normalized integrated square error defined as follows [7]:

$$ERR^{(k)} = \left\| T^{(k)}(r) \right\|^2 / \left\| X^{(0)}(r) \right\|^2$$

is the error criterion used in order to stop the iterative process which is proved to be always convergent [6].

3 Computational Scheme

Since the electrodes of the physical grounding structure are modelled by a thin-wire skeleton, the longitudinal currents I(r) along the conductors are the unknowns which must be computed for each of the frequencies employed in order to obtain the time behaviour of the interesting electromagnetic quantities [1]. Hence the computational scheme can be summarized as follows:

```
            { Frequencies loop }
            DO J=1,16000
                {Conjugate Gradient method }
                First residual vector T;
                First integrated square error ERR;
                WHILE ERR>=10⁻ᵛ DO
                        Gradient Vector S;
                        Square norm a;
                        Variational vector H;
                        Variational deviation F;
                        Square norm b;
                        Variational parameter η;
                        Residual vector T;
                        Currents unknown I;
                        Integrated square error ERR;
            END DO.
```

By discretizing the problem [3], the integrals involved in the computation of the first residual T, of the gradient vector S and of the variational deviation F are replaced by matrix vector products which represent the heaviest computational tasks of the iterative process.

From now on, N will mean the number of parts in which the physical structure is partitioned, i.e. the problem size.

4 Parallel Analysis

In a multiprocessor environment two different parallel solutions can be characterized in the computational scheme just reported. An intrinsic parallelism is in the algorithm at frequency level and a second one, a data parallelism, can be obtained by fixing a frequency value and subdividing the computations of the Conjugate Gradient method among the processors. Hence, a coarse grain and a fine grain parallel strategies will be examined.

4.1 Computational Environment

The parallel experiments have been carried out on the Cray T3D equipped with 64 processors and working in space sharing modality. Each processor has a local memory which can be addressed by the others. Hence, data objects not distributed across the processors are *private*, i.e. each task that references a private object uses its own private version. Whilst *shared* data objects, in contrast, are accessible by all processors and if the object is an array it can be distributed across multiple processors. In this case, the size of the shared array must be a power of 2 in order to optimize the load balancing, by considering that in space sharing modality only $p=2^r$ $(r=1,...,6)$ processors are allocable.

The PVM (Parallel Virtual Machine) constructs working only with private data realize the explicit parallelism; the implicit one can be implemented by taking into account the work sharing paradigm which makes use of the compiler directives for data and statements distribution. In this last context an automatic parallelism can be performed by using the data parallel paradigm which is mostly based on FORTRAN 90 array syntax and intrinsics.

All these programming styles may be combined in the same code, but when possible the use of the work sharing parallel paradigm must be preferred.

4.2 Parallel Solutions

As just underlined a coarse grain parallelism can be obtained by dividing up the frequencies loop j among the processors: each of them, having a set of iterations to execute, generates N currents for each iteration.

Hence, the vectors involved in the iterative method are private data, whilst the matrix R of the geometric configuration is the only global data shared by all processors (Fig.1). The communication task required in this parallel solution is extremely limited and the implementation is trivial.

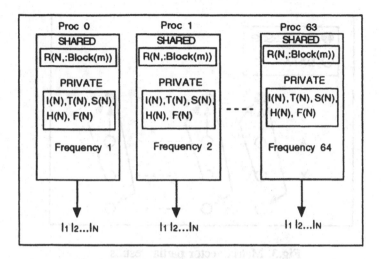

Fig.1. Coarse grain data distribution.

By fixing a frequency value, a parallel solution with a finer grain can be carried out by distributing the computations of each step k of the Conjugate Gradient method among the processors. In order to successfully apply the work sharing programming model, the matrix and all the vectors involved in the computations are shared by all processors.

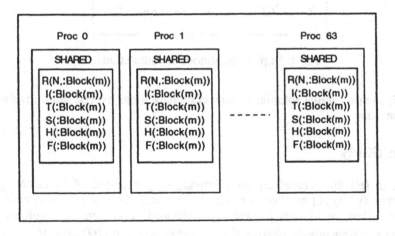

Fig.2. Fine grain data distribution.

By indicating with A*X=Y a generic matrix vector product, each processor computes partial results by using the local m=N/p matrix columns and the local vector entries (Fig.3). Hence, only local loads and local stores are involved in this computational phase. The elaboration is completed by realizing an implicit multinode accumulation, i.e. each processor adds the local and remote partial results and stores the final result in the local data Y:

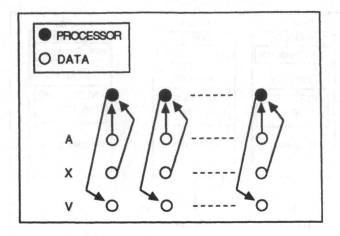

Fig.3. Matrix vector partial results.

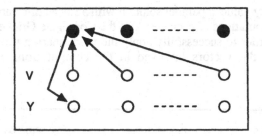

Fig.4. Implicit multinode accumulation.

In this fine grain parallel solution implicit communications exceed those involved in the coarse grain.

5 Case Study

In order to test the parallelism the following easy example of ElectroMagnetic Interference (EMI) [1] has been examined. Two metallic (copper) non connected electrodes belonging to a same plane and embedded in homogeneous soil with the following electromagnetic characteristics: $\varepsilon=5\varepsilon_0$; $\mu=\mu_0$; $\rho=100$ $\Omega \cdot m$ (ε_0, μ_0 as the permeattivity and the magnetic permeability of vacuum, ρ as the soil resistivity). The electrodes are 100 m long with 0.5 cm of radius.

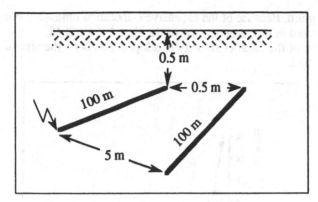

Fig.5. Two electrodes in homogeneous soil.

Size	p	Fine grain	Coarse grain
256	2	28.7	16.2
	4	16.1	8.1
	8	9.5	4.0
	16	6.4	2.1
	32	4.8	1.1
	64	4.0	0.6
512	2	212.3	125.2
	4	110.9	62.9
	8	61.8	31.8
	16	36.2	16.1
	32	24.3	8.2
	64	18.1	4.2

Table 1. CPU times.

In Table 1 the CPU times (min.) of the two parallel solutions have been reported. The measurements have been carried out by using the *IRTC* routine and the CF77 compiler has been used. Only 128 frequencies, instead of those required in a real problem, have been elaborated because just this number is able to give answers on the behaviour of the parallel schemes adopted.

As shown for the problem sizes analyzed, the coarse grain solution is always better than the fine grain and because of the limited communication cost it is also more scalable than the fine grain.

By increasing the problem size the fine grain improves the performances. In fact, the following graph (Fig. 6) reporting the behaviour of the ratio between the fine and the coarse grain, shows that when the problem size is about 2000 the fine grain is better than the coarse grain. Because of the expensive elaboration time, only p=64 processors have been interested in the computations of the sizes 1024 e 2048.

Hence the choice of the parallel strategy to adopt depends on the size of the problem that must be studied.

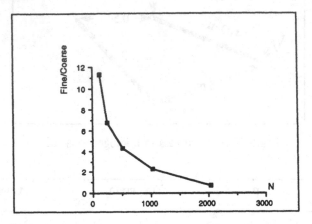

Fig. 6. Behaviour of the ratio between the fine and the coarse grain.

This work has been supported by Italian Ministry of University and Scientific and Tecnological Research.

References

[1]   Ala, G., Buccheri, P., Campoccia, A., High frequency coupling among buried electrodes, paper n. 300-05, *Proceedings of CIGRE' Symposium on Power System Electromagnetic Compatibility*, Lausanne, Switzerland, 1993.

[2]   Ala, G., Buccheri, P., Francomano, E. Tortorici, A., An advanced algorithm for transient analysis of grounding systems by Moments Method, pp 363-366, *Proceedings of IEE 2nd Int. Conf. on Computation in Electromagnetics*, Nottingham, United Kingdom, 1994.

[3]   Ala, G., Francomano, E., Tortorici, A., Parallel approach for transients in grounding systems, pp 3-8, *Proceedings of Fourth International Conference ASE 95*, Milano, Italy, 1995.

[4]   Bertsekas, D.P., Tsitsiklis, J.N., *Parallel and distributed computation - Numerical methods*, Prentice Hall, Cliffs, New Jersey, 1989.

[5]   Van den Berg, P.M., Iterative computational techniques in scattering based upon the integrated square error criterion,*IEEE Transaction on antennas and propagation*, vol.AP-32, No.10, October 1984.

[6]   Wang, J.J.H., *Generalized moment methods in electromagnetics*, Wiley Interscience Publication, New York,1991.

[7]   Wang, J.J.H., Dubberley, J.R., Computation of electromagnetic fields in large biological bodies by an iterative moment method with a restart technique, *IEEE Transaction on microwave theory and techniques*, vol. 37, No 12, pp 1918 - 1923, 1989.

Analysis of Crystalline Solids by Means of a Parallel FEM Method

B. S. Andersen[1] and N. J. Sørensen[2]

[1] UNI-C, DTU Building 304,
DK-2800 Lyngby, Denmark
[2] Materials Department, Risø National Laboratory,
Risø Denmark

Abstract. A parallel finite element method suitable for the analysis of 3D crystal plasticity problems on parallel computers using the PVM environment is presented. The method is based on a division of the original mesh into a number of substructures which are treated as isolated finite element models related via the interface conditions. The resulting interface equations are solved using a direct parallel solution method.

1 Introduction

The finite element analysis of elastic-plastic behaviour of crystalline solids is a subject of great importance for fundamental and practical reasons (see e.g. Teodosiu *et al.* [1], Beaudoin *et al.* [2] and Bassani [3] for an overview). However, the quantitative analysis is complicated by the large CPU time and memory requirements in a realistic three dimensional analysis. The new parallel machines seems to honour both of these demands, by the expense of a more complicated program, see Dagum *et al.* [4], Law [5], Carter *et al.* [6], Babuska and Elman [7] and Yagawa *et al.* [8].

The computational environment assumed in the present work is a cluster of processors connected in a communication network which allows each of the processors in the cluster to communicate with all the others.

The finite element mesh is divided into smaller non-overlapping meshes called *substructures*, which are mapped onto the cluster of processors. In order to ensure static and kinematic continuity on the *interface* between the substructures the processors have to communicate.

Basically, either a direct or an iterative method can be used to solve both the substructure equations and the interface equations. In the present work both sets of equations are solved by means of direct methods, as described by Sørensen and Andersen [9].

2 Non-linear Crystal Plasticity Problems

The analysis of finite strain crystal behaviour involves the solution of highly non-linear problems which can be solved using a general iterative finite element method, (see Peirce *et al.* [10], Hill and Rice [11] and Needleman and Tvergaard [12]).

The method used in the present work is based on the principle of virtual work in the Lagrangian convected coordinate form

$$\int_{V_0} \tau^{ij} \delta\eta_{ij} dV = \int_{S_0} T^i \delta u_i dS \tag{1}$$

In (1) V_0 and S_0 denote the volume and surface of the region analysed in the reference configuration. The Kirchhoff stress is τ^{ij}, the displacements are u_i and the nominal tractions are T^i. The deformation measure is the Lagrangian strain tensor η_{ij} which can be expressed in term of the displacement components by the non-linear relation

$$\eta_{ij} = \frac{1}{2}(u_{i,j} + u_{j,i} + u^k_{,i} u_{k,j}) \tag{2}$$

A polar decomposition of the deformation into elastic and plastic parts is used

$$\mathbf{F} = \mathbf{F}^* \cdot \mathbf{F}^P \tag{3}$$

The plastic part of the deformation gradient is

$$\dot{\mathbf{F}}^P = \sum_\alpha \dot{\gamma}^{(\alpha)} \mathbf{s}^{(\alpha)} \mathbf{m}^{(\alpha)} \cdot \mathbf{F}^P \tag{4}$$

where $\dot{\gamma}^{(\alpha)}$ is the slip rate on the slip system α and $\mathbf{s}^{(\alpha)}$ and $\mathbf{m}^{(\alpha)}$ are the slip direction and slip plane normal, respectively.

The crystal elasticity is taken to be of the form

$$\hat{\tau}^* = \mathbf{L} : \mathbf{D}^* \tag{5}$$

where $\hat{\tau}^*$ is the Jaumann rate of the Kirchhoff stress with respect to the lattice, $\mathbf{L}$ is the instantaneous elastic moduli and the elastic rate of stretching is D^* (i.e. the symmetric part of the elastic rate of deformation $\dot{\mathbf{F}}^* \cdot \mathbf{F}^{*-1}$).

A specification of the slip rate is needed

$$\dot{\gamma}^{(\alpha)} = \dot{\gamma}_0 \left[\frac{\tau^{(\alpha)}}{g^{(\alpha)}} \right] \left[\left| \frac{\tau^{(\alpha)}}{g^{(\alpha)}} \right| \right]^{(1/m)-1} \tag{6}$$

where $g^{(\alpha)} = g^{(\alpha)}(\gamma)$ is a viscoplastic flow stress, $\dot{\gamma}_0$ is a reference slip rate and m is a strain rate hardening exponent and the Schmid stress is $\tau^{(\alpha)}$. An evolution for $g^{(\alpha)}$ is used in (6). This involves a hardening matrix which describes the coupling between the various slip systems and the variation of the hardening with the accumulated slip, (further details can be found in Peirce *et al.* [10]).

The numerical solution of the crystal plasticity problems are obtained by an incremental solution of (1) which involves updating of the various history dependent quantities involved in the solution of linear algebraic equations

$$\mathbf{A} \cdot \dot{\mathbf{U}} = \dot{\mathbf{B}} \tag{7}$$

where $\mathbf{A}$ is the *instantaneous* stiffness matrix, $\dot{\mathbf{U}}$ is the unknown displacement increments and the incremental load vector is $\mathbf{B}$. In the present case the final form of the incremental constitutive equations leads to a *banded* and *non-symmetric* instantaneous stiffness matrix.

3 Parallel Finite Element Solution

The global mesh, representing the discretized field equations for the structure to be analysed, can be decomposed in a number of non-overlapping meshes each representing a *substructure*. The set of degrees-of-freedom that are shared by two or more substructures is called the *interface*, see Bjørstad and Widlund [13].

The present method for parallel solution of (7) is,

1. Parallel factorization of the stiffness matrix of each of the substructures.
2. Parallel computation of each substructures contribution to the interface stiffness matrix and the interface load rate vector.
3. Assembly of the interface equations followed by a parallel decomposition and solution to give the interface displacement rate vector.
4. Parallel solution of the substructure equations to give the substructure displacement rate vector of each of the substructures.

The instantaneous interface stiffness matrix and the interface load rates needed in the task 2. are computed by superimposing contributions from each substructure. For the J'th substructure the equations has the form,

$$\mathbf{A}^J \dot{\mathbf{D}}^J = \dot{\mathbf{B}}^J \tag{8}$$

By a suitable permutation of the equations, they can also be written as,

$$\begin{bmatrix} \mathbf{A}^J_{ss} & \mathbf{A}^J_{sk} & \mathbf{A}^J_{si} \\ \mathbf{A}^J_{ks} & \mathbf{A}^J_{kk} & \mathbf{A}^J_{ki} \\ \mathbf{A}^J_{is} & \mathbf{A}^J_{ik} & \mathbf{A}^J_{ii} \end{bmatrix} \left\{ \begin{array}{c} \dot{\mathbf{D}}^J_s \\ \dot{\mathbf{D}}^J_k \\ \dot{\mathbf{D}}^J_i \end{array} \right\} = \left\{ \begin{array}{c} \dot{\mathbf{B}}^J_s \\ \dot{\mathbf{R}}^J_k \\ \dot{\mathbf{B}}^J_i \end{array} \right\} \tag{9}$$

where the sub indices s, k and i refers to the static (or free), kinematic and interface degrees-of-freedom, respectively.

The contributions to the interface stiffness matrix and the interface load rate vector from the J'th substructure can be computed by introducing *trial* displacement rate vectors $\dot{\mathbf{d}}^J_i$, $\dot{\mathbf{d}}^J_k$ and $\dot{\mathbf{d}}^J_s$ and *trial* load rate vectors $\dot{\mathbf{b}}^J_i$, $\dot{\mathbf{r}}^J_k$ and $\dot{\mathbf{b}}^J_s$ corresponding to the i-, k- and s-categories of DOF's.

$$\begin{bmatrix} \mathbf{A}^J_{ss} & \mathbf{A}^J_{sk} & \mathbf{A}^J_{si} \\ \mathbf{A}^J_{ks} & \mathbf{A}^J_{kk} & \mathbf{A}^J_{ki} \\ \mathbf{A}^J_{is} & \mathbf{A}^J_{ik} & \mathbf{A}^J_{ii} \end{bmatrix} \left\{ \begin{array}{c} \dot{\mathbf{d}}^J_s \\ \dot{\mathbf{d}}^J_k \\ \dot{\mathbf{d}}^J_i \end{array} \right\} = \left\{ \begin{array}{c} \dot{\mathbf{b}}^J_s \\ \dot{\mathbf{r}}^J_k \\ \dot{\mathbf{b}}^J_i \end{array} \right\} \tag{10}$$

Using the trial vectors $\dot{\mathbf{d}}_k^J$ and $\dot{\mathbf{d}}_i^J$ as prescribed vectors (to be specified below) the solution to (10) can be expressed as

$$\left\{ \begin{array}{c} \dot{\mathbf{d}}_s^J \\ \dot{\mathbf{d}}_k^J \\ \dot{\mathbf{d}}_i^J \end{array} \right\} = \left[\begin{array}{ccc} \mathbf{A}_{ss}^{J-1} & 0 & 0 \\ 0 & \mathbf{I} & 0 \\ 0 & 0 & \mathbf{I} \end{array} \right] \left\{ \begin{array}{c} \dot{\mathbf{b}}_s^J - \mathbf{A}_{sk}^J \dot{\mathbf{d}}_k^J - \mathbf{A}_{si}^J \dot{\mathbf{d}}_i^J \\ \dot{\mathbf{d}}_k^J \\ \dot{\mathbf{d}}_i^J \end{array} \right\} \tag{11}$$

where $\mathbf{A}_{ss}^{J-1}$ is the inverse of $\mathbf{A}_{ss}^J$. The reactions are (see [9]),

$$\dot{\mathbf{r}}_s^J = 0 \tag{12}$$

$$\dot{\mathbf{r}}_k^J = \mathbf{A}_{ks}^J \mathbf{A}_{ss}^J \dot{\mathbf{b}}_s^J + (\mathbf{A}_{kk}^J - \mathbf{A}_{ks}^J \mathbf{A}_{ss}^{J-1} \mathbf{A}_{sk}^J)\dot{\mathbf{d}}_k^J + (\mathbf{A}_{ki}^J - \mathbf{A}_{ks}^J \mathbf{A}_{ss}^{J-1} \mathbf{A}_{si}^J)\dot{\mathbf{d}}_i^J \tag{13}$$

$$\dot{\mathbf{r}}_i^J = \mathbf{A}_{is}^J \mathbf{A}_{ss}^J \dot{\mathbf{b}}_s^J + (\mathbf{A}_{ik}^J - \mathbf{A}_{is}^J \mathbf{A}_{ss}^{J-1} \mathbf{A}_{sk}^J)\dot{\mathbf{d}}_k^J + (\mathbf{A}_{ii}^J - \mathbf{A}_{is}^J \mathbf{A}_{ss}^{J-1} \mathbf{A}_{si}^J)\dot{\mathbf{d}}_i^J - \dot{\mathbf{b}}_i^J \tag{14}$$

Specifying zero interface displacements in (14), using the actual values for the kinematic prescribed displacement rates $\dot{\mathbf{V}}_k^J$ and the actual load rates $\dot{\mathbf{B}}_s^J$ and $\dot{\mathbf{B}}_i^J$ gives

$$\dot{\mathbf{r}}_i^{\prime J} = \mathbf{A}_{is}^J \mathbf{A}_{ss}^{J-1} \dot{\mathbf{B}}_s^J + (\mathbf{A}_{ik}^J - \mathbf{A}_{is}^J \mathbf{A}_{ss}^{J-1} \mathbf{A}_{sk}^J)\dot{\mathbf{V}}_k^J - \dot{\mathbf{B}}_i^J \quad \text{for} \quad \dot{\mathbf{d}}_i^J \equiv 0 \tag{15}$$

Specifying zero trial load rates and zero trial displacement rates for the k-DOF's in (14) gives

$$\dot{\mathbf{r}}_i^{\prime\prime J} = (\mathbf{A}_{ii}^J - \mathbf{A}_{is}^J \mathbf{A}_{ss}^{J-1} \mathbf{A}_{si}^J)\dot{\mathbf{d}}_i^J \quad \text{for} \quad \dot{\mathbf{b}}_s^J \equiv 0, \dot{\mathbf{b}}_i^J \equiv 0 \text{ and } \dot{\mathbf{d}}_k^J \equiv 0 \tag{16}$$

The reactions $\dot{\mathbf{r}}_i^{\prime J}$ given by (15)—in this case with the interface displacement rates fixed to zero—is equal to the J'th contribution to the interface load rates. Furthermore, taking the load rates and the displacement rates at the kinematic restricted DOF's to zero, as in (16) allows the J'th contribution to the interface stiffness matrix to be computed. This contribution, i.e. the matrix $\mathbf{A}_{ii}^J - \mathbf{A}_{is}^J \mathbf{A}_{ss}^{J-1} \mathbf{A}_{si}^J$ is established by calculating the column vector $\dot{\mathbf{r}}_i^{\prime\prime J}$ as many times as there are degrees of freedom on the substructure interface, with $\dot{\mathbf{d}}_i^J$ set to $\{1, 0, 0, \cdots\}, \{0, 1, 0, \cdots\}$ etc.

In this work, where the interface system is solved by a direct method, the interface equations are *assembled* by superposing all the substructure contributions into one system. Then a parallel LU-decomposition as described in Van de Velde [14] is used in the solution for the interface displacement rates. Finally, the unknown displacement rates for the s-category of DOF's are obtained at the substructure level by solving the finite element equations (10). (The factorization needed in this solution is already done).

4 Results

The parallel finite element method presented here is implemented in a Fortran program with calls to the PVM message passing library, see Geist *et al.* [15]. The timings are measured while running on an IBM SP2 with a High Performance Switch, and calling the PVMe library, see IBM [16].

The global finite element mesh, i.e. the unsubstructured mesh used for the time measurements consists of an 8×8×9 block of isoparametric 20 nodes brick elements. Table 1 shows the properties of the single substructures, when the global mesh is divided in 1, 3, 6 and 12 parts. The number of equations per substructure and—even more important—the bandwidth of the coefficient matrix is considerably reduced when the number of substructures and thus the number of processors is increased, but of course the number of equations in the interface system is also increased with the number of substructures. In summary, the total number of floating point operations is slightly increased with the number of substructures, but this is greatly surpassed by the advantage of parallel processing. It is also very important to note, that a big finite element model can be distributed to the memory of many processors.

Number of processors	1	3	6	12
Elements per proc.	576	192	96	48
DOF per proc.	8937	3429	1857	1005
Halfbandwidth per proc.	1002	432	252	252
Interface DOF's per proc.	0	675–1350	597–909	429–573
Interface DOF's in total	0	1350	2001	2601

Table 1. Substructuring examples for the 8×8×9 block of elements

In table 2 the timings for 3, 6 and 12 substructures are shown. The finite element model for the complete structure i.e. *one* substructure cannot be accommodated in the memory of a single processor node. The long overall time for the *three* substructures case, reflects that the load balancing for this case is poor, with the number of interface degrees-of-freedom per substructure ranging from 675 to 1350. The load balance becomes better for more substructures but will never become even as it could be for another choice of the initial finite element mesh, see Sørensen and Andersen [9].

5 Summary

A parallel finite element method suitable for the analysis of crystal-plasticity problems on parallel computers using the PVM-environment has been presented. The method shows good overall speedup by substructuring an 8×8×9 mesh into

no. of processors	input	smLU	interf. contrib.	interf. solution	final	overall
	s	s	s	s	s	s
3	0.03	20.8	96.4	67.9	0.35	189.7
6	0.27	7.5	29.2	37.3	0.53	74.9
12	1.09	3.7	10.3	39.7	0.04	53.5

Table 2. Average increment times used in the different computational tasks based on substructuring of the initial 8×8×9 mesh. The "input" time is needed only in the first increment. The term "smLU" denotes the the computational work needed to calculate the instantaneous stiffness matrix for a substructure and the LU-factorization of this quantity.

3–12 substructures. Apart from the speedup the use of the parallel method allows for the solution of crystal- plasticity problems whose size is much too large for the effective solution on a single workstation (node) in the parallel machine. The use of direct parallel equation solvers for the interface makes the method very stable and allows for the treatment of problems involving many right-hand sides with much smaller computational costs than indirect solutions would require.

Acknowledgment

The work of N. J. Sørensen was carried out within the Risø Engineering Science Centre for Structural Characterization and Modelling of Materials.

References

1. C. Teodosiu, J. L. Raphanel, and L. Tabourot. Finite element simulation of the large elastoplastic deformation of multicrystals. In C. Teodosiu, J. L. Raphanel, and F. Sidoroff, editors, *Proceedings of the International Seminar Mecamat'91*, pages 153–168, Fontainebleau France, 7-9 August 1991, (1993). Balkema, Rotterdam.

2. A. J. Beaudoin, K. K. Mathur, P.R. Dawson, and C. G. Johnson. Three-dimensional deformation process simulation with explicit use of polycrystallne plasticity models. *Int. Jour. Plasticity*, 9:833–860, (1993).

3. J. L. Bassani. *Advances in Applied Mechanics*, pages 192–258. Academic Press, Inc., (1994).

4. L. Dagum D. H. Bailey, E. Barszcz and H. D. Simon. *NAS Parallel Benchmark Results 10-94*. NASA Ames Research Center, Mail Stop T27A-1, Moffett Field, CA 94035-1000, (1994).

5. K. H. Law. A parallel finite element solution method. *Comput. Struct.*, 23:845–858, (1986).

6. W. T. Carter Jr., T.-L. Sham, and K. H. Law. A parallel finite element solution method. *Comput. Struct.*, 31:921–934, (1989).

7. I. Babuska and H.C. Elman. Some aspects of papallel implementation of the finite-element method on message passing architectures. *Journal of Computational and Applied Mathematics*, 27:157–187, (1989).
8. G. Yagawa, A. Yoshioka, S. Yoshimura, and N. Soneda. A parallel finite element method with a supercomputer network. *Comput. Struct.*, 47:407–418, (1993).
9. N. J. Sørensen and B. S. Andersen. A parallel finite element method for the analysis of crystalline solids. *To be published*, , (1995).
10. D. Peirce, R.J. Asaro, and A. Needleman. Material rate dependence and localized deformation in crystalline solids. *Acta Metall.*, 31:1951–1976, (1983).
11. R. Hill and J. R. Rice. Constitutive analysis of elastic-plastic crystals at arbitrary strain. *J. Mech. Phys. Solids*, 20:401–413, (1972).
12. A. Needleman and V. Tvergaard. Finite elements – special problems in solid mechanics. In J. Tinsdley Oden and Graham F Carey, editor, *Finite Element Analysis of Localization in Plasticity*, volume volume V, pages 94–157. Prentice-Hall, Inc., Englewood Cliffs, New Jersey 07632, (1984).
13. P.E. Bjørstad and O.B. Widlund. Iterative methods for the solution of elliptic problems on regions partitioned into substructures. *Journal of Computational and Applied Mathematics*, 38:1097–1120, (1986).
14. Eric F. Van de Velde. *Concurrent Scientific Computing*. Springer Verlag 1994, (1994).
15. A. Geist, A. Beguelin, J. Dongarra, W. Jiang, R. Manchek, and V. Sunderam. *PVM Parallel Virtual Machine - A Users' Guide and Tutorial for Networked Parallel Computing*. MIT Press, Cambridge, Massachusetts, (1994).
16. IBM Corporation. *IBM AIX PVMe User's Guide and Subroutine Reference, Release 3.1*. IBM Corporation, (1994).

Parallelization Strategies for Tree N-body Codes

Vincenzo Antonuccio-Delogu[1], Ugo Becciani[1] and Fabrizio Magugliani[2]

[1] Osservatorio Astrofisico di Catania, Viale A. Doria 6, I-95125 Catania, ITALY
[2] Convex Computers S.p.A.,Viale Colleoni, 17 - 20041 Agrate Brianza (Milano), ITALY

Abstract. Particle codes which simulate the evolution of self-gravitating systems are a popular tool of contemporary cosmological research. Here we compare two different approach to the parallelization of *Tree N-body* codes, namely *work-sharing* and *message passing*. We also discuss a *Dynamical Load Balancing* algorithm which we have applied to the work-sharing code. preliminary results show merits and problems of these two approaches.

1 Introduction

Among the current algorithms devised to simulate N-body systems of particles interacting through long-range forces, the one based on the *oct-tree* decomposition devised by J. Barnes and P.Hut ([2]) has a very interesting computational feature: its complexity scales as $O(N_b \log N_b)$, N_b being the number of particles. This scaling has been verified in *serial* implementations of the algorithm ([6]), but it is not at all obvious that the same scaling will hold for *parallel* implementations of the same algorithm. Indeed, the parallelization of the Barnes-Hut algorithm has been attained either by modifying the original Barnes-Hut algorithm in order to parallelize only the most time-consuming part of it ([8], [1], [4]), or by directly porting the serial version and exploiting software and hardware features of Massively Parallel systems ([8], [9]). Both approaches have merits and disadvantages. Starting from the observation that the most time-consuming part of the serial BH algorithm is the tree's analysis, Salmon [8] has introduced a strategy of message passing based on the introduction of *Locally Essential Trees*, where each parallel "task" builds up a local reduced oct-tree necessary to compute the evolution of its own particles. This strategy results in a *spatial* decomposition of the workload among the tasks, and it is easy to implement under many popular parallelization environments like PVM ([5]) and MPI. With this approach the average timestep execution time scales as $T_{step} \propto N^\alpha$, with α ranging from 1.05 to 1.4 ([8]; [1]). This rather large variance depends on *intrinsic* factors (e.g. communication overheads, latency bandwith) and on the *complexity* of the tree. It also affects the critical issue of load balancing.

The alternative strategy for parallelization is to start from the serial BH algorithm and to exploit the available compiler features on some MPP systems which allow the programmer to distribute work and/or data among the available Processing Elements (hereafter PEs). In this latter case the programmer will *mostly* rely on the compiler to optimize the code. Strategies which give good results for a given architecture will not necessarily do the same on a different architecture. On the other hand, one can choose to adopt explicitly message-passing software (like PVM), and this choice has the advantage that the parallel code can be ported on *many* different architectures (Massively Parallel Systems, mixed Networks and Clusters, etc...). Different merits and disadvantages of these strategies have often been discussed in the literature, mostly for benchmarks. Here we will report some first results of our tests of these two different strategies on a complex application: a *cosmological Tree N-body code*. We have developed a PVM parallel treecode and we have also ported a *serial* code to the CRAFT environment of the Cray T3D. We will first shortly discuss the structures common to both codes in section 2, then we will give some results concerning the two approaches in section 3, and we will finally give some comments on present results and some prospects for future work in the last section.

The simulations we report here have been performed on a Cray T3D at the Cineca (Casalecchio di Reno (BO) - Italy), a 64 DEC Alpha processor with 8 Mword (64 bits) memory per processor and 9.6 GigaFLOPS peak performance, while the preliminary results for the Convex-PVM version have been obtained from runs on the Convex Exemplar-SPP 1000, a 16 HP-R6000 processors system at Cilea, Italy.

2 Serial and parallel Treecodes

The *Treecode* is based on an algorithm for force calculation which was introduced (in the context of cosmological N-body calculations) by J. Barnes and P.Hut ([2]; see Salmon [8] for a deeper description of the algorithm). In this algorithm the calculation of the force for each particle of the system is done by "inspecting" a 2^{ndim} tree (where $ndim$ is the dimensionality of the space where the particles are moving). The *tree* is made of nested (hyper)cubic cells: starting from the parent cell which contains all the particles, each cell is further divided into 2^{ndim} cells (e.g. 8 cells for $ndim = 3$), until all the cells contain n_{min} or 0 cells (typically one chooses $n_{min} = 1$).

Each particle within the system will then "look" inside the tree in order to compute the force due to the gravitational action of all the remaining particles. We do not describe in detail here this procedure, we only mention the fact that it has been shown in the references quoted above that the complexity grows as $O(N_b \log N_b)$, and that the coefficient of proportionality is not easy to calculate because it depends on the "clumpiness' of the configuration and on two parameters (the *opening angle* and the *softening parameter*).

In the serial version of this algorithm the part which takes about 73% of the CPU time is precisely the *tree walk*: for this reason we have concentrated our efforts toward the parallelization of this part of the code. Other parts can be parallized as well: for instance the initial input phase during which particles positions and velocities are read from input files.

We will now describe in some detail two different strategies we have adopted to

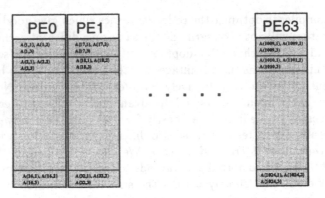

Fig. 1. Data sharing. The figure shows how an array is divided among the PEs.

parallelize the serial BH algorithm within the CRAFT environment of the Cray T3D and with PVM.

2.1 Parallel porting on the T3D

The CRAFT parallel environment on the Cray T3D allows the programmer to choose among three distinguishable, but cooperating, programming styles([3]).

1. *Data Sharing.* Data (e.g.arrays) are distributed over the memory of the PEs concurring to solve the program execution. All data are accessible from any PE, but the goal is to give the PEs the highest priority to perform operations on their own data (i.e. data residing in the PE's local memory). This because local data can be accessed faster than remote data. Attempts of a given PE of accessing data residing on another PE (remote memory) can add extra execution time due to the following factors:
   (a) Latency time: the associated time with getting on and off the PE node;
   (b) Contention of communication links: this happens when some PEs are trying to access the same communication links;
   (c) Distance between PE and remote memory.
2. *Work Sharing.* When these directives are given the program enters a parallel region code, and all PEs execute the same code. Using some specific directives a code region can be executed as a serial region: only one PE at a time (or only a master PE) executes this region of code. The work sharing paradigm allows one to distribute the statements of the code among the PEs with the goal of executing them in parallel. For example it is possible to divide a DO-LOOP among PEs so that each PE executes a subset of the whole iteractions.
3. *Message Passing.* This is supported by some low-level explicit message passing directives.

All these three styles can be mixed together within the same program.

In our case we have used the Data Sharing and the Work Sharing programming styles to parallelize the original serial Barnes & Hut algorithm. A detailed description of the algorithm is found in [2] and [1]. In the following we shortly describe the main phases of the algorithm steps, and we describe the aspects concernig the parallelizzation issue.

Whe can distinguish three phases in our code structure. a) **Data distribution**; b) **Tree formation and cell proprerties calculation**; c) **Force evaluation**. Let us then analyse the main paralleization directives during each of these phases.

Data Distribution. This is a very crucial phase to attain high performance of our code on Cray T3D. As a guiding principle we have tried to let the arrays containing the bodies properties to be shared among the available PEs, and to let each PE work, primarily, on bodies resident in its own local memory. At the same time, we would like all PEs' to have almost the same workload. The CDIR\$ SHARED directive of the CRAFT allows data to be shared among all available PEs. If the data distribution method were not correctly tuned to this strategy, the total performance of the run would be greatly affected.

In our case it is very important that each PE works prevalently on local data, in order to reduce the terms depending on data access.

At the start, the arrays containing bodies' properties (positions, mass, velocity, acceleration), are spread in contiguos block, using the CDIR\$ SHARED A(:BLOCK,:) directive, as shown in Fig. 1, showing the data distribution of arrays containing bodies' properties. Due to the coarse-grainy nature of the software, when the number of particles is enough large each PE mostly works on its local bodies.

The goal of this kind of data distribution, is to let as many PEs as possible work mainly on their allocated data, rather than on data allocated on other PEs.

The arrays containing the tree properties (cells' number, geometric and physical characteristics) are shared with CDIR\$ SHARED B(:BLOCK(2),:) in a way

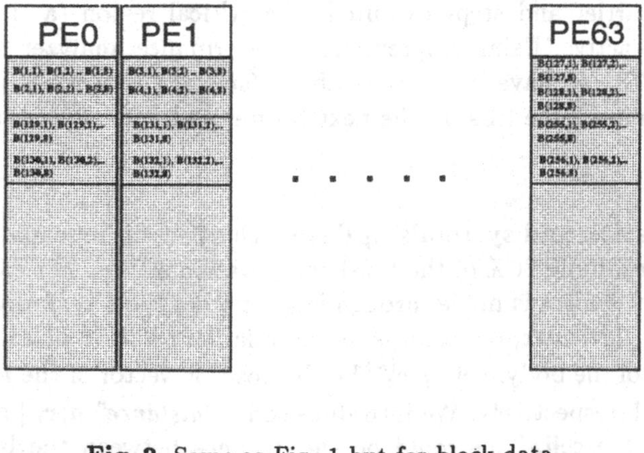

Fig. 2. Same as Fig. 1 but for block-data.

similar to the one shown in Fig. 1. Using this kind of tree data distribution we have reached the best performances with many runs.

Tree formation and cell properties. This phase is repeated at each time step. The spatial domain containing the system is divided into a set of nested cubic cells by means of an oct-tree decomposition. For those cells of the tree containing more than one body (*internal cells*, herafter icells) one stores the position, size, total mass and quadrupole moment in corresponding arrays. For cells containing only one body (*terminal cells*, hereafter fcells), on the other hand, one stores only the positions of the bodies.

Using the work sharing model, all the available PEs contribute to tree formation and to the calculation of icells' properties. The parallelism is performed sharing the DO-LOOPS structures among the PEs. The CRAFT directive CDIR\$ DO SHARED (ind1,[ind2, ind3,...]) mechanism allows to share a DO-LOOP: inside the shared DO-LOOPs each PE executes its assigned loop iteration. The following example shows a do loop shared operation.

```
DIMENSION B(1024,8), C(1024,8)
CDIR$ SHARED B(:BLOCK(2),:), C(:BLOCK(2),:)

CDIR$ DO SHARED (k,p) on B(p,k)
DO k=1,8
DO p=1,1024
B(p,k)=B(p,k)+C(p,k)*const
ENDDO
ENDDO
```

The B and C arrays are shared as shown in Fig. 1. Each K,P iteration acts on local elements and is executed only by the PE's owning the local B(p,k) elements. Access to local data is very fast. If a task must go to different PE to get the data it needs for the iteration, it does so, but a considerably longer time for the iteration execution will be needed. Syncronization mechanisms, as implicit and explicit barrier and steps executed in a critical region (a single PE at a time) are also active. Using Apprentice, a performance analyzer tool designed for CRAY MPP, we have noted that about 65% of the work is performed in parallel by the available PEs. In the next future, we hope to reach up to 90% of parallelism.

Force calculation and system's update. This phase is repeated at each time step and is consuming 80% of the total computational time. For each body, the calculation of the force is made through inspection of the *tree*, forming an *interaction list*: in particular, one compares the relative position of each cell of the tree with that of the body. Let $r_i, r_c^{(k)}$ be the position vector of the i-th body and of the k-th cell, respectively. We introduce some *"distance"* $d(r_i \mid r_c^{(k)})$ between the body and the cell. This could be the distance between the body and the center of mass of the given cell. Considering the ratio $z_{OC} = l^{(k)}/d(\mathbf{r}_i \mid \mathbf{r}_c^{(k)})$ for $1 \leq k \leq N_{cell}$, where $l^{(k)}$ is the size of the k-th cell (assuming that $k = 1$ is the root cell), those cells for which $\theta < z_{OC}$, where $0 \leq \theta \leq 1$ are considered too "near" to the body, and are not added to the interaction cell list: this means that the particle "sees" these cells as extended objects and one needs to "look inside" the smaller cells contained within them. For each of these one recalculates z_{OC} and one checks whether it is larger or lesser than θ. At the end, cells which do not satisfy the criterion and terminal cells (*fcell*) are included in the interaction cell list. Then the calculation of the force is made and the position is updated. Each PE executes the routines for this job separately (in a parallel region code) and, generally, for all the bodies that are resident in the local memory, as described in the data distribution phase. Using this calculation scheme, no synchronization mechanisms are needed. A dynamical load balancing is active in this phase and can re-distribute the loads among PEs if an unbalanced situation arises. Using Apprentice, we noticed that 100% of the work is performed in parallel by the available PEs. At the end of this phase a specific synchronization mechanism (CDIR$ BARRIER) acts before starting the next step (*tree formation*).

2.2 Dynamical Load Balancing

In the ideal case all PEs do the same work consuming the same time. Load imbalance strongly depends on how much data are *uniformly* distributed: ultimately on the geometry and mass distribution of the particles within the system, and can greatly vary during the run when clusters are formed and data become unevenly distributed.

The Dynamical Load Balancing (DLB) routines that we have implemented, attempt to avoid that during the run, an imbalance situation arises. The algorithms on which they are based assign to each body a PE executor (PEX), the PE that executes the *force calculation* phase for the body. At the beginning, using the data distribution shown in Fig. 1, the PEX assigned to each body is the PE where the body's properties are residing. If an unbalanced situation arises, a load re-distribution is performed. PEXs are reassigned taking into account the medium load (WHLD) for all the availbale PEs.

The use of this tecnique of Load Balancing, for several run's, allows a gain from 10% up to 25% in a uniform data distribution. Higher gains are expected in a "not uniform" data distribution (an evolved simulation).

2.3 Message Passing on The Convex SPP

The Convex Exemplar SPP-1000 is a Massively Parallel System which also allows a variety of programming models to the programmer, although it has a different hardware architecture in comparison to the Cray T3D (see Michielse [7] for an introduction to some of the hardware and software features). PEs are grouped into *hypernodes*, and a 5 class memory structure is superimposed on top of this,

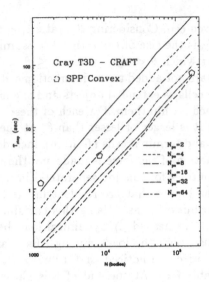

Fig. 3. Scaling with N_{bodies}. Open pentagons are for runs on the Convex SPP-1000. Values are averages over 10 timesteps. The points at $N_{bodies} = 216.000$ for the SPP-1000 is for only one timestep.

so that data can be accessed with different speeds according to the relative memory class which they belong to. In addition, message pasing primitives act within a shared memory environment, so that exchanging data amounts only to copy pointers and not to send/receive packets ([7]). Message passing is then greatly simplifies and accelerated in comparison to the TCP/IP case. As for the T3D, also on the SPP-1000 the programmer can choose between work-sharing and message passing styles. We have tested a PVM version of the treecode based on the *Orthogonal Recursive Bisection* (hereafter ORB) and *Locally Essential Tree* (hereafter LET) modifications of the original BH serial code inroduces by J. Salmon and S.M. Warren ([8], [9]), as we have shortly described in the Introduction and in more detail in [1]. An explicit message passing scheme allows a more direct control of the programmer over the optimization issues, although at the price of introducing an overhead due to synchronization and the structure of message passing primitives. We will see in the next Section that, despite these conjectures, the performance of the PVM version on the Convex Exemplar SPP is comparable to the one on the Cray T3D for high values of N_b.

3 Results

We have tested our codes with an initially homogeneous (in space) configuration for a small number of time steps ($10 T_{step}$). For such small number of timesteps

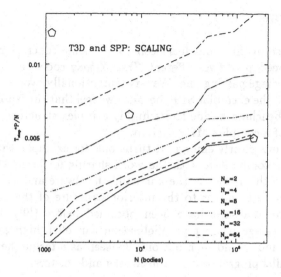

Fig. 4. Scaling with number of processors. In the ideal scaling case $T_{step} \propto 1/N_{proc}$ and curves should be horizontal lines. Symbols as in Fig. 3.

the system does not evolve very much: this should be kept in mind when considering these results.

On the other hand, this choice allows us to compare our results with those found by other authors ([4], [8], [9]), whose results refer also to almost homogeneous configurations. The results shown in Figs. 3 and 4 for the runs on the Cray T3D have been taken by averaging over 10 runs for each set of initial data. We have performed runs for different numbers of particles ($1000 \leq N_{bodies} \leq 216.000$) and for different numbers of processors ($4 \leq N_{proc} \leq 64$). The performance of the work- and data-sharing code scales self-similarly both with N_{proc} and with N_{bodies} (i.e. with the size of the job). The apparent "jumps" at some points in Fig. 4 are caused by the more or less efficient data and work redistribution which occurs when the number of processors is more or less far from a power of 2. As it is clear from the plots, this is the main factor controlling the speedup.

The open pentagons in Figs. 3 and 4 are the results of the runs of the message passing PVM code on the Convex SPP-1000. The points for $N_{bodies} = 1000$ and 8000 are averages over 10 timesteps. For $N_{bodies} = 216.000$ at the time of this Conference we have been successfully run only one timestep. The behaviour of the scaling parameter $T_{step} * N_{proc}/N_{bodies}$ for this code in Fig. 4 is similar to the one found by Salmon & Warren in their message passing implementation on the CM-5 ([8]), although the absolute values are obviously different.

From these preliminary results it is clear that the message passing code has an efficiency which scales almost linearly with N_{proc}, and the scaling seems to saturate already at $N_{bodies} \approx 200.000$.

4 Conclusions

Our main purpose in this work has been to get some concrete results on thye efficiency and speedup of two different *Tree N-body* codes: a work- and data-sharing and a message passing one. We have intentionally avoided to apply both to the T3D and to the exemplar SPP, because we feel that this application should not be seen as abenchmark, due to its highly complex structure which exploits all the features of a complex MPP system.

However, some hints clearly emerge. It turns out, for example, that the message passing version is competitive with the work-sharing for large jobs. We do not know yet whether this is due to the particular hardware and software platform where we run the test and/or to the interior structure of the algorithm. Note however that similar results have been obtained before ([9]). It seems clear. however, that message passing parallelization can offer high performance on MPP platforms, and should then not only be seen as a technique to be adopted only within parallel programming on networks and clusters.

References

1. Antonuccio-Delogu, V. and Becciani, U., "A Parallel Tree N-Body Code for Heterogeneous Clusters" *Proceedings of* PARA94, Dongarra, J. and Wasniewsky, J. eds., Lecture Notes in Computer Science **879**, 17 (Springer Verlag: 1994)
2. Barnes, J, and Hut, P., *Nature* **324**, 446 (1986)
3. Cray Research Inc., *"Cray MPP Fortran Refeference Manual* SR-2504 6.1 (1994)
4. Gouhong Xu, *"A new parallalel N-body gravity solver: TPM"*, *Astrophys. J. Supp.* **97**, 884 (1995)
5. Geist, A., Beguelin, A., Dongarra, J., Jiang, W., Manchek, R. and Sunderam, V., *"PVM 3 User's Guide and Reference Manual"*, ORNL/TM-12187, September 1994
6. Hernquist, L., *Astrophys. J. Suppl.* **64**, 715 (1987)
7. Michielse, P., "Programming the Convex Exemplar Series SPP System, *Proceedings of* PARA94, Dongarra, J. and Wasniewsky, J. eds., Lecture Notes in Computer Science **879** Lecture Notes in Computer Sciences **879** (Springer Verlag: 1994)
8. Salmon, J.K., *"Parallel hierarchical N-body methods"*, Ph. D.. Thesis, *unpublished* (California Institute of Technology: 1991)
9. Salmon, J.K., Warren, M. *"A portable Parallel N-body code"*, *unpublished* , (California Institute of Technology: 1995)
10. Stiller, L., Daemene, L.L. and Gubernatis, J.E., *J. Comp. Phys.* **115**, 550 (1994)

Numerical Solution of Stochastic Differential Equations on Transputer Network

S.S. Artemiev, S.A. Gusev, O.G. Monakhov

Computing Center, Sibirian. Division of Russian Academy of Sciences,
Pr. Lavrentiev, 6, Novosibirsk, 630090, Russia
e-mail: monakhov@comcen.nsk.su

Abstract. The problems of a design of computations by statistical simulation of trajectories of a the solution of systems of stochastic differential equations on transputer net are considered. The efficiency of the parallel programs for various numerical methods and calculated functionals of the solutions of stochastic differential equations is discussed. Results of numerical experiments are demonstrated.

1 Introduction

An equation of kind

$$dy(t) = f(t, y)dt + \sigma(t, y)dw(t), \quad t_0 \leq t \leq T, \quad y \in R^N, \tag{1}$$

where $w(\cdot)$ - M-dimensional standard Winner process, is called a stochastic differential equation (SDE) . Depending on the second integral in (1) is understood SDE's are understood in Stratonovitch or Ito sense.

Mathematical models, containing SDE's, are used in chemistry, medicine, radio physics, bank business and in many other areas of science and engineering.

The numerical methods which we consider are based on simulation of trajectories of SDE solution. It permits to receive the various probability characteristics of research object at nodes of given uniform grid on $[0, T]$. In this report we consider calculation of following functionals: (I) a mathematical expectation; (II) a covariance matrix; (III) a matrix of second moments; (IV) a correlation function of chosen component of SDE solution at the given time point; (V) a density function of a chosen component at the given time point; (VI) a density function of first times of SDE solution going out on the boundary of a given region; (VII) a joint density function of two chosen components of SDE solution at a given time point; (VIII) two-dimensional density function of a chosen component at two given time points .

Five methods [1] for numerical solving SDE's are offered in this report. They differ in order of convergence, properties of stability and choice of an integration step. Such are: a generalization of Euler's method (EULER), Milstein's method (MILSH), generalized one stage (STABLE) and two stage (ROSSDE) Rosenbrock type methods and generalized Runge-Kutta-Felberg method (RKFSDE).

First four methods simulate trajectories by a constant integration step, but RKFSDE – by variable random one. Each of this methods can be considered as mathematical expression of form

$$y_{n+1} = \Phi(y_n, t_n, h, f, \sigma) \tag{2}$$

where y_n, y_{n+1} are values of modeling trajectory at points $t_n, t_n + h \in [0, T]$, h is integration step.

Since the separate trajectories of random process are simulated independently, it is possible to calculate mentioned functionals by using a well-known method of organizing of network of transputers - processor farm.

However, there is a limit on volume of transmitted data $LMAX$ between "master" and "workers" for one access to commands $SEND$ and $RECEIVE$ in such parallel programs. In accord with the transmitted data it was convenient to partition functionals (I) – (VIII) on three groups: (I) – (IV), (V) – (VI) and (VII) – (VIII).

Independent simulation of the trajectories permit also to create programs with maximum saturation of transputer network by working packages. That gives high efficiency of parallel programs.

2 Calculation of functionals (I) – (IV)

We discuss here design of the parallel program for calculation of functionals (I) – (IV) in more detail. These tasks require to process data for each node of the grid. So as the number of nodes can reach a few thousands in many real problems, our program is based on simultaneous simulation of segments of different trajectories. The size of each segment does not exceed LMAX.

In this case the "worker" program does following operations: it accepts a working package containing necessary information about simulated segment; then it simulates the segment of trajectory and stories values of the trajectory at each node in the segment; after that the "worker" program sends to "master" program the package of results.

The "master" program at initial stage of work is made loading in transputer network k working packages, representing tasks for modelling of the first segments of k trajectories. The number k is a parameter of the program. It is chosen such that the most total saturation of the transputer network reaches.

Now we consider the sequence of operations of "master" program for calculation of mathematical expectation of SDE solution at the each node of the grid:

1) *to generate k working packages for simulation of the first segments of k trajectories and send its to the routing program;*

2) *to receive a package of results, containing simulated segment of trajectory and summarize the received vectors with appropriate elements of the array for calculating mathematical expectation;*

3) *If the received segment is not last in it's trajectory then go to 4) , else go to 5);*

4) *to generate a working package for a following segment of the trajectory and send it to the routing program; Go to 2);*

5) *If the number of begun trajectories less given volume of sample then put $k = 1$ and go to 1) , else go to 6);*

6) *If the number of received packages of results less number of sent working packages then go to 2) , else end of work of the program.*

For calculation of functionals (II) - (IV) appropriate programs the "owner" and the "worker" differ from adduced programs of calculation of mathematical expectation by small changes.

We consider valuation of efficiency for offered of programs for calculation of functionals (I) – (IV). Let L - quantity of segments of all simulated trajectories, T_s - time of transfer of one working package from the "master" to "worker", T_r - time of transfer of one package of results from the "worker" to "master", T_w - time, required to "master" program for processing of one packages of results, T_c - time, which is required to "worker" for simulation of one segment of a trajectory. Then for simulation L segments of trajectories at one working processor it will be required to the program time

$$T_1 = L(T_s + T_r + T_w) + LT_c \tag{3}$$

Let assume the program gives approximately identical load for each of N working in parallel processors. This assumption is valid when a sample volume is sufficiently large. Then the time for N working processors is

$$T_N = L(\mu_s T_s + \mu_r T_r + \mu_w T_w) + \frac{LT_c}{N}, \tag{4}$$

where $0 < \mu_s \le 1, 0 < \mu_r \le 1, 0 < \mu_w \le 1$, factors, meaning that the transferring of data and the processing of packages of results take place simultaneously with simulation of segments of trajectories. So we receive the efficiency

$$E = \frac{T_s + T_r + T_w + T_c}{N(\mu_s T_s + \mu_r T_r + \mu_w T_w) + T_c}, \tag{5}$$

Taking into account minimum and maximum significances μ_s, μ_r, μ_w the top and lower bounds for E will be

$$\frac{T_s + T_r + T_w + T_c}{T_c} > E \ge \frac{T_s + T_r + T_w + T_c}{N(T_s + T_r + T_w) + T_c} \tag{6}$$

¿From (5), (6) it is visible, that for $T_c \gg N(T_s + T_r + T_w)$ the significance E will be close to one and it will negligible depend of sample volume. When μ_s, μ_r, μ_w are sufficiently small it is possible $E > 1$. That sometimes is observed in practice.

3 Calculations of one-dimensional and two-dimensional densities

Functionals (V) – (VI) are one-dimensional densities, received as result of statistical simulation of trajectories. They are evaluated by means of histograms. For the majority of practical tasks volume of memory, required for the histograms, less $LMAX$. Therefore the parallel program can be made on the basis partition of all sample volume between processors, working in parallel. In this case the "master" distributes general number of simulated trajectories among "workers". Each the "worker" calculates histogram for that volume of sample, which to it is intended, and transmits the histogram to "master". "Master" summarizes received results from "workers" to general histogram. It is possible to show, that the efficiency of program will be higher, if the whole sample volume will be divided among "workers" equally.

We consider valuation of efficiency for this case. Let is formed k working packages in the program, T_s, T_r and T_w have such a sense as in previous case. T_c and Ntr is present time of simulation of one trajectory and the sample volume respectively. Then

$$T_1 = T_s + T_r + T_w + T_c \cdot Ntr,$$

$$T_N = k(\mu_s T_s + \mu_r T_r + \mu_w T_w) + \frac{T_c \cdot Ntr}{N}$$

So we receive

$$E = \frac{T_s + T_r + T_w + T_c \cdot Ntr}{Nk(\mu_s T_s + \mu_r T_r + \mu_w T_w) + T_c \cdot Ntr}, \tag{7}$$

The top and low bounds for E have sight

$$\frac{T_s + T_r + T_w + T_c \cdot Ntr}{T_c \cdot Ntr} > E \geq \frac{T_s + T_r + T_w + T_c \cdot Ntr}{Nk(T_s + T_r + T_w) + T_c \cdot Ntr} \tag{8}$$

¿From (7), (8) follows, that $E \to 1$ for fixed N, k at $Ntr \to \infty$.

The functionals (VII) – (VIII) are two-dimensional densities, which also are evaluated by means of histograms. However, in majority of cases the data of such volume are not placed in a buffer of length $LMAX$. For reception of the histogram it is necessary to receive two numbers from each simulated trajectory. In this case "worker" program is designed so, that it simulates not more $l = [LMAX/8]$ trajectories and sends to "master" l pairs of numbers for histogram.

For distribution of whole sample volume equaled Ntr among " workers" it will be required $k_1 = [(Ntr-1)/l] + 1$ working packages. If the saturation of the transputer network by working packages is equal k, at $k_1 \leq k$ the valuation of efficiency will such as well as for calculation of fuctionals (V) – (VI).

Let assume $k_1 > k$, then

$$T_1 = k_1(T_s + T_r + T_w) + T_c \cdot Ntr,$$

$$T_N = k_1(\mu_s T_s + \mu_r T_r + \mu_w T_w) + \frac{T_c \cdot Ntr}{N}$$

So we have

$$E = \frac{k_1(T_s + T_r + T_w) + T_c \cdot Ntr}{N k_1(\mu_s T_s + \mu_r T_r + \mu_w T_w) + T_c \cdot Ntr}, \qquad (9)$$

The top and low bounds for E have are

$$\frac{k_1(T_s + T_r + T_w) + T_c \cdot Ntr}{T_c \cdot Ntr} > E \geq \frac{k_1(T_s + T_r + T_w) + T_c \cdot Ntr}{N k_1(T_s + T_r + T_w) + T_c \cdot Ntr} \qquad (10)$$

¿From (9), (10) follows, that the given valuation of efficiency has not the asymptotic property of one for functionals (V) – (VI). However at

$$T_c \cdot Ntr \gg N k_1(T_s + T_r + T_w)$$

its efficiency will be close to one.

4   Numerical results

¿From adduced valuations of efficiency clearly, that the efficiency becomes more at increase of costs on simulation of trajectories T_c. The numerical experiments on networks for one and five transputers confirm that.

In the table below we demonstrate a speed up for some test problems. The speed up here is the ratio of calculation time for one transputer to calculation time for five transputers. The numerical methods are placed in order of increase T_c. From this table it is visible, that the speed up grows at increase of T_c. Exception is a method RKFSDE as it makes simulation of trajectories by variable random step.

Prog-ram	functional number							
	1	2	3	4	5	6	7	8
EULER	4.77	2.82	3.42	5.35	4.86	4.91	4.85	4.64
MILSH	5.16	4.93	4.92	5.10	4.88	5.00	4.85	4.72
STABLE	5.62	4.36	4.93	5.74	4.92	5.00	4.92	5.10
ROSSDE	5.83	5.05	5.11	5.71	4.93	5.03	4.93	5.10
RKFSDE	19.06	5.82	5.81	5.43	4.85	4.74	4.88	2.61

References

1. S.S. Artemiev. *The numerical solution of the Cauchy Problem for Systems of Ordinary and Stochastic Differential Equations*, Computing Center, Sib. Div. Russian Academy of Sciences, 1993.

Development of a Stencil Compiler for One-Dimensional Convolution Operators on the CM-5

Roch Bourbonnais

Thinking Machines Corporation,
245 First Street, Cambridge, MA 02142-1214, USA

Abstract. Finite difference methods are ideally suited to modern computer architectures and in particular parallel ones. Convolution operators or "stencils" are commonly used to solve a wide variety of problems. These operators can be elegantly written in data-parallel language but the performances obtained can be lower than expected.

With its cache-less vector node architecture, the CM-5 is particularly well suited to this type of operation on large volumes of data. We have developed a code generator that can improve performance over Connection-Machine Fortran (CMF) by a factor of sometimes 3 and reach computation rates in excess of 100 Mflops/node.

We will first describe the performance characteristics of stencil operators expressed in Connection Machine Fortran. Then, based on a performance model of the CM-5 processors , we will describe the three vectorization strategies that were implemented in the code generator.

1 Introduction

1.1 Justification

Finite difference methods are ideally suited to modern computer architectures and in particular parallel ones. In many cases, the integration scheme can be written in terms of convolution operators in which the source data are shifted then multiplied by a coefficient. These operators are collectively called "stencils".

In the case of petroleum reservoir modelling, it is common to see stencil expressions with a large ($\gtrsim$ 6) number of terms and in which the source data are shifted along a *single* axis of the 2-D or 3-D data set [1, 2, 3].

Despite the fact that these operators can be simply and elegantly written in data-parallel language such as HPF, the performances obtained are usually quite far from peak performance of the machines. The problem is that data-parallel constructs introduce extra temporaries and data-movement which are not strictly required to implement the stencils. Current compilers are unable to recognize stencils or produce inefficient code.

With its cache-less vector node architecture, the CM-5 is particularly well suited to this type of operation on large volumes of data. We have developed a code generator that can improve performance over Connection-Machine Fortran

(CMF) by a factor of sometimes 3 and reach computation rates in excess of 100 Mflops/node. The low-level CDPEAC (vector units assembler) generated code minimizes the within processor data movements keeping only the inevitable off-processor communication.

To reach good performance on a variety of cases, we developed three underlying strategies, each one striding memory in a particular order. The fastest vectorization strategy is selected at run-time using a semi-heuristic method that takes into account the data distribution in processor memory.

This work has led to the creation of a stencil utility capable of generating code for one-dimensional stencils of up to 9 points in either single or double precision, with either scalar or array coefficients.

This work distinguishes itself from other efforts, notably from a stencil compiler developed in the framework of the Connection Machine Scientific Software Library [4] or other [5], in that it aims to provide the highest possible performance reachable on a CM-5 system but only on a restricted set of stencils to be defined in the following section.

After the definitions, we will be discussing the implications of writing stencils using a high level language such as Connection-Machine Fortran [6]. We will then discuss in detail the implementation of our stencil generator for the CM-5.

1.2 Definitions

A stencil is formed by the successive accumulation into a destination array of a number of *terms* or *points* where each term is the product of a coefficient with a CSHIFT-ed [7, 6] source array. Each coefficient can be either full size arrays with the same rank and dimensions as the source data or more simply a scalar value. We differentiate variable (or array) coefficients to fix (or scalar) coefficient stencils. One-dimensional stencils are characterized by the fact that all terms of the expression are shifted relative to one another along what is called the *stencil* axis. A compact one-dimensional stencil is one in which the successive terms involve shifted data with consecutive shift counts. To specify an n-point one-dimensional compact stencil, one also needs to give the starting shift value which we call the *from* parameter; For example a 5 point compact stencil in which the first shift count is -3 will be formed of terms with shift counts of -3,-2,-1, 0, and 1 (see fig. 1).

If n is odd, then the stencil can be *centered* which means that for each term with a negative shift $(-s)$ value there will be a term with the positive shift value $(+s)$. So from here on, we will consider a one-dimensional compact stencil to be uniquely specified at compile time by :

n	:	the number of terms or points in the expression,
from	:	the leftmost (or smallest) shift value,
prec	:	the precision of all floating point data involved,
coeff type	:	the kind of coefficients used (fix or variable).

Furthermore, at runtime, it will be necessary to specify :

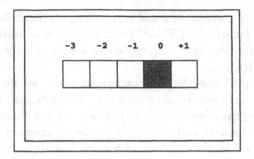

Fig. 1. 5-point compact stencil with *from* = -3.

data   :   the specific source, destination and coefficients to be used
  in the computation,
axis   :   which axis to use in the shift operations.

2   Writing Stencils in Connection-Machine Fortran

Consider a 5-point centered stencil. It could be written in CMF as in Fig. 2. To
analyze this expression the CMF compiler would split this into communication
and computation primitives [8]. The generated code would then be equivalent to
the code seen in Fig. 3. Lets now analyze in turn the performance characteristics
of each of the two components of this stencil.

```
    dest =  c1 * CSHIFT(source,dim=1,shift=-2) +
!           c2 * CSHIFT(source,dim=1,shift=-1) +
!           c3 *          source                +
!           c4 * CSHIFT(source,dim=1,shift=+1) +
!           c5 * CSHIFT(source,dim=1,shift=+2)
```

Fig. 2. 5-point centered stencils expressed in CMF.

2.1   Communication

In a CSHIFT [7, 6] , some data motion occurs internally in each processor and
only a restricted amount of data will be communicated to neighboring processors. Consider an array that is distributed block-wise in processor memory; each
processor holds a portion of the array which we refer to as the *subgrid*. For a
$sg1 \times sg2$ subgrid and a shift count x along dimension 1 the portion of data
that needs to be moved off the processor is $x/sg1$. Conversely, $(sg1 - x)/sg1$ of
the data needs to be moved inside the memory of each processor. The ability of
processors to move data in local memory, the *on-processor bandwidth*, is usually
orders of magnitude larger than the off-processor bandwidth. The consequence
is that, in many cases, the on-node data movement associated with CSHIFT,

```
        tmp1 = CSHIFT(source,dim=1,shift=-2)
        tmp2 = CSHIFT(source,dim=1,shift=-1)
        tmp3 = CSHIFT(source,dim=1,shift=+1)
        tmp4 = CSHIFT(source,dim=1,shift=+2)
        dest =  c1 *  tmp1    +
     !          c2 *  tmp2    +
     !          c3 *  source  +
     !          c4 *  tmp3    +
     !          c5 *  tmp4
```

Fig. 3. Equivalent CMF representation for 5-point centered stencils. CMF alternate the work between communication and computation phases

which strictly speaking is unnecessary, is also not too expensive. Nevertheless, if $sg1$ is very large, then on-node data movement can significantly impact the performance of a stencil.

Consider tmp1 and tmp2 in Fig. 3. Both are the result of a CSHIFT with different shift counts. A CSHIFT with a shift count of -2 will be sending $2/sg1$ of the data to neighboring processors and will naturally be more expensive than a CSHIFT with a shift count of -1 which sends only half as much data off-processor $(1/sg1)$. It is easy to see that one can express both tmp1 and tmp2 using two CSHIFT each using the "cheap" shift count of -1 as done in Fig. 4.

```
        tmp2 = CSHIFT(source,dim=1,shift=-1)
        tmp1 = CSHIFT(tmp2,  dim=1,shift=-1) ! this is faster
C       tmp1 = CSHIFT(source,dim=1,shift=-2) ! than this.
```

Fig. 4. Use shift count of ± 1 when possible

2.2 Efficiency of Code Generated by CMF.

We now consider the efficiency of the generated code when writing stencils in CMF. The computation part of a stencil is basically a sequence of mult-add operations and should lead to quite efficient code. When the coefficients are scalar values they can be loaded into scalar registers and reused during the computation process. Using the -S option to the CMF compiler, one can extract performance information about the generated code. We gathered this information in table 1; the cycle count refers to SPARC cycles while the flop count refers to the number of flops per VU. In these units the peak performance of the CM-5 is reached at 1 flop / cycle. We note that for stencils of more that 7 points the maximum performance of the generated code decreases significantly. This is because CMF allocates a maximum of 7 registers to hold scalar data.

From this information, we can infer a Mflops count for the computation part of each stencil (table 1). The actual performance will necessarily be smaller and one of the main reason for this will be DRAM page faults which stalls the processing for 5 SPARC cycles [9]. Taking into account these page faults one can refine table 1 into a new table that gives expected performance (not taking into account the communication phase) for each of the stencils. This is done in table 2.

	![cyc , flop]	% Mach. Peak	Mflops (32MHz)
2pt	![48, 24]	50 %	64
3pt	![64, 40]	63 %	81
4pt	![80, 56]	70 %	90
5pt	![96, 72]	75 %	96
6pt	![112, 88]	79 %	101
7pt	![128, 104]	81 %	103
8pt	![208, 120]	57 %	73
9pt	![232, 136]	58 %	74

Table 1. Peak stencil performance not taking into account possible DRAM page faults

	![cyc , flop]	% Mach. Peak	Mflops (32MHz)
2pt	![63, 24]	38 %	49
3pt	![84, 40]	48 %	61
4pt	![105, 56]	53 %	68
5pt	![126, 72]	57 %	73
6pt	![147, 88]	60 %	77
7pt	![168, 104]	62 %	79
8pt	![253, 120]	47 %	60
9pt	![282, 136]	48 %	61

Table 2. Peak performance taking into account DRAM page faults

Even for these operators that are very compute intensive, one can see that the performance that can be obtained from a high level language are limited. Note that these numbers do not take communication time into consideration.

3 Hand Coded Solution

3.1 Justification of Handwritten Solution

The main advantage of writing a low-level stencil package is to be able to avoid on-processor data movement. A performance gain can also be obtained by careful organization of data movement between memory and registers. Another advantage of writing a low-level stencil will be memory requirements. In CMF each CSHIFT requires a full-size temporary array. This can be a great obstacle to code that needs to work with very large data sets. A low-level approach will use only the amount of temporary space that is strictly necessary.

Also, a low-level approach can be implemented as a single subroutine which avoids the synchronization steps required by CMF between each communication steps as well as between the communication and computation step. One can also avoid weaknesses of the global compiler such as the 7 scalar register limitation mentioned in section 2.2

3.2 Communication

We decided at the onset of this project to use a model that required no in-processor memory moves. This means that whenever some data element is required in some computation it will be fetched directly from its location in the source array. For the data that is not locally available, a communication phase needs to be performed before running the computation phase. For a 2-d data set temporary *slabs* of data orthogonal to the shift axis will be used as in Fig. 5. These slabs are filled by calls to the Connection Machine Run-Time System (CMRTS). Today, there are no restriction on array sizes that can be treated with the stencil generator.

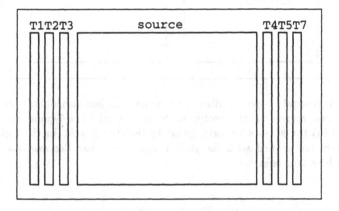

Fig. 5. Layout of data in every processing element after communication phase

3.3 The Three Strategies of Computation

Having brought the data inside the local memory of each processing element (in this case in each vector unit) one is ready to start the computation phase of the stencil.

As we have said, one-dimensional stencil operators are characterized by a specific user selected axis. We will use this axis to orient our view of the problem. The data on which this stencil is to be performed is kept in memory which has a linear organization. The data of a multidimensional array thus has an axis that varies fastest in memory (the axis is *aligned* with memory) and other axes that vary in memory each with their own non-unit stride.

Therefore two situation can occur. Either the axis of the stencil is aligned with the axis that varies fastest in memory or it is not. The CM-5 vector units introduce a third direction which is the axis along which vectors will be aligned. So a given user-selected stencil axis, can be either aligned with the axis that varies fastest in memory or not, and in either case one can perform the computation using vectors that are either aligned with the stencil axis or orthogonal to it (see Fig. 6). It is expected that the performance characteristic of either vectorization of this problem will be different in each case.

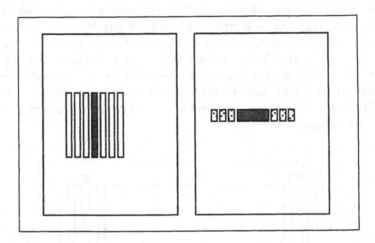

Fig. 6. Two vectorization possibilities. The stencil axis lies horizontally in both cases. The shaded box represents the vector to be computed in a 7-point stencil. In the leftmost case (a) the vectors are orthogonal to the stencil axis. In the rightmost case (b) the vectors are aligned with the stencil axis. Note that the axis varying first in memory can be either axis.

The orthogonal vectorization (fig. 6a) works its way through the stencil from left to right computing each vector in succession while keeping previously loaded data in registers as long as they are needed. In principle this is the most efficient method since a piece of data is loaded once and only once into the vector registers and used as many times as there are terms in the stencil. The number of terms that can be implemented using this method is limited by the number of registers available to hold the data.

The aligned vectorization (fig. 6b) works somewhat differently in that it works on one vector at a time, loading all of the data needed to compute a given vector. It does not try to re-use loaded data in further computations. For example, to compute a 7-point stencil with a *from* value of -3 (with shifts of -3,-2,...+2,+3), for a vector of 8 elements (dark shaded box), requires loading an *extended* vector of $8 + 7 - 1 = 14$ elements total. Discarded data will eventually need to be reloaded for the computation of some other vector but this inefficiency is sometimes compensated by a better alignment of vectors in memory which leads to faster loads or store instructions.

The fact that each vector is computed independently of previous and future computations means that we can compute different vectors in any order. Since memory accesses via the vector units are not cached there is no cache miss ratio to be optimized. However the memory of the vector units are part of the SPARC virtual memory system that is accessed through a Translation Look-aside Buffer (TLB) [10]. The TLB maps the memory between the virtual address space and the physical address space using 64 entry table of 4k-bytes pages. The TLB cost will come into play only if the data needed for a given problem does not all fit using the 64 entries. At this point, it becomes important to minimize the TLB cost by using a memory access pattern that makes maximal use of each entry. Fig. 7 shows two variations on the aligned vectorization strategy that behave differently in terms of TLB usage.

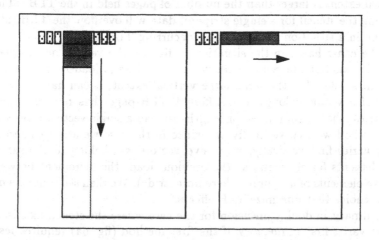

Fig. 7. Two alternatives for Aligned Vectorization. In the leftmost case (a) the successive computations are ordered by stepping orthogonally to the stencil axis while in the rightmost case (b) successive computations lie along the stencil axis.

The Selection. Each one of the three strategies has its own performance characteristic as a function of the data layout in the processors. For single precision, one important point affecting the performance of a given strategy, is whether the vectors are aligned with memory or not : Single precision vectors aligned with memory can be stored (as doubles) 7 times faster than unaligned ones [9]. Other factors that come into play when selecting the optimal strategy are DRAM page faults and TLB misses.

Orthogonal vectorization will be used when memory varies first in the direction that is not aligned with the stencil axis (vertically in fig. 6a) , and aligned vectorization(fig. 7) will be used when memory varies first in the same direction as the stencil axis. In this case, the TLB effect will favor the second case (fig. 7b) of aligned vectorization which will be used in preference.

For double precision, the selection is more complex because there are no big gains to be expected from aligning vectors with memory. If the memory varies first along the same axis as the stencil axis then the method showed in figure 6b will be a method that shows the best behavior in terms of memory access since all elements of the array are accessed sequentially with unit stride. This will be the method used when the stencil axis is aligned with the memory axis. However if memory varies first along the axis that is not the stencil axis the analysis needs to be pushed further.

In the first case (fig. 6a), the vectors are aligned with memory, but successive computed vectors will correspond to memory addresses that are far apart (the increment being proportional to the vertical extent). For a very large vertical extent, each vector of data will belong to a different TLB page. Now, if the horizontal extent is larger than the number of pages held in the TLB table, then computing the stencil for a single stripe of data will overflow the TLB table and we will be in a situation with regularly occurring TLB misses.

On the other hand, in the situation of fig. 7a, the vectors are aligned with the stencil axis but not with memory (i.e. the data is loaded in vectors using a non-unit stride). For the same large vertical extent, it can happen that *each* element of a vector belongs to a different TLB page thus requiring up to 16 TLB entries (the maximum vector length) to load a single vector. However since in this method we step vertically and since in the case we are considering that memory varies first vertically, successive vectors will belong to the same TLB pages (elements for elements) as the previous load (the increment between two successive elements of a column is here unit stride). We then see that this method can be a choice that minimize TLB effects.

In summary, in double precision for the case where the stencil axis is not the axis that varies first in memory, if the first method (fig. 6a) requires less TLB pages than the table can hold (64), then we use this method. In other cases we use the method shown in fig. 7a.

It should be clear at this point that the selection process is based on a semi-heuristic mechanism that, if well tested, is certainly not fool-proof which means that the best method will not be selected in every case. A mechanism was put in place to allow the user if he chooses so, to select the vectorization strategy.

Another possible solution to this problem, would be to develop an accurate model of performance of each strategy and use it to select the best method at run-time. Preliminary remarks on this problem will be presented in the next section.

3.4 Performance Model

As a general rule, any vector instruction on n elements takes $2 * n$ cycles but stores on of n singles take $7 * n$ cycles. One must also take into account DRAM page faults. Finally, the vector units will sometimes introduce delay cycles or "bubbles" between successive instructions that are not "pipeline" compatible.

When working on very large data sets, TLB effects will come into play. We measured the time taken *per* load instruction as a function of the number of pages accessed (Fig. 8). When working with less that 64 vectors, the time per load is constant at about 21 cycles (16 regular cycles plus 5 cycles for the DRAM page fault). If N_p is the number of entries in the TLB table (64) and p_{miss} is the probability of incurring a TLB miss and if TLB pages are replaced at random [10] in the 64 entry table, we can express a self-consistent equation between n and p_{miss} as :

$$p_{miss} = 1 - ((N_p - 1)/N_p)^{n*p_{miss}}$$

This equation establishes a relation between n and p_{miss} that is plotted alongside measurements in Figure 8.

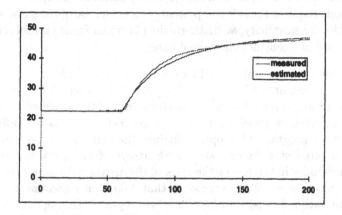

Fig. 8. Performance model of vector load instruction

While the current model under development gives satisfactory results for medium size arrays, there are still many unresolved question regarding the effect of striding and TLB misses on large arrays. Nevertheless the model was used as a guide in the development of the heuristic selection process. It would be premature at this point to rely on it for the full vectorization strategy selection problem.

3.5 Stencil Generator Product

We have developed a stencil generator that takes into account the considerations we have developed in this paper. The generator is a argument driven UNIX command written in C to be used on a CM-5 compile server. The arguments fully specify the stencil to be generated. The files are then generated and compiled leaving only the two object files (one for the control processor and one for the nodes of the CM-5) to be linked with data-parallel code. From the standpoint of the user, the stencil becomes a subroutine which takes as arguments the source and destination arrays, the axis along which to apply the stencil, and the coefficients involved in the computation.

Initially the first version of the generator required that arrays should be distributed in CM-5 node memory without the use of garbage masks. Currently an option can be specified to remove this restriction. Since this option does not seem to impact the performance of the initial restricted case, we expect that it will disappear in the future releases.

We also have implemented the option for the user to *accumulate* the result of a stencil to a given destination addresses rather than to overwrite that destination. This allows the combination of multiple stencils to eventually generate more general multidimensional stencils.

The generated code works with arrays of any rank. For simplicity reason, the generated code requires that all components of a stencil be of the same arithmetic precision. Also for simplicity, we distinguished between fix or variable coefficients stencils but did not consider the mixed case.

Finally, the generator also produces a fortran user code and Makefile that can serve as an example of usage of the generated stencil. This example code will report a timing comparison of the performance of the specified stencil versus an equivalent fortran expression. Additional generator options will influence this example fortran program. One option defines the rank of the test arrays (note that the generated stencils will work with arrays of any rank but the fortran program hard-codes in the source the rank of the arrays it manipulates). Another option will produce a CMF expression that takes into account the shift=1 optimization outlined in section 2.1. This will report a timing comparison that accurately reflects the capability of straight CMF for writing stencils. By default the generated code will rely of the stencil package to select the best vectorization for the tested stencils. However an option can be used to make the example code try all three of the vectorization strategies.

As an example case, we will consider the 9-point, single precision centered stencil applied on a 2048×2048 data-set on a 32-node CM-5. The global stencil performance (communication + computation) is 87 and 73 Mflops/node depending on the axis along which the stencil is applied. The best CMF code (version 2.2.11) is respectively 3.2 and 2.4 times slower. On a CM-5E, the numbers are 112 and 89 Mflops/nodes with improvements over CMF of 3.2 and 2.3.

For a 9 point stencil with a *from* value of -4, then 4 leading slabs and 4 trailing slabs would be needed to compute the stencil. One additional slab is used for an optimization in the communication phase so that a total of 9 slabs ($9/sg1$ of a full-sized temp) are required. It is common for the subgrid to be in the hundreds or even thousands which means that only a fraction of a full-size temp is needed to compute a stencil using this method. This is to be contrasted with a CMF approach which requires a full-sized temporary variable for each CSHIFT in an expression.

4 Conclusions and Prospects

We have developed a code generator for one-dimensional compact stencils to be used on CM-5 equipped with vector units. For each case, three different vectorization strategies were developed. A heuristics selection mechanism was tuned to select the fastest of the strategies at runtime depending on the parameters of the stencil. The generated code demonstrate levels of performances not reached using the standard data-parallel fortran compiler specially when the number of terms involved is large.

In many cases, we believe that the generated code will not allow many more performance improvements. However some work can still be done to reduce the latency of stencil calls improving overall performance on small array size. An area of improvement lies in the development of stencil with still higher number of terms. Also it would be interesting to be able to combine the currently supported stencils to form higher dimension 2-D cross- or block- stencils.

Finally, if this product proves to be of high enough quality then its incorporation into Thinking Machines' stencil compiler software would be the logical step. Further down the line, the ideal situation from the users standpoint would be to incorporate this directly and transparently into data-parallel compilers.

References

1. Holberg O. Computational aspects of the choice of operator and sampling interval for numerical differentiation in large-scale simulation of wave phenomena. *Geophysical prospecting*, (35):629–655, 1987.
2. Rodrigues D. *Large scale modeling of seismic wave propagation.* PhD thesis, Ecole Centrale de Paris, 1993.
3. Igel H., Djikpesse H., and Tarantola A. Waveform inversion of marine reflection seismograms for p-impedance and poisson's ratio. *Geophysical Journal International*, 1994. in press.
4. Ralph G. Brickner, Kathy Holian, Balaji Thiagarajan, S. Lennart Johnsson, and Palle Pederson. A stencil compiler for the connection machine model cm-5. *unknown*, 1995.
5. Andreas Pirklbauer. Stengen- array aliasing based stencil generator. private communication.
6. Thinking Machines Corp., Cambridge, MA. *CM Fortran Language Reference Manual*, 2.2 edition, October 1994.
7. International Organisation for Standardization and International Electrotechnical Commission, editors. *Fortran 90 [ISO/IEC 1539:1991 (E)]*. 1991.
8. Thinking Machines Corp., Cambridge, MA. *CM-5 CM Fortran Performance Guide*, 2.2 edition, October 1994.
9. Thinking Machines Corp., Cambridge, MA. *DPEAC Reference Manual*, 7.1 edition, October 1992.
10. ROSS Technology, Inc, Austin, TX. *SPARC RISC users's guides*, second edition, February 1990.

Automatic Parallelization
of the AVL FIRE Benchmark
for a Distributed-Memory System*

Peter Brezany[a], Viera Sipkova[a], Barbara Chapman[b], Robert Greimel[c]

[a]Institute for Software Technology and Parallel Systems
[b]European Centre for Parallel Computing at Vienna
University of Vienna, Liechtensteinstrasse 22, A-1092 Vienna, Austria
[c]AVL List GmbH, Kleiststrasse 48, A-8020 Graz, Austria

Abstract. Computational fluid dynamics (CFD) is a Grand Challenge discipline whose typical application areas, like aerospace and automotive engineering, often require enormous amount of computations. Parallel processing offers very high performance potential, but irregular problems like CFD have proven difficult to map onto parallel machines. In such codes, access patterns to major data arrays are dependent on some runtime data, therefore runtime preprocessing must be applied on critical code segments. So, automatic parallelization of irregular codes is a challenging problem. In this paper we describe parallelizing techniques we have developed for processing irregular codes that include irregularly distributed data structures. These techniques have been fully implemented within the Vienna Fortran Compilation System. We have examined the AVL FIRE benchmark solver GCCG, to evaluate the influence of different kinds of data distributions on parallel-program execution time. Experiments were performed using the Tjunc dataset on the iPSC/860.

1 Introduction

Computational Fluid Dynamics (CFD) is a Grand Challenge discipline which has been applied as a successful modelling tool in such areas as aerospace and automotive design. CFD greatly benefits from the advent of massively parallel supercomputers [12]. On the close cooperation between researchers in CFD and specialists in parallel computation and parallel programming, successful CFD modelling tools were developed. This paper deals with the automatic parallelization of the FIRE solver benchmark developed at AVL Graz, Austria.

The CFD software **FIRE** is a general purpose computational fluid dynamics program package. It was developed specially for computing compressible and incompressible turbulent fluid flows as encountered in engineering environments.

* The work described in this paper was carried out as part of the European ESPRIT project PPPE and CEI-PACT project funded by the Austrian Science Foundation (FWF) and the Austrian Ministry for Science and Research (BMWF).
Authors' e-mail addresses: {brezany,barbara,sipka}@par.univie.ac.at, rg@avl.co.at

Two- or three-dimensional unsteady simulations of flow and heat transfer within arbitrarily complex geometries with moving or fixed boundaries can be performed.

For the discretization of the computational domain a finite volume approach is applied. The resulting system of strongly coupled nonlinear equations has to be solved iteratively. The solution process consists of an outer non-linear cycle and an inner linear cycle. The matrices which have to be solved in the linear cycle are extremely sparse and have a large and strongly varying bandwidth. In order to save memory, only the non-zero matrix elements are stored in linear arrays and are accessed by indirect addressing.

Automatic parallelization of irregular codes like FIRE is a challenging problem. In irregular codes, the array accesses cannot be analyzed at compile time to determine either independence of these or to find what data must be pre-fetched and where it is located. Therefore, the appropriate language support is needed, as well as compile time techniques relying on runtime mechanisms. The **Vienna Fortran** (VF) language [14] provides several constructs to deal efficiently with irregular codes. These include constructs for specifying irregular distributions and for explicitly specifying asynchronous parallel loops (FORALL loops).

We have transformed one benchmark solver (**GCCG**, orthomin with diagonal scaling [1]) of the AVL FIRE package to VF and parallelized it by the **Vienna Fortran Compilation System** (VFCS). Because, access patterns to data arrays of the inner cycle are not known until runtime, a parallelization method based on the combination of compile time and runtime techniques has been applied.

In Section 2 the method and data structures of the GCCG program are discussed. Section 3 describes the data structures as they are specified by VF. We focus on the selection of the appropriate data and work distribution for irregular code parts. We elaborated two versions of the program. In the first version, the INDIRECT data distribution was specified for the data arrays accessed in the irregular code parts, and in the second one, all arrays got BLOCK data distributions. The mapping array used for the irregular distribution was determined by an external partitioner. Section 3 also describes the automatic parallelization strategy which the VFCS applies to irregular codes and the interface to runtime support. Performance results for the Tjunc dataset achieved for both GCCG program versions on the Intel iPSC/860 system are discussed in Section 4. The rest of the paper deals with related work (Section 5), followed by the conclusion (Section 6). Concepts described in this paper can be used in the compiler of any language that is based on the same programming model as VF.

2 Survey of the GCCG solver

The FIRE benchmark consists of solving the linearized continuity equation with selected input datasets using the unstructured mesh linear equation solver GCCG. Each dataset contains coefficients, sources, and the addressing array including linkage information. The numerical solution to differential equation describing the transport of a scalar variable Φ is performed with the finite volume method. For the discrete approximation the computational domain is subdivided into a finite number of hexahedral elements, the control volumes or *internal* cells. A compass notation is used to identify the interconnections between the centres of the control volumes (E=East, W=West, etc.). Integrating over the control volumes results in a system of non-linear algebraic equations:

$$A_p \, \Phi_p = \sum_{c=\{E,S,N,W,SE,...\}} A_c \Phi_c + S_\Phi \qquad (1)$$

where the pole coefficient A_p can always be written as the sum of the neighbour coefficients in different directions. Due to the locality of the discretisation method equation (1) represents an extremely sparse system of algebraic equations which has to be solved iteratively. The solution process consists of an outer cycle dealing with the non-linearities and an inner cycle dealing with the solution to the linearized equation systems. In particular, the outer cycle is designed to update the coefficients and sources from the previous iterations in order to achieve strong coupling between momentum and pressure. The innermost cycle consists of solving the linear equation systems:

$$A \, \Phi = S_\Phi \qquad (2)$$

for every flow variable Φ. The main diagonal of A consists of the pole coefficient A_p, and the sidebands of A are obtained from the neighbouring node coefficients A_c of (1). Both, the coefficients and the sources are kept constant during the iterative solution steps. To solve the linear equation system the truncated Krylov subspace method Orthomin is used [13].

The solution process stops when the ratio of the residual vector R^n at iteration n and the residual vector R^0 at iteration 0 falls below a small value ϵ :

$$\| \, R^n \, \| = \| \sum_{c=\{E,S,...\}} A_c \Phi_c^n + S_\Phi - A_p \Phi_p^n \, \| < \epsilon \cdot \| \, R^0 \, \| \qquad (3)$$

For boundary conditions an additional layer of infinitely thin cells is introduced around the computational domain, called *external* cells.

Flow variables, coefficients and sources associated with each cell are held in 1-dimensional arrays of two different sizes: one size corresponds to the number of internal cells (parameter NNINTC), and the second size corresponds to the total number of cells, internal plus external cells (parameter NNCELL). These two kinds of arrays are related with each other in such a way that the first portion of the bigger arrays (including internal cells), is aligned with the smaller arrays. To determine the interconnection of cells (links cell-centre to cell-centre) an indirect addressing scheme is used whereby a unique number, the *address*, is associated with each cell. These linkage information is stored in the 2-dimensional array LCC, where the first index stands for the actual cell number and the second index denotes the direction to its neighbouring cell. The value of LCC(cell number, direction) is the cell number of the neighbouring cell. All calculations in the solver GCCG are carried out over the internal cells. External cells can only be accessed from the internal cells, no linkage exists between the external cells.

3 Automatic Parallelization of GCCG

Initially, the sequential code of the GCCG solver has been ported to the VF code. Declarations of arrays with distribution specifications and the main loop within the outer iteration cycle is outlined in Figure 1. This code fragment will be used as a running example to illustrate our parallelization method.

```
        PARAMETER :: NP = ...
        PARAMETER :: NNINTC = 13845, NNCELL = 19061
        PROCESSORS P1D(NP)
        DOUBLE PRECISION, DIMENSION(NNINTC),                &
           DISTRIBUTED(BLOCK) :: BP, BS, BW, BL, BN, BE, BH,    &
                                 DIREC2, RESVEC, CGUP
        INTEGER, DISTRIBUTED(BLOCK, :) :: LCC(NNINTC, 6)
S1      DOUBLE PRECISION, DIMENSION(NNCELL), DYNAMIC ::  &
                                 DIREC1, VAR, DXOR1, DXOR2
S2      INTEGER, DISTRIBUTED(BLOCK) :: MAP(NNCELL)
        ...
S3      READ ( u) MAP
S4      DISTRIBUTE DIREC1, VAR, DXOR1, DXOR2 :: INDIRECT (MAP)
        ...
F1      FORALL nc = 1, NNINTC ON OWNER(DIREC2(nc))
                 DIREC2(nc) =  BP(nc) * DIREC1(nc)          &
                             - BS(nc) * DIREC1(LCC(nc, 1))  &
                             - BW(nc) * DIREC1(LCC(nc, 4))  &
                             - BL(nc) * DIREC1(LCC(nc, 5))  &
                             - BN(nc) * DIREC1(LCC(nc, 3))  &
                             - BE(nc) * DIREC1(LCC(nc, 2))  &
                             - BH(nc) * DIREC1(LCC(nc, 6))
        END FORALL
        ...
```

Fig. 1. Kernel loop of the GCCG solver

Arrays DIREC1, VAR, DXOR1 and DXOR2 are distributed irregularly using the INDIRECT distribution function with the mapping array MAP which is initialized from a file referred to as mapping file; the remaining arrays have got the BLOCK distribution.

The mapping file has been constructed separately from the program through the

Domain Decomposition Tool (DDT) [8] employing the recursive spectral bisection partitioner. The DDT inputs the computational grid corresponding to the set of internal cells (see Figure 2), decomposes it into a specified number of partitions, and allocate each partition to a processor. The resulting mapping array is then extended for external cells in such a way that the referenced external cells are allocated to that processor on which they are used, and the non-referenced external cells are spread in a round robin fashion across the processors. Arrays VAR, DXOR1 and DXOR2 link DIREC1 to the code parts that don't occur in Figure 1.

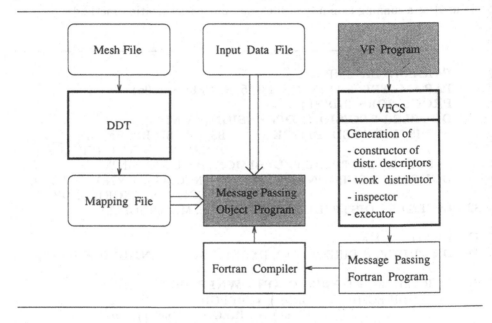

Fig. 2. Flow of Data in Compilation and Execution Phases

The code version that only includes regular distributions is simply derived from Figure 1 by replacing the keyword **DYNAMIC** in statement S1 by the distribution specification: **DISTRIBUTED(BLOCK)**, and removing statements S2, S3 and S4. The main loop (F1) within the outer iteration cycle computes a new value for every cell by using the old value of this cell and of its six indirectly addressed neighbored cells. This loop has no dependences and can be directly transformed to the parallel loop. Work distribution **ON OWNER**(DIREC2(nc)) ensures that communications are caused only by the array DIREC1. The remaining loops, with exception of the one that updates the DIREC1, perform either calculations on local data only or global operations (global sums).
In the rest of this section, we first describe the parallelization strategy applied to the code version from Figure 1 and then briefly discuss simplifications applied when processing the code containing regularly distributed arrays only. Our ap-

proach is based on a model that is graphically sketched in Figure 2. The method applied by DDT is outlined on the previous page. The strategy used by the VF-CS generates four code phases for the above kernel, called the **constructor of data distribution descriptors (CDDD), work distributor, inspector**, and **executor** [10]. CDDD constructs a runtime distribution descriptor for each irregularly distributed array. The work distributor determines how to spread the work (iterations) among the available processors. The inspector analyzes the communication patterns of the loop, computes the description of the communication, and derives translation functions between global and local accesses, while the executor performs the actual communication and executes the loop iterations. All the phases are supported by the appropriate runtime library that is based on the PARTI library developed by Saltz and his coworkers [7].

INTEGER :: MAP(local^{MAP}_size)
INTEGER, DIMENSION(:), **POINTER** :: local^{DIREC1}, local^{DIREC1}_size
TYPE (TT_EL_TYPE), **DIMENSION**(:), **POINTER** :: tt^{DIREC1}
TYPE (DIST_DESC) :: dd$_{DIREC1}$
...
    C--**Constructing local index set for DIREC1 using the MAP values**
        CALL build_Local(dd$_{MAP}$, MAP, local^{DIREC1}, local^{DIREC1}_size)
    C--**Constructing translation table for DIREC1**
        tt^{DIREC1} = build_TRAT(local^{DIREC1}_size, local^{DIREC1})
    C--**Constructing runtime distribution descriptor for DIREC1**
        dd$_{DIREC1}$%local = local^{DIREC1}
        dd$_{DIREC1}$%local_size = local^{DIREC1}_size
        dd$_{DIREC1}$%tt = tt^{DIREC1}
        ... initialization of other fields of dd$_{DIREC1}$ ...

Fig. 3. CDDD for array DIREC1

3.1 Constructor of Data Distribution Descriptors (CDDD)

For each irregularly distributed array A and in each processor p, CDDD constructs the runtime distribution descriptor $dd_A(p)$ which includes information about: shape, alignment and distribution, associated processor array, and size of the local data segment of A in processor p. In particular, it includes the local index set $local^A(p)$ and the translation table tt^A. The set $local^A(p)$ is an ordered set of indices designating those elements of A that have to be allocated in a particular processor p. The translation table tt^A is a distributed data structure which records the home processor and the local address in the home processor's memory for each array element of A. Only one distribution descriptor is constructed for a set of arrays having the same distribution. The construction of

dd_{DIREC1} (it also describes the distribution of VAR, $DXOR1$ and $DXOR2$) is illustrated by Figure 3. On each processor p, the procedure *build_Local* computes the set $local^{DIREC1}(p)$ and its cardinality denoted by variables $local^{DIREC1}$ and $local^{DIREC1}\_size$ respectively. These two objects are input arguments to the PARTI function *build_TRAT* that constructs the translation table for $DIREC1$ and returns a pointer to it.

3.2 Work Distributor

On each processor p, the work distributor computes the *execution set* $exec(p)$, i.e. the set of loop iterations to be executed on processor p.

For the GCCG kernel loop in Figure 1, the execution set can be determined by a simple set operation (see [2]):

$$exec(p) = [1 : NNINTC] \cap local^{DIREC2}(p)$$

VF provides a wide spectrum of possibilities for the work distribution specification. The appropriate techniques for processing individual modes are described in [3, 4, 5, 6].

3.3 Inspector, Executor, and PARTI support

The **inspector** performs a dynamic loop analysis. Its task is to describe the necessary communication by a set of so called *schedules* that control runtime procedures moving data among processors in the subsequent executor phase. The dynamic loop analysis also establishes an appropriate addressing scheme to access local elements and copies of non-local elements on each processor. The inspector generated for the GCCG kernel loop is introduced in Figure 4.

Information needed to generate a schedule, to allocate a communication buffer, and for the global to local index conversion for the rhs array references can be produced from the appropriare *global reference* lists, along with the knowledge of the array distributions. Each global reference list $globref_i$ is computed from the subscript functions and from $exec(p)$[2] (see lines G1–G16 of Figure 4). The list $globref_i(p)$, its size and the distribution descriptor of the referenced array are used by the Parti procedure *localize* to determine the appropriate *schedule*, the size of the communication buffer to be allocated, and the local reference list $locref_i(p)$ which contains results of global to local index conversion. The declarations of rhs arrays in the message passing code allocate memory for the local segments holding the local data and the communication buffers storing copies of non-local data. The buffers are appended to the local segments. An element from $locref_i(p)$ refers either to the local segment of the array or to the buffer. The procedure ind_conv performs the global to local index conversion for array references that don't refer non-local elements.

The **executor** is the final phase in the execution of the FORALL loop; it performs communication described by schedules, and executes the actual computations for all iterations in $exec(p)$. The schedules control communication in such a

---

[2] In Figure 4, $exec(p)$ and its cardinality are denoted by the variables *exec* and *exec_size* respectively.

```
C--INSPECTOR code

C--Constructing global referece list for the 1st dimension of LCC
G1     n1 = 1
G2     DO k=1, exec_size
G3         globref1(n1) = exec(k); n1 = n1 + 1
G4     END DO
C--Index Conversion for LCC having the data distribution desc. dd_1
       CALL  ind_conv (dd_1, globref1, locref1, n1-1)
C--Constructing global reference list for DIREC1 and
C--DIREC2, BP, BS, BW, BL, BN, BE, BH
G5     n1 = 1; n2 = 1; n3 = 1
G6     DO k=1, exec_size
G7         globref2(n2) = exec(k); n2 = n2 + 1
G8         globref3(n3) = exec(k)
G9         globref3(n3+1) = LCC(locref1(n1),1)
G10        globref3(n3+2) = LCC(locref1(n1),4)
G11        globref3(n3+3) = LCC(locref1(n1),5)
G12        globref3(n3+4) = LCC(locref1(n1),3)
G13        globref3(n3+5) = LCC(locref1(n1),2)
G14        globref3(n3+6) = LCC(locref1(n1),6)
G15        n3 = n3 + 7; n1 = n1 + 1
G16    END DO
C--Index Conversion for DIREC2, BP, BS, BW, BL, BN, BE, BH
C--having the common data distribution descriptor dd_2
       CALL  ind_conv (dd_2, globref2, locref2, n2-1)
C--Computing schedule and local reference list for DIREC1
       CALL  localize(tt^{DIREC1}, sched3, globref3, locref3, n3-1, nonloc3)


C--EXECUTOR code

C--Gather non-local elements of DIREC1
       CALL  gather(DIREC1(1), DIREC1(local^{DIREC1}_size+1), sched3)
C--Transformed forall loop
       n2 = 1; n3 = 1
       DO k=1, exec_size
          DIREC2(locref2(n2)) = BP(locref2(n2))  * DIREC1(locref3(n3))      &
             - BS(locref2(n2)) * DIREC1(locref3(n3+1))      &
             - BW(locref2(n2))  * DIREC1(locref3(n3+2))      &
             - BL(locref2(n2))  * DIREC1(locref3(n3+3))      &
             - BN(locref2(n2))  * DIREC1(locref3(n3+4))      &
             - BE(locref2(n2))  * DIREC1(locref3(n3+5))      &
             - BH(locref2(n2))  * DIREC1(locref3(n3+6))
          n2 = n2 + 1; n3 = n3 + 7
       END DO
```

Fig. 4. Inspector and Executor for the GCCG kernel loop

way that execution of the loop using the local reference list accesses the correct data in local segments and buffers. Non-local data needed for the computations on processor p are gathered from other processors by the runtime communication procedure **gather**[3]. It accepts a schedule, an address of the local segment, and an address of the buffer as input arguments.

4 Performance Results

This section presents the performance of the automatically parallelized benchmark solver GCCG for two types of data distribution: the regular BLOCK, and the irregular distribution according to the mapping array. The generated code was slightly optimized by hand in such a way that the inspector was moved out of the outer iteration cycle, since the communication patterns do not change between the solver iterations.

Dataset: TJUNC		
Number of internal cells: 13845		
Number of total cells: 19061		
Number of solver iterations: 338		
Sequential code		
Number of Processors	Time (in secs)	
1	43.47	
Parallel code		
Number of Processors	Time (in secs)	
	Data Distr: BLOCK	Data Distr: INDIRECT
4	30.25	27.81
8	24.66	19.50
16	22.69	13.50

Table 1. Performance results of the GCCG solver

We examined the GCCG using the input dataset Tjunc, with 13845 internal cells and 19061 total number of cells. Calculation stopped after the 338 iterations. The program has been executed on the iPSC/860 system, the sequential code on 1 processor, the parallel code on 4, 8, and 16 processors. The results are sumarized in the Table 1. Figure 5 shows the speedup defined by the ratio of the

---

[3] If necessary, space for the local segment and buffer is reallocated prior to the **gather** call, depending on the current size of the segment and the number of non-local elements computed by localize.

single processor execution time T_s to the multi-processor execution time T_{np}, where np stands for the number of processors.

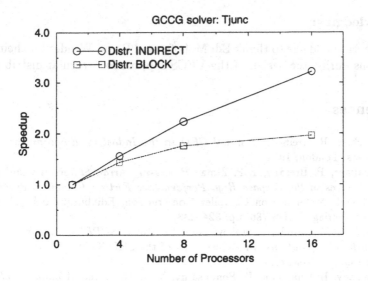

Fig. 5. Speedup of the GCCG solver

5 Related Work

Solutions of many special cases of the problems discussed in this paper appeared in the literature. Koelbel [10] introduces the term inspector/executor for processing irregular data-parallel loops. PARTI [7] has been the first runtime system constructed to support the handling of irregular computations on massively parallel systems on the basis of the inspector/executor paradigm. Van Hanxleden [9] developes a method for optimizing communication placement in irregular codes. Techniques developed for automatic coupling parallel data and work partitioners are described in [6, 11].

6 Conclusions

Our experiments have shown that in irregular codes like the AVL FIRE Benchmark it is advantageous to use irregular distributions. There are many heuristic methods to obtain irregular data distribution based on a variety of criteria; different partitioners that have been developed in the last years offer the implementation of those methods. VFCS automatically generates interface to external and on-line partitioners on the basis of the information provided by the user and derived by the compile time program analysis. This paper described the automatic parallelization method based on the utilization of the external partitioner.

Acknowledgment

The authors would like to thank Edi Mehofer and Bernd Wender for their helpful discussions during the design of the VFCS support for irregular distributions.

References

1. G. Bachler, R. Greimel. *Parallel CFD in the Industrial Environment.* Unicom Seminars, London, 1994.
2. S. Benkner, P. Brezany, H.P. Zima. *Processing Array Statements and Procedure Interfaces in the Prepare High Performance Fortran Compiler.* Proc. 5th International Conference on Compiler Construction, Edinburgh, U.K., April 1994, Springer-Verlag, LNCS 786, pp. 324–338.
3. P. Brezany, M. Gerndt, V. Sipkova, and H.P. Zima. *SUPERB Support for Irregular Scientific Computations.* In Proceedings of the SHPCC '92, Williamsburg, USA, April 1992, pp.314-321.
4. P. Brezany, B. Chapman, R. Ponnusamy, V. Sipkova, and H Zima, *Study of Application Algorithms with Irregular Distributions,* tech. report D1Z-3 of the CEI-PACT Project, University of Vienna, April 1994.
5. P. Brezany, O. Chéron, K. Sanjari, and E. van Konijnenburg, *Processing Irregular Codes Containing Arrays with Multi-Dimensional Distributions by the PREPARE HPF Compiler,* HPCN Europe'95, Milan, Springer-Verlag, 526–531.
6. P. Brezany, V. Sipkova. *Coupling Parallel Data and Work Partitioners to VFCS.* Submitted to the Conference EUROSIM – HPCN Challenges 1996, Delft.
7. R. Das, and J. Saltz. *A manual for PARTI runtime primitives - Revision 2.* Internal Research Report, University of Maryland, Dec. 1992.
8. N. Floros, J. Reeve. *Domain Decomposition Tool (DDT).* Esprit CAMAS 6756 Report, University of Southampton, March 1994.
9. R. van Hanxleden. *Compiler Support for Machine-Independent Parallelization of Irregular Problems.* Dr. Thesis, Center for Research on Parallel Computation, Rice University, December 1994.
10. C. Koelbel. *Compiling Programs for Nonshared Memory Machines.* Ph.D. Dissertation, Purdue University, West Lafayette, IN, November 1990.
11. R. Ponnusamy, J. Saltz, A. Choudhary, Y-S. Hwang, G. Fox. *Runtime Support and Compilation Methods for User-Specified Data Distributions.* Internal Research Report, University of Maryland, University of Syracuse, 1993.
12. H.D. Simon. *Parallel CFD.* The MIT Press, Cambridge, 1992.
13. P.K.W. Vinsome. *ORTHOMIN, an iterative method for solving sparse sets of simultaneous linear equations.* In Proc. Fourth Symp. on Reservoir Simulation, Society of Petroleum Engineers of AIME, pp. 149–159.
14. H. Zima, P. Brezany, B. Chapman, P. Mehrotra, A. Schwald. *Vienna Fortran - A language Specification Version 1.1.* University of Vienna, ACPC-TR 92-4, 1992.

2-D Cellular Automata and Short Range Molecular Dynamics Programs for Simulations on Networked Workstations and Parallel Computers

Marian Bubak[1,2]

[1] Institute of Computer Science, AGH, al. Mickiewicza 30, 30-059, Kraków, Poland
[2] Academic Computer Centre – CYFRONET, ul. Nawojki 11, 30-950 Kraków,Poland
email: bubak@uci.agh.edu.pl

Abstract. In the paper parallel algorithms for two popular simulation methods: cellular automata and short range molecular dynamics are presented and their implementation on networked workstations and CONVEX Exemplar SPP1000/XA is reported. Programs are equipped with load balancing procedures based on local self-timings and they are useful for simulation of large systems wit 10^7+ lattice sites and 10^5+ particles.

1 Introduction

Cellular automata and molecular dynamics are recently applied for investigation of fluid mechanics problems at microscopic level [1, 2] as well as for simulation of macroscopic phenomena with intrinsically discontinuous nature like friction and fracture associated with earthquakes [3] or deformation and penetration in materials of non-liquid properties [4]. These new fields of research need large number of lattice sites or molecules (above 1 million) to be involved in simulations. This can be achieved on massively parallel computers (see e.g. [5]) or on networked workstations.

In this kind of parallel programs master-slave scheme is obvious (see Figs. 1 and 2). Since master programs only read system layout, decompose it into domains, gather data for the load-balancer and receive and store intermediate and final data, the processor which hosts the master process may also run the slave one.

2 Cellular automata programs

Two cellular automata programs have been developed. The first one is meant as a tool for lattice gas (LGA) investigations of fluid flows while the second one is appropriate for modelling of immune systems (CA).

Lattice gas consists of unit mass particles moving with unit velocities on a hexagonal lattice and colliding at lattice sites according to energy and momenta conservation rules. Macroscopic values are obtained by averaging after

```
read computational box layout
decompose box into domains
send layout of domains to all nodes
while ( not end of simulation )
        receive values from nodes
        calculate new global values
        send new global data to nodes
        if load balancing required
                compute new boundaries of domains
                send new boundaries to nodes
        store values
end    while
final report
```

Fig. 1. Host program structure for cellular automata and molecular dynamics simulation.

```
receive domain layout
generate initial state of domain
while ( not end of simulation )
        compute new state of domain
        exchange boundary layers of domains with neighbors
        if global values calculation required
                store or send domain state to host
                send timings to host
                receive new global data from host
                receive new domain boundaries from host
                rearrange domain data
end    while
final report
```

Fig. 2. Slave program structure for cellular automata and molecular dynamics simulation.

each specified number of steps. The evolution of the lattice consist of: absorption, collision, injection, and free particle streaming [6].

In the immune system model [7] the following cells are considered at each lattice site: virus V, macrophages M, cytotoxic suppresing cells K and helpers H. Time evolution at a given site is described by a set of local rules which take into account the state of nearest neighbor sites.

Parallelization of cellular automata simulation consists in division of a lattice into domains assigned to different processes (processors, workstations). Only border rows of domains are transmitted between neighbor domains after each timestep. Using multispin coding one can represent cellular automata with binary arrays assigned to different automaton variables (directions of motion in the case of LGA, cells for CA). In this way parallelization is introduced on the instruction level.

Wall-clock execution time (per one lattice site and one timestep) for parallel LGA program is presented in Figs 3 and 4, and for CA program – in Fig. 5. Computation on CONVEX Exemplar SPP1000 were done on dedicated subcomplex with 7 processors in it.

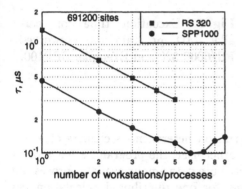 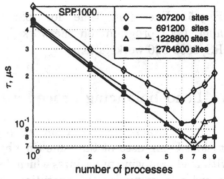

Fig. 3. Execution time for parallel LGA program on networked RS/6000-320 and Exemplar SPP100.

Fig. 4. Execution time for parallel LGA program on Exemplar SPP100 for different number of lattice sites.

3 Molecular dynamics program

In the short-range 2-D *MD* program Verlet neighbor list (N-list) (reconstructed automatically) is used to speed up the evaluation of interactions between molecules and Hockney linked-list is applied to build the neighbor list. Equations of motion are solved with leap-frog algorithm [8].

The parallel program is based on division of a computational box into domains with a number of layers of Hockney cells in each. Domains are assigned to different processes. Each process keeps the complete data set of its own particles, local N-list and positions of particles which are in the boundary layer of cells in one of the two adjacent domains. Communication is necessary only for particles in boundary Hockney layer.

Wall-clock execution time per particle and timestep for homogenous cluster of IBM RS/6000 and CONVEX Exemplar SPP1000/XA (on dedicated subcomplex with 7 processors in it) is presented in Fig. 6.

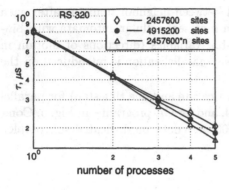 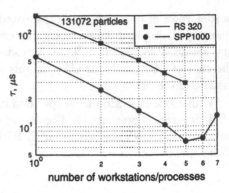

Fig. 5. Execution time for parallel CA program on networked RS/6000-320 workstations.

Fig. 6. Execution time for parallel MD program on networked RS/6000-320 and Exemplar SPP100.

4 Load balancing procedure

As the programs have been devised to run on heterogenous clusters of workstations being used at the same time by other user, the programs are equipped with load balancing procedures which distribute data among available processors in order to minimize the execution time for one distributed calculation (so called restricted LB problem [9]). Load balance is based on local self-timings and is accomplished by changing the partitioning of the computational box at chosen timesteps in order to obtain the same computation time for all processes. Changes of particle density in domains are taken into account and a mechanism is included to reduce unstable behavior introduced by an operating system scheduler [10]. If the elements (sites, particles) were distributed on p processors according to the current speed of each of them n_i/T_i, (n_i is number of elements on i-th processor and T_i is wall-clock time of computation in the last time interval on i-th processor), the time of computation would be

$$t_o = \frac{n}{\sum_{i=1}^{p} \frac{n_i}{T_i}} \quad , \qquad n = \sum_{i=1}^{p} n_i \quad . \tag{1}$$

so in the next interval the number of elements on i-th workstation should be

$$n_i' = (1 - \lambda_i) \cdot \frac{n_i}{T_i} \cdot t_o + \lambda_i \cdot n_i \quad , \qquad \lambda_i \in [0, 1] \quad , \tag{2}$$

where λ_i is the dumping parameter.
More efficient load balancing algorithm is based on exponential distribution:

$$\theta_i = (1 - \lambda_i)\tau_i^{comp} + \lambda_i\theta_i' \quad , \tag{3}$$

$$\lambda_i = exp(-\frac{t_i^{cpu}}{T_0}), \tag{4}$$

$$\theta_o = \frac{1}{\sum_{i=1}^{p} \frac{1}{\theta_i}}, \tag{5}$$

$$n_i' = \frac{n}{\theta_i} \cdot \theta_o, \tag{6}$$

where τ_i^{comp} is the average wall-clock computation time per one timestep and one lattice site, θ_i' is the previous value of θ_i, t_i^{cpu} is cpu time used since the last LB action, and T_0 is a parameter.

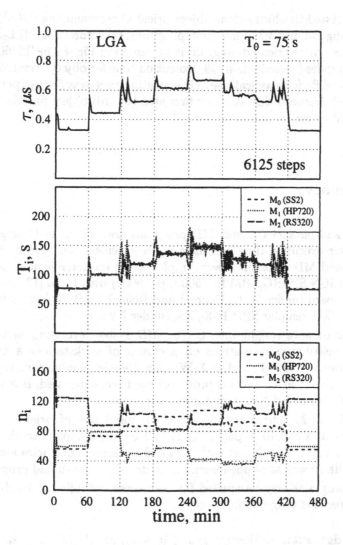

Fig. 7. Performance of load balancing procedure for LGA program.

Performance of the load-balancer for LGA program on 3 workstation intentionally loaded is shown in Fig. 7. The additional programs were running on M_1 (one program from 60 to 120 and 300 to 420, two programs - from 180 to 300 minute of the experiment) and on $M - 2$ (one program from 120 to 240 and from 360 to 420, two programs - from 240 to 360). n_i in Fig 7 is the number of stripes; one strip comprises b rows, where b is the side of averaging box.

5 Molecular dynamics object oriented program

It was also tested to which extent object-oriented programming (OOP) is useful for developing *CA* and *MD* simulation programs. Structure of a 3-D L-J molecular dynamics object-oriented program is presented in Fig. 8 On RS/6000-520H, with *gcc* compiler (optimization O3) execution time for object-oriented program was $1200\mu s$ while for classical one [11] execution time was $560\mu s$ (per timestep and particle). Nethertheless it is our first step towards object parallelism which is considered as very promising approach [13].

6 Summary

Sequential and parallel *CA* and *MD* programs are written in C language and optimized for execution time and memory requirements on workstations. Parallel LGA and MD algorithms were implemented on heterogenous network of computers (SUN SPARC, IBM RS/6000, HP 9000) with PVM [12] as well as on transputer board (under 3L C, Express and Helios), on iPSC/860 (with NX-2), and CONVEX Exemplar SPP 1000/XA (under PVM).

In order to have insight into *CA* or *MD* system evolution in the course of parallel distributed simulations on a cluster of workstations a distributed graphical system – *DEG* (based on X-Window libraries, nonblocking asynchronic sockets and UDP communication protocol) has been elaborated. It may be used with any of distributed programming environments.

The *CA* and *MD* programs are useful for simulation of large systems with 10^7+ lattice sites and 10^5+ particles. They have benn applied to the following simulations: a HIV infection at early stage, flows around obstacles and through porous media as well as of phenomena in materials of non-liquid properties.

These programs may be applied for comparison of different hardware platforms and parallel programming tools.

Acknowledgments. J. Mościński, J. Kitowski, M. Pogoda and R. Słota are acknowledged for enlightening discussions and collaboration.
The work was partially supported by KBN under grant 8 S503 006 07.

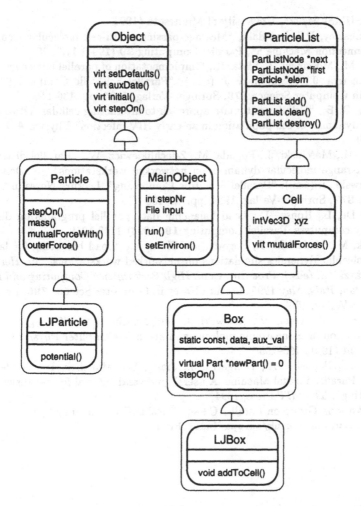

Fig. 8. Structure of molecular dynamics object oriented program.

References

1. M. Mareschal, and B.L. Holian, eds., *"Microscopic simulations of complex hydro-dynamic phenomena"*, *NATO ASI Series, Series B: Physics VOL.292, Plenum Press, New York and London*, 1992
2. Bhattacharya, D.K., G.C. Lie, G.C., Clementi, E., "Molecular dynamics simulations of fluid flows", in Enrico Clementi, ed., *Modern Techiques in Computational Chemistry, MOTECC-91*, ESCOM Leiden, 1991
3. Mora, P., Place, D., " A lattice solid model for the nonlinear dynamics of earth-quakes", International Journal of Modern Physics C 4 (1993) 1059-1074
4. Alda, W., Bubak, M., Dzwinel, W., Kitowski, J., Mościński, J., "Computational molecular dynamics at the Institute of Computer Science, AGH - Cracow, Poland", International Symposium on Computational Molecular Dynamics, Minneapolis, University of Minnesota, 23-16 October 1994; also Supercomputer Institute Research

Report UMSI 94/224, University of Minnesota (1994).

5. D.M. Beazley, P.S. Lomdahl, "Message-passing multi-cell molecular dynamics on the Connection Machine 5", Parallel Computing **20** (1994) 173-195

6. Bubak, M., Mościński, J., Słota, R., "Implementation of parallel lattice gas program in: Dongarra, J., Waśniewski, J. (eds.): "Parallel Scientific Computing", Lecture Notes in Computer Science **879**, Springer-Verlag, 1994, pp. 136-146

7. Pandey, R. B., "Cellular automata approach to interacting cellular network models for the dynamics of cell population in an early HIV infection" Physica **A 179** (1990) 442-470

8. Bubak, M., Mościński, J., Pogoda, M., Zdechlikiewicz, W., "Parallel distributed 2-D short-range molecular dynamics on networked workstations", in: Dongarra, J., Waśniewski, J. (eds.): "Parallel Scientific Computing", Lecture Notes in Computer Science **879**, Springer-Verlag, 1994, pp.127-135

9. Keyser De, L., Roose., D.: Load balancing data parallel programs on distributed memory computers. Parallel Computing **19** (1993) 1199-1219

10. Bubak, M., Mościński, J., Pogoda, M., Słota, R., "Load balancing for lattice gas and molecular dynamics simulations on networked workstations", in: *Hertzberger, B., Serazzi, G., (eds.), Proc. Int. Conf. High Performance Computing and Networking, Milan, Italy, May 1995*, Lecture Notes in Computer Science **796**, pp. 329-334, Springer-Verlag, 1995

11. Bargieł, M., Dzwinel, W., Kitowski, J., and Mościński, J., "C-language program for simulation of monoatomic molecular mixtures", *Computer Physics Communication*, **64** (1991) 193-205

12. Geist, A., Beguelin, A., Dongarra, J., Jiang, W., Manchek, R., Sunderam, V., "PVM: Parallel Virtual Machine. A users' guide and tutorial for networked parallel computing", *The MIT Press*, 1994

13. The Working Group on Parallel C++, "Parallel C++ roadmap", http://www.lpac.ac.uk/europa/index.html

Pablo-Based Performance Monitoring Tool for PVM Applications

Marian Bubak[1,2], Włodzimierz Funika[1], Jacek Mościński[1,2] and Dariusz Tasak[1]

[1] Institute of Computer Science, AGH, al. Mickiewicza 30, 30-059, Kraków, Poland
[2] Academic Computer Centre – CYFRONET, ul. Nawojki 11, 30-950 Kraków,Poland
email: bubak@uci.agh.edu.pl

Abstract. In the paper we present a tool for off-line performance monitoring of parallel applications running under PVM. A trace file collected by XPVM is converted into the pure SDDF one and then it is analysed and visualized using a tool built of the Pablo environment modules. In this way we have avoided developing our own instrumentation. Our monitoring tool is easily extensible and ongoing work focuses on implementing most of the ParaGraph functions. Sample monitoring sessions of a 3-D molecular dynamics program and conjugate gradient benchmark are presented and obtained results are interpreted.

1 Introduction

Recent development of parallel distributed computing and increase of its acceptance have resulted in demand for monitoring systems which should provide information necessary to get insight to behavior of parallel programs in order to improve their performance [1]. The main attributes of a distributed program behavior users are interested in are: state of CPU utilization, network utilization, load balancing over the whole system and overhead induced by communication between interacting processes. Tools should measure and present parameters such as speed of data transmission, amount of message traffic, length of messages, errors occured in the network traffic, total number of calls to the communication system, the time spent in communication, the number of probes in the message queue. According to Malony [2] a general performance visualization environment architecture should consist of a *database* formed of performance data, *views* and *displays*. A *view* is generated from performance data through any combination of filtering, reduction and analysis. Each *display* should provide a set of *resources* and *methods*, where *resources* are the configuration parameters defining a display instance and *methods* are used to control display operations.

Our goal was to have got an off-line (post mortem) performance monitoring tool suitable for parallel programs running under PVM on networked workstations. We would like to avoid developing our own instrumentation and the idea was to build the tool with elements available in the public domain.

In the next section we present a brief overview of performance monitoring tools for distributed programs. In section 3 we present our way to get a tool that could meet the requirements of portability, extensibility and reusability.

The tool is constructed with facilities provided by XPVM and Pablo so data transformation from the form generated by XPVM into the form meaningful for the Pablo analysis and visualization modules is described. In section 4 we supply a short hints on how the tool could be used. In section 5 two sample sessions are presented. The paper concludes with an outlook on future work.

2   An overview of performance monitoring tools

In Tab. 1 we have collected main features of available monitoring tools [3]. The tools are divided into three categories according to their usefulness for PVM applications. Symbols used in Tab. 1 are explained in Tab. 2.

Table 1. Monitoring tools for distributed programs.

Name	type	trace generation	mode	visualization	instrumentation
for PVM and PVM based programs					
Xab	d, pt	au	on	an	au
XPVM	m, d, pt	au	on, off	g, an, sm, p	au
EASYPVM	ctl	au	off	ParaGraph	au
HeNCE	env, d, pt	au	on, off	g, an, sm, p	au
environment independent					
Pablo	pt	u	off	g, an	g, i-a
IPS-2	pt	au, u	off	sm, g, p	au
JEWEL	pt	u	on, off	sm, g, p, an	src
AIMS	pt	au, tpt	off	g	g, i-a, src
ParaGraph	pt	–	off	g, an, sm, p	PICL
environment dependent					
PICL	ctl	au, u	off	ParaGraph	au, src
Express-tools	pt	au, u	off	g, sm, p	au, src
TOPSYS-tools	d, pt	au	on, off	g, an, sm, p	au
Linda-tools	d, pt	au	on, off	g, an, sm, p	au

As one can see the PVM users have few monitoring facilities with rather limited possibilities. Formerly there was Xab [4], now replaced by XPVM [5].

Table 2. Explanation of symbols used in Tab.1.

Symbol	Meaning
ctl	communication & tracing library
pt	performance tool
m	monitoring program
d	debugging tool
env	environment
au	automatically generated
tpt	post-mortem trace processing tool
u	user-defined
off, on	off-line, on-line
g	graphical
an	animated
sm	summaries
p	performance
i-a	interactive
src	manual source instrumentation
PICL	executed by means of PICL
ParaGraph	visualized by means of ParaGraph

XPVM is an on-line PVM graphical interface that allows to start, kill and monitor tasks. It provides a space-time diagram, utilization count, network animation display, network queue and two textual displays: call trace and task output. Unfortunately, it gives no statistics, cumulative displays and traces of user events. EASYPVM [6] provides a set of communication functions built on the top of PVM and enables the use of conventional PVM primitives in a program. The instrumentation in EASYPVM is provided only for its own communication library and it does not allow to customize easily EASYPVM for pure PVM program. Moreover, tachyons often occur in EASYPVM trace files and they are the source of problems at visualization stage. HeNCE [7] is an environment that assists users in developing parallel programs. The drawback of environment-independent tools like IPS-2 [8], JEWEL [9] and AIMS [10] is that they are difficult to install and maintain and, moreover, it is not sure that they will be supported for a longer time. The environment-dependent tools are included in parallel programming environments: *Express* [11], *Linda* [12], *TOPSYS* [13] and their tracing facilities use intrinsic mechanisms of individual environments.

The features mentioned in section 1 have been implemented in the Pablo environment [14] which is based on semantically-free SDDF (Self Defining Data Format) for handling trace data, the graphical interactive instrumentation facility and modules which may be used for building up a graph of analysis and visualization.

ParaGraph [15] is an off-line visualization tool with about 25 displays and it allows a refined insight into program behavior, especially in communication between processes. For a PVM user the drawback is that ParaGraph consumes only trace data files produced by PICL [16] which uses its own (non-PVM) communication primitives and data created by EASYPVM[1]. Paragraph is considered by the distributed computing community as the most instructive and informative performance data visualization system, therefore we intend to create a tool that can provide most of ParaGraph functionality.

Taking all above into account we have decided to create the tool which uses trace files generated by XPVM and Pablo modules for analysis and visualization. As the both products will be supported in the future we do not need worry about instrumentation of new mechanisms which may be included in PVM.

3 Monitoring using XPVM trace file and Pablo

We have found the XPVM trace files did not meet all requirements of SDDF and Pablo (the sequence of record descriptors in XPVM trace files differed from that requested by the input module, time-stamps in XPVM required taking into an account the local clocks synchronization, the range of some data in the XPVM trace-file records was not acceptable by Pablo), therefore a program module for transforming trace files produced by XPVM into the Pablo acceptable input files has been developed. Its general structure is shown in Fig. 1.

The *error-pruner* removes a few minor errors from XPVM trace file while the *record data processor* puts the packets into the required order, tunes values of local time stamps, reshapes data into the valid ranges, generates new information: task utilization and execution time between barriers, computes statitistics indispensable for setting display parameters. The most important fields in XPVM's Record Descriptors (valid in every record) are: time stamp of event and PVM task identifier. Other fields depend on the type of event described by the individual record. A number of new fields has been added to record descriptors (see Tab. 3).

dtPablo extracts record descriptors from the incoming data and puts them into the record dictionary (RD) of XPVM descriptors. Then it transforms the data to the format defined by the record dictionary (RDNew) according to the rules specified in the trans.rules file. Transformed data are recorded using format from the rec.descr.proto file. *dtPablo* has been developed with methods of Pablo system SDDF library. Now we handle a set of records resulting from the applications developed in the field of computer simulation using lattice gas, molecular dynamics and cellular automata methods.

Graphical user interface for analysis and visualization was created using the Pablo modules (see Fig. 2). It enables presentation of the following performance

---

[1] When this paper was completed we learnt that this approch was applied also in TAPE/PVM available from *ftp.imag.fr*.

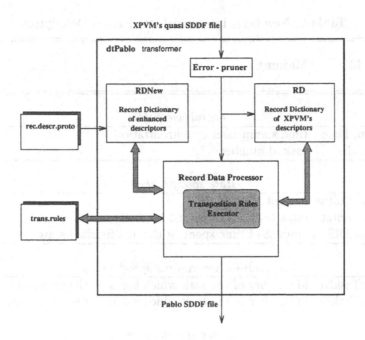

Fig. 1. Dataflow in the dtPablo transformer.

data: synchronized time of every event in a parallel program (Time_Stamp), matrix of current connections with the trace of recent message volume (Active_Transmissions), current computation time spent by every task between two consecutive barriers (Barrier_Mean_Time), cumulated time spent in each state by every task (Task_Utilization), cumulated number of messages sent between tasks (Summ_of_Msg_Counts), cumulated volume of messages sent between tasks (Summ_of_Msg_Volume), current total volume of messages in a program as a function of time (Msg_Volume_vs_Time), state (busy, system, idle) of every task (Task_Activity). There are three layers of processing: input pipe which consists of data file submitted by transformer and `Fileinput` Pablo module, analysis level which is built of modules synthesizing data into structures required by displays, and presentation level (displays).

4 How to use the tool

In order to make the tool working one should get and install Pablo Analysis Environment and SDDF library (available from *cs.bugle.uiuc.edu*), install XPVM, get and compile transformer dtPablo using *makeProgram* script from SDDF package.
Monitoring of parallel program behavior consists of the following steps:
- run XPVM and monitor the program; option WHERE must be active and point to the machine where XPVM is started and in order to minimize XPVM's

Table 3. New fields introduced into Record Descriptors.

Field	Meaning
	general use fields
SyncTime	time stamp after synchronization
TaskNr	task id number
	state specific fields
TaskStatus	current state of task
PosOfStat	state that has been just finished
TimeOfStat	amount of time spent in the just finished state
	communication records specific fields
SrcTaskNr	id number of the task which has sent the message
DstTaskNr	id number of the task which has to receive the message
	barrier specific field
BarrTime	amount of time between two consecutive barriers

effect on the application being monitored all XPVM views should be inactive,
- transform the trace file obtained from XPVM's monitoring with *dtPablo* transformer; the new file gets ".Pablo" extension,
- start Pablo and load pvmGraph ("Load Configuration" command from "File" menu); as an input data the file obtained from *dtPablo* should be used,
- configure display parameters ("Module Parameters" command from "Configure" menu) and insert the data statistics produced by *dtPablo* into proper fields of the emerged dialog box,
- configure parameters for all the modules suggested by *dtPablo*,
- run visualization.

5 Sample sessions

In Fig. 3 and in Fig. 4 we present examples of monitoring sessions of two parallel programs running on networked workstations. Time_Stamp is presented as a dial, Active_Transmissions and Task_Activity are shown as matrix displays, Msg_Volume_vs_Time has form of chart diagram, and the rest of displays are bar graphs.

In Fig.3 a snapshot of a molecular dynamics (MD) program monitoring session[17] is presented. The program is executed on 3 SUN SPARCStation2

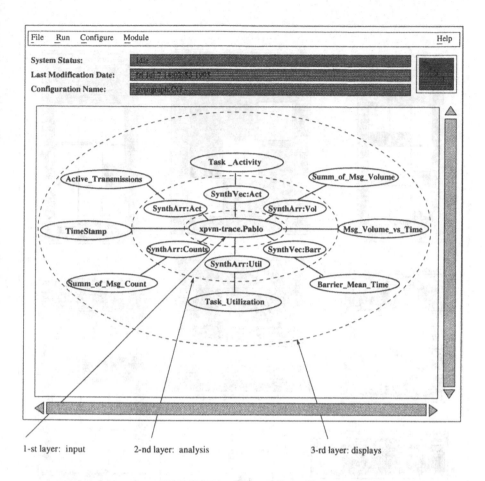

Fig. 2. An enhanced trace file-based analysis and visualization graph.

workstations. *Task_Utilization* display shows that *Busy* bar graph, i.e. computation time is much greater than that of communication indicated by the *Idle* bar graph. Communication and computation times for all tasks are almost the same. *Summ_of_Msg_Counts* shows that the number of messages passed between tasks 0 and 1, 0 and 2 are almost equal, whereas communication between tasks 1 and 2 is lower. *Msg_Volume_vs_Time* shows that along with the short messages there are single long messages. We can conclude that this program is well suited for networked workstations.

In Fig.4 a snapshot from the conjugate gradient(CG) benchmark [18] monitoring session is shown. In CG, the power method is used to find an estimate of the smallest eigenvalue of a symmetric positive definite sparse matrix with a random pattern of nonzeros. This program must execute on n^2 processors where $n = 2, 3, 4...$ due to the nature of its parallelization. In our case study it was executed on 4 RS/6000-320 workstations. *Task_Utilization* display shows domination of the time spent awaiting for messages while the time of computing is much

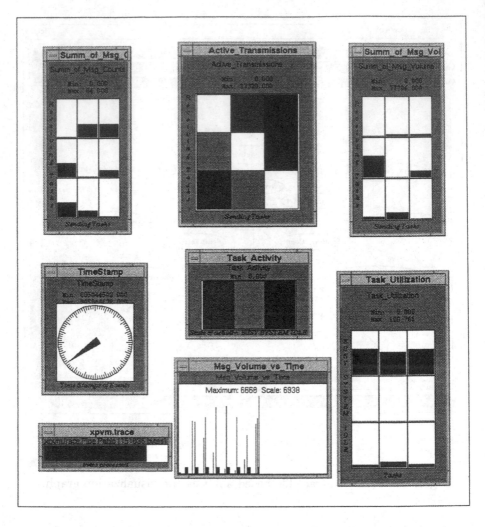

Fig. 3. Monitoring session of molecular dynamics program.

lower. Task 1 spent much more time in the *System* state (i.e. PVM routines) than the other tasks. The first bar graphs in all the displays present the data from the host task that has spawned node tasks and exited. *Msg_Volume_vs_Time* reveals almost even distribution of communication volume in the system over time. It is clear that this program is not suitable for computing on networked workstations.

6 Conclusions

We have elaborated the tool for post mortem performance analysis and visualization using trace files from XPVM and the Pablo environment modules. This approach allowed us to avoid developing our own instrumentation of application

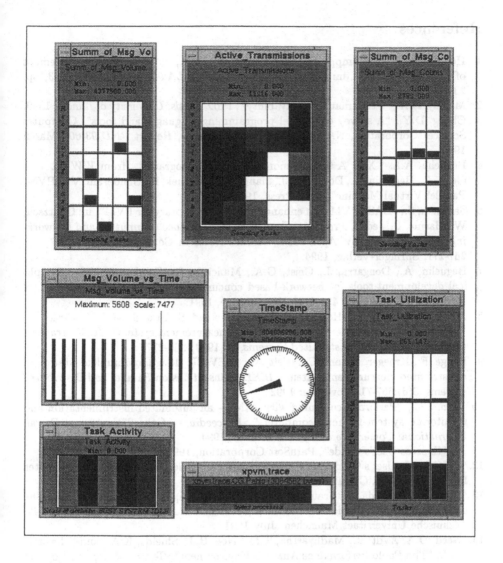

Fig. 4. Monitoring session of conjugate gradient benchmark program.

source codes. The tool is easy extensible and our final goal is to implement most of ParaGraph functionality.

Acknowledgments. We are very grateful to Jim Kohl of the PVM group for his frequent and almost immediate help as well as to Daniel Reed and Jon Reid for relieving our first steps in Pablo. We are also indebted to Marek Pogoda for enlightening discussions.
The work was partially supported by KBN under grant 8 S503 006 07.

References

1. Hollingsworth, J., Lumpp, J.E., Jr., and Miller, B.P., "Performance measurement of parallel programs", in: Tvrdik, P., *Proc. of ISIPALA'93*, Prague, July 1993, pp. 239-254.
2. Malony, D.A., "Performance observability", *PhD Thesis, University of Illinois*, 1990.
3. Cheng, D.Y., "A survey of parallel programming languages and tools", Computer Sciences Corporation, NASA Ames Research Center, *Report RND-93-005* March 1993.
4. Beguelin, A.L. " Xab: A tool for monitoring PVM programs", (from *WWW*).
5. Geist, A., Beguelin, A., Dongarra, J., Jiang, W., Manchek, R., Sunderam, V., "PVM: Parallel Virtual Machine", MIT Press, 1994.
6. Saarinen, S., "EASYPVM – an enhanced subroutine library for PVM", in: Gentzsch, W., Harms, U., (eds.), *Proc. Int. Conf. High Performance Computing and Networking, Munich, Germany, April 1994*, Lecture Notes in Computer Science **797**, pp. 267-271, Springer-Verlag, 1994.
7. Beguelin, A., Dongarra, J., Geist, G.A., Manchek, R., Sunderam, V.S., "Graphical development tools for network-based concurrent supercomputing", in *Proc. of Supercomputing 91* pp. 435-444, Albuquerque, 1991.
8. Miller, B.P., Clark, M., Hollingsworth, J., Kierstead, S., Lim, S., and Torzewski, T., "IPS-2: The second generation of a parallel measurement system", *IEEE Transactions on Parallel and Distributed Systems*, **1** (1990) 206-217.
9. Lange, F., Kroeger, R., and Gergeleit, M., "JEWEL: design and implementation of a distributed measurement system", *IEEE Transactions on Parallel and Distributed systems*, 3(6):657-71, November 1992.
10. Yan, J.C., "Performance tuning with AIMS - an automated instrumentation and monitoring system for multicomputers", *Proceedings of the 27th Annual Hawaii International Conference on System Science*, 1994.
11. –, "Express User's Guide", ParaSoft Corporation, 1990.
12. –, "C-Linda User's Guide and Reference Manual", Scientific Computing Associates Inc., (New Haven, CT, 1993).
13. Bemmerl, T., Bode, A., Braum, P., Hansen, O., Tremi, T., and Wismuller, R., "The design and implementation of TOPSYS", *Technical Report TUM-INFO-07-71-440*, Technische Universitaet Muenchen, July 1991
14. Reed, D.A. Aydt, R., Madhyastha, T.M., Noe, R.J., Shields, K.A., and Schwartz, B.W., "The Pablo Performance Analysis Environment", *Technical Report, Dept. of Comp. Sci., University of Illinois*, 1992.
15. Heath, M.T. Etheridge, J.A., "Visualizing the performance of parallel programs", *IEEE Software* September 1991, pp. 29-39.
16. Worley, P.H., "A new PICL trace file format", *ORNL, Oak Ridge*, September 1992.
17. Boryczko, K., Bubak, M., Kitowski, J., Mościński, J. and Pogoda, M., "Molecular dynamics and lattice gas parallel algorithms for transputers and networked workstations", *Transport Theory and Statistical Physics* **23** (1994) 297-311.
18. –, "The NAS parallel benchmarks", *Report RNR-91-002 revision 2*, NASA Ames Research Center, 22 August 1991.

Linear Algebra Computation on Parallel Machines

Marian Bubak[1,2], Wojciech Owczarz[3], Jerzy Waśniewski[3]

[1] Institute of Computer Science, AGH, al. Mickiewicza 30, 30-059, Kraków, Poland
[2] Academic Computer Centre – CYFRONET, ul. Nawojki 11, 30-950 Kraków, Poland
[3] UNI•C, Building 305, DTU, DK-2800 Lyngby, Denmark

Abstract. In the paper we present the results of running DGETRF, DPOTRF and DGEQRF linear algebra subprograms on new three parallel computers: IBM SP-2, SGI Power Challenge and CONVEX Exemplar SPP1000/XA. Computing time, speed in MFLOPS, speedup and efficiency are presented.

1 Introduction

Linear algebra - and in particular, the solution of linear systems of equations - is the central part of most calculations in scientific computing. It is important to carry this kind of computations in the most efficient way especially for large-scale problems arising in science and engineering. Linear algebra routines are the fundamental building blocks for most of numerical software.

The LAPACK library [1] provides software for solving systems of linear equations, linear least square problems, eigenvalue problems and singular value problems. LAPACK routines are written in a way which allows most of computations to be performed by calls to the Basic Linear Algebra Subprograms (BLAS) which are usually efficiently implemented on machine specific level.

The ScaLAPACK library [2] is a subset of LAPACK routines redesigned for distributed memory parallel computers. Currently it is written in a SPMD style using explicit message passing for interprocessor communications. ScaLAPACK relies on the Parallel Basic Linear Algebra Subprograms (PBLAS) and Basic Linear Algebra Communication Subprograms (BLACS).

The main purpose of many ongoing projects is to develop and implement algorithms for numerical linear algebra on parallel computers. The most popular comparison of computers is based on their performance in solving systems of linear equations [3]. We have tested how efficiently basic linear algebra subroutines have been implemented on new three parallel computers: IBM SP-2, SGI Power Challenge and CONVEX Exemplar SPP1000/XA.

2 Parallel computers used in tests

Experiments were run on three parallel computers: an IBM SP-2, Silicon Graphics Power Challenge and CONVEX Exemplar SPP1000/XA. A short description of these relatively new and very powerful computers is given below.

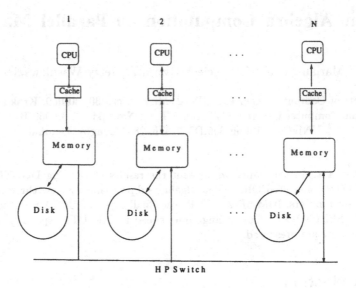

Fig. 1. The architecture of an IBM-SP2 computer.

The essential features of architecture of IBM SP-2 parallel computer is shown in Fig. 1. The peak performance of each processor is 266 MFLOPS. Number of processors may be large - even a few hundreds. Each processor has its own 256 Kbytes cache. More details about this computer can be found in [4].

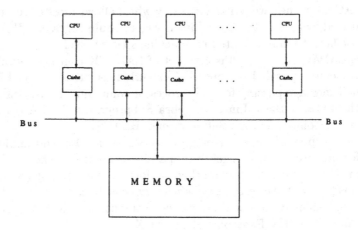

Fig. 2. The architecture of an SGI-PC computer.

The architecture of SGI Power Challenge is presented in Fig. 2. It contains up to 18 processors and the peak performance of each processor is 300 MFLOPS. Each processor has relatively large cache: 32 Mbytes. The total peak performance of fully configured SGI Power Challenge is 5.4 GFLOPS; the total memory can vary from 64 Mbytes to 16 GBytes [5].

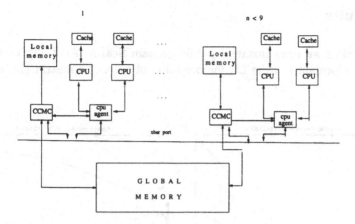

Fig. 3. The architecture of an CONVEX-SPP computer.

The CONVEX Exemplar SPP100 parallel computer consists of one to sixteen hypernodes each of them with up to 8 HP 7100 (or 7200 in the model SPP1200) processors. The peak performance of one processor is 200 MFLOPS. The peak performance of the fully configured Exemplar SPP1000 is 25.6 GFLOPS and its total memory is 32 Mbytes. The architecture of a SPP-1 node is shown in Fig. 3. Each processor has its own cache (1 Mbyte instruction cache and 1 Mbyte data cache). The communication from the cache to its processor is the fastest and therefore it is important to exploit the data that are in cache as long as possible. More details about this computer can be found in [6]. Our experiments has been done on one SPP1000 hypernode.

3   Performace parameters

Performace of parallel computers running basic linear algebra routines are described with the following parameters:

- computing time T_i in seconds; (i is the number of processors),
- speed in MFLOPS calculated from computing times using the theoretical number of floating point operation to be performed for a given algorithm,
- speedup defined as $S_i = \frac{T_1}{T_i}$,
- efficiency $\epsilon_i = 100 \times \frac{S_i}{PP_i}$, where PP_i is the peak performance on i processors of a given parallel computers.

In the next Section all above defined parameters are presented as functions of the size of matrix, N.

4 Results

On IBM SP-2 we have installed public domain ScaLAPACK [2]. In Figs. 4,5, 6 and 7 performance of the LU factorization of a general matrix (DGETRF) is shown.

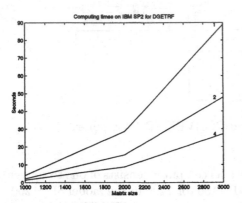

Fig. 4. Computing times on IBM SP2 for DGETRF.

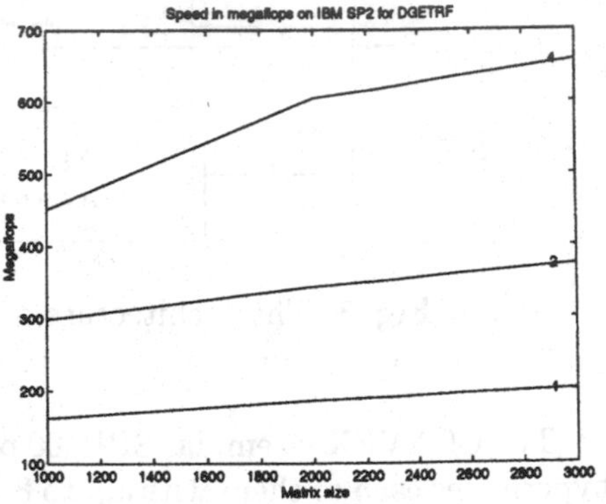

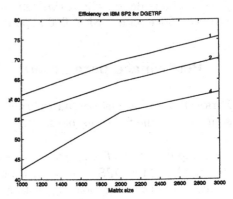

Fig. 5. Speed in MFLOP on IBM SP2 for DGETRF.

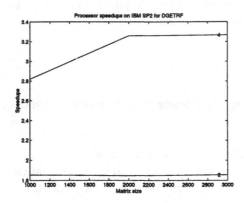

Fig. 6. Speedup on IBM SP2 for DGETRF.

Fig. 7. Efficiency on IBM SP2 for DGETRF.

In Figs. 8, 9, 10 and 11 we present performace on IBM SP-2 of the subprogram for computing of the Cholesky factorization of a positive definite matrix (DPOTRF).

In Figs. 12, 13, 14 and 15 performance of DGEQRF on IBM SP-2 is shown.

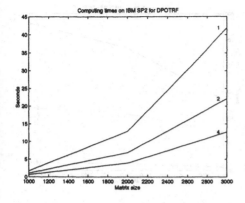

Fig. 8. Computing times on IBM SP2 for DPOTRF.

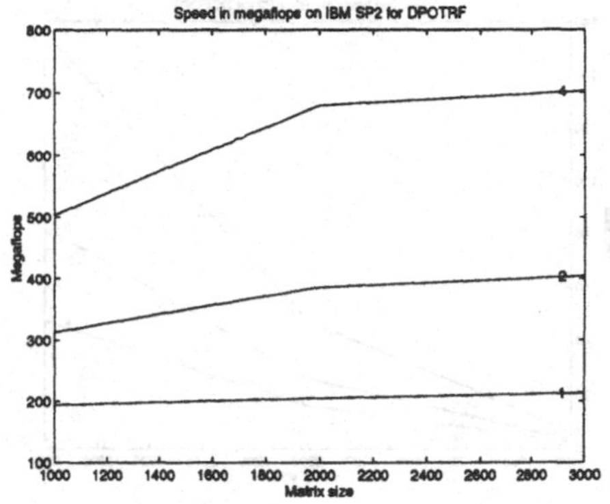

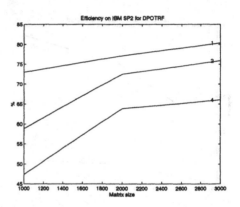

Fig. 9. Speed in MFLOP on IBM SP2 for DPOTRF.

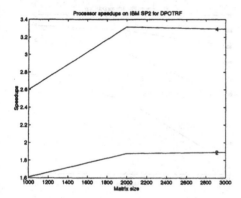

Fig. 10. Speedup on IBM SP2 for DPOTRF.

Fig. 11. Efficiency on IBM SP2 for DPOTRF.

On SGI Power Challenge we have used manufacture developed LAPACK subprograms. In Figs. 16, 17 , 18 and 19 performance of the LU factorization of a general matrix (DGETRF) is shown.

On CONVEX Exemplar SPP1000 we have tested both manufacture developed LAPACK subprograms and public domain LAPACK routines. In Figs. 20, 21 , 22 and 23 performance of the LU factorization of a general matrix (DGETRF) taken from the CONVEX LAPACK library is shown.

In Figs. 24, 25 , 26 and 27 performance of the LU factorization of a general matrix (DGETRF) for model-implemented public domain LAPACK routines are shown.

Results presented in this paper may be treated as a complimentary to performance data given in [3] and may be used for comparison of these new parallel computers.

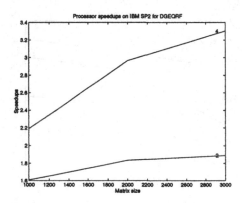

Fig. 12. Computing times on IBM SP2 for DGEQRF.

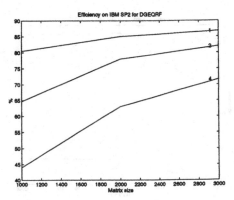

Fig. 13. Speed in MFLOP on IBM SP2 for DGEQRF.

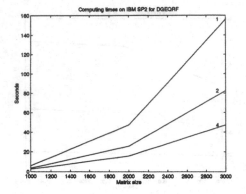

Fig. 14. Speedup on IBM SP2 for DGEQRF.

Fig. 15. Efficiency on IBM SP2 for DGEQRF.

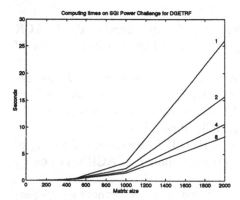

Fig. 16. Computing times on SGI Power Challenge for DGETRF.

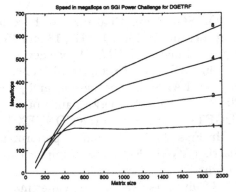

Fig. 17. Speed in MFLOP on SGI Power Challenge for DGETRF.

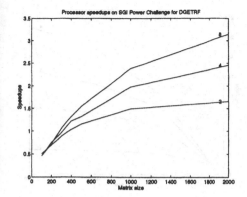

Fig. 18. Speedup on SGI Power Challenge for DGETRF.

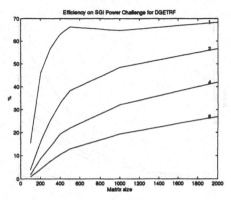

Fig. 19. Efficiency on SGI Power Challenge for DGETRF.

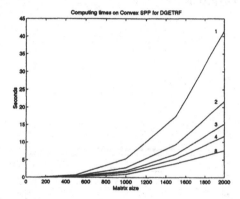

Fig. 20. Computing times on CONVEX Exemplar SPP1000 for DGETRF.

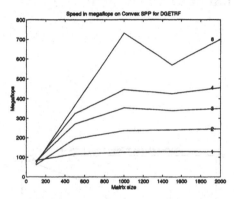

Fig. 21. Speed in MFLOP on CONVEX Exemplar SPP1000 for DGETRF.

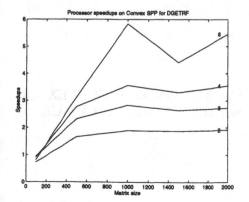

Fig. 22. Speedup on CONVEX Exemplar SPP1000 for DGETRF.

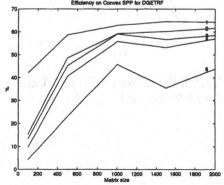

Fig. 23. Efficiency on CONVEX Exemplar SPP1000 for DGETRF.

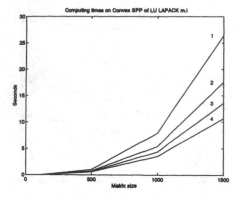

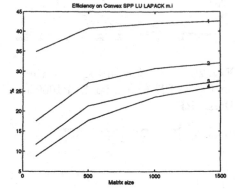

Fig. 24. Computing times on CON-VEX Exemplar SPP1000 for LU LAPACK.

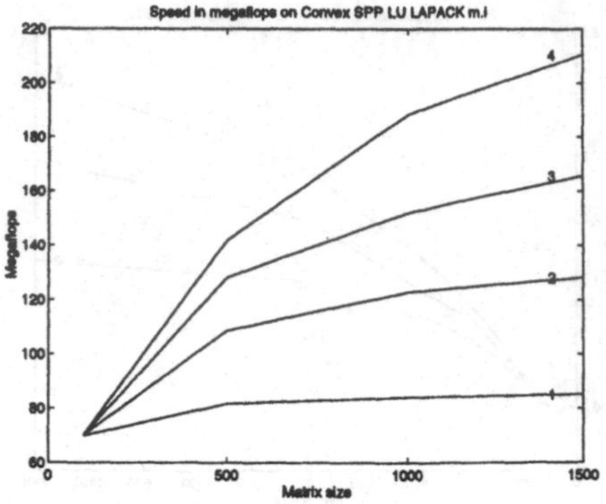

Fig. 25. Speed in MFLOP on CON-VEX Exemplar SPP1000 for LU LAPACK.

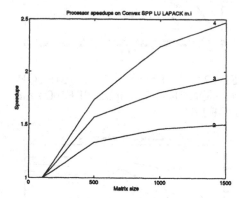

Fig. 26. Speedup on CONVEX Exemplar SPP1000 for LU LAPACK.

Fig. 27. Efficiency on CONVEX Exemplar SPP1000 for LU LAPACK.

References

1. Andrson, E., Bai, Z., Bischof, C.H., Demmel, J., Dongarra, J.J., Du Croz, J., Greenbaum, A., Hammarling, S., McKenney, A., Ostrouchov, S., and Sorensen, D.C., "LAPACK Users' Guide Release 2.0", SIAM, Philadelphia, 1995.
2. Choi, J., Demmel, J., Dhillon, I., Dongarra, J., Ostrouchov, S., Petitet, A., Stanley, K., Walker, D., and Whaley, R. C., "ScaLAPACK: a portable linear algebra library for distributed memory computers – design issues and performance", LAPACK Working Note 95.
3. Dongarra, J.J., "Performance of various computers using standard linear equation software", Computer Science Department, University of Tennessee, Knoxville, CS – 89 – 85, August 29, (1995).
4. Tengwall, C. G., "SP2 architecture and performance", in: Dongarra, J. and Waśniewski, J., eds., "Parallel Scientific Computing", pp. 481-492, Springer Verlag, Berlin, (1994).
5. Silicon Graphics Computer Systems, "POWER CHALLENGE Supercomputing Servers", Silicon Graphics Computer Systems, Marketing Department, Supercomputing Systems Division, 485 Central Avenue, Mountain View, California CA 94043, USA, (1994).
6. CONVEX Computer Corporation, "CONVEX SPP1000 Programming Model", CONVEX Computer Corporation, 3000 Waterview Parkway, P.O. Box 833851, Richardson, Texas 75083-3851, USA, (1993).

A Neural Classifier for Radar Images

T.Casinelli[1], M.La Manna[2], A.Starita[1]

[1] Dept. of Informatics, University of Pisa, Italy
[2] Alenia, Pisa, Italy

Abstract. Goal of our work has been to design a Neural Network, based on the Counterpropagation Network, able to solve the classification problem in a multi-radar system.

The final aim is to implement the network on chip, and embed it in an integrated environment, for the automatic detection and control of vessels in the proximity of ports, based on radar sensors and signal/data processing equipment. The network has been implemented on the *TMC CM200 8k processor* system, which is a SIMD (Single Instructions and Multiple Data) parallel computer.

1 Introduction

Aim of our work is automating the classification of vessels in a multi-radar system, the Vessel Traffic Service (VTS).

This task is currently performed through radio-cooperation between radar operators and ships' personnel. Even if it is a difficult task and needs to be automated, it is not automated yet, due to the severe difficulties to perform real time classification in a classical manner, and/or to the problems encountered in real time image processing.

On the other side, as resulted from numerous studies carried out during the last decades, the Neural Networks (NN) approach has shown to be meant for solving problems difficult to be understood in a mathematical/procedural manner [9].

The common feature of all NN paradigms is the ability to produce consistent classifications, self-adjusting itself, that is adapting weights during the training phase, with a well fitted training set. Then, it is possible to realize a system that can be, to certain degree, insensitive to minor variations, and able to see through noise and distortion of the input pattern.

Since neural network architecture is inherently parallel, general purpose parallel computers turn out to be an excellent choice for the high speed and for a good mapping of the computations, required for the learning and processing phases. For these reasons our NN has been implemented on a *TMC CM200 8k processor* system, based on a SIMD (Single Instruction Multiple Data) architecture.

The paper is organized as follows: Section 2 shows the architecture of the network and its functionalities; Section 3 introduces the main characteristics of the *TMC CM200* system, and addresses to the programmability issues; Section 4 reports the classification results.

2 The Network Tassonomy

Our network is based on the Counterpropagation network developed by Hecht-Nielsen [1], [2], which is a combination of two well-known algorithms: the self-organizing map of Kohonen [3], and the Grossberg outstar [4], [5], [6], [7], possessing together properties not available if considered separately.

Our proposed network provides some modifications to the original architecture of Counterpropagation, as described in the following (see figure below):

Layer 0 Buffers the input vector, and send it to all the neurons of the first layer; it can also perform some additional computation (for example a slight rotation on the input image) in case of erroneous classification.

Layer 1 MaxNN, it is composed by two sub-layers of N neurons, where N is the number of the examples (we will call them attractors) presented to the network during the training phase. The network is a modified MAXNET [8], which implements a "few-winners-take-all" learning rule, i.e. more than one neuron can win the competition.

Layer 2 C*NN, it is constituted by two sub-layers with a total of M+1 neurons, where M is the number of classes of the objects to be classified. It can distinguish between a successful classification or an insuccessful one, and, in this case, the network propagates the error back to Layer 0, where some additional image processing takes place.

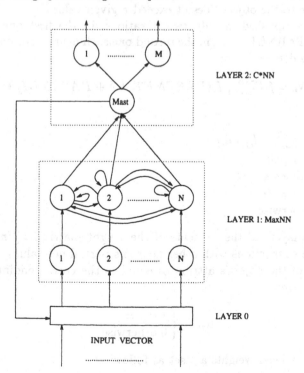

3 Learning and Processing Procedures

We have chosen to have all the weights predetermined, except for the C*NN layer, which is trained as the standard Grossberg layer is, so the training phase result to be very easy to perform. This solution overcomes some typical NN problems, and reduces the training time essentially to that necessary to make assignaments.

As already said, Layer 0 buffers the object, and tests the connection with the third layer to see if there is any need to perform additional processing on the image representing the object.

The MaxNN layer selects the winners of the competition, in particular in the first sub-layer each neuron calculates the "distance matrix", $D=[d_{ij}]$, between the object and all the attractors, using the logical equivalence:

$$d_{ij} = x_{ij} \; eq \; w_{ij}$$

where:

d_{ij} is the ij-th component of the distance matrix;
x_{ij} is the ij-th component of the matrix representing the object;
w_{ij} is the ij-th component of the weight matrix, set to be equal to the ij-th binary component of the corresponding attractor;
eq is the logical equivalence.

The second sub-layer selects the neurons corrisponding to the attractors whose distance to the object doesn't exceed a given value δ.

This is accomplished in only two iterations, in the first one the neurons evaluate their PARNET value, in the second one each neuron produces its output value, which is derived from:

$$WIN_i = f_1(\sum_{j=1}^{N} f_o(PARTNET_i + \delta + PARTNET_j * \gamma_{ji})$$

where:

$PARNET_i = \sum_{i,j=1}^{N} l_{ij} * d_{ij}$

$f_o(x) = \begin{cases} 1 \text{ if } x \geq 0 \\ 0 \text{ otherwise} \end{cases}$

$f_1(x) = \begin{cases} 1 \text{ if } x = N \\ 0 \text{ otherwise} \end{cases}$

In this sub-layer, all the elements of the weight matrix $L=[lin_{ij}]$ are set to 1, whereas the connections with other neuron's partial net-values, $PARTNET_i$ (the number of the object's attributes equal to the corresponding ones of the i-th attractor) are:

$$\alpha_{ji} = \begin{cases} 1 \text{ if } i = j \\ 0 \text{ otherwise} \end{cases}$$

In the C*NN layer weights are set as following:

$$\beta_{ji} = \begin{cases} 1 \text{ if } i = j \\ 0 \text{ otherwise} \end{cases} \qquad \gamma_{ji} = \begin{cases} 0 \quad \text{ if } i = j \\ -1 \text{ otherwise} \end{cases}$$

where β_{ji} is the ji-th weight connected with the ji-th element of the input vector from the previous layer, and γ_{ji} is the ji-th weight connected with the ji-th element of the partial net-value vector.

The C*NN layer gives the classification result. To do so, it first checks if there is more than one winner, if it is the case, it sends a codified message back to Layer 0, which starts an additional processing on the image, i.e. a slight rotation; in this case a message is also sent to the operator , in order to trace the evolution of the classification phase, and allow him to take further decisions.

4 The Connection Machine System

We have chosen a SIMD machine after having carefully examined the kind of functions that are usually involved in a NN computation.

The hardest problems a parallel programmer has to deal with is how to define and distribute data elements and how to partition the calculation among the processors.

These problems disappear for the programmer if she/he is able to treat the data in an object oriented fashion as matrices, vectors, scalars, and these objects are managed by a vendor specific memory management library. In that case the programmer is only required to express typical functions in a matrix and vector algebra context; whereas the memory manager distributes the data of the object in a well defined fashion.

The *TMC CM200 8k processor* is a data parallel computing system, which associates one processor with each data element. This computing style exploits the natural computational parallelism inherent in many data-intensive problems, significantly decreasing the execution time of a problem, and even simplifying its programming.

Execution time is generally reduced in proportion to the number of data elements; whereas the programming effort is reduced proportionally to the complexity involved in trying to express an intrinsically parallel problem in a serial manner.

The hardware elements include front-end computers, that provide the development and execution environment for the users' software, connected with a parallel processing unit of 8k processors that executes the data parallel operations.

Software elements include a standard operating system and a program development environment of the front-end computer, and provide also an useful enhancement of the environment, with extensions to the standard languages and tools that facilitate data parallel program development.

4.1 The Parallel Implementation

Writing NN models requires a good initial design and technique of software engineering, particularly to solve some severe problems such as memory management when the amplitude of the network appreciably increases.

A good method involves the use of data structures and dynamic allocation, which are realised with variable and function pointers and their dynamical allocation, constructs we find in Standard C, for example.

The languages currently supported for the Connection Machine system are CM Fortran, which implements the Fortran 90 array features directly, C^*, and *Lisp.

We have chosen, for the reasons above explained, the C^* language, which is based on the standard version of C specified by the American National Standard Institute.

C language constructs such as data types, operators, structures, pointers, and functions are all maintained in C^*, extended so as to allow them to work also with parallel data; C^* further introduces a small set of new features to help writing programs for massively parallel computers.

5 Testing results

We have trained and tested the network simulating the radar images with patterns closely resembling the real images we expect will be soon available from an advanced radar extraction system.

We have used images represented by 256×256 matrices, trained the network with specific patterns, which should be compliant with the most significant commercial vessel's models, such as ferry-boat, transatlantic, oil-tanker, and tug boats. We have only considered the dimension proportions of the different models (see Fig.1), without taking into account the real ships' dimensions, even because a good radar extraction system is able to extract ships' images and enclose them in a window, this ability to zoom in on an image realizes the natural dimension-tolerance of the classification system.

Fig. 1. On the left is shown our simulation of a ferry-boat; on the right of a transatlantic

The network has then been tested with images after having added a certain amount of noise and/or distortion. The classification has been successful, even when noise and distortion were both introduced in a massive way, on condition that the proportion dimensions of the ships' images didn't differ more than a given tolerance value from the learned ones (we have chosen to realize a static classifyer) this limit can be overcome designing the network as a dynamic classifyer); under the above mentioned conditions, the network has proved to have a good ability to differentiate between objects, distinguishing even when the only available feature was a different proportion (see Fig.2).

Fig. 2. An example of a correct recognition; a transatlantic on the left, and a ferry-boat on the right

6 Conclusion

In this paper we proposed a network able to classify efficiently simulated radar images of vessels.

The intrinsic parallelism of neural networks has been fully exploited, implementing the network on the *TMC CM200 8k processor* system, which is a highly parallel machine, based on a SIMD (Single Instruction Multiple Data) architecture.

The high speed has enabled us to get answers in few milliseconds, which meets our requirement to get significant results in a short time, even if the speed and the accurancy should be further increased realizing the network on a dedicated parallel hardware, as it is in our purpose.

7 Acknowledgements

The work has been performed in the frame of a cooperation between Alenia and the University of Pisa.

Alenia has mainly played the role of the final user and research manager, whereas the University of Pisa has carried out the technical work of research, exploiting its know-how in the NN field.

We wish to thank the *Scuola Normale Superiore of Pisa* , and in particular its C.E.D., for allowing us to use its *TMC CM200* system, and for giving us all the support needed in a serene and friendly atmosphere; and finally, we whish to thank Antonio Storino, for his helpfulness and kindness.

References

1. Hetch-Nielsen,R.: Nearest Matched Filter Classification of Spatiotemporal Patterns. Applied Optics 26, no.10, 1987, 1892-1899
2. Hetch-Nielsen,R.: Neurocomputing. Addison-Wesley, 1990
3. Kohonen,T.: Self-Organization and Associative Memory. 2d ed. Springer-Verlag, 1988
4. Grossberg, S.: Neural Networks and Natural Intelligence. MIT Press, Cambridge MA, 1988
5. Carpenter,G., Grossberg,S.: A massively parallel architecture for a Self-Organizing neural pattern recognition machine. Computer Vision, Graphics, and Image Processing, 37, 1987, 54-115
6. Carpenter,G., Grossberg,S.: ART 2: Self organization of stable Category Recognition Codes for Analog Input Patterns. Applied Optics 26, 1987, 4919-4930
7. Carpenter,G., Grossberg,S., and Reynolds, J.: ARTMAP: A Self-Organizing Neural Network Architecture for Fast Supervised Learning and Pattern Recognition. International Joint Conference on Neural Networks, Seattle WA, July 8-12, 1991, I-863 to I-868
8. Carpenter,G., Grossberg,S., and Reynolds, J.: ARTMAP: Supervised real time learning and classification of nonstationary data by a self-organizing neural network. Neural Networks, 4, 565-588, 1991
9. Pao,Y.,H.: Adaptive Pattern Recognition and Neural Network. Addison-Wesley, Mass.,1989
10. Dobnikar,A. et al.: Invariant Pattern Classification - Neural Network versus Ft approach. International Summer School and Workshop on Neurocomputing, Dubrovnik, 1990

ScaLAPACK: A Portable Linear Algebra Library for Distributed Memory Computers - Design Issues and Performance *

J. Choi[1], J. Demmel[2], I. Dhillon[1], J. Dongarra[1,3], S. Ostrouchov[1], A. Petitet[1], K. Stanley[1], D. Walker[3] and R.C. Whaley[1]

[1] Department of Computer Science, University of Tennessee
Knoxville, TN 37996-1301, USA
[2] Computer Science Division, University of California
Berkeley, CA 94720, USA
[3] Mathematical Sciences Section, Oak Ridge National Laboratory
Oak Ridge, TN 37831, USA

Abstract. This paper outlines the content and performance of ScaLA-PACK, a collection of mathematical software for linear algebra computations on distributed memory computers. The importance of developing standards for computational and message passing interfaces is discussed. We present the different components and building blocks of ScaLAPACK. This paper outlines the difficulties inherent in producing correct codes for networks of heterogeneous processors. Finally, this paper briefly describes future directions for the ScaLAPACK library and concludes by suggesting alternative approaches to mathematical libraries, explaining how ScaLAPACK could be integrated into efficient and user-friendly distributed systems.

1 Overview and Motivation

ScaLAPACK is a library of high performance linear algebra routines for distributed memory MIMD computers. It is a continuation of the LAPACK project, which designed and produced analogous software for workstations, vector supercomputers, and shared memory parallel computers. Both libraries contain routines for solving systems of linear equations, least squares problems, and eigenvalue problems. The goals of both projects are efficiency (to run as fast as possible), scalability (as the problem size and number of processors grow), reliability (including error bounds), portability (across all important parallel machines), flexibility (so users can construct new routines from well-designed parts), and ease-of-use (by making LAPACK and ScaLAPACK look as similar

* This work was supported in part by the National Science Foundation Grant No. ASC-9005933; by the Defense Advanced Research Projects Agency under contract DAAL03-91-C-0047, administered by the Army Research Office; by the Office of Scientific Computing, U.S. Department of Energy, under Contract DE-AC05-84OR21400; and by the National Science Foundation Science and Technology Center Cooperative Agreement No. CCR-8809615.

as possible). Many of these goals, particularly portability, are aided by developing and promoting *standards*, especially for low-level communication and computation routines. We have been successful in attaining these goals, limiting most machine dependencies to two standard libraries called the BLAS, or Basic Linear Algebra Subroutines [5, 6, 12, 14], and BLACS, or Basic Linear Algebra Communication Subroutines [7, 9]. LAPACK and ScaLAPACK will run on any machine where the BLAS and the BLACS are available.

This paper presents the design of ScaLAPACK. After a brief discussion of the BLAS and LAPACK, the block cyclic data layout, the BLACS, the PBLAS (Parallel BLAS), and the algorithms used are discussed. We also outline the difficulties encountered in producing correct code for networks of heterogeneous processors; difficulties we believe are little recognized by other practitioners.

Finally, the paper discusses the performance of ScaLAPACK. Extensive results on various platforms are presented. One of our goals is to model and predict the performance of each routine as a function of a few problem and machine parameters. One interesting result is that for some algorithms, speed is *not* a monotonic increasing function of the number of processors. In other words, speed can be increased by letting some processors remain idle.

2 Design of ScaLAPACK

2.1 Portability, Scalability and Standards

In order to be truly portable, the building blocks underlying parallel software libraries must be *standardized*. The definition of computational and message-passing standards [10, 12] provides vendors with a clearly defined base set of routines that they can optimize. From the user's point of view, standards ensure portability. As new machines are developed, they may simply be added to the network, supplying cycles as appropriate.

From the mathematical software developer's point of view, portability may require significant effort. Standards permit the effort of developing and maintaining bodies of mathematical software to be leveraged over as many different computer systems as possible. Given the diversity of parallel architectures, portability is attainable to only a limited degree, but machine dependences can at least be isolated.

Scalability demands that a program be reasonably effective over a wide range of numbers of processors. The scalability of parallel algorithms over a range of architectures and numbers of processors requires that the granularity of computation be adjustable. To accomplish this, we use block algorithms with adjustable block sizes. Eventually, however, polyalgorithms (where the actual algorithm is selected at runtime depending on input data and machine parameters) may be required.

Scalable parallel architectures of the future are likely to use physically distributed memory. In the longer term, progress in hardware development, operating systems, languages, compilers, and communication systems may make it possible for users to view such distributed architectures (without significant loss of efficiency) as having a shared memory with a global address space. For the near term, however, the distributed nature of the underlying hardware will continue to be visible at the programming level; therefore, efficient procedures for explicit communication will continue to be necessary. Given this fact, standards for basic message passing (send/receive), as well as higher-level communication constructs (global summation, broadcast, etc.), are essential to the development of portable scalable libraries. In addition to standardizing general communication primitives, it may also be advantageous to establish standards for problem-specific constructs in commonly occurring areas such as linear algebra.

2.2 ScaLAPACK Software Components

The following figure describes the ScaLAPACK software hierarchy. The components below the line, labeled Local, are called on a single processor, with arguments stored on single processors only. The components above the line, labeled Global, are synchronous parallel routines, whose arguments include matrices and vectors distributed in a 2D block cyclic layout across multiple processors. We describe each component in turn.

ScaLAPACK Software Hierarchy

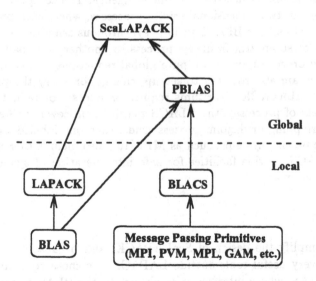

2.3 Processes versus Processors

In ScaLAPACK, algorithms are presented in terms of *processes*, rather than physical processors. In general there may be several processes on a processor, in which case we assume that the runtime system handles the scheduling of processes. In the absence of such a runtime system, ScaLAPACK assumes one process per processor.

2.4 Local Components

The **BLAS** (Basic Linear Algebra Subprograms) [5, 6, 14] include subroutines for common linear algebra computations such as dot-products, matrix-vector multiplication, and matrix-matrix multiplication. As is well known, using matrix-matrix multiplication tuned for a particular architecture can effectively mask the effects of the memory hierarchy (cache misses, TLB misses, etc.), and permit floating point operations to be performed at the top speed of the machine.

As mentioned before, **LAPACK**, or Linear Algebra PACKage [1], is a collection of routines for linear system solving, least squares, and eigenproblems. High performance is attained by using algorithms that do most of their work in calls to the BLAS, with an emphasis on matrix-matrix multiplication. Each routine has one or more *performance tuning parameters*, such as the sizes of the blocks operated on by the BLAS. These parameters are machine dependent, and are obtained from a table at run-time.

The LAPACK routines are designed for single processors. LAPACK can also accommodate shared memory machines, provided parallel BLAS are available (in other words, the only parallelism is implicit in calls to BLAS). Extensive performance results for LAPACK can be found in the second edition of the manual [2].

The **BLACS** (Basic Linear Algebra Communication Subprograms) [7, 9] are a message passing library designed for linear algebra. The computational model consists of a one or two dimensional grid of processes, where each process stores matrices and vectors. The BLACS include synchronous send/receive routines to send a matrix or submatrix from one process to another, to broadcast submatrices to many processes, or to compute global reductions (sums, maxima and minima). There are also routines to set up, change, or query the process grid. Since several ScaLAPACK algorithms require broadcasts or reductions among different subsets of processes, the BLACS permit a processor to be a member of several overlapping or disjoint process grids, each one labeled by a *context*. Some message passing systems, such as MPI [10], also include this context concept. The BLACS provide facilities for safe interoperation of system contexts and BLACS contexts.

2.5 PBLAS

In order to simplify the design of ScaLAPACK, and because the BLAS have proven to be very useful tools outside LAPACK, we chose to build a Parallel BLAS, or PBLAS, whose interface is as similar to the BLAS as possible. This decision has permitted the ScaLAPACK code to be quite similar, and sometimes nearly identical, to the analogous LAPACK code. Only one substantially new routine was added to the PBLAS, matrix transposition, since this is a complicated operation in a distributed memory environment [3].

We hope that the PBLAS will provide a distributed memory standard, just as the BLAS have provided a shared memory standard. This would simplify and encourage the development of high performance and portable parallel numerical software, as well as providing manufacturers with a small set of routines to be optimized. The acceptance of the PBLAS requires reasonable compromises among competing goals of functionality and simplicity. These issues are discussed below.

The PBLAS operate on matrices distributed in a 2D block cyclic layout. Since such a data layout requires many parameters to fully describe the distributed matrix, we have chosen a more object-oriented approach, and encapsulated these parameters in an integer array called an *array descriptor*. An array descriptor includes

(1) the number of rows in the distributed matrix,
(2) the number of columns in the distributed matrix,
(3) the row block size, r
(4) the column block size, c
(5) the process row over which the first row of the matrix is distributed,
(6) the process column over which the first column of the matrix is distributed,
(7) the BLACS context, and
(8) the leading dimension of the local array storing the local blocks.

For example, here is an example of a call to the BLAS double precision matrix multiplication routine DGEMM, and the corresponding PBLAS routine PDGEMM; note how similar they are:

```
CALL DGEMM ( TRANSA, TRANSB, M, N, K, ALPHA,
                     A( IA, JA ), LDA,
                     B( IB, JB ), LDB, BETA,
                     C( IC, JC ), LDC )


CALL PDGEMM( TRANSA, TRANSB, M, N, K, ALPHA,
                     A, IA, JA, DESC_A,
                     B, IB, JB, DESC_B, BETA,
                     C, IC, JC, DESC_C )
```

DGEMM computes C = BETA * C + ALPHA * op(A) * op(B), where op(A) is either A or its transpose depending on TRANSA, op(B) is similar, op(A) is M-by-K, and op(B) is K-by-N. PDGEMM is the same, with the exception of the way in which submatrices are specified. To pass the submatrix starting at A(IA,JA) to DGEMM, for example, the actual argument corresponding to the formal argument A would simply be A(IA,JA). PDGEMM, on the other hand, needs to understand the global storage scheme of A to extract the correct submatrix, so IA and JA must be passed in separately. DESC_A is the array descriptor for A. The parameters describing the matrix operands B and C are analogous to those describing A. In a truly object-oriented environment matrices and DESC_A would be the synonymous. However, this would require language support, and detract from portability.

Our implementation of the PBLAS emphasizes the mathematical view of a matrix over its storage. In fact, it is even possible to reuse our interface to implement the PBLAS for a different block data distribution that would not fit in the block-cyclic scheme.

The presence of a context associated with every distributed matrix provides the ability to have separate "universes" of message passing. The use of separate communication contexts by distinct libraries (or distinct library invocations) such as the PBLAS insulates communication internal to the library from external communication. When more than one descriptor array is present in the argument list of a routine in the PBLAS, it is required that the individual BLACS context entries must be equal. In other words, the PBLAS do not perform "intra-context" operations.

We have not included specialized routines to take advantage of packed storage schemes for symmetric, Hermitian, or triangular matrices, nor of compact storage schemes for banded matrices.

3 Performance

An important performance metric is *parallel efficiency*. Parallel efficiency, $E(N, P)$, for a problem of size N on P processors is defined in the usual way [11] as

$$E(N, P) = \frac{1}{P} \frac{T_{\text{seq}}(N)}{T(N, P)} \tag{1}$$

where $T(N, P)$ is the runtime of the parallel algorithm, and $T_{\text{seq}}(N)$ is the runtime of the best sequential algorithm. An implementation is said to be *scalable* if the efficiency is an increasing function of N/P, the problem size per processor (in the case of dense matrix computations, $N = n^2$, the number of words in the input).

We will also measure the *performance* of our algorithm in Megaflops/sec (or Gigaflops/sec). This is appropriate for large dense linear algebra computations, since floating point dominates communication. For a scalable algorithm with N/P held fixed, we expect the performance to be proportional to P.

We seek to increase the performance of our algorithms by reducing overhead due to load imbalance, data movement, and algorithm restructuring. The way the data are distributed over the memory hierarchy of a computer is of fundamental importance to these factors. We present in this section extensive performance results on various platforms for the ScaLAPACK factorization and reductions routines. Performance data for the symmetric eigensolver (PDSYEVX) are presented in [4].

3.1 Choice of Block Size

In the factorization or reduction routines, the work distribution becomes uneven as the computation progresses. A larger block size results in greater load imbalance, but reduces the frequency of communication between processes. There is, therefore, a tradeoff between load imbalance and communication startup cost, which can be controlled by varying the block size.

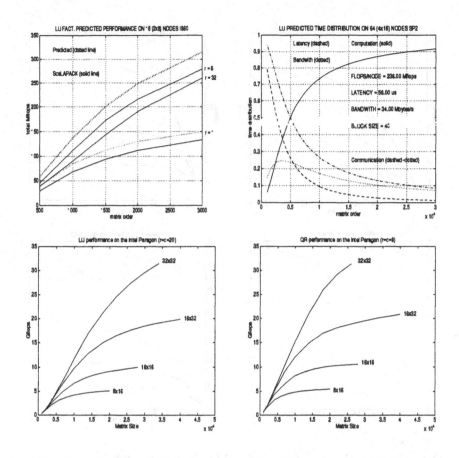

Most of the computation of the ScaLAPACK routines is performed in a blocked fashion using Level 3 BLAS, as is done in LAPACK. The computational blocking factor is chosen to be the same as the distribution block size. Therefore, smaller distribution block sizes increase the loop and index computation overhead. However, because the computation cost ultimately dominates, the influence of the block size on the overall communication startup cost and loop and index computation overhead decreases very rapidly with the problem size for a given grid of processes. Consequently, the performance of the ScaLAPACK library is not very sensitive to the block size, as long as the extreme cases are avoided. A very small block size leads to BLAS 2 operations and poorer performance. A very large block size leads to computational imbalance.

The chosen block size impacts the amount of workspace needed on every process. This amount of workspace is typically large enough to contain a block of columns or a block of rows of the matrix operands. Therefore, the larger the block size, the greater the necessary workspace, i.e the smaller the largest solvable problem on a given grid of processes. For Level 3 BLAS blocked algorithms, the smallest possible block operands are of size $r \times c$. Therefore, it is good practice to choose the block size to be the problem size for which the BLAS matrix-multiply GEMM routine achieves 90 % of its reachable peak.

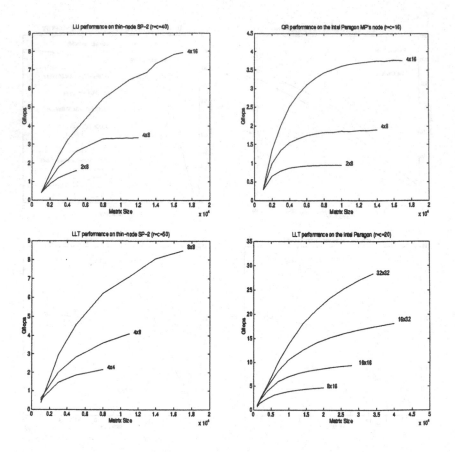

Determining optimal, or near optimal block sizes for different environments is a difficult task because it depends on many factors including the machine architecture, speeds of the different BLAS levels, the latency and bandwidth of message passing, the number of process available, the dimensions of the process grid, the dimension of the problem, and so on. However, there is enough evidence and expertise for automatically and accurately determining optimal, or near optimal block sizes via an enquiry routine. Furthermore, for small problem sizes it is also possible to determine if redistributing n^2 data items is an acceptable cost in terms of performance as well as memory usage. In the future, we hope to calculate the optimal block size via an enquiry routine.

3.2 Choice of Grid Size

The best grid shape is determined by the algorithm implemented in the library and the underlying physical network. A one link physical network will favor $P_r = 1$ or $P_c = 1$. This affects the scalabilty of the algorithm, but reduces the overhead due to message collisions. It is possible to predict the best grid shape given the number of processes available. The current algorithms for the factorization or reduction routines can be split into two categories.

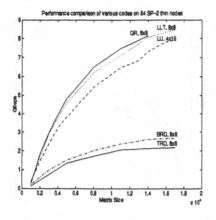
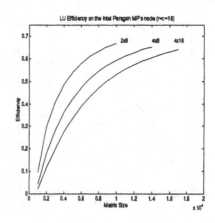

If at every step of the algorithm a block of columns and/or rows needs to be broadcast, as in the LU or QR factorizations, it is possible to pipeline this communication phase and overlap it with some computation. The direction of the pipeline determines the shape of the grid. For example, the LU, QR and QL factorizations perform better for "flat" process grids ($P_r < P_c$). These factorizations share a common bottleneck of performing a reduction operation along each column (for pivoting in LU, and for computing a norm in QR and QL). The first implication of this observation is that large latency message passing perform better on a "flat" grid than on a square grid. Secondly, after this reduction has been performed, it is important to update the next block of columns as fast as possible. This is done by broadcasting the current block of columns using a ring topology, i.e, feeding the ongoing communication pipe. Similarly, the performance of the LQ and RQ factorizations take advantage of "tall" grids ($P_r > P_c$) for the same reasons, but transposed.

The theoretical efficiency of the LU factorization can be estimated by:

$$E_{LU}(N, P) = \frac{1}{1 + \frac{3P \log P_r}{n^2} \frac{\alpha}{\gamma_3} + \frac{3}{4n}(2P_c + P_r \log P_r)\frac{\beta}{\gamma_3}}$$

For large n, the last term in the denominator dominates, and it is minimized by choosing a P_r slightly smaller than P_c. $P_c = 2P_r$ works well on Intel machines. For smaller n, the middle term dominates, and it becomes more important to choose a small P_r. Suppose that we keep the ratio P_r/P_c constant as P increases, thus we have $P_r = u\sqrt{P}$ and $P_c = v\sqrt{P}$, where u and v are constant [8]. Moreover, let ignore the $\log_2(P_r)$ factor for a moment. In this case, P_r/n and P_c/n are proportional to $\sqrt{P}/n$ and n^2 must grow with P to maintain efficiency. For sufficient large P_r, the $\log_2(P_r)$ factor cannot be ignored, and the performance will slowly degrade with the number of processors P. This phenomenon is observed in practice as shown in the plot above showing the efficiency of the LU factorization on the Intel Paragon.

The second group of routines physically transpose a block of columns and/or rows at every step of the algorithm. In these cases, it is not usually possible to maintain a communication pipeline, and thus square or near square grids

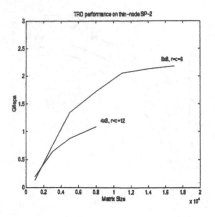

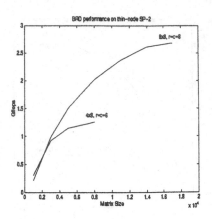

are more optimal. This is the case for the algorithms used for implementing the Cholesky factorization, the matrix inversion and the reduction to bidiagonal form (BRD), Hessenberg form (HRD) and tridiagonal form (TRD). For example, the update phase of the Cholesky factorization of a lower symmetric matrix physically transposes the current block of columns of the lower triangular factor.

Assume now that at most P processes are available. A natural question arising is: could we decide what process grid $P_r \times P_c \leq P$ should be used? Similarly, depending on P, it is not always possible to factor $P = P_r.P_c$ to create the appropriate grid. For example, if P is prime, the only possible grids are $1 \times P$ and $P \times 1$. If such grids are particularly bad for performance, it may be beneficial to let some processors remain idle, so the remainder can be formed into a "squarer" grid [13]. These problems can be analyzed by a complicated function of the machine and problem parameters. It is possible to develop models depending on the machine and problem parameters which accurately estimate the impact of modifying the shape of the grid on the total execution time, as well as predicting the necessary amount of extra memory required for each routine.

4 Future Directions

4.1 Future addition to ScaLAPACK

Basic building blocks like the BLAS, the BLACS and the PBLAS have been made publically available. At the time this paper was written, the current version of the PBLAS was being extended by removing alignment assumptions made on the operands. Moreover, the PBLAS package is being internally restructured to facilitate its maintenance and reinforce its robustness. Concurrently, many of the LAPACK functions missing in ScaLAPACK are being assembled and integrated. These include condition estimation, iterative refinement of linear solutions and linear least square solvers. We are planning improved versions of the symmetric eigenvalue routine. SVD and nonsymmetric eigenvalue routines are also in preparation. More elaborate testing and timing programs are being developed to ensure the robustness and the efficiency of the library. Finally, banded, general sparse, and out-of-core prototype routines are being investigated.

4.2 Alternative Approaches to Libraries

Traditionally, large, general-purpose mathematical software libraries on uniprocessors and shared memory machines have tried to hide much of the complexity of data structures and performance issues from the user. For example, the LAPACK project incorporates parallelism in the Level 3 BLAS, where it is not directly visible to the user. Unfortunately, it is not possible to hide these details as neatly on distributed memory machines. Currently, the data structures and data decomposition must be specified by the user, and it may be necessary to explicitly transform these structures in between calls to different library routines. These deficiencies in the conventional user interface have prompted extensive discussion of alternative approaches for scalable parallel software libraries of the future. Here are some possibilities.

1. Traditional function library (i.e., minimum possible change to the status quo in going from serial to parallel environment). This will allow one to protect the programming investment that has been made. More aggressive use of performance models may permit us to choose the best layout and redistribute the input data structure automatically. This is attractive for dense linear algebra since for large problems the $O(n^3)$ floating point operations will dominate the $O(n^2)$ cost of redistribution.

2. Reactive servers on the network. A user would be able to send a computational problem to a server that was specialized in dealing with the problem. This fits well with the concepts of a networked, heterogeneous computing environment with various specialized hardware resources (or even the heterogeneous partitioning of a single homogeneous parallel machine). Again, this is attractive for dense linear algebra since $O(n^3)$ flops are performed on a data structure of size $O(n^2)$.

3. Interactive environments like Matlab or Mathematica, perhaps with "expert" drivers (i.e. knowledge-based systems) for special domains, such as structural analysis. Such environments have proven to be especially attractive for rapid prototyping of new algorithms and systems that may subsequently be implemented in a more customized manner for higher performance. With the growing popularity of the many integrated packages based on this idea, this approach would provide an interactive, graphical interface for specifying and solving scientific problems. Both the algorithms and data structures are hidden from the user, because the package itself is responsible for storing and retrieving the problem data in an efficient, distributed manner. In a heterogeneous networked environment, such interfaces could provide seamless access to computational engines that would be invoked selectively for different parts of the user's computation according to which machine is most appropriate for a particular subproblem.

4. Reusable templates (i.e., users adapt "source code" to their particular applications). A template is a description of a general algorithm rather than the executable object code or the source code more commonly found in a conventional software library. Nevertheless, although templates use generic

versions of key data structures, they offer whatever degree of customization the user may desire. We have constructed such a set of template for interactive linear system solvers, and are currently constructing one for eigenvalue problems.

References

1. E. Anderson, Z. Bai, C. Bischof, J. Demmel, J. Dongarra, J. Du Croz, A. Greenbaum, S. Hammarling, A. McKenney, S. Ostrouchov, and D. Sorensen. *"LAPACK Users' Guide"*. SIAM, Philadelphia, PA, 1992.
2. E. Anderson, Z. Bai, C. Bischof, J. Demmel, J. Dongarra, J. Du Croz, A. Greenbaum, S. Hammarling, A. McKenney, S. Ostrouchov, and D. Sorensen. *"LAPACK Users' Guide, Second Edition"*. SIAM, Philadelphia, PA, 1995.
3. J. Choi, J. Dongarra, and D. Walker. "Parallel Matrix Transpose Algorithms on Distributed Concurrent Computers". Technical Report UT CS-93-215, LAPACK Working Note #65, University of Tennessee, 1993.
4. J. Demmel and K. Stanley. "The Performance of Finding Eigenvalues and Eigenvectors of Dense Symmetric Matrices on Distributed Memory Computers". In *Proceedings of the Seventh SIAM Conference on Parallel Proceesing for Scientific Computing*. SIAM, 1994.
5. J. Dongarra, J. Du Croz, I. Duff, and S. Hammarling. "A Set of Level 3 Basic Linear Algebra Subprograms". *ACM Transactions on Mathematical Software*, 16(1):1–17, 1990.
6. J. Dongarra, J. Du Croz, S. Hammarling, and R. Hanson. "Algorithm 656: An extended Set of Basic Linear Algebra Subprograms: Model Implementation and Test Programs". *ACM Transactions on Mathematical Software*, 14(1):18–32, 1988.
7. J. Dongarra and R. van de Geijn. "Two dimensional Basic Linear Algebra Communication Subprograms". Technical Report UT CS-91-138, LAPACK Working Note #37, University of Tennessee, 1991.
8. J. Dongarra, R. van de Geijn, and D. Walker. "A Look at Scalable Dense Linear Algebra Librairies". Technical Report UT CS-92-155, LAPACK Working Note #43, University of Tennessee, 1992.
9. J. Dongarra and R. C. Whaley. "A User's Guide to the BLACS v1.0". Technical Report UT CS-95-281, LAPACK Working Note #94, University of Tennessee, 1995.
10. Message Passing Interface Forum. "MPI: A Message-Passing Interface standard". *International Journal of Supercomputer Applications*, 8(3/4), 1994.
11. G. Fox, M. Johnson, G. Lyzenga, S. Otto, J. Salmon, and D. Walker. *"Solving Problems on Concurrent Processors"*, volume 1. Prentice Hall, Englewood Cliffs, N.J, 1988.
12. R. Hanson, F. Krogh, and C. Lawson. "A Proposal for Standard Linear Algebra Subprograms". *ACM SIGNUM Newsl.*, 8(16), 1973.
13. W. Hsu, G. Thanh Nguyen, and X. Jiang. "Going Beyond Binary". http://www.cs.berkeley.edu/ xjiang/cs258/project_1.html, 1995. CS 258 Class project.
14. C. Lawson, R. Hanson, D. Kincaid, and F. Krogh. "Basic Linear Algebra Subprograms for Fortran Usage". *ACM Transactions on Mathematical Software*, 5(3):308–323, 1979.

A Proposal for a Set of Parallel Basic Linear Algebra Subprograms[*]

Jaeyoung Choi[1][**], Jack Dongarra[2,3], Susan Ostrouchov[2], Antoine Petitet[2],
David Walker[3] and R. Clinton Whaley[2]

[1] School of Computing, Soongsil University, Seoul 156-743, Korea
[2] Department of Computer Science, University of Tennessee, Knoxville,
TN 37996-1301, USA
[3] Mathematical Sciences Section, Oak Ridge National Laboratory, Oak Ridge,
TN 37831, USA

Abstract. This paper describes a proposal for a set of Parallel Basic
Linear Algebra Subprograms (PBLAS) for distributed memory MIMD
computers. The PBLAS are targeted at distributed vector-vector, matrix-
vector and matrix-matrix operations with the aim of simplifying the parallelization of linear algebra codes, especially when implemented on top
of the sequential BLAS.

1 Introduction

In 1987 Dongarra, Du Croz, Duff and Hammarling wrote an article in the *ACM
Trans. Math. Soft.* (Vol. 16, no. 1, page 1) defining and proposing a set of Level 3
Basic Linear Algebra Subprograms. That proposal logically concluded a period
of reflection and discussion among the mathematical software community [8, 9] to
define a set of routines that would find wide application in software for numerical
linear algebra and provide a useful tool for implementors and users. Because
these subprograms and their predecessors – the Levels 1 and 2 BLAS – are an
aid to clarity, portability, modularity and maintenance of software, they have
been embraced by the community and have become a *de facto* standard for
elementary linear algebra computations.

Many of the frequently used algorithms of numerical linear algebra can be
implemented so that a majority of the computation is performed within the
BLAS. It is thus possible to develop portable and efficient software across a
wide range of architectures, with emphasis on workstations, vector-processors
and shared-memory computers, as has been done in LAPACK [2].

[*] This work was supported in part by the National Science Foundation Grant No.
ASC-9005933; by the Defense Advanced Research Projects Agency under con-
tract DAAL03-91-C-0047, administered by the Army Research Office; by the Of-
fice of Scientific Computing, U.S. Department of Energy, under Contract DE-AC05-
84OR21400; and by the National Science Foundation Science and Technology Center
Cooperative Agreement No. CCR-8809615.
[**] The author's research was performed at the Department of Computer Science of
University of Tennessee and Oak Ridge National Laboratory.

The large variety of existing distributed-memory systems motivated the development of a set of portable communication subprograms well suited for linear algebra computations: the Basic Linear Algebra Communication Subprograms (BLACS) [10]. The BLACS define a portable interface and provide the correct level of abstraction.

There has been much interest recently in developing parallel versions of the BLAS for distributed memory computers [1, 3, 11, 12]. Some of this research proposed parallelizing the BLAS, and some implemented a few important BLAS routines, such as matrix-matrix multiplication. Almost ten years after the very successful BLAS were proposed, we are in a position to define and implement a set of Basic Linear Algebra Subprograms for distributed-memory computers with similar functionality as their sequential predecessors. Internally, the local computations within a process are performed by the BLAS, while the communication operations are handled by the BLACS.

The scope of this proposal is limited. First, the set of routines described in this paper constitutes an extended proper subset of the BLAS. For instance, this proposal does not contain vector rotation routines or dedicated subprograms for banded or packed matrices. A matrix transposition routine has been added to the Level 3 subprograms since this operation is much more complex to perform and implement on distributed-memory computers. Second, this proposal does not include routines for matrix factorizations or reductions; these are covered by the ScaLAPACK (Scalable Linear Algebra PACKage) project [4]. A reference implementation version of the PBLAS is available on netlib (http://www.netlib.org). Vendors can then supply system optimized versions of the BLAS, the BLACS and eventually the PBLAS. It is our hope that this proposal will initiate discussions among the computer science community so that this project will best reflect its needs.

The details of this proposal are concerned with defining a set of subroutines for use in FORTRAN 77 and C programs. However, the essential features of this standard should be easily adaptable to other programming languages. We have attempted to pave the way for such a future evolution by respecting the driving concepts of the HPF [14] and MPI [13] projects.

2 Conventions of the PBLAS

The name of a PBLAS routine begins with a 'P' and then follows the conventions of the BLAS [8, 9]. The matrix transposition routine is called P□TRAN□. The collection of routines can be thought of as being divided into four separate parts, *real, double precision, complex and complex*16*. These routines can be written in C or FORTRAN 90 for example; their implementation takes advantage of dynamic memory management features present in these programming languages. However, the local storage convention of the distributed matrix operands in every process's memory is assumed to be FORTRAN like, i.e., "column major". Thus, it is possible to rely on the BLAS to perform the local computations within a process.

2.1 Storage Conventions

The current model implementation of the PBLAS assumes the matrix operands to be distributed according to the block-cyclic decomposition scheme. This allows the routines to achieve scalability, well balanced computations and to minimize synchronization costs. It is not the object of this paper to describe in detail the data mapping onto the processes, for further details see [4]. Let us simply say that the set of processes is mapped to a virtual mesh, where every process is naturally identified by its coordinates in this $P \times Q$ grid. This virtual machine is in fact part of a larger object defined by the BLACS and called a context [10].

An M_ by N_ matrix operand is first decomposed into MB_ by NB_ blocks starting at its upper left corner. These blocks are then uniformly distributed across the process mesh. Thus every process owns a collection of blocks, which are locally and contiguously stored in a two dimensional "column major" array. The local entries of every matrix column are contiguously stored in the processes' memories. It follows that a general M_ by N_ distributed matrix is defined by its dimensions, the size of the elementary MB_ by NB_ block used for its decomposition, the coordinates of the process having in its local memory the first matrix entry {RSRC_, CSRC_}, and the BLACS context (CTXT_) in which this matrix is defined. Finally, a local leading dimension LLD_ is associated with the local memory address pointing to the data structure used for the local storage of this distributed matrix.

These pieces of information are grouped together into a single 8 element integer array, called the descriptor, DESC_. Such a descriptor is associated with each distributed matrix. The entries of the descriptor uniquely determine the mapping of the matrix entries onto the local processes' memories. Moreover, with the exception of the local leading dimension, the descriptor entries are global values characterizing the distributed matrix operand. Since vectors may be seen as a special case of distributed matrices or proper submatrices, the larger scheme just defined encompasses their description as well.

2.2 Argument Conventions

The order of the arguments of a PBLAS routine is as follows:

1. Arguments specifying matrix options
2. Arguments defining the sizes of the distributed matrix or vector operands
3. Input-Output scalars
4. Description of the input distributed vector or matrix operands
5. Input scalar (associated with the input-output distributed matrix or vector operand)
6. Description of the input-output distributed vector or matrix operands

Note that every category is not present in each of the routines. The arguments that specify options are character arguments with the names SIDE, TRANS, TRANSA, TRANSB, UPLO and DIAG. These arguments have similar values and meanings as for the BLAS [8, 9].

The distributed submatrix operands of the Level 3 PBLAS are determined by the arguments M, N and K, which specify their size. These numbers may differ from the two first entries of the descriptor (M_ and N_), which specifies the size of the distributed matrix containing the submatrix operand. Also required are the global starting indices IA, JA, IB, JB, IC and JC.

The description of the distributed matrix operands consists of a pointer in every process to the local array (let say A) containing the local pieces of the corresponding distributed matrix, the global starting indices in row column order (IA, JA), and the descriptor of the distributed matrix as declared in the calling (sub)program (DESCA).

The description of a distributed vector operand is similar to the description of a distributed matrix (X, IX, JX, DESCX) followed by a global increment INCX, which allows the selection of a matrix row or a matrix column as a vector operand. Only two increment values are currently supported by our model implementation, namely 1 to select a matrix column and DESCX(1) (i.e INCX=MX) specifying a matrix row.

The input scalars always have the dummy argument names ALPHA and BETA. Output scalars are only present in the Level 1 PBLAS and are called AMAX, ASUM, DOT, INDX and NORM2.

We use the description of two distributed matrix operands X and Y to describe the invalid values of the arguments:

- Any value of the character arguments SIDE, TRANS, TRANSA, TRANSB, UPLO, or DIAG, whose meaning is not specified,
- $M < 0$ or $N < 0$ or $K < 0$,
- $IX < 1$ or $IX+M-1 > M\_(=DESCX(1))$, (if $X(IX:IX+M-1,*)$ is an operand),
- $JX < 1$ or $JX+N-1 > N\_(=DESCX(2))$, (if $X(*,JX:JX+N-1)$ is an operand),
- $MB\_(=DESCX(3)) < 1$ or $NB\_(=DESCX(4)) < 1$,
- $RSRC\_ (=DESCX(5)) < 0$ or $RSRC\_ \geq P$ (number of process rows),
- $CSRC\_ (=DESCX(6)) < 0$ or $CSRC\_ \geq Q$ (number of process columns),
- $LLD\_ (=DESCX(8)) <$ local number of rows in the array pointed to by X,
- $INCX \neq 1$ and $INCX \neq M\_ (= DESCX(1))$ (only for vector operands),
- $CTXT\_X (=DESCX(7)) \neq CTXT\_Y (=DESCY(7))$.

If a routine is called with an invalid value for any of its arguments, then it must report the fact and terminate the execution of the program. In the model implementation, each routine, on detecting an error, calls a common error-handling routine PBERROR(), passing to it the current BLACS context, the name of the routine and the number of the first argument that is in error. If an error is detected in the j-th entry of a descriptor array, which is the i-th argument in the parameter list, the number passed to PBERROR() has been arbitrarily chosen to be $100 * i + j$. This allows the user to distinguish an error on a descriptor entry from an error on a scalar argument. For efficiency purposes, the PBLAS routines only perform a local validity check of their argument list. If an error is detected in at least one process of the current context, the program execution is stopped.

A global validity check of the input arguments passed to a PBLAS routine must be performed in the higher-level calling procedure. To demonstrate the need and cost of global checking, as well as the reason why this type of checking is not performed in the PBLAS, consider the following example: the value of a global input argument is legal but differs from one process to another. The results are unpredictable. In order to detect this kind of error situation, a synchronization point would be necessary, which may result in a significant performance degradation. Since every process must call the same routine to perform the desired operation successfully, it is natural and safe to restrict somewhat the amount of checking operations performed in the PBLAS routines.

Specialized implementations may call system-specific exception-handling facilities, either via an auxiliary routine PBERROR or directly from the routine. In addition, the testing programs [5] can take advantage of this exception-handling mechanism by simulating specific erroneous input argument lists and then verifying that particular errors are correctly detected.

3 Specifications of the PBLAS

The PBLAS subroutine specifications can easily be deduced from their BLAS equivalent. The matrix specification A, LDA (resp. X, INCX for a vector) becomes A, IA, JA, DESCA (resp. X, IX, JX, DESCX, INCX). For example, the BLAS double precision real matrix-vector product:

```
DGEMV( TRANS, M, N, ALPHA, A, LDA, X, INCX, BETA, Y, INCY )
```

becomes

```
PDGEMV( TRANS, M, N, ALPHA, A, IA, JA, DESCA, X, IX, JX,
        DESCX, INCX, BETA, Y, IY, JY, DESCY, INCY ).
```

The double precision real PBLAS matrix transposition routine has the following interface:

```
PDTRAN( M, N, ALPHA, A, IA, JA, DESCA, BETA, C, IC, JC, DESCC )
```

The Level 1 BLAS functions have become subroutines in the PBLAS, for further details see [5].

4 Rationale and Questions to the Community

In the design of all levels of the PBLAS, as with the BLAS, one of the main concerns is to keep both the calling sequences and the range of options limited, while at the same time maintaining sufficient functionality. This clearly implies a compromise, and a good judgement is vital if the PBLAS are to be accepted as a useful standard. In this section we discuss some of the reasoning behind the decisions we have made.

A large amount of sequential linear algebra software relies on the BLAS. Because software reusability is one of our major concerns, we wanted the BLAS and PBLAS interfaces to be as similar as possible. Consequently, only one routine, the matrix transposition, has been added to the PBLAS, since this operation is much more complex to perform in a distributed-memory environment [6].

The scalar values returned by the Level 1 PBLAS, such as the dot-product routine, are only correct in the scope of their operands and zero elsewhere. For example, when INCX is equal to one, only the column of processes having part of the vector operands gets the correct results. This decision was made for efficiency purposes. It is, however, very easy to have this information broadcast across the process mesh by directly calling the appropriate BLACS routine. Consequently, these particular routines do not need to be called by any other processes other than the ones in the scope of their operands. With this exception in mind, the PBLAS follow an SPMD programming model and need to be called by every process in the current BLACS context to work correctly.

Nevertheless, there are a few more exceptions in the current model implementation, where some computations local to a process row or column can be performed by the PBLAS, without having every process in the grid calling the routine. For example, the rank-1 update performed in the LU factorization [4] involves data which is contained by only one process column. In this case, to maintain the efficiency of the factorization it is important to have this particular operation performed only within one process column. In other words, when a PBLAS routine is called by every process in the grid, it is required that the code operates successfully as specified by the SPMD programming model. However, it is also necessary that the PBLAS routines recognize the scope of their operands in order to save useless communication and synchronization costs when possible. This specific part of the PBLAS specifications remains an open question.

Internally, the PBLAS currently rely on routines requiring certain alignment properties to be satisfied [7]. These properties have been chosen so that maximum efficiency can be obtained on these restricted operations. Consequently, when redistribution or re-alignment of input or output data has to be performed some performance will be lost. So far, the PBLAS do not perform such redistribution or alignment of the input/output matrix or vector operands when necessary. However, the PBLAS routines would provide greater flexibility and would be more similar in functionality to the BLAS if these operations where provided. The question of making the PBLAS more flexible remains open and its answer largely depends on the needs of the user community.

A few features supported by the PBLAS underlying tools [7] have been intentionally hidden. For instance, a block of identical vectors operands are sometimes replicated across process rows or columns. When such a situation occurs, it is possible to save some communication and computation operations. The PBLAS interface could provide such operations, for example, by setting the origin process coordinate in the array descriptor to -1 (see Sect. 2.1). Such features, for example, would be useful in the ScaLAPACK routines responsible for applying a block of Householder vectors to a matrix. Indeed, these Householder vectors need to be broadcast to every process row or column before being applied. Whether or not this feature should be supported by the PBLAS is still an open question.

We have adhered to the conventions of the BLAS in allowing an increment argument to be associated with each distributed vector so that a vector could, for example, be a row of a matrix. However, negative increments or any value other than 1 or $DESC\_(1)$ are not supported by our current model implementation. The negative increments -1 and $-DESC\_(1)$ should be relatively easy to support. It is still unclear how it would be possible to take advantage of this added complexity and if other increment values should be supported.

The presence of BLACS contexts associated with every distributed matrix provides the ability to have separate "universes" of message passing. The use of separate communication contexts by distinct libraries (or distinct library invocations) such as the PBLAS insulates communication internal to the library execution from external communication. When more than one descriptor array is present in the argument list of a routine in the PBLAS, it is required that the BLACS context entries must be equal (see Sect. 2.2). In other words, the PBLAS do not perform "intra-context" operations.

We have not included specialized routines to take advantage of packed storage schemes for symmetric, Hermitian, or triangular matrices, nor of compact storage schemes for banded matrices. As with the BLAS no check has been included for singularity, or near singularity, in the routines for solving triangular systems of equations. The requirements for such a test depend on the application and so we felt that this should not be included, but should instead be performed outside the triangular solve.

For obvious software reusability reasons we have tried to adhere to the conventions of, and maintain consistency with, the sequential BLAS. However, we have deliberately departed from this approach by explicitly passing the global indices and using *array descriptors*. Indeed, it is our experience that using a "local indexing" scheme for the interface makes the use of these routines much more complex from the user's point of view. Our implementation of the PBLAS emphasizes the mathematical view of a matrix over its storage. In fact, other block distributions may be able to reuse both the interface and the descriptor described in this paper without change. Fundamentally different distributions may require modifications of the descriptor, but the interface should remain unchanged.

The model implementation in its current state provides sufficient functionality for the use of the PBLAS modules in the ScaLAPACK library. However, as we mentioned earlier in this paper, there are still a few details that remain open questions and may easily be accommodated as soon as more experience with these codes is reported. Hopefully, the comments and suggestions of the user community will help us to make these last decisions so that this proposal can be made more rigorous and adequate to the user's needs.

Finally, it is our experience that porting sequential code built on the top of the BLAS to distributed memory machines using the PBLAS is much simpler than writing the parallel code from scratch [4]. Taking the BLAS proposals as our model for software design was in our opinion a way to ensure the same level of software quality for the PBLAS.

References

1. Aboelaze, M., Chrisochoides, N., Houstis, E.: The parallelization of Level 2 and 3 BLAS Operations on Distributed Memory Machines. Technical Report CSD-TR-91-007, Purdue University, West Lafayette, IN, 1991.
2. Anderson, E., Bai, Z., Bischof, C., Demmel, J., Dongarra, J., Du Croz, J., Greenbaum, A., Hammarling, S., McKenney, A., Ostrouchov, S., Sorensen, D.: LAPACK Users' Guide, Second Edition. SIAM, Philadelphia, PA, 1995.
3. Brent, R., Strazdins, P.: Implementation of BLAS Level 3 and LINPACK Benchmark on the AP1000. Fujitsu Scientific and Technical Journal, 5(1):61–70, 1993.
4. Choi, J., Demmel, J., Dhillon, I., Dongarra, J., Ostrouchov, S., Petitet, A., Stanley, K., Walker, D., Whaley, R.C.: ScaLAPACK: A Portable Linear Algebra Library for Distributed Memory Computers - Design Issues and Performance. Technical Report UT CS-95-283, LAPACK Working Note #95, University of Tennessee, 1995.
5. Choi, J., Dongarra, J., Ostrouchov, S., Petitet, A., Walker, D., Whaley, R.C.: A Proposal for a Set of Parallel Basic Linear Algebra Subprograms. Technical Report UT CS-95-292, LAPACK Working Note #100, University of Tennessee, 1995.
6. Choi, J., Dongarra, J., Walker, D.: Parallel matrix transpose algorithms on distributed memory concurrent computers. In Proceedings of Fourth Symposium on the Frontiers of Massively Parallel Computation (McLean, Virginia), pages 245–252. IEEE Computer Society Press, Los Alamitos, California, 1993. (also LAPACK Working Note #65).
7. Choi, J., Dongarra, J., Walker, D.: PB-BLAS: A Set of Parallel Block Basic Linear Algebra Subroutines. In Proceedings of the Scalable High Performance Computing Conference, pages 534–541, Knoxville, TN, 1994. IEEE Computer Society Press.
8. Dongarra, J., Du Croz, J., Duff, I., Hammarling, S.: A Set of Level 3 Basic Linear Algebra Subprograms. ACM Transactions on Mathematical Software, 16(1):1–17, 1990.
9. Dongarra, J., Du Croz, J., Hammarling, S., Hanson, R.: Algorithm 656: An extended Set of Basic Linear Algebra Subprograms: Model Implementation and Test Programs. ACM Transactions on Mathematical Software, 14(1):18–32, 1988.
10. Dongarra, J., Whaley, R.C.: A User's Guide to the BLACS v1.0. Technical Report UT CS-95-281, LAPACK Working Note #94, University of Tennessee, 1995.
11. Elster, A.: Basic Matrix Subprograms for Distributed Memory Systems. In D. Walker and Q. Stout, editors, Proceedings of the Fifth Distributed Memory Computing Conference, pages 311–316. IEEE Press, 1990.
12. Falgout, R., Skjellum, A., Smith, S., Still, C.: The Multicomputer Toolbox Approach to Concurrent BLAS. submitted to Concurrency: Practice and Experience, 1993. (preprint).
13. Message Passing Interface Forum. MPI: A Message Passing Interface Standard. International Journal of Supercomputer Applications and High Performance Computing, 8(3–4), 1994.
14. Koebel, C., Loveman, D., Schreiber, R., Steele, G., Zosel, M.: The High Performance Fortran Handbook. The MIT Press, Cambridge, Massachusetts, 1994.

Parallel Implementation of a Lagrangian Stochastic Particle Model of Turbulent Dispersion in Fluids

Lianne G.C. Crone[1], Han van Dop[2] and Wim Lourens[1]

[1] University of Utrecht, Department of Physics
[2] Institute for Marine and Atmospheric Research Utrecht

Abstract. In Lagrangian models, trajectories of particles are generated to describe turbulent dispersion. The parallel implementation of such a model appears to be straightforward by distributing the particles over the processors. The model itself is fully parallel, but communicating the results to the outside world leads to a parallel output problem. We discuss different implementation methods of this model on message passing machines. In this paper we show performance characteristics of the different implementations on a Parsytec GCel-512 and a PowerXplorer-32.

1 Introduction

The turbulent transport of pollutants in fluids can be modelled using either a Eulerian or a Lagrangian approach. In a Eulerian frame-work, the transport and chemical behaviour of species is described in a fixed frame of reference by an advection-diffusion-reaction equation where the diffusion term reflecting the turbulence is derived by applying K-theory neglecting the concentration fluctuations (Seinfeld (1975)). However, concentration fluctuations are in the order of the mean concentration itself and can not be disregarded. Currently, also second order approximations, which include concentration fluctuations, are being successfully applied (Vilà-Guerau and Duynkerke (1992)). Numerically solving the resulting equations offers numerous problems and usually requires large computational effort (CPU-time and memory).

Alternatively, the statistical Lagrangian approach consists of calculating the displacement and velocity of fluid as a random process. This approach is more intuitive because a close look at fluid elements in turbulent flows shows that their trajectories are generally smooth, apart from 'sudden' changes in small, concentrated regions of the fluid with high vorticity (van Dop et al. (1985)). Van Dop et al (1985) state that Lagrangian models are able to include more turbulence properties than the first order K-models.

Here, we shall give an outline of an application of the Lagrangian description of turbulent diffusion with an emphasis on the numerical aspects and advantages of the method.

Nearly all flows occurring in nature are turbulent, for example atmospheric flows (Stull (1988)). Atmospheric turbulence occurs within the convective boundary layer (CBL) originating primarily from heating of the land surface in the daytime. This leads to the formation of random up-draft and down-draft thermals called 'eddies'. The size of these eddies determines the typical size of the turbulence and the associated time-scale. Transport of pollutants in the atmosphere is dominated horizontally by the mean wind and vertically by turbulence. Hence, when modelling atmospheric dispersion, one often assumes turbulence to be homogeneous in the horizontal directions, meaning that it has the same structure at every horizontal location.

In this paper, we first give a brief outline of the Lagrangian method describing turbulent dispersion, where we focus on the application of the model to the atmosphere. In section 3 an outline is given of the random walk model. Numerical implementation questions are handled in section 4. The model is demanding with respect to computer time. Therefore, in section 5, we discuss the possibility of implementing the model on a parallel system and study the performance of the implementation strategies in section 6. We present results of a specific test on two multiprocessor platforms in section 7.

2 Lagrangian method

The Lagrangian method for description of transport in turbulent fluids is based on statistical relations involving 'marked particles'. By 'particles' we mean here a small volume of fluid, that is small enough to follow all turbulent eddies but large enough to contain many molecules. Such a particle can be 'marked' by adding a tracer. We will not consider molecular processes and assume the particle's concentration to be constant.

The mean concentration can be expressed in terms of these particles by

$$\overline{C}(\mathbf{x}, t) = \int_{-\infty}^{t} \int_{\mathbb{R}^3} P(\mathbf{x}, t; \mathbf{x}', t') S(\mathbf{x}', t') d\mathbf{x}' dt', \qquad (1)$$

where $P(\mathbf{x}, t; \mathbf{x}', t')$ is the probability that particles leaving $\mathbf{x}'$ at time t' arrive at $\mathbf{x}$ at time t, $S(\mathbf{x}', t')$ is the initial concentration field of the particles.

If we use N_p particles to approximate the concentration, the discrete form of the mean concentration in a region $\Omega_{x,\Delta x} = [x - \frac{\Delta x}{2}, x + \frac{\Delta x}{2}]$ reads

$$\overline{C}(\Omega_{x,\Delta x}, t) = \frac{1}{\Delta x} \int_{x - \frac{\Delta x}{2}}^{x + \frac{\Delta x}{2}} \sum_{i=1}^{N_p} C_i \cdot \delta(x' - X_i(t)) dx' \qquad (2)$$

where C_i equals the weight of the ith particle and δ is the Dirac δ-function. $X_i(t)$ is the ith realisation of the stochastic process $X(t)$ with probability density function P. For clarity, we only presented the one-dimensional case, noting that the one-dimensional theory can easily be adapted to the more general 3-dimensional case.

To compute the concentration we need either to determine the probability density function, $P(x, t; x', t')$ or the particle's positions at time t.

3 Calculating particle trajectories

We assume stationary turbulence to be homogeneous in the horizontal directions (x- and y-directions). The Langevin equations describing the horizontal velocities U_1, U_2 and positions X_1, X_2 of a particle are

$$dU_i = -\frac{U_i}{T_L^H}dt + d\omega_i, \quad dX_i = U_i dt \text{ for } i = 1, 2. \tag{3}$$

The Lagrangian time-scales in both horizontal directions are considered to be equal and constant and will be denoted by T_L^H. The stochastic velocity increments $d\omega_i$, $i = 1, 2$ have moments

$$\overline{d\omega_i} = 0, \quad \overline{d\omega_i(t')d\omega_j(t'')} = \frac{2\,\sigma_i^2}{T_L^H}\,\delta_{t't''}\delta_{ij}\,dt \text{ for } j = 1, 2, 3, \tag{4}$$

where σ_i^2 is the Eulerian velocity variance in the ith direction and is constant in homogeneous stationary turbulence. In equation 4, the Kronecker δ-functions, $\delta_{t't''}$ and δ_{ij} indicate that velocity increments are uncorrelated.

To describe the transport in the vertical direction, where turbulence is considered inhomogeneous, we use a normalised formulation of the Langevin equation with $\tilde{U}_3 = \frac{U_3}{\sigma_3}$ (see Sawford (1986)):

$$d\tilde{U}_3 = -\frac{\tilde{U}_3}{T_L^V}dt + d\omega_3, \quad dX_3 = \sigma_3\tilde{U}_3 dt, \tag{5}$$

where T_L^V is the vertical Lagrangian time-scale and $d\omega_3$ is a stochastic process with a Gaussian distribution and moments

$$\overline{d\omega_3} = \frac{\partial\sigma_3}{\partial x_3}dt, \quad \overline{d\omega_3(t')d\omega_j(t'')} = \frac{2}{T_L^V}\delta_{t't''}\delta_{3j}\,dt. \tag{6}$$

The Eulerian velocity variance in the z-direction varies with height depending on the mixing height h and the convective velocity scale w_*:

$$\frac{\sigma_3^2(z)}{w_*^2} = f_h(z). \tag{7}$$

4 Random-walk model

The trajectories of the particles are calculated using discrete formulations of the stochastic differential equations described above. The velocity and position of a particle at time $t + \Delta t$ are calculated from the position and velocity of the particle at time t.

$$
\begin{aligned}
U_i(t + \Delta t) &= \left(1 - \frac{\Delta t}{T_L^H}\right) U_i(t) + \sigma_i\sqrt{\frac{2\Delta t}{T_L^H}}\nu_i, \\
X_i(t + \Delta t) &= X_i(t) + \frac{\Delta t}{2}\left[U_i(t) + U_i(t + \Delta t)\right] \text{ for } i = 1, 2
\end{aligned} \tag{8}
$$

and

$$\tilde{U}_3(t + \Delta t) = \left(1 - \frac{\Delta t}{T_L^V}\right)\tilde{U}_3(t) + \sqrt{\frac{2\Delta t}{T_L^V}}\nu_3 + \Delta t\frac{\partial \sigma_3}{\partial z},$$
$$X_3(t + \Delta t) = X_3(t) + \frac{\Delta t}{2}\left[\sigma_3(X_3(t))\tilde{U}_3(t) + \sigma_3(X_3(t + \Delta t))\tilde{U}_3(t + \Delta t)\right], \tag{9}$$

where we will use the approximation

$$\sigma_3(X_3(t + \Delta t)) = \sigma_3(X_3(t)) + U_3(t)\Delta t\frac{\partial \sigma_3}{\partial z}. \tag{10}$$

In formula's 9 and 10 the ν_i, $i = 1, 2, 3$, are normally distributed independent random variables with zero mean and unit variance.

The next algorithm describes the calculation of a particle's trajectory from time $start$ to time $stop$ with time steps Δt.

Algorithm *Trajectory* $(start, stop, \Delta t)$

    The position $\mathbf{X}(start)$ and velocity $\mathbf{U}(start)$ of the particle are known

    (1) $t = start$

    (2) While $t \leq stop$ do

            Generate 3 random numbers with distribution N(0,1)

            Calculate $U_1(t + \Delta t)$, $U_2(t + \Delta t)$, $\tilde{U}_3(t + \Delta t)$

            calculate $\sigma_3(X(t))$ and $\sigma_3(X(t + \Delta t))$

            Calculate $\mathbf{X}(t + \Delta t)$

            $t = t + \Delta t$

    (3) end

Now that we know the positions of all particles at discrete times, we can compute the mean concentration in a small domain. The mean concentration in a volume equals the sum of the concentrations of the particles located in the specific volume. We give an outline of one of the possible implementations of a program that predicts the concentration evolution of a gas from time t_0 to T with sub-results every $\Delta save$ seconds, using a particle model with time-step Δt.

Algorithm *Concentration prediction* $(t_0, T, \Delta save)$

    (0) input parameters and initialisation

    (1) $t = t_0$, compute the initial position and velocity

        of the particles depending on the source-type, compute

        the initial concentration profile and store it in a file

    (2) While $t + \Delta save \leq T$ do

            for all particles compute

                Trajectory $(t, t + \Delta save, \Delta t)$

            compute the concentration at $t + \Delta save$ and file it

            $t = t + \Delta save$

    (3) end

The accuracy of the concentration-prediction depends on the number of particles located within the volume. The accuracy is in the order of $\frac{1}{\sqrt{N}}$. Chock and Winkler 1994 remark that at least 2000 particles per volume are needed to predict the mean concentration with very good agreement with the numerical solutions of the corresponding Eulerian model, i.e. within two percent.

5 Parallelisation

To make good concentration predictions, we have to calculate the trajectories of many particles which requires much computing time. Therefore, the need for parallelisation is obvious. One way of introducing parallelism is to use several processors each including a program control unit, an arithmetic logic unit and several memory and I/O modules: multiprocessors. The processors in a shared-memory machine communicate via a common memory, which forms a bottle-neck if all processors want to access the memory concurrently. A message-passing multiprocessor system consists of several independent modules and an interconnection network. Properties such as simplicity and scalability make message-passing multiprocessors prime candidates for very large systems. For our application we purchase massive parallelism, therefore, we focus on the implementation of the particle-model on a message-passing system.

First, we notice that the particles move independently of each other. One possible approach for parallelisation is to distribute the particles evenly over all the processors and compute on each processor the trajectories of its particles. This part is fully parallel. Whenever the concentration has to be calculated, each processor computes the distribution of its own particles. This leads to two problems: how to combine these local results and how to store them in a file. There are different methods to deal with these parallel I/O problems.

The simplest solution to the I/O problems is to let each processor write its own results to disc. This method encounters problems due to hardware restrictions: often it is not possible to access file-systems from multiple processors simultaneously. On most multi-processor systems the disc-access will be handled more or less sequentially. We will refer to this method as the direct filing method.

We could also gather the results in one processor and let this processor access the disc. The costs of this operation depend on the network-topology and the communication-algorithm. We will look at several topologies: a hypercube, a 2d-grid and a linear array with P processors. On the hypercube the gathering-algorithm costs at least $^2 \log P$ neighbour-neighbour communications no matter which processor gathers the result. In a 2d-grid topology gathering in a middle processor costs at least $\sqrt{P}$ communications and gathering in a corner $2(\sqrt{P}-1)$ communications. A global gathering at the end processor of a linear array costs at least $P-1$ neighbour communications.

A third method to file the concentration fields, is using a master-slave hierarchy where the master gathers all the results and files them while all the other processors calculate particle trajectories and send their local-concentrations to the master one after the other.

6 Performance analysis

To measure the performance of a certain implementation on P processors, we will use the notion of speed-up,

$$\frac{\text{time on 1 processor}}{\text{time on P processors}}. \tag{11}$$

The speed-up would be perfectly equal to P, if we only considered the calculation of the particle trajectories, but the I/O requirements will lower the performance.

The parallel algorithm can be split into three parts: a fully parallel part (the calculation of the particle trajectories), a sequential part (the filing of the concentration fields) and an extra part due to the implementation of the model on a multiprocessor system, the parallelisation overhead. The first two implementation methods differ from each-other in the overhead part. The parallelisation overhead depends on the possibilities of overlapping communication with computation and on some features of the application itself.

When running the application, at first the processors operate synchronously, but after the first filing operation they fall out of step because processors can not access the disc simultaneously or processors do not participate in the same communication actions. The processors will stay out of step until the next filing operation where in most cases it is favourable that the processors are not working synchronously. In the best circumstances (if the computational work is large enough) all processors can start their filing actions immediately without delay.

For each implementation method we can derive the maximum and the minimum parallelisation overhead depending on the number of processors, P and the number of filings, N_s. Other factors influencing the parallelisation overhead are $T_{disc}(N)$, the time needed to file N bytes, $T_{send}(N)$ the time needed to send a message of N bytes to a neighbour and $N_p \cdot \frac{N_t}{N_s} \cdot T_{traject}$, the time needed to compute the trajectories of N_p particles for the $\frac{N_t}{N_s}$ time-steps between two filing actions (N_t equals the total number of time-steps). In table 1 the maximum and minimum overhead are presented for some implementation strategies.

Implementation method	Parallelisation overhead	
	Maximum	Minimum
Direct filing	$(P-1)N_s T_{disc}(N)$	$(P-1)T_{disc}(N)$
Hypercube	$^2\log P \, N_s \, T_{send}(N)$	$^2\log P N_s T_{send}(N)$
2d-grid mid-gathering	$\sqrt{P}N_s T_{send}(N)$	$(\sqrt{P}+4(N_s-1))T_{send}(N)$
2d-grid corner-gathering	$2(\sqrt{P}-1)N_s T_{send}(N)$	$(2(\sqrt{P}-1)+2(N_s-1))T_{send}(N)$
Linear array end-gathering	$(P-1)N_s T_{send}(N)$	$(P+N_s-2)T_{send}(N)$

Table 1. Minimal and Maximal parallelisation overhead for different filing methods

For the direct filing method, the minimum overhead is attained if all processors are able to file their results within the time needed to compute the trajectories for the next field, hence if $P \cdot T_{disc} < N_p \cdot \frac{N_t}{N_s} \cdot T_{traject}$. All gathering methods perform best if the total task (gathering and filing) of the gather-processor is larger than the gathering tasks of the other processors.

We see that the direct-filing method has the largest overhead but it is independent of the filing frequency, which can be very large. For long simulations where the concentration has to be filed very often, the linear array will attain the best performance. If, the application is short or if the user is only interested in the final result, the best performance will be obtained by using a hypercube.

The master-slave method is different. A slave needs for the application at least $\frac{N_p N_t}{P-1} T_{traject} + (N_s + P) T_{send}(N)$ time and the master $((P-1)T_{send}(N) + T_{disc})N_s$. The best performance is attained if the load of the master and the slaves are nearly equal.

7 The single particle model on the Parsytec GCel and the PowerXplorer

To illustrate the parallel possibilities of a particle model on a multiprocessor machine, we simulated the dispersion of a cloud of tracer in the atmosphere.

7.1 Test-case

The convective velocity scales, the Eulerian velocity variances and the Lagrangian time scales, can only be determined by observations. Chock and Winkler (1994) adopted some choices for their three dimensional test-case. In our paper we will use the same test-model for comparison.

Turbulence is not stationary in the atmospheric boundary layer, but we will assume it is. From studies of data sets, one may conclude that the horizontal velocity variances show no significant dependence on height and can be chosen constant such that

$$\frac{\sigma_u^2}{w*^2} = \frac{\sigma_v^2}{w*^2} = 0.3, \tag{12}$$

where the convective velocity scale, $w*$ is chosen equal to 1.5 m/s. The profile used for the vertical velocity variance was obtained by fitting it to the AMTEX data set,

$$\frac{\sigma_w^2}{w*^2} = 1.8 \left(\frac{z}{h}\right)^{\frac{2}{3}} \left(1 - 0.8\frac{z}{h}\right)^2, \tag{13}$$

where h is the mixing height equal to 1000 m.

The Lagrangian time-scales show a dependence on the altitude. We will assume constant time scales for all heights $T_L^H = 1000$ s and $T_L^V = 100$ s.

We calculate the mean concentration following equation 2 in grid-points of a $21 \times 21 \times 25$ grid and for regions of 5km by 5km by 50 meters.

The initial concentration is a 2-dimensional cosines hill located just above the ground in the bottom cells and centred in the middle of the computational grid. The concentration in the bottom cell $C_B(x, y)$ at $t = 0$ is

$$C_B(x, y) = \begin{cases} 50(1 + \cos \frac{\pi\sqrt{x^2+y^2}}{2000}) & \text{if } \sqrt{x^2 + y^2} \leq 2000 \\ 0 & \text{else.} \end{cases} \tag{14}$$

The cloud of pollutant will dissipate slowly in the horizontal directions and spread evenly within the height of the convective boundary layer.

The time-step needed to make accurate predictions for the inhomogeneous turbulence in the vertical direction is equal to 10 seconds. Although, a larger time-step size could be used to describe the transport in the horizontal directions (in homogeneous turbulence), we used the same time-step of 10 seconds for all directions. All tests cover a 10 hour time-period.

For further explanation of this test-case, we refer to Chock and Winkler (1994).

7.2 Results

We implemented the single particle model on two message-passing systems located at the Interdisciplinary Centre for Complex Computer facilities Amsterdam: a Parsytec GCel with 512 T805 transputers and a PowerXplorer with 32 nodes, consisting of a PowerPC as computation unit and a transputer for the communication. We coded the simulation in the ANSI-C language in combination with PARIX library routines to handle the communication. The network topologies were set up with routines from the virtual topology library. More information about these two systems can be found in Hoekstra et al. (1995).

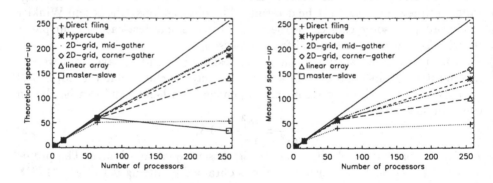

Fig. 1. Theoretical and measured speed-ups on the Parsytec GCel

The T805 transputer has a low sustained floating point performance hence, the simulation with only 2560 particles costed nearly eight hours on one single processor. Therefore, we compared the performance of multiple nodes with the performance of a single processor for this small application. The concentration fields are calculated every hour. We compared the theoretical and measured speed-ups on 4, 16, 64 and 256 processors for the different implementation methods described earlier. It was not possible to make a master-slave hierarchy with more than 32 processors. In figure 1, we see that even for this small application the performance is good. The speed-up deteriorates on 256 processors because the computational work is very small: only 10 particles per processor. For accurate predictions, about 2.5 million particles should be used, making the parallel part of the application much bigger, which leads to even better speed-ups.

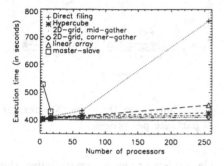 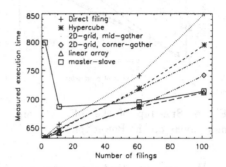

Fig. 2. Timings on the Parsytec GCel for 100 particles per processor

Fig. 3. Timings of the application ran on 16 processors

We studied the scalability of the problem by measuring the execution times for a constant number of particles per processor, thus for a constant computational load per processor. In figure 2 the results are presented.

We see that the methods using the interconnection network scale very good. The direct filing method performs well as long as the computational work is larger than the time needed for all processors to access the disc. On 256 processors the computational load is not large enough and we see that the performance decreases rapidly. The master-slave implementation on only 4 processors suffers a load-imbalance, the master's task is much smaller than the tasks of the slaves. We used the theoretical analysis to predict the performance of the master-slave method for larger numbers of processors. When using 256 processors the performance will deteriorate due to the increasing task of the master.

In figure 4, the speedup on the PowerXplorer is shown. The PowerXplorer is computationally faster than the Parsytec but has comparable speed for communication and disc-access. For a specific application the PowerXplorer has a worse speed-up than the Parsytec, because the parallelisation overhead is relatively expensive compared to the computational work. On the PowerXplorer, we studied the scalability with 1000 particles per processor, figure 5. We, again, see that the direct filing and master-slave methods do not scale as well as the gathering methods.

8 Conclusion

Transport of pollutants in turbulent fluids can be described by Eulerian and Lagrangian models. Both methods are computer time consuming, hence only parallel computers seem capable of handling them in an effective way.

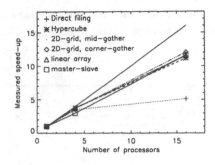 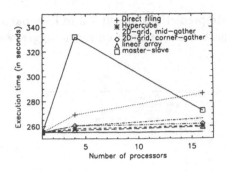

Fig. 4. Speedup of the different implementations on the PowerXplorer

Fig. 5. Timings on the PowerXplorer for 1000 particles per processor

We have seen that the Lagrangian particle models are easy to parallelise: the only problem is filing the concentration fields. The parallel implementations on message passing multi-processor systems are efficient and scale very well.

The direct filing and the master-slave strategies are good, but they scale not as good as the gathering methods. Best performances are attained when the filing of the concentration fields is done by a gathering operation using the interconnection network. The choice of the network topology depends on the number of filings in one application, for low numbers the hypercube is best and for many filings the linear array.

References

Chock, D. P. and S. L. Winkler (1994), A particle grid air quality modeling approach, 1. The dispersion aspect, *J. Geophys. Res.*, **99**, pp. 1019–1031.

Dop van, H., F. T. M. Nieuwstadt and J. C. R. Hunt, (1985), Random walk models for particle displacements in inhomogeneous unsteady turbulent flows, *Phys. Fluids*, **28** (6), pp. 1639–1653.

Hoekstra, A. G. , F. van der Linden, M. van Muiswinkel, J. J. J. Vesseur, L. O. Hertzberger and P. M. A. Sloot, (1995), *Native and generic parallel programming environments on a transputer and a powerpc platform*, Concurrency, Practice and Experience, in press.

Sawford B. L., (1986), Generalized random forcing in random-walk turbulent dispersion models, *Phys. Fluids*, **29**, p. 3582–3585.

Seinfeld, J. H., (1975), **Air pollution, physical and chemical fundamentals**, McGraw-Hill, Inc., New York.

Stull, R. B., (1988), **An introduction to boundary layer meteorology**, Kluwer Academic Publishers, Dordrecht.

Vilà-Guerau de Arellano, J. and P. G. Duynkerke, (1993), Second order closure study of the covariance between chemically reactive species in the surface layer, *J. atmos. Chem.*, **16**, pp. 145–155

Reduction of a Regular Matrix Pair (A, B) to Block Hessenberg-Triangular Form

Krister Dackland and Bo Kågström

Department of Computing Science, University of Umeå, S–901 87 Umeå, Sweden.
Email addresses: `dacke@cs.umu.se` and `bokg@cs.umu.se`.

Abstract. An algorithm for reduction of a regular matrix pair (A, B) to block Hessenberg-triangular form is presented. This condensed form $Q^T(A, B)Z = (H, T)$, where H and T are block upper Hessenberg and upper triangular, respectively, and Q and Z orthogonal, may serve as a first step in the solution of the generalized eigenvalue problem $Ax = \lambda Bx$. It is shown how an elementwise algorithm can be reorganized in terms of blocked factorizations and higher level BLAS operations. Several ways to annihilate elements are compared. Specifically, the use of Givens rotations, Householder transformations, and combinations of the two. Performance results of the different variants are presented and compared to the LAPACK implementation DGGHRD, which indeed is unblocked.

1 Introduction

A reliable way to solve the generalized eigenvalue problem $Ax = \lambda Bx$, where A and B are square real $n \times n$ matrices, is to reduce (A, B) to generalized Schur (GS) form (S, T), where S is upper quasi-triangular and T is upper triangular. A quasi-triangular matrix is block upper triangular with 1×1 and 2×2 blocks on the diagonal. The 2×2 blocks on the diagonal of $S - \lambda T$ correspond to pairs of complex conjugate eigenvalues. The real eigenvalues are given by $(s_{ii}, t_{ii}) \neq (0, 0)$ (i.e., the 1×1 diagonal blocks of (S, T)). The finite eigenvalues are s_{ii}/t_{ii}, where $t_{ii} \neq 0$. If $(s_{ii}, t_{ii}) \neq (0, 0)$ for all i, then (A, B) is a regular matrix pair.

Typically, the GS form is computed in two steps. First, (A, B) is reduced to condensed form (H, T) by using orthogonal transformations such that H is upper Hessenberg and T is upper triangular (e.g. see [5]). In the second step the QZ algorithm [9] is iteratively applied to the condensed pencil (H, T) to obtain the desired GS form (S, T).

One way to achieve high performance on modern advanced architectures is to restruct well-known elementwise algorithms in terms of level 2 (matrix-vector) and level 3 (matrix-matrix) operations. These blocked algorithms permit reuse of data in the higher levels of a memory hierarchy, which make them run closer to peak machine speed.

In this contribution we focus on the first step in the process of solving the generalized eigenvalue problem. An implementation of an elementwise algorithm for the Hessenberg-triangular reduction (called DGGHRD for double precision real data) is included in LAPACK [1]. We show how the elementwise algorithm can be reorganized in terms of blocked factorizations and higher level (matrix-matrix) operations to reduce A to upper *block* Hessenberg form and B to upper triangular form. In order to make the algorithm portable and efficient most operations are expressed in terms of BLAS [8], [4], [6], [7] and LAPACK.

Several ways to annihilate elements are evaluated. We compare the use of Givens rotations, Householder transformations, and combinations of the two. Performance results of the different variants are presented and compared to the unblocked LAPACK implementation DGGHRD.

The rest of the paper is organized as follows. Section 2 outlines the elementwise and level 3 (blocked) algorithms. In Section 3 different variants of the blocked algorithm are presented. Section 4 presents some measured performance results of the implementations for an IBM RS6000 workstation. Finally, in Section 5 we give a brief summary and outline future work.

2 Reduction of (A, B) to Hessenberg-Triangular Form

We present algorithms for reducing a regular matrix pair (A, B) to condensed form in terms of an orthogonal equivalence transformation such that

$$Q^T A Z = H, \quad Q^T B Z = T,$$

where H is block upper Hessenberg, T is upper triangular and Q and Z are orthogonal. First, we review the elementwise algorithm that reduces A to upper Hessenberg form. The second algorithm is based on level 3 operations and reduces A to block upper Hessenberg form. In both cases the final form of B is upper triangular.

2.1 Elementwise Reduction to Hessenberg-Triangular Form

The elementwise algorithm reduces a regular matrix pair (A, B) to upper Hessenberg-triangular form [5]. In the first step of the reduction an orthogonal matrix U is determined such that $U^T B$ is upper triangular. To preserve the generalized eigenvalues of the matrix pair U is also applied to A: $A \leftarrow U^T A$. Now A is dense and B is triangular. To further reduce A to upper Hessenberg form while preserving B upper triangular, Givens rotations or 2×2 Householder reflections are used. Orthogonal transformations applied from left annihilate elements in A but destroy the structure of B. The triangular form of B is restored by right hand side orthogonal transformations. We illustrate one step of the reduction using Givens rotations (A and B are of size 4×4):

$$Q_{34}^T A = \begin{bmatrix} \cdot & \cdot & \cdot & \cdot \\ \cdot & \cdot & \cdot & \cdot \\ \cdot & \cdot & \cdot & \cdot \\ & \cdot & \cdot & \cdot \end{bmatrix}, \quad Q_{34}^T B = \begin{bmatrix} \cdot & \cdot & \cdot & \cdot \\ & \cdot & \cdot & \cdot \\ & & \cdot & \cdot \\ & \star & \cdot & \end{bmatrix},$$

$$BZ_{34} = \begin{bmatrix} \cdot & \cdot & \cdot & \cdot \\ & \cdot & \cdot & \cdot \\ & & \cdot & \cdot \\ & & & \cdot \end{bmatrix}, \quad AZ_{34} = \begin{bmatrix} \cdot & \cdot & \cdot & \cdot \\ \cdot & \cdot & \cdot & \cdot \\ \cdot & \cdot & \cdot & \cdot \\ & & \cdot & \cdot \end{bmatrix}$$

The rotation Q_{34} eliminates a_{41} and when applied to B introduces a non-zero element b_{43}. This fill-in is set to zero by a right rotation Z_{34}. Since Z_{34} only operates on columns 3 and 4, no new non-zero elements are introduced in A. To annihilate the remaining elements below the sub-diagonal similar two-sided rotations are used.

2.2 Level 3 Reduction to Block Hessenberg-Triangular Form

We reorganize the elementwise algorithm to a blocked variant that mainly perform level 3 (matrix-matrix) operations and some level 2 (matrix-vector) and level 1 (vector-vector) operations. The final forms of H and T, the transformed A and B, are block upper Hessenberg and upper triangular, respectively.

Initially, B is reduced to upper triangular form using a blocked (level 3) QR factorization algorithm [1], [3]. The transformations are also applied to A, resulting in an equivalence transformation of (A, B). From now on we focus on the reduction of A to upper block Hessenberg form, while preserving B upper triangular.

The matrices A and B are partitioned into $M \times M$ square blocks $A_{i,j}$ and $B_{i,j}$, respectively, of size $nb \times nb$, i.e., $M = n/nb$ and the block size nb is chosen with respect to the underlying architecture. Without loss of generality, we assume that nb is a multiple of n.

The following example used to present the algorithm shows A and B in block form after one iteration of the block reduction algorithm. The symbol □ shows a dense block and the symbol ◺ shows an upper triangular block. All blocks are of size $nb \times nb$ and $M = 6$.

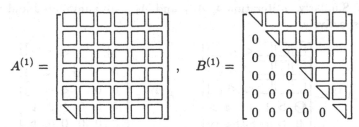

This form is obtained by first reducing $A_{M,1}$ to triangular form. The resulting (left) orthogonal transformation U (of size $nb \times nb$), where $U^T A_{M,1}$ is upper triangular, is also applied to B, which destroys the structure of $B_{M,M}$. The triangular form of $B_{M,M}$ is restored via an RQ factorization of the diagonal block, giving the right hand side transformation V (of size $nb \times nb$). The application of U and V onto the remaining blocks in the last block row of A and the last block columns of A and B are performed using the level 3 BLAS routine DGEMM for general matrix multiply and add: $A_{M,2:M} \leftarrow U^T A_{M,2:M}$, $A_{:,M} \leftarrow A_{:,M} V$, $B_{1:M-1,M} \leftarrow B_{1:M-1,M} V$.

Below follows a sequence of illustrations that describe the operations performed on different blocks of A and B during the remaining $M - 2(= 4)$ iterations to reduce the first block column to block upper Hessenberg form. The symbols $\triangleleft$, $\triangleright$ and $\triangleright\!\triangleleft$ illustrate level 3 updates from the right, left and both hand sides, respectively. Lower level reductions and updates are shown by the symbols $\otimes$ (two sided) and $\oslash$ (left hand side). $A^{(i)}$ and $B^{(i)}$ are the reduced A and B after iteration $i(= 1 : M - 1)$.

$$
A^{(2)} = \begin{bmatrix}
\square & \square & \square & \square & \triangleleft & \triangleleft \\
\square & \square & \square & \square & \triangleleft & \triangleleft \\
\square & \square & \square & \square & \triangleleft & \triangleleft \\
\square & \square & \square & \square & \triangleleft & \triangleleft \\
\oslash & \triangleright & \triangleright & \triangleright & \triangleright\!\triangleleft & \triangleright\!\triangleleft \\
0 & \triangleright & \triangleright & \triangleright & \triangleright\!\triangleleft & \triangleright\!\triangleleft
\end{bmatrix}, \quad
B^{(2)} = \begin{bmatrix}
\boxtimes & \square & \square & \square & \triangleleft & \triangleleft \\
0 & \boxtimes & \square & \square & \triangleleft & \triangleleft \\
0 & 0 & \boxtimes & \square & \triangleleft & \triangleleft \\
0 & 0 & 0 & \boxtimes & \triangleleft & \triangleleft \\
0 & 0 & 0 & 0 & \otimes & \otimes \\
0 & 0 & 0 & 0 & 0 & \otimes
\end{bmatrix}
$$

The second iteration starts by annihilating the triangular block $A_{M,1}$ and reducing $A_{M-1,1}$ to upper triangular form. When applied to B, these (left) orthogonal transformations destroy the structure of the diagonal blocks $B_{M-1,M-1}$ and $B_{M,M}$ and introduce a non-zero subdiagonal block $B_{M,M-1}$. The triangular structure of B is restored by computing an RQ factorization of

$$
\begin{bmatrix}
B_{M-1,M-1} & B_{M-1,M} \\
B_{M,M-1} & B_{M,M}
\end{bmatrix}.
$$

The accumulated left and right orthogonal transformations U and V (now both of size $2nb \times 2nb$) from these two steps are applied to the remaining parts of block columns and rows $M - 1$ and M: $A_{M-1:M,2:M} \leftarrow U^T A_{M-1:M,2:M}$, $A_{:,M-1:M} \leftarrow A_{:,M-1:M}V$, $B_{1:M-2,M-1:M} \leftarrow B_{1:M-2,M-1:M}V$.

Similar operations are performed in the remaining iterations as illustrated below. In iteration 3, $A_{5,1}$ and $A_{4,1}$ are annihilated and triangularized, respectively. Notice that some B-blocks are updated from the left hand side using level 3 routines. Similarly, in iteration 4, $A_{4,1}$ and $A_{3,1}$ are annihilated and triangularized, respectively.

$$
A^{(3)} = \begin{bmatrix}
\square & \square & \square & \triangleleft & \triangleleft & \square \\
\square & \square & \square & \triangleleft & \triangleleft & \square \\
\square & \square & \square & \triangleleft & \triangleleft & \square \\
\oslash & \triangleright & \triangleright & \triangleright\!\triangleleft & \triangleright\!\triangleleft & \triangleright \\
0 & \triangleright & \triangleright & \triangleright\!\triangleleft & \triangleright\!\triangleleft & \triangleright \\
0 & \square & \square & \triangleleft & \triangleleft & \square
\end{bmatrix}, \quad
B^{(3)} = \begin{bmatrix}
\boxtimes & \square & \square & \triangleleft & \triangleleft & \square \\
0 & \boxtimes & \square & \triangleleft & \triangleleft & \square \\
0 & 0 & \boxtimes & \triangleleft & \triangleleft & \square \\
0 & 0 & 0 & \otimes & \otimes & \triangleright \\
0 & 0 & 0 & 0 & \otimes & \triangleright \\
0 & 0 & 0 & 0 & 0 & \boxtimes
\end{bmatrix}
$$

$$
A^{(4)} = \begin{bmatrix}
\square & \square & \triangleleft & \triangleleft & \square & \square \\
\square & \square & \triangleleft & \triangleleft & \square & \square \\
\oslash & \triangleright & \triangleright\!\triangleleft & \triangleright\!\triangleleft & \triangleright & \triangleright \\
0 & \triangleright & \triangleright\!\triangleleft & \triangleright\!\triangleleft & \triangleright & \triangleright \\
0 & \square & \triangleleft & \triangleleft & \square & \square \\
0 & \square & \triangleleft & \triangleleft & \square & \square
\end{bmatrix}, \quad
B^{(4)} = \begin{bmatrix}
\boxtimes & \square & \triangleleft & \triangleleft & \square & \square \\
0 & \boxtimes & \triangleleft & \triangleleft & \square & \square \\
0 & 0 & \otimes & \otimes & \triangleright & \triangleright \\
0 & 0 & 0 & \otimes & \triangleright & \triangleright \\
0 & 0 & 0 & 0 & \boxtimes & \square \\
0 & 0 & 0 & 0 & 0 & \boxtimes
\end{bmatrix}
$$

After $M-1$ iterations the first block column of A is reduced to block upper Hessenberg form with $A_{2,1}$ upper triangular, while preserving B upper triangular.

$$
A^{(5)} = \begin{bmatrix} \square & \triangleleft & \triangleleft & \square\square\square \\ \oslash & \bowtie\bowtie & \triangleright & \triangleright & \triangleright \\ 0 & \bowtie\bowtie & \triangleright & \triangleright & \triangleright \\ 0 & \triangleleft & \triangleleft & \square\square\square \\ 0 & \triangleleft & \triangleleft & \square\square\square \\ 0 & \triangleleft & \triangleleft & \square\square\square \end{bmatrix}, \quad
B^{(5)} = \begin{bmatrix} \searrow & \triangleleft & \triangleleft & \square\square\square \\ 0 & \otimes & \otimes & \triangleright & \triangleright & \triangleright \\ 0 & 0 & \otimes & \triangleright & \triangleright & \triangleright \\ 0 & 0 & 0 & \searrow\square\square \\ 0 & 0 & 0 & 0 & \searrow\square \\ 0 & 0 & 0 & 0 & 0 & \searrow \end{bmatrix}
$$

To complete the block reduction similar operations are performed on the trailing $M-2$ block columns.

3 Annihilation of elements

There are several ways to annihilate block-subdiagonal elements of A and fill-in elements in B. We describe five variants, where two of them use Givens rotations, two of them use Householder reflections and one uses a combination of Givens and Householder transformations. All five are based on the level 3 reduction algorithm outlined in Section 2.2.

3.1 Givens 1

In the implementation called Givens 1, elements are annihilated by use of Givens rotations following the elementwise algorithm in Section 2.1. All the other blocks involved are updated by use of the level 3 BLAS DGEMM.

Specifically, when an element of A is annihilated by a left hand side rotation, the diagonal block of B is immediately updated. The resulting (single) fill-in element is then annihilated by a right hand side rotation. This rotation, when applied to A do not affect the A-block under annihilation and may therefore be accumulated and applied later.

3.2 Givens 2

The variant Givens 2 differs from Givens 1 in the following way. Instead of eliminating the fill-in immediately all rotations from the left hand side (to kill one block of A) are accumulated and applied to the whole current block row of B in one single DGEMM operation. This will destroy the structure of the involved diagonal blocks of B as illustrated in the following example:

$$
U^T B^{(2)}_{M-1:M,M-1:M} = \begin{bmatrix}
\cdot & \cdot & \cdot & \cdot & \cdot & \cdot & \cdot & \cdot \\
f_1 & \cdot & \cdot & \cdot & \cdot & \cdot & \cdot & \cdot \\
f_2 & f_1 & \cdot & \cdot & \cdot & \cdot & \cdot & \cdot \\
f_3 & f_2 & f_1 & \cdot & \cdot & \cdot & \cdot & \cdot \\
f_4 & f_3 & f_2 & f_1 & \cdot & \cdot & \cdot & \cdot \\
 & f_4 & f_3 & f_2 & f_2 & \cdot & \cdot & \cdot \\
 & & f_4 & f_3 & f_3 & f_3 & \cdot & \cdot \\
 & & & f_4 & f_4 & f_4 & f_4 & \cdot
\end{bmatrix}.
$$

The example shows the fill-in introduced in B resulting from iteration 2 of the example in Section 2.2. The fill-in elements are labeled f_j, where j tells what column of A introduced the fill-in. To restore B to upper triangular form $U^T B^{(2)}_{M-1:M,M-1:M}$ is RQ-factorized using a routine based on Givens rotations that takes the new structure of $B^{(2)}_{M-1:M,M-1:M}$ into account.

3.3 Householder 1

Householder 1 follows the same operation pattern as Givens 2, but here Householder reflections are used in all annihilations. Refering to the last example, $U^T B^{(2)}_{M-1:M,M-1:M}$ will now be a full matrix. The RQ factorization that restores B to triangular form is also based on Householder reflections.

3.4 Householder 2

In all previous described implementations the orthogonal transformations U and V have been explicitly constructed before applying the DGEMM updates (which require a temporary array of size $2nb \times n$). In Householder 2, the compact WY representation of Householder matrices is used to reduce the amount of extra storage and copying [11]. We compute Y (of size $2nb \times nb$) and S (of size $nb \times nb$) such that U (or V) $= I - Y S Y^T$. When calling the LAPACK routines for computing a QR or RQ factorization, the Householder vectors Y are returned as the lower part of the factorized block. To extract the triangular matrix S the LAPACK routine DLARFT is called and to apply the reflections to the remaining part of the matrices A and B the LAPACK routine DLARFB is used.

3.5 Givens-Householder

In this implementation both Givens and Householder transformations are used. We extend techniques for reducing a non-symmetric matrix A to condensed form [10], [2] to matrix pairs. First the current block column of A to be reduced is QR-factorized block by block using Householder reflections. This step results in a shark fin shaped block column.

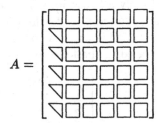

Next all triangular blocks (except the subdiagonal block) are annihilated using Givens rotations. The lower of two subsequent upper triangular blocks is annihilated column by column from left to right as follows: to kill an element $a_{i,j}$, $a_{i-1,j}$ is used as pivot. If $a_{i-1,j}$ refer to an element in the upper of the two triangles, then $a_{j,j}$ becomes the pivot element. We illustrate the elimination of the first two columns of two 4×4 subsequent triangular blocks.

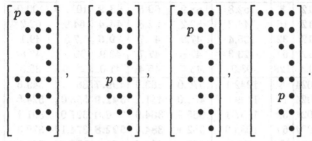

The transformations are of course applied to the associated remaining blocks of A and B. The update of B with respect to the QR factorizations of the first block column and the annihilation (with Givens rotations) of the bottom triangular block $A_{M,1}$ results in the following structure ($M = 6$):

$$
B = \begin{bmatrix} \searcc \end{bmatrix} \rightarrow \begin{bmatrix} \end{bmatrix},
$$

i.e. all diagonal blocks of B are full matrices. Moreover, one non-zero subdiagonal block of B is introduced. To restore $B_{M-1:M,M-1:M}$ to triangular form the same operations as in Householder 1 are performed.

4 Performance Results

We present performance results of our implementations of the five variants of the level 3 reduction algorithm. The target machine is an IBM RS/6000, model 25T, 66MHz. The results are compared to the elementwise implementation DGGHRD from LAPACK [1]. DGGHRD is modified to compute the same condensed form as the level 3 implementations, i.e., H is block upper Hessenberg and T is upper triangular.

The results in Table 1 show that the implementation Householder 1 outperforms the Givens-based implementations independent of the block size nb. Moreover, for small block sizes it is advantageous to explicitly construct the orthogonal matrices Q and Z while for larger block sizes the compact WY approach is to prefer. An explanation to this is that in the WY approach DLARFB is called to perform the update [1]. In DLARFB the update of a block is split into two parts one using the level 3 BLAS DTRMM (triangular matrix multiply) and one using DGEMM. If Q and Z are small the gain in this strategy is lost resulting in poorer performance.

Note that, when the the block size increases, the amount of work decreases and thereby the execution time of DGGHRD.

Table 1. Performance results in seconds on IBM RS/6000, model 25T, 66MHz

n	nb	DGGHRD	Gvns 1	Gvns 2	Hsh 1	Hsh 2	Gvn-Hsh
512	8	152.8	68.9	69.6	**63.9**	101.4	100.0
512	16	144.2	52.2	52.8	**44.4**	64.9	62.2
512	32	135.4	47.9	47.7	**39.6**	47.3	49.1
512	64	123.3	47.6	47.7	40.9	**39.8**	43.8
512	128	93.0	46.5	45.4	41.0	**27.6**	35.3
1024	8	1232.1	570.0	563.9	**499.7**	882.1	849.6
1024	16	1199.6	408.0	411.1	**341.9**	553.0	524.6
1024	32	1167.1	365.9	364.9	**296.0**	397.9	391.1
1024	64	1089.8	382.1	384.5	**322.8**	374.1	379.3
1024	128	975.8	449.9	404.6	364.4	**303.7**	365.5

5 Summary

Five variants of a level 3 algorithm for reducing a regular matrix pair (A, B) to block Hessenberg-triangular form have been presented. It has been shown through computational experiments on a high-performance workstation that the new level 3 based algorithms reduce the execution time by a factor 2–4 compared to an elementwise algorithm. Furthermore, our computational experiments show that algorithms based on Householder reflections achieve the best performance among the blocked (level 3) algorithms. Notably, for large enough block sizes the compact WY approach gives the best performance.

The results reported here is ongoing work and we currently extend different blocked variants to parallel algorithms for scalable MIMD computers.

References

1. E. Anderson, Z. Bai, J. Demmel, J. Dongarra, J. Du Croz, A. Greenbaum, S. Hammarling, A. McKenney, S. Ostrouchov, and D. Sorensen. LAPACK Users' Guide, Second Edition. *SIAM Press, Philadelphia*, 1995.
2. M. Berry, J. Dongarra, and Y. Kim. A Highly Parallel Algorithm for the Reduction of a Nonsymmetric Matrix to Block Upper-Hessenberg Form. *Technical Report, LAPACK Working Note 68, University of Tennessee, CS-94-221*, 1994.
3. K. Dackland, E. Elmroth, and B. Kågström. Parallel Block Matrix Factorizations on the Shared Memory Multiprocessor IBM 3090 VF/600J. *The International Journal of Supercomputer Applications*, 6:69–97, 1992.
4. J. Dongarra, J. Du Croz, S. Hammarling, and Richard J. Hanson. An Extended Set of FORTRAN Basic Linear Algebra Subroutines. *ACM Transactions on Mathematical Software*, 14(1):1–17, March 1988.
5. Gene H. Golub and Charles F. Van Loan. Matrix Computations, 2nd ed. *The John Hopkins University Press, Baltimore, Maryland*, 1989.
6. S. Hammarling J. Dongarra, J. Du Croz and I. Duff. A Set of Level 3 Basic Linear Algebra Subprograms. *ACM Trans. on Mathematical Software*, 16:1–17, 1990.

7. B. Kågström and C. Van Loan. GEMM-Based Level 3 BLAS. *Tech. rep. CTC91TR47, Department of Computer Science, Cornell University*, 1989.
8. C. Lawson, R. Hanson, D. Kincaid, and F. Krogh. Basic Linear Algebra Subprograms for Fortran Usage. *ACM Transactions on Mathematical Software*, 5:308–323, 1979.
9. C. B. Moler and G. W. Stewart. An Algorithm for Generalized Matrix Eigenvalue Problems. *SIAM J. Num. Anal.*, 10:241–256, 1973.
10. A. Pothen and P. Raghavan. Distributed Orthogonal Factorizations: Givens and Householder Algorithms. *SIAM J. Sci. Statist. Comput.*, 10:1113–1134, 1989.
11. R. Schreiber and C. Van Loan. A Storage Efficient WY Representation for Products of Householder Transformations. *SIAM J. Sci. Statist. Comput.*, 10:53–57, 1989.

Parallelization of Algorithms for Neural Networks *

Beniamino Di Martino

Dipartimento di Informatica e Sistemistica
Università di Napoli "Federico II"
Via Claudio 21, 80125 Napoli - Italy
E-mail: dimartin@nadis.dis.unina.it

Abstract. In this paper we present the strategies adopted in the parallelization of two algorithms for the simulation of two classes of neural networks: the Hopfield Network and the Error BackPropagation Network.
Although the parallel algorithms have been developed within the (loosely synchronous) SPMD parallel programming model, the particular nature of the strategies adopted make the final parallel algorithms not expressible within the HPF-like programming paradigm; therefore a more flexible programming model, the message passing programming paradigm, has been adopted, and the final development has been carried out in the PVM environment.

1 Introduction

Artificial Neural Networks represent a promising area of research in computer science and engineering. They appear to complement conventional approaches to symbolic computation by suggesting solutions to problems requiring associative searches, pattern recognition, classification, and optimization, even on incomplete or unpredictable data.

The fundamental characteristics of this novel approach to computation (inherently collective computational model, distributed representation and memorization of data, high interconnectivity, massive parallelism, fault tolerance) make it very well suited to parallel and distributed implementation of their simulations.

In this paper we present the strategies adopted in the parallelization of two algorithms for the simulation of two classes of neural networks: the Hopfield network and the Error BackPropagation Network; the Hopfield network is widely used for classification purposes, while the Error BackPropagation network is a typical example of "learning" network.

Although the execution scheme selected for both the algorithms can be classified as (loosely synchronous) SPMD, the particular nature of the computational

---

* This work has been supported in part by the Italian Ministry of University and Scientific and Technological Research within the "40%" Project.

kernels, and the incompatibilities among them in the access patterns of the data structures, have made the parallelization strategy not expressible in the context of HPF-like programming models.

Therefore, a more flexible programming model, the Message Passing programming paradigm, has been adopted, and the final algorithmic development has been carried out in the PVM environment.

The remainder of the paper proceeds as follows: a conceptual description of the characteristics of the two networks is provided in section 2. Section 3 describes the techniques used in the parallelization process, and section 4 gives results of tests on a local area network (LAN) of workstations.

2 Characteristics of the Networks

The *Hopfield Network* [1] is widely used for classification purposes, such as Pattern Matching, when an exact binary representation of the input elements is available: this is the typical case of binary images, or for ASCII text (binary coded to 8 bit).

Given an input pattern, it evolves (always) towards a stable state, that represents the network output for the input pattern, that can be codified in such a way to represent the class to which that pattern belongs.

It is an associative neural network: that is, composed of only one neurons layer. The output signals of the layer elements are brought back to the inputs: for this feature it is a *feedback* network, and, besides, an *autoassociative* one, because the input and output vectors have the same dimensionality.

Named **x** the vector of the neurons' outputs (coincident with the inputs, because of the feedback connection), and **w** the matrix of the connections among the neurons, the convergence law of the system is given by:

$$x_i^{new} = x_i^{old} + \frac{1}{\tau} \left(\tanh \left(\beta \left(\sum_{j<i} w_{ij} x_j^{new} + \sum_{j \geq i} w_{ij} x_j^{old} \right) + \alpha s_i \right) - x_i^{old} \right) \quad (1)$$

This network cannot *learn*: the matrix of the neuron's weight is fixed.

The *Error BackPropagation Network* [1] is instead a typical example of learning network: the typical operation performed by this network is the *mapping* of a function $f : A \in \mathcal{R}^n \to \mathcal{R}^m$, by means of *training* on examples of such a mapping.

It is a *multilayer, eteroassociative* network. The layers' elements (one input layer, one output layer, and one or more *hidden* layers) are connected in the *feedforward* direction, for what concerns the output transmission through the network. During the training phase, the layers are connected in the *feedback* direction too, for which concerns the *errors* transmission through the network. These latter are a measure of the deviations between the output values and the values of the examples in the training set. They are propagated through the network during the training phase to permit the correction of the connections' weights.

Named z_{lj} the output of the j-th neuron of the l-th layer, and w_{lji} the weight of the connection between the j-th neuron of the l-th layer and the i-th neuron of the $(l-1)$-th layer, the updating law is given by:

$$z_{li} = s\left(\sum_i w_{lji} z_{(l-1)i}\right) \tag{2}$$

where $s(I) = 1/(1 + e^{-I})$ is the *sigmoidal function*.

During the training phase, the correction of the connection weight w_{lji}, after the k-th trial, is given by:

$$w_{lji}^{new} = w_{lji}^{old} + \eta \delta_{lj}^k z_{(l-1)i}^k \tag{3}$$

The error δ_{lj}, associated to the j-th neuron of the l-th layer, propagates in backward direction, and is given by:

$$\delta_{lj} = s'(I_{lj}) \sum_i w_{(l+1)ij} \delta_{(l+1)i} \tag{4}$$

where:

$$s'(I) = \frac{e^{-I}}{(1 + e^{-I})^2} = s(I) * (1 - s(I))$$

$$I_{lj} = \sum_i w_{lji} z_{(l-1)i}$$

$$s'(I_{lj}) = z_{lj}(1 - z_{lj})$$

3 Description of the parallelization strategy

The parallel execution model utilized for both the networks is the *Processor Farm* [2]: a coordinator process spawns and coordinate a set of worker processes, which perform computation, and manages knowledge sharing between them. In our case, the role of the coordinator is limited, for both the network, to the I/O management; thus the execution model can be restricted to the SPMD (loosely synchronous) paradigm [3].

The work decomposition and data distribution among workers has been driven by the blockwise subdivision of the neurons of each layer among workers, chosen for both the networks.

This choice has had different consequences in the development of the two algorithms, due to the different computational kernels of the two networks.

3.1 Hopfield Network

With regard to the Hopfield network, this choice corresponds to a partitioning of the array **x**, updated by the convergence rule 1, into P blocks, and its distribution to the P workers. The update formula 1 induces a serialization in the computation, since an element x_j can only be updated after all elements x_h, $1 \leq h < j$ have been updated, and so elements assigned to different processors cannot be updated in parallel. This is a typical *stencil computation*, as can appear in SOR iterative solvers for systems of linear equations [4], but here the stencil overlaps the whole array, that is the interaction is not only among neighbors, but among all elements.

One way of working with this constraint is to make processors work in pipeline, so that already updated elements can be passed over and used for updating elements with higher indices. But, if the *owner computes rule* has to be followed, that is each worker can perform computations to update only elements belonging to it, the "all to all" interaction stencil of the elements lead to a complete serialization of the pipeline.

To overcome this difficulty, we have adopted a computational workload distribution strategy which follows an "altruistic" approach, and overcomes the owner compute rule, relying on the associative properties of the operators in the updating law 1. After updating its own elements, each processor computes partial sums for updating elements assigned to subsequent and preceding processors in the pipeline, using elements assigned to it only. As soon as these partial sums are computed, they are sent to the processors that will need them, so that these processors can start a new update step. This step now requires, for each element to be updated, only the computation of the partial sum of the elements assigned to that processor, and a final sum of the partial results received from the other stages of the pipeline.

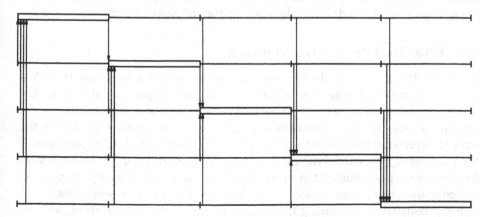

Fig. 1. Gantt diagram of the parallel execution scheme for the Hopfield algorithm

Figure 1 shows the Gantt diagram of the execution strategy just described.

We have adopted the following conventions. The white rectangles represent the update of the elements assigned to each processor, whereas the thin horizontal lines represent the computation of partial sums to be sent to other processors. Arrows represent communications of partial sums. To avoid too many symbols for each iteration, we have only reported one arrow to represent all communications between each pair of processors. The arrow is then placed at the beginning of the segment representing the computation of a group of partial sums to mean that such sums are communicated individually as soon as their evaluation has been completed (i.e. no message vectorization is performed). Finally, communications are assumed to be ideal, i.e. without overhead (the length of segments representing computations is not modified by the presence of communications) and with zero latency (arrows are drawn vertically) to simplify an informal analysis of the execution scheme just described.

Figure 1 shows that in ideal conditions the outlined strategy lead to very high efficiency because all idle times can be reduced to a negligible fraction of the overall iteration time. In the more realistic assumptions that communications have an associated overhead and a non-zero latency, communications of partial results have been aggregated (*message vectorization*) to reduce communication overhead. This has introduced an idle time for processors, with the result of a decreased efficiency with respect to the ideal case.

The other computational data structure of the Hopfield algorithm, the weight matrix **w**, has been block decomposed in stripes, and each stripe assigned to a process, so that each process owns those weights representing interactions of elements $x(j)$ of its own with the interacting others. It is worth noting that, with our computational distribution strategy (that is, to distribute computational workload for updating an element to all processes that own elements concurring in this update) there is no need to overlap parts (or all) this data structure among processes; this is of crucial importance when also the matrix elements have to be updated: if there was overlap, the overlapping elements should be communicated at each update, with a dramatic increase in communication overhead.

3.2 Error BackPropagation Network

The main difficulty in the development of the algorithm for the Error BackPropagation Network has not been the structure of the computational kernels (the computations schematized in 2, 3 and 4), which are essentially matrix-vector multiplications, but the differences among them in the access patterns on the data structures, combined with the particular sequence of their application. If the most straightforward data distribution (block distribution of the outermost loop-accessed dimension of the arrays involved) and work decomposition (owner computes rule) were applied to each of the kernels, this would lead to a series of redistributions among the various phases of the computation, with an unacceptable increase of the communication overhead.

For instance, the updating law 2 (matrix-vector multiplication) is sequentially applied to each of the layers of the network, so that the computed vector of one application becomes the input vector of the successive application: if a block distribution for the computed vector, and a replication for the input vector, is applied, then each layer vector has to be scattered among the processes after its computation (each process performs a broadcast, and then a gather communication).

Even more crucial, for the communication overhead issue, is the different access pattern of the weight matrix **w** between the forward and the backward propagation. In the eqs. 3 (l-th forward step) and 4 (l-th backward step, the access of the l-th page of **w** is the transpose mode of each other (the indices i and j are inverted in the two equations): this means that, if a block distribution for one dimension of each page is chosen (distribution driven by the use of this page in the updating law 2), then this page has to be scattered among the processes in the backward step, if the owner computes rule has to be chosen for the work distribution of the backward computation 4. This redistribution is by far heavier than the previous, because of the bidimensionality of the data structure.

The approach chosen to optimize the communication overhead has been that of applying the straightforward data distributions and work decompositions in the forward phase (the neurons' update), thus leaving the redistributions of the neurons' arrays; but avoiding the redistribution of the weight matrix pages, by relaxing the owner computes rule in the work decomposition of the backward error computations 4. This has been possible due to the associative properties of the operators in those computations. The work decomposition adopted is the "altruistic" strategy, described before for the Hopfield net. In this case, each process make use of its own elements of the weight matrix page w_l, to compute partial results for updating its own δ array elements, and also partial results for updating the elements belonging to other processors. At the end of the computation of all partial results, each process sends the results needed by the others, and receives those that it needs; then it combines them with the partial results computed by itself, thus updating its own elements of the δ array.

The difference with the execution scheme of the Hopfield net is that here there is no serialization effect in the computational kernel, thus no pipelining is introduced in the parallel execution scheme; moreover, processes can aggregate the communication of the partial results, and perform it at the end of the computation, without introducing idle time to the other processes.

Although the execution scheme for both the algorithms can be classified as (loosely synchronous) SPMD, the introduction of the described optimizations in the work distribution makes the parallelization strategy not expressible in the context of HPF-like programming models. Therefore, a more flexible programming model, the Message Passing programming paradigm, has been adopted, and the final algorithmic development has been carried out in the PVM environment. PVM environment [5] allows for an high degree of flexibility in concurrent software development: application programs written using PVM primitives can hold arbitrary control and dependence structures, and this allows for the most general forms of MIMD computation. In our case, has been very easy to program, making use of the non-blocking send and receive PVM primitives, the scatter and gather communication pattern needed in the case of the Error BackPropagation Network, and the "generalized pipeline" communication pattern of the Hopfield Network.

4 Results

The two algorithms have been tested by executing them on a bus-connected cluster of 4 workstations (each process executed on one processor).

dim.	Speed-up	Efficiency
64	0.3	0.07
256	1.6	0.39
1024	3.7	0.92

dim.	Speed-up	Efficiency
128	1.2	0.29
256	1.9	0.48
512	2.9	0.73
1024	3.2	0.80

The two tables show the measured speed-ups and efficiencies of the two networks (left: Error BackPropagation; right: Hopfield), for different dimensions of the input data. These results make apparent that, in the case of the Error Backpropagation Network, when the problem size is sufficiently large (inputs by far greater than 1024 bits are of common use in text or image Pattern Recognition, for instance) the communication overhead induced by the redistributions is masked, and the efficiency reach a very high value. In the case of the Hopfield Network, the lower value of the efficiency is due to the waiting time introduced for the processors when the communications are aggregated.

Acknowledgments

I wish to thank G. Iannello and P. Sguazzero for their precious hints and suggestions.

References

1. R.H.Nielsen, *Neurocomputing*, Addison-Wesley.
2. J. N. Magee and S. C. Cheung, "Parallel Algorithm Design for Workstation clusters", *Software - Practice and Experience*, 21(3), pp. 235-250, Mar. 1991.
3. A. H. Karp, "Programming for Parallelism", *IEEE Computer*,20, pp. 43-57, May 1987.
4. J.R. Westlake, *A handbook of numerical matrix inversion and solution of linear equations*, Wiley, New York, 1968.
5. A. Geist V.S.Sunderam, "Network-Based concurrent computing on the PVM system", *Concurrency: practice and experience*, vol. 4, n. 4, pp. 293 − 311, Jun 1992.

Paradigms for the Parallelization of Branch&Bound Algorithms *

Beniamino Di Martino, Nicola Mazzocca and Stefano Russo

Dipartimento di Informatica e Sistemistica
Università di Napoli "Federico II"
Via Claudio 21, 80125 Napoli - Italy
E-mail: {dimartin,mazzocca,russo}@nadis.dis.unina.it

Abstract. Branch&Bound (B&B) algorithms represent a typical example of techniques used to solve irregularly structured problems. When porting sequential B&B applications to a network of workstations, a very popular class of MIMD distributed memory machines, several issues have to be coped with, such as sharing a global computation state and balancing workload among processors. The parallel programming paradigm to adopt has to be chosen as a compromise between simplicity and efficiency. In this paper we discuss issues in the parallelization of B&B algorithms according to two paradigms: *coordinator/workers* and *SPMD* (Single Program Multiple Data). The implementation according to the message-passing mechanisms provided by the PVM parallel programming environment is presented. The two approaches are compared qualitatively, with respect to the solutions adopted for knowledge sharing, communication, load balancing, and termination condition. Comparison is also performed quantitatively, by evaluating the performances of the two algorithms on a local area network of workstations.

1 Introduction

Branch&Bound (B&B) is a technique widely used to solve combinatorial optimization problems, which occur frequently in physics and engineering science. B&B is a typical example of techniques for irregularly structured problems. The exploitation of parallelism is clearly a promising approach to derive efficient solutions to this class of scientific problems. There are however many complex issues to be addressed when defining new parallel algorithms for such problems, or adapting existing sequential algorithms to parallel computers. Examples of such issues are load balancing, global knowledge sharing, scheduling, mapping, exploitation of system heterogeneity.

The choice of the paradigm to follow in the parallelization of existing sequential B&B programs involves finding a compromise between simplicity of implementation, in order to reuse as much of the existing code as possible, and efficiency. For distributed memory multiprocessor architectures, which are well

---

* This work has been supported in part by the Italian Ministry of University and Scientific and Technological Research within the "40%" Project.

suited for coarse grain parallelization strategies, the *task parallelism* approach fits well for these problems; within this class, several paradigms can be adopted to manage the needed interactions among executing tasks.

In this paper we discuss issues in the parallelization of B&B algorithms according to two paradigms. The *coordinator/workers* (also called *host/node* or *master/slaves*) paradigm, is one which provides straightforward guidelines for the parallelization of existing programs [1]. As the coordinator becomes a bottleneck for communications, in the case of a large number of workers or of frequent interactions, a different paradigm is considered. This is the *loosely synchronous* SPMD paradigm [2], where multiple instances of a single process perform the same computations on different data. An algorithm is presented, which uses a *token-passing* strategy to manage load balancing and termination detection.

The remainder of the paper proceeds as follows: a conceptual description of B&B technique, together with advantages and problems arising within a parallel framework is presented in Section 2. Sections 3 and 4 describe the characteristics of the two algorithms, which have been implemented with the message-passing mechanisms provided by the PVM parallel programming system, currently considered a *de facto* standard by the community of high-performance scientific application developers. In Section 5 we discuss results of tests on a local area network (LAN) of workstations. Finally we give some concluding remarks.

2 The Branch&Bound technique

A *discrete optimization problem* consists in searching the optimal value (maximum or minimum) of a function $f : \mathbf{x} \in \mathcal{Z}^n \to \mathcal{R}$, and the solution $\mathbf{x} = \{x_1, \ldots, x_n\}$ in which the function's value is optimal. $f(\mathbf{x})$ is said *cost function*, and its domain is generally defined by means of a set of m constraints on the points of the definition space. Constraints are generally expressed by a set of inequalities:

$$\sum_{i=1}^{n} a_{i,j} x_i \leq b_j \qquad \forall j \in \{1, \ldots, m\} \tag{1}$$

and they define the set of feasible values for the x_i variables (the *solutions space* of the problem).

Branch&Bound is a class of methods solving such problems according to a *divide&conquer* strategy. The initial solution space is recursively divided in subspaces, until attaining to the individual solutions; such a recursive division can be represented by a (abstract) tree: the nodes of this tree represent the solution subspaces obtained by dividing the parent subspace, the leaf nodes represent the solutions of the problem, and the tree traversal represents the recursive operation of dividing and conquering the problem.

The method is enumerative, but it aims to a non-exhaustive scanning of the solutions space. This goal is achieved by estimating the best feasible solution for each subproblem, without expanding the tree node, or trying to prove that there are no feasible solutions for a subproblem, whose value is better than the *current* best value. (It is assumed that a best feasible solution *estimation function* has been devised, to be computed for each subproblem.) This latter situation corresponds to the so called *pruning* of a search subtree.

B&B algorithms can be parallelized at a *fine* or *coarse grain* level. The fine grain parallelization involves the computations related to each subproblem, such as the computation of the estimation function, or the verification of constraints defining feasible solutions. The coarse grain parallelization involves the overall tree traversal: there are several computation processes concurrently traversing a different branch of the search tree. In this case the effects of the parallelism are not limited to a speed up of the algorithmic steps. Indeed, the search tree explored is generally different from the one traversed by the sequential algorithm. As a result, the resolution time can be lower, even though the number of explored nodes can be greater than in the sequential case. As it has been demonstrated in [5], this approach exhibits *anomalies*, so that the parallelization does not guarantee an improvement in the performance. For most practical problems, however, the bigger the problem size, the larger are the benefits of the parallel traversal of the search tree.

Parallel B&B algorithms can be categorized on the basis of four features [4]:

1. how information on the global state of the computation is shared among processors (we refer to such information as the *knowledge* generated by the algorithm);
2. how the knowledge is utilized by each process;
3. how the workload is divided among the processes;
4. how the processes communicate and synchronize among them.

The knowledge is *global* if there is a single common repository for the state of the computation at any moment, accessed by all processes, otherwise it is *local*. In this latter case, processes have their own knowledge bases, which must be kept consistent to a certain degree to speed up the tree traversal and to balance the workload.

With respect to the knowledge use, the algorithms are characterized by: (1) the *reaction strategy* of processes to the knowledge update (it can range from an instantaneous reaction, to ignore it until the next decision is to be taken); (2) the *dominance rule* among nodes (a node *dominates* another if its solutions best value is better than the lower bound on the solutions of the other): it can be *partial* or *global*, if a node can be eliminated only by a dominant node belonging to the same subtree traversed by a process, or by any other dominant node; (3) the *search strategy*, that can be *breadth*, *depth* or *best* first.

With regard to the workload division, if all generated subproblems are stored in a common knowledge base, to each process that becomes idle the most promising subproblem is assigned (on the basis of an heuristic evaluation). If the state of the computation is distributed among processes (local knowledge), then a *workload balancing* strategy has to be established, consisting of a relocation of subproblems not yet traversed.

3 The coordinator/workers parallel algorithm

The two parallel algorithms we present solve the $(0-1)$ *knapsack problem*, which can be stated as follows [6]:

$$maximize \sum_{i=1}^{n} c_i x_i \tag{2}$$

$$with \quad x_i \in \{0, 1\} \quad \forall i \in \{1, \ldots, n\}$$

$$subject \ to \ \sum_{i=1}^{n} a_{i,j} x_i \leq b_j \quad \forall j \in \{1, \ldots, m\} \tag{3}$$

where $a_{i,j}$ and b_j are positive integers. They have been designed for distributed memory parallel architectures, such as a network of workstations. The implementation of both algorithms has been carried out in the PVM parallel programming system [7]. In particular, the algorithms exploit the PVM primitives for interrupting sends and multicast communication.

The characteristics of both algorithms, with respect to the categorization of the B&B algorithms presented above, are the following: (1) each processor has its own knowledge base (local knowledge); (2) processes react instantaneously to knowledge updates (the *signal* primitives of the PVM environment are used for this purpose); (3) a global dominance rule is adopted, since the current optimal value is broadcasted to all processes as soon as it is updated; (4) the search strategy is *depth-first*, as it is more suited in the case of local knowledge; (5) with respect to workload sharing between processes, load balancing is provided, activated in presence of an idle process, since there is a local knowledge sharing; (6) with regard to synchronization, the algorithms are asynchronous, since there is no synchronous exchange of information and workload between executors, but this exchange is performed on the basis of executor's local events, and so asynchronously with respect to the other executors.

The first algorithm follows the *coordinator/workers* concurrent programming paradigm. In this model, a coordinator process spawns a set of worker processes, which perform the actual computation; the coordinator also manages the sharing of the global knowledge among the workers. The structure of the application is depicted in Fig. 1a.

The algorithm consists of the following phases:

- a *coordinator* process produces P instances of a *worker* process, decomposes the assigned problem in P disjoint subproblems and assigns them to the workers;
- each worker explores its own subtree with a *depth first* strategy, and updates its local current best value, sending it to the coordinator, when it generates a feasible solution;
- when the coordinator receives a local current best value, it compares it with the global current best value and eventually updates and broadcasts it to the workers;

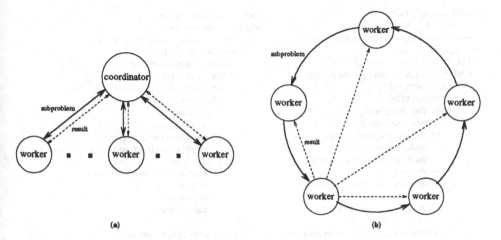

Fig. 1. (a) Structure of the coordinator/worker algorithm; (b) structure of the SPMD token-passing algorithm.

The load balancing among the workers is accomplished with the following strategy:

- when a worker completes the exploration of its subtree, it sends a message to the coordinator, and waits for a new load share from it;
- the coordinator manages a list of idle workers: when it is not empty, it polls active workers for a share of the load, until it receives a positive answer, then it assigns the share to the first idle process in the list.

Thus, the coordinator is in charge of managing the updates and the broadcast of the current best value, and of balancing the load among workers. The presence of a coordinator also allows the detection of the *termination condition*. This is verified when all workers are idle at the same time, and there are no messages carrying work units, which have been sent but not received yet. Since the coordinator holds the global state of the computation (lists of idle workers and of workload messages), it is able to detect this situation.

Fig. 2 shows the pseudo-code of the coordinator/workers algorithm.

4 The SPMD parallel algorithm

This algorithm aims at eliminating the coordinator presence, by having multiple instances of the same process, that perform a distributed workload-balancing strategy and termination detection. This approach has the advantage of not presenting the bottleneck on communication, represented by the coordinator process; the drawback is the need of a more elaborate management of the load balancing and termination detection phases. The structure of the application is depicted in Fig. 1b, while Fig. 3 shows the pseudo-code of the processes.

```
spawn(worker,NPROC);
for (i=0; i < NPROC; i++)
  send(initial_subproblem,worker[i]);
idles_list = {};
while (num_idles < NPROC)
  nonblocking_recv(any_message,any);
  if (msg_received)
    switch(msgtype)
      case new_supposed_best:
        update(num_sup_best_rcv);
        if (supposed_best < best)
          update_best();
          multicast(best,workers);
      case worker_idle:
        update(num_sup_best_snt);
        update(idles_list);
        if (num_idles == NPROC)
          exit_from_while;
        update(num_shareload_request);
  if (num_shareload_request > 0)
    worker = next_worker(!in idles_list);
    send(shareload_request,worker);
    recv(shareload_answer,worker);
    if (shareload_answer == positive)
      worker = next_worker(in idles_list);
      update(idles_list);
      send(load,worker);
      update(num_shareload_request);
while (num_sup_best_rcv < num_sup_best_snt)
  recv(supposed_best,any);
  update(num_sup_best_rcv);
  if (supposed_best < best)
    update_best();
notify(best);
multicast(terminate,workers);
```

(a)

```
N=initial_subproblem;
while (!terminate)
  if "not other nodes to explore"
    send(worker_idle,coor);
    send(num_sup_best_snt,coor);
    while (!active)
      recv(any_message,coor);
      switch(msgtype)
        case best:
          update_best();
        case shareload_request:
          send(shareload_answer_neg,coor);
        case load:
          N=new_subproblem(load);
          active = true;
        case terminate:
          terminate=true;
  else
    x = select_next_node(N);
    verify_feasibility(x);
    if "x is feasible solution"
      send(new_supposed_best,coor);
      update(num_sup_best_snt);
  if (!terminate) & (signaling_message)
    recv(any_message,coor);
    switch(msgtype)
      case best:
        update_best();
      case shareload_request:
        if "no other nodes to explore"
          send(share_load_answer_neg,coor);
        else
          load=select_next_node(N);
          send(shareload_answer_pos,coor);
          send(load,coor);
```

(b)

Fig. 2. (a) Pseudo-code of the coordinator (a) and worker (b) processes.

In this second algorithm, processes perform the exploration of the search tree at the same way of the worker processes in the previous algorithm. In both cases, the sequential code is fully reused.

The current best value update phase is slightly different with respect to the previous algorithm:

– when a process produces a local current best value, broadcasts it to the other processes;
– when a process receives a current best value from another process, it compares that with its current best, which is eventually updated.

The termination detection phase adds complexity, in the absence of a coordinator, and influences planning of the workload balancing phase. As for the previous algorithm, in fact, the termination condition is satisfied when all processes are at the same time idle, and there are no messages carrying workload, sent but not received yet. In the absence of a coordinator, every idle process

```
if "I have no parents"                         case token:
  spawn(process,NPROC-1);                        recv(token,prev_proc_on_ring);
  N=initial_subproblem;                          if my_token
  for(i=1; i<NPROC; i++)                           if token_full
    send(initial_subproblem,process[i]);            N=new_subproblem(load);
else                                                active=true;
  recv(initial_subproblem,parent);               else /* token_empty */
  N=initial_subproblem;                            if no_terminate
while (!terminate)                                   send(my_token,next_proc_on_ring);
  if "not other nodes to explore"                 else /* no_terminate=false */
    send(my_token,next_proc_on_ring);               multicast(terminate,processes);
    while (!active)                                  terminate=true;
      probe(any_message,any);                   else /* not my_token */
      switch(msgtype)                             if token_full
        case new_supposed_best:                     no_terminate=true;
          recv(new_supposed_best,any);              send(token,next_proc_on_ring);
          if (supposed_best < best)             else /* other nodes to explore */
            update_best();                        x = select_next_node(N);
        case terminate:                           verify_feasibility(x);
          recv(terminate,any);                    if "x is feasible solution"
          terminate=true;                           multicast(new_supposed_best,processes);
                                                if (!terminate) & (signaling_message)
                                                  probe(any_message,any);
                                                  switch(msgtype)
                                                    case new_supposed_best:
                                                      recv(new_supposed_best,any);
                                                      if (supposed_best < best)
                                                        update_best();
                                                    case token:
                                                      recv(token,prev_proc_on_ring);
                                                      if token_empty
                                                        load_token=select_next_node(N);
                                                      send(token,next_proc_on_ring);
```

Fig. 3. Pseudo-code for the token-passing SPMD algorithm

could ask the other processes for a share of the load. In a distributed system with asynchronous communications, however, if answers were all negative, an idle process could only conclude that every other process was idle at the moment of the reception of its request, but not necessarily they were simultaneoulsy idle. Even in this latter case, there is no guarantee that there is no messages in the system, still to be delivered. Hence, algorithm termination cannot be detected in this way. There is the need to check for the presence of messages in the asynchronous communication system.

The strategy adopted in our algorithm is based on a *token-passing* mechanism. Processes form a (purely virtual) ring topology, where every process has a clockwise and a counterclockwise neighbor. Every process can send messages only to the neighbor process which follows it in the circular ordering, and can receive only from the neighbor which precedes it. Furthermore, communications are "sequential", that is messages are received in the order they are sent.

When a process P_i becomes idle, it puts a *token* on the ring, labeled with its identification number, which represents its request for workload sharing. The first active process, in the sense of the circular ordering, that receives the token, will load on it a node to expand, and will put the token on the ring again. Process idle P_i which have launched the token, will wait for it; in the meanwhile, it checks if tokens, not addressed to it, that it receives, carry workload to share. If none of them carries workload, and the token addressed to it has no workload, process P_i can conclude that all processes are idle at that time, and no activating message is traveling on the ring, so satisfying termination condition: P_i can now broadcast a termination message. On the contrary, if a token not addressed to P_i carries workload, then another idle process, P_j, will become active, and it will become after it have received token addressed to P_i; so, if no other process is active, P_i will be returned an empty token, but it cannot conclude, like earlier, that all processes are idle and none of them will return active: so it can not terminate computation, and it will limit itself to send another token with workload share request.

5 Experimental results

The two algorithms have been experimentally evaluated and compared, executing them on a local area Ethernet network of 9 IBM RS6000 workstations (each process run on a different processor).

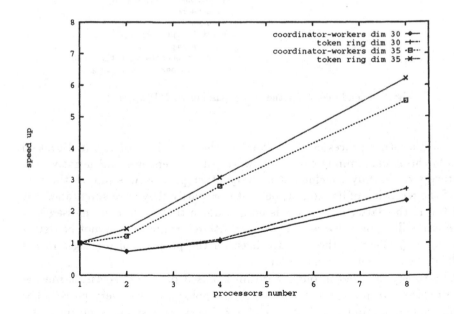

Fig. 4. Speed-up of the two algorithms

Fig. 4 plots the speed up obtained for the two algorithms for problem sizes of 30 and 35 variables, as the number of processors varies (the parallel execution times, for eight processors, are on the order of half minute for problem size 30, and 3 minutes for problem size 35). The curves show that the SPMD algorithm

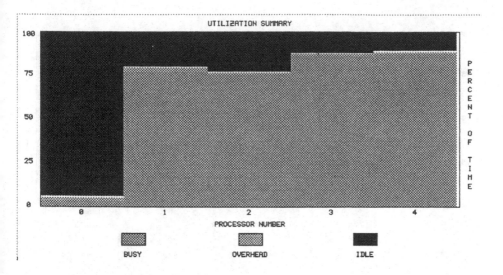

Fig. 5. CPU utilization for the coordinator/worker algorithm (process number 0 is the coordinator).

performs better than the coordinator/worker implementation. The difference increases as the number of processes increases. Although the maximum number of workstations which were available is too small to experimentally prove the scalability of this feature, we argue that the difference in performance is due to the presence of the coordinator process in the first algorithm, that acts as a bottleneck for communications.

This characteristic is confirmed by the comparison of Fig. 5 and Fig. 6, that show that percentage of CPU usage of all processors on a whole program run. The diagrams have been obtained profiling the execution of the algorithms, and using the performance analysis and visualization tool ParaGraph [8]. Fig. 5 show that the workers in the first algorithm are idle for about 20% of the time. In the second algorithm (Fig. 6), CPU utilization raises up to the 95%.

6   Conclusions

We have presented two strategies for the parallelization of Branch&Bound algorithms for distributed memory MIMD architectures. They are based on a coordinator/worker and on a SPMD paradigm, respectively. In synthesis, The coordinator/worker paradigm allows an easy parallelization of existing sequential algorithms. The SPMD program has an additional complexity in the token-passing mechanism to manage load balancing and program termination. However, the experimental results on a workstation LAN under the PVM parallel programming system show that the SPMD algorithm can perform better, because of the elimination of the bottleneck represented by the coordinator, even

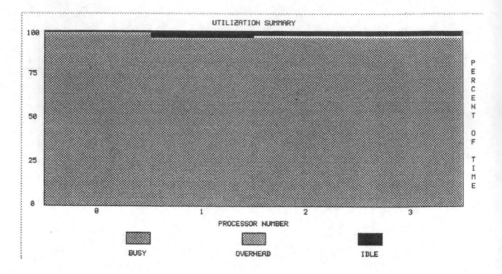

Fig. 6. CPU utilization for the SPMD token-passing algorithm.

for a small number of processors, and despite of the more elaborate workload balancing and termination detection procedure.

References

1. J.N. Magee and S.C. Cheung, "Parallel Algorithm Design for Workstation clusters", *Software - Practice and Experience*, 21(3), pp. 235-250, Mar. 1991.
2. A.H. Karp, "Programming for Parallelism", *IEEE Computer*, 20, pp. 43-57, May 1987.
3. A. Mazzeo N. Mazzocca A. Sforza and S. Russo, "Algoritmi Branch&Bound paralleli per sistemi multicomputer", *Ricerca Operativa*, vol. 24 n. 71, 1994.
4. H.W.J. Trienekens, "Parallel Branch&Bound Algorithms", *Ph.D. Thesis at Erasmus Universiteit-Rotterdam*, Nov. 1990.
5. H.T. Lai and S.Sahni, "Anomalies in Parallel Branch&Bound Algorithms", *Communications of the ACM*, vol. 27, n. 6, pp. 594 − 602, Jun. 1984.
6. C. Ribeiro, "Parallel Computer Models and Combinatorial Algorithms", *Annals of Discrete Mathematics*, North-Holland, pp. 325 − 364, 1987.
7. A. Geist and V.S.Sunderam, "Network-Based concurrent computing on the PVM system", *Concurrency: Practice and Experience*, vol. 4, n. 4, pp. 293 − 311, Jun 1992.
8. M.T. Heath, J.A. Etheridge, "ParaGraph: A Tool for Visualizing Performance of Parallel Programs", Tech. Rep. Oak Ridge National Laboratory, Oak Ridge, TN, March 1994.

Three-Dimensional Version of the Danish Eulerian Model

I. Dimov[1], K. Georgiev[1], J. Waśniewski[2], and Z. Zlatev[3]

[1] Institute of Informatics and Computer Technology
Bulgarian Academy of Sciences
G. Bonchev street, Bl. 25-A, BG-1113 Sofia, Bulgaria
Email: dimov@iscbg.acad.bg and georgiev@iscbg.acad.bg respectively
[2] UNI•C, The Danish Computing Centre for Research and Education
DTU, Bldg. 304, DK-2800 Lyngby, Denmark
Email: jerzy.wasniewski@uni-c.dk
[3] National Environmental Research Institute
Frederiksborgvej 399, P.O. Box 358, DK-4000 Roskilde, Denmark
Email: luzz@sun2.dmu.dk

Abstract. The two-dimensional Danish Eulerian Model has been developed during the 80'ies. More than 70 chemical reactions (some of them photochemical) are involved in the model. The space domain of the model covers the whole of Europe. It has been discretized by using a (32 × 32) equidistant spatial grid. The concentrations and the species calculated by the model were compared both with measurements taken over land and with measurements taken over sea. The model has also been run by using finer grids; as, for example, a (96 × 96) grid. The experiments indicated that in general the model calculates rather reliable results. However, it is also clear that the results might be improved if a three-dimensional version of the model is developed. Three-dimensional air pollution models are very time-consuming. Therefore the development of a reliable three-dimensional version of an air pollution model is a very challenging task. The efforts to solve some of the numerical problems arising during the development of a large three-dimensional air pollution model (with non-linear chemical reactions) will be discussed in this paper.

1 Need for large air pollution models

Air pollutants emitted by different sources can be transported, by the wind, on long distances. Several physical processes (diffusion, deposition and chemical transformations) take place during the transport. Regions that are very far from the large emission sources may also be polluted. It is well-known that the atmosphere must be kept clean (or, at least, should not be polluted too much). It is also well-known that if the concentrations of some species exceed certain acceptable (or critical) levels, then they may become dangerous for plants, animals and humans.

Mathematical models are needed in the efforts to predict the optimal way of keeping the air pollution under acceptable levels. It should be emphasized here that the mathematical models are the only tool by the use of which one can predict the results of many different actions and, moreover, one can attempt to choose the best solution (or, at least, a solution which is close to the best one) in the efforts to reduce the air pollution in an optimal way. The mathematical models are necessarily large (the transport of air pollutants is carried out over long distances and, thus, the space domains are very large).

2 The Danish Eulerian Model

The work on the development of the Danish Eulerian Model has been initiated in 1980. First a simple transport scheme was developed (one pollutant only and without chemical reactions). The next step was the development of a simple model with two pollutants (and linearized chemical reactions). An experimental model containing ten pollutants and non-linear chemical reactions (including here photochemical reactions) was the third step. The first operational version of the Danish Eulerian Model was based on a chemical scheme with 35 pollutants. Some experiments with chemical schemes containing 56 and 168 pollutants are carried out at present. Different versions of the Danish Eulerian Model are discussed in [3] and [6].

The reliability of the operational model with 35 pollutants has been tested by comparing model results both with measurements taken over land ([4], [5]) and with measurements taken over sea ([2]). Test-problems, where the analytical solution is known, have been used to check the accuracy of the numerical algorithms ([3]).

3 Mathematical description of the Danish Eulerian Model

The Danish Eulerian model is described by a system of PDE's (the number of equations in this system being equal to the number of pollutants involved in the model):

$$\frac{\partial c_s}{\partial t} = -\frac{\partial(uc_s)}{\partial x} - \frac{\partial(vc_s)}{\partial y} - \frac{\partial(wc_s)}{\partial z} + \frac{\partial}{\partial x}\left(K_x\frac{\partial c_s}{\partial x}\right) + \frac{\partial}{\partial y}\left(K_y\frac{\partial c_s}{\partial y}\right) + \frac{\partial}{\partial z}\left(K_z\frac{\partial c_s}{\partial z}\right)$$

$$-(\kappa_{1s} + \kappa_{2s})c_s + E_s + Q_s(c_1, c_2, \ldots, c_q), \qquad\qquad s = 1, 2, \ldots, q. \text{(1)}$$

The different quantities that are involved in the mathematical model have the following meaning:

- the concentrations are denoted by c_s;
- u, v and w are wind velocities;
- K_x, K_y and K_z are diffusion coefficients;
- the emission sources in the space domain are described by the functions E_s;
- κ_{1s} and κ_{2s} are deposition coefficients;

- the chemical reactions are described by the non-linear functions $Q_s(c_1, c_2, \ldots, c_q)$.

The non-linear functions Q_s are of the form:

$$Q_s(c_1, c_2, \ldots, c_q) = -\sum_{i=1}^{q} \alpha_{si} c_i + \sum_{i=1}^{q} \sum_{j=1}^{q} \beta_{sij} c_i c_j, \qquad s = 1, 2, \ldots, q. \qquad (2)$$

This is a special kind of non-linearity, but it is not clear how to exploit this fact during the numerical treatment of the model.

It is clear from the above description of the quantities involved in the mathematical model that all five physical processes (advection, diffusion, emission, deposition and chemical reactions) can be studied by using the above system of PDE's.

The mathematical model described by (1) must be considered together with appropriate initial and boundary conditions ([3]).

The Danish Eulerian Model has mainly been run as a two-dimensional model until now, which can formally be obtained from (1) by removing the derivatives with regard to z. The computational difficulties connected with the development of a three-dimensional version will be discussed in the following sections. Some numerical results will be presented.

4 Numerical algorithms used in the 3-D version of the Danish Eulerian Model

The mathematical model described by the system of PDE's (1) is split to five sub-models according to the physical processes involved in the model. The sub-models are: horizontal advection sub-model, horizontal diffusion sub-model, deposition sub-model, chemistry sub-model (the emission being included here) and vertical exchange sub-model. The splitting procedure leads to a series of parallel tasks. In the first four sub-models, each horizontal plane can be treated as a parallel task. In the fifth sub-model, each vertical line can be treated as a parallel task. This shows that the splitting procedure leads to a lot of natural parallelism. More details about the splitting procedure and the possibilities to run the model in parallel are given in [3] and [6].

Various numerical algorithms have been tried in the treatment of the different sub-models. The numerical algorithms currently used are listed below; the algorithms used in the numerical treatment of the model are discussed in [3].

- A pseudospectral discretization is used in the horizontal advection sub-model. Experiments with other methods (finite elements, corrected flux transport algorithms and semi-Lagrangian methods) are presently carried out.
- A semi-analytical approach based on expansions of the unknown functions in Fourier series is used in the horizontal diffusion sub-model. Some experiments with finite elements are also carried out.

- The splitting procedure leads to a deposition sub-model consisting of independent linear ODE's; these are solved exactly.

- The QSSA (the quasi-steady-state-approximation) is used when the chemical sub-model is handled numerically. Experiments with several classical time-integration algorithms are also carried out.

- Linear finite elements are used in the treatment of the vertical exchange sub-model.

5   Computational difficulties connected with 3-D version of the Danish Eulerian Model

The development of three-dimensional air pollution models leads to huge computational tasks. The computational tasks are very large even if the model is considered as a two-dimensional model. Assume that some splitting procedure has been applied and that the spatial derivatives are discretized by using an appropriate numerical algorithm (see the previous section or [3], [6]). Then the system of PDE's (1) is transformed into five systems of ODE's corresponding to the five sub-models discussed in the previous section. The five ODE systems have to be treated successively at every time-step. The number of equations in each of these ODE systems is equal to the product of the number of grid-points and the number of pollutants. In the case where the model is considered as two-dimensional, the numbers of equations in the ODE systems for different space discretizations and for different chemical schemes are given in Table 1. The sizes of the grid-squares for the three grids used in Table 1 are given in Table 2. The total size of the space domain of the Danish Eulerian Model is normally (4800 $km \times 4800$ km); the space domain contains the whole of Europe together with parts of Asia, Africa and the Atlantic Ocean.

Number of pollutants	(32 × 32)	(96 × 96)	(192 × 192)
1	1024	9216	36864
2	2048	18432	73728
12	12288	110592	442368
35	35840	322560	1290240
56	57344	516096	2064384
168	172032	1548288	6193152

Table 1
Numbers of equations per ODE system in the two-dimensional versions of the Danish Eulerian Model.

1	(32 × 32)	(150 km × 150 km)
2	(96 × 96)	(50 km × 50 km)
3	(192 × 192)	(25 km × 25 km)

Table 2

The sizes of the grid-squares for the three grids used in Table 1.

It should be mentioned here that (i) if a three-dimensional model with ten vertical layers is used, then all figures in Table 1 must be multiplied by ten, (ii) 3456 time-steps have been used for a typical run (with meteorological data for one month + five days to start up the model; [3]) when the (96 × 96) grid is used and (iii) the chemical reactions lead to a very stiff (and also very badly scaled) ODE systems; this fact causes extra computational difficulties.

6 Need for high-speed computers

The size of the ODE systems shown in Table 1 (these systems have to be treated during several thousand time-steps) explains why the use of high-speed computers is absolutely necessary when large air pollution problems are to be handled. It is even more important to perform carefully the programming work in order to try to exploit better the great potential power of the modern high-speed computers. As a rule, this is not an easy task (especially on the newest parallel computers). Finally, it should be emphasized that even when the fastest computers are available and even when the programming work is very carefully done, it is still not possible, at present, to solve some of the biggest problems listed in Table 1 (especially in the case where the model is treated as a three-dimensional model). Therefore faster and bigger computers are needed in order to be able to treat successfully big air pollution models.

Different versions of the Danish Eulerian Model have been used in runs on several high-speed computers ([1], [3], [6]). Some results obtained by running the three-dimensional version on a CRAY Y-MP C90A will be presented in the next section.

7 Runs on a CRAY Y-MP C90A computer

The three-dimensional version of the Danish Eulerian Model has until now been run only on a CRAY Y-MP C90A computer. The performance achieved in the different parts of the model (corresponding to the five physical processes) is shown in Table 3. The total computing time and the overall speed of computations, measured in MFLOPS (millions of floating point operations, additions and multiplications, per second), are given in Table 4.

Physical process	(32x32x10)		(96x96x10)	
	time	percent	time	percent
Advection	4.00	30.5	45.83	36.3
Diffusion	1.21	9.2	10.70	8.5
Deposition	0.38	2.9	4.27	3.4
Chemistry	6.90	52.5	59.50	47.1
Vertical exchange	0.18	1.3	2.60	2.0
Overhead	0.46	3.6	3.43	2.7

Table 3

Computing times (in minutes and in percent) for the different parts of the model obtained on a CRAY Y-MP C90A computer.

Quantity measured	(32x32x10)	(96x96x10)
Computing time	13.13	126.33
Speed in MFLOPS	383	471

Table 4

The total computing time (in minutes) and the overall computational speed obtained in a run with one-month meteorological data (plus five days to start up the model) on a CRAY Y-MP C90A computer.

8   Concluding remarks

It is seen that the chemistry is the most time-consuming part of the model. Therefore improvements of the chemical subroutines are most desirable (however, also the performance of the mathematical modules describing the other physical processes are to be improved). It should be emphasised that the three-dimensional model can be treated numerically at present only on a relatively coarse spatial grid, the (32 × 32) grid, and only for the relatively simple chemical scheme with 35 pollutants. It is necessary to improve both the numerical algorithms and the computer programs for the different modules in order to be able to run also some of the more complicated cases (refined grids and chemical schemes with more pollutants). It is crucial to select (or develop) algorithms that perform well on the new high-speed computers. Some work in these directions is carried out at present.

Acknowledgments

This research was partially supported by NMR (Nordic Council of Ministers), EMEP (European Monitoring and Evaluating Programme), NATO (North Atlantic Treaty Organisation), SMP (Danish Strategic Environmental Programme) and the Danish Natural Sciences Research Council.

References

1. BROWN, J., WASNIEWSKI, J., ZLATEV, Z.: Running air pollution models on massively parallel machines; Parallel Computing, **21** (1995), 971-991.
2. HARRISON, R. M., ZLATEV, Z., OTTLEY, C. J.: A comparison of the predictions of an Eulerian atmospheric transport-chemistry model with measurements over the North Sea; Atmos. Environ., **28** (1994), 497-516.
3. ZLATEV, Z.: Computer treatment of large air pollution models; Kluwer Academic Publishers, Dordrecht-Boston-London, 1995.
4. ZLATEV, Z., CHRISTENSEN, J., ELIASSEN, A.: Studying high ozone concentrations by using the Danish Eulerian Model; Atmos. Environ., **27A** (1993), 845-865.
5. ZLATEV, Z., CHRISTENSEN, J., HOV, Ø.: A Eulerian air pollution model for Europe with nonlinear chemistry; J. Atmos. Chem., **15** (1992), 1-37.
6. ZLATEV, Z., DIMOV, I., GEORGIEV, K.: Studying long-range transport of air pollutants; Computational Science and Engineering, **1**, No. 3 (1994), 45-52.

A Proposal for a Fortran 90 Interface for LAPACK

Jack J. Dongarra[1], Jeremy Du Croz[2], Sven Hammarling[2],
Jerzy Waśniewski[3] and Adam Zemła[4]

[1] Department of Computer Science, University of Tennessee
107 Ayres Hall, Knoxville, TN 37996-1301, USA
and
Mathematical Sciences Section, Oak Ridge National Laboratory
P.O.Box 2008, Bldg. 6012, Oak Ridge, TN 37831-6367, USA
Email: dongarra@cs.utk.edu
[2] Numerical Algorithms Group Ltd, Wilkinson House
Jordan Hill Road, Oxford OX2 8DR, UK
Email: jeremy@nag.co.uk or sven@nag.co.uk respectively
[3] UNI•C, The Danish Computing Centre for Research and Education
DTU, Bldg. 304, DK-2800 Lyngby, Denmark
Email: jerzy.wasniewski@uni-c.dk
[4] Institute of Mathematics, Polish Academy of Sciences
Śniadeckich 8, 00-950 Warsaw, Poland
Email: adamz@impan.gov.pl

1 Introduction

The purpose of this paper is to initiate discussion of the design of a Fortran 90 interface to LAPACK [1]. Our emphasis at this stage is on the design of an improved *user-interface* to the package, taking advantage of the considerable simplifications which Fortran 90 allows (see [3]).

The new interface can be implemented initially by writing Fortran 90 jackets to call the existing Fortran 77 code.

Eventually we hope that the LAPACK code will be rewritten to take advantage of the new features of Fortran 90, but this will be an enormous task. We aim to design an interface which can persist unchanged while the underlying code is rewritten.

For convenience we use the name "LAPACK 77" to denote the existing Fortran 77 package, and "LAPACK 90" to denote the new Fortran 90 interface which we are proposing.

2 LAPACK 77 and Fortran 90 Compilers

2.1 Linking LAPACK 77 to Fortran 90 programs

LAPACK 77 can be called from Fortran 90 programs in its present form — with some qualifications. The qualifications arise only because LAPACK 77 is not written entirely in *standard* Fortran 77; the exceptions are the use of the

COMPLEX*16 data type and related intrinsic functions, as listed in Section 6.1 of [1]; these facilities are provided as extensions to the standard language by many Fortran 77 and Fortran 90 compilers. Equivalent facilities are provided in standard Fortran 90, using the parameterized form of the COMPLEX data type (see below).

To link LAPACK 77 to a Fortran 90 program (which must of course be compiled by a Fortran 90 compiler), one of the following approaches will be necessary, depending on the compilers available.

1. Link the Fortran 90 program to an existing LAPACK 77 library, compiled by a Fortran 77 compiler. This approach can only work if the compilers have designed to allow cross-linking.
2. If such cross-linking is not possible, recompile LAPACK 77 with the Fortran 90 compiler, provided that the compiler accepts COMPLEX*16 and related intrinsics as extensions, and create a new library.
3. If these extensions are not accepted, convert the LAPACK 77 code to standard Fortran 90 (see below), before recompiling it.

The conversions needed to create standard Fortran 90 code for LAPACK 77 are:

$$
\begin{array}{rcl}
\text{COMPLEX*16} & \Rightarrow & \text{COMPLEX(KIND=Kind(0.0D0)} \\
\text{DCONJG(z) for COMPLEX*16 z} & \Rightarrow & \text{CONJG(z)} \\
\text{DBLE(z) for COMPLEX*16 z} & \Rightarrow & \text{REAL(z)} \\
\text{DIMAG(z) for COMPLEX*16 z} & \Rightarrow & \text{AIMAG(z)} \\
\text{DCMPLX(x,y) for DOUBLE PRECISION x, y} & \Rightarrow & \text{CMPLX(x,y,KIND=Kind(0.0D0))}
\end{array}
$$

One further obstacle may remain: it is possible that if LAPACK 77 has been recompiled with a Fortran 90 compiler, it may not link correctly to an optimized assembly-language BLAS library that has been designed to interface with Fortran 77. Until this is rectified by the vendor of the BLAS library, Fortran 77 code for the BLAS must be used.

2.2   Interface blocks for LAPACK 77

Fortran 90 allows one immediate extra benefit to be provided to Fortran 90 users of LAPACK 77, without making any further changes to the existing code: that is a *module* of *explicit interfaces* for the routines. If this module is accessed by a USE statement in any program unit which makes calls to LAPACK routines, then those calls can be checked by the compiler for errors in the numbers or types of arguments.

The module can be constructed by extracting the necessary specification statements from the Fortran 77 code, as illustrated by the following example (in fixed-form source format) containing an interface for the single routine CBDSQR:

```
MODULE LAPACK77_INTERFACES
INTERFACE
SUBROUTINE CBDSQR( UPLO, N, NCVT, NRU, NCC, D, E, VT, LDVT, U,
$                  LDU, C, LDC, RWORK, INFO )
CHARACTER          UPLO
INTEGER            INFO, LDC, LDU, LDVT, N, NCC, NCVT, NRU
REAL               D( * ), E( * ), RWORK( * )
COMPLEX            C( LDC, * ), U( LDU, * ), VT( LDVT, * )
END
END INTERFACE
END MODULE LAPACK77_INTERFACES
```

A single module containing interfaces for all the routines in LAPACK 77 (over 1000 of them) may be too large for practical use; it may be desirable to split it (perhaps, one module for single precision documented routines, one for double precision documented routines, and similarly for auxiliary routines).

3 Proposals for the Design of LAPACK 90

In the design of a Fortran 90 interface to LAPACK, we propose to take advantage of the features of the language listed below.

1. **Assumed-shape arrays:** All array arguments to LAPACK 90 routines will be assumed-shape arrays. Arguments to specify problem-dimensions or array-dimensions will not be required.
   This implies that the actual arguments supplied to LAPACK routines *must* have the *exact* shape required by the problem. The most convenient ways to achieve this are:
   − using allocatable arrays, for example:
   ```
   REAL, ALLOCATABLE :: A(:,:), B(:)
   . . .
   ALLOCATE( A(N,N), B(N) )
   . . .
   CALL LA_GESV( A, B )
   ```
   − passing array sections, for example:
   ```
   REAL :: A(NMAX,NMAX), B(NMAX)
   . . .
   CALL LA_GESV( A(:N,:N), B(:N) )
   ```
   Zero dimensions (empty arrays) will be allowed.
   There are some grounds for concern about the effect of assumed-size arrays on performance, because compilers cannot assume that their storage is contiguous. The effect on performance will of course depend on the compiler, and may diminish in time as compilers become more effective in optimizing compiled code. This point needs investigation.

2. **Automatic allocation of work arrays:** Workspace arguments and arguments to specify their dimensions will not be needed. In simple cases, *automatic arrays* of the required size can be declared internally. In other cases, allocatable arrays may need to be declared and explicitly allocated. Explicit allocation is needed in particular when the amount of workspace required depends on the block-size to be used (which is not passed as an argument).

3. **Optional arguments:** In LAPACK 77, character arguments are frequently used to specify some choice of options. In Fortran 90, a choice of options can sometimes be specified naturally by the presence or absence of optional arguments: for example, options to compute the left or right eigenvectors can be specifed by the presence of arguments VL or VR, and the character arguments JOBVL and JOBVR which are required in the LAPACK 77 routine DGEEV, are not needed in LAPACK 90.

In other routines, a character argument to specify options may still be required, but can itself be made optional if there is a natural default value: for example, in DGESVX the argument TRANS can be made optional, with default value 'N'.

Optional arguments can also help to combine two or more routines into one: for example, the functionality provided by the routine DGECON can be made acessible by adding an optional argument RCOND to DGETRF.

4. **Generic Interfaces:** The systematic occurrence in LAPACK of analogous routines for real or complex data, and for single or double precision lends itself well to the definition of generic interfaces, allowing four different routines to be accessed through the same generic name.

Generic interfaces can also be used to cover routines whose arguments differ in *rank*, and thus provide a slight increase in flexibility over LAPACK 77. For example, in LAPACK 77, routines for solving a system of linear equations (such as DGESV), allow for multiple right hand sides, and so the arrays which hold the right hand sides and solutions are always of rank 2. In LAPACK 90, we can provide alternative versions of the routines (covered by a single generic interface) in which the arrays holding the right hand sides and solutions may *either* be of rank 1 (for a single right hand side) *or* be of rank 2 (for several right hand sides).

5. **Naming:** For the generic routine names, we propose:

   (a) the initial letter (S, C, D or Z) is simply omitted.

   (b) the letters LA_ are prefixed to all names to identify them as names of LAPACK routines.

In other respects the naming scheme remains the same as described in Section 2.1.3 of [1]: for example, LA_GESV.

It would also be possible to define longer, more meaningful names (which could co-exist with the shorter names), but we have not attempted this here. We have *not* proposed the use of any *derived types* in this Fortran 90 interface. They could be considered — for example, to hold the details of an *LU* factorization and equilibration factors. However, since LAPACK routines are so frequently used as building blocks in larger algorithms or applications, we feel that there are advantages in keeping the interface simple, and avoiding possible loss of efficiency through the use of array pointers (which such derived types would require).

6. **Error-handling:**

In LAPACK 77, all documented routines have a diagnostic output argument INFO. Three types of exit from a routine are allowed:

successful termination: the routine returns to the calling program with INFO set to 0.

illegal value of one or more arguments: the routine sets INFO < 0 and calls the auxiliary routine XERBLA; the standard version of XERBLA issues an error message identifying the first invalid argument, and stops execution.

failure in the course of computation: the routine sets INFO > 0 and returns to the calling program without issuing any error message. Only some LAPACK 77 routines need to allow this type of error-exit; it is then the resposibility of a user to test INFO on return to the calling program.

For LAPACK 90 we propose that the argument INFO becomes *optional*: if it is not present and an error occurs, then the routine *always* issues an error message and stops execution, even when INFO > 0 (in which case the error message reports the value of INFO). If a user wishes to continue execution after a failure in computation, then INFO must be supplied and tested on return.

This behaviour simplifies calls to LAPACK 90 routines when there is no need to test INFO on return, and makes it less likely that users will forget to test INFO when necessary.

If an invalid argument is detected, we propose that routines issue an error message and stop, as in LAPACK 77. Note however that in Fortran 90 there can be different reasons for an argument being invalid:

illegal value : as in LAPACK 77.

invalid shape (of an assumed-shape array): for example, a 2-dimensional array is not square when it is required to be.

inconsistent shapes (of two or more assumed-shape arrays): for example, arrays holding the right hand sides and solutions of a system of linear equations must have the same shape.

The specification could be extended so that the error-message could distinguish between these cases.

4 Prototype Implementation of LAPACK 90 Procedures

We have implemented Fortran 90 jacket procedures to the group of LAPACK 77 routines concerned with the solution of systems of linear equations $AX = B$ for a general matrix A — that is, the driver routines xGESV and xGESVX, and the computational routines xGETRF, xGETRS, xGETRI, xGECON, xGERFS and xGEEQU.

In Appendix of [2], we give detailed documentation of the proposed interfaces. Here we give examples of calls to each of the proposed routines, the first without using any of the optional arguments, the second using all the arguments. For the time being and for ease of comparison between LAPACK 77 and LAPACK 90, we have retained the same names for the corresponding arguments, although of course Fortran 90 offers the possibility of longer names (for example, `IPIV` could become `PIVOT_INDICES`).

In this prototype implementation, we have assumed that the code of LAPACK 77 is not modified.

`LA_GESV` (simple driver):

```
CALL LA_GESV( A, B )

CALL LA_GESV( A, B, IPIV, INFO )
```

Comments:

- The array B may have rank 1 (one right hand side) or rank 2 (several right hand sides).

`LA_GESVX` (expert driver):

```
CALL LA_GESVX( A, B, X )

CALL LA_GESVX( A, B, X, AF, IPIV, FACT, TRANS, EQUED, R, C, &
               FERR, BERR, RCOND, RPVGRW, INFO )
```

Comments:

- The arrays B and X may have rank 1 (in which case `FERR` and `BERR` are scalars) or rank 2 (in which case `FERR` and `BERR` are rank-1 arrays).
- `RPVGRW` returns the reciprocal pivot growth factor (returned in `WORK(1)` in LAPACK 77).
- the presence or absence of `EQUED` is used to specify whether or not equilibration is to be performed, instead of the option `FACT = 'E'`.

`LA_GETRF` (LU factorization):

```
CALL LA_GETRF( A, IPIV )

CALL LA_GETRF( A, IPIV, RCOND, NORM, INFO )
```

Comments:

- instead of a separate routine `LA_GECON`, we propose that optional arguments `RCOND` and `NORM` are added to `LA_GETRF` to provide the same functionality in a more convenient manner. The argument `ANORM` of xGECON is not needed, because `LA_GETRF` can always compute the norm of A if required.

LA_GETRS (solution of equations using *LU* factorization):

     `CALL LA_GETRS( A, IPIV, B )`

     `CALL LA_GETRS( A, IPIV, B, TRANS, INFO )`

Comments:

    − The array B may have rank 1 or 2.

LA_GETRI (matrix inversion using *LU* factorization):

     `CALL LA_GETRI( A, IPIV )`

     `CALL LA_GETRI( A, IPIV, INFO )`

LA_GERFS (refine solution of equations and optionally compute error bounds):

     `CALL LA_GERFS( A, AF, IPIV, B, X )`

     `CALL LA_GERFS( A, AF, IPIV, B, X, TRANS, FERR, BERR, INFO )`

Comments:

    − The arrays B and X may have rank 1 (in which case FERR and BERR are
      scalars) or rank 2 (in which case FERR and BERR are rank-1 arrays).

LA_GEEQU (equilibration):

     `CALL LA_GEEQU( A, R, C )`

     `CALL LA_GEEQU( A, R, C, ROWCND, COLCND, AMAX, INFO )`

5   Documentation

In the Appendix of [2], we give a first attempt at draft documentation for these
routines. The style is somewhat similar to that of the LAPACK Users' Guide, but
with various obvious new conventions introduced to handle the generic nature
of the interfaces.

6   Test Software

Additional test software will be needed to test the new interfaces.

7 Timings

We have done some timings to measure the extra overhead of the Fortran 90 interface. We timed LA_GETRF on a single processor of an IBM SP-2 (in double precision) and a single processor of a Cray YMP C90A (in single precision). All timings are given in megaflops.

IBM results:
1. Speed of LAPACK 90 calling LAPACK 77 and BLAS from the ESSL library.
2. Speed of LAPACK 77, using BLAS from the ESSL library.

Array size	600	700	800	900	1000	1100	1200	1300	1400	1500
LAPACK90	187	180	182	170	172	172	176	177	181	182
LAPACK77	191	181	182	171	172	173	176	179	180	182

CRAY results:
1. Speed of LAPACK 90 calling LAPACK 77 as provided by CRAY in LIB-SCI.
2. Speed of LAPACK 77 as provided by CRAY in LIBSCI.

Array size	600	700	800	900	1000	1100	1200	1300	1400	1500
LAPACK90	723	828	646	841	822	855	789	857	846	868
LAPACK77	778	834	649	845	825	860	794	864	848	873

The above tables show the LAPACK 90 results are a little slower (1 or 2%) than the LAPACK 77 results.

8 Acknowledgments

Jerzy Waśniewski's research is partly supported by the Danish Project, Efficient Parallel Algorithms for Optimization and Simulation (EPOS).

We thank very much Dr. Christian de Polignac. He ran the test programs on an IBM RS/6000 and using the NAG compiler on an HP workstation.

References

1. E. Anderson, Z. Bai, C. H. Bischof, J. Demmel, J. J. Dongarra, J. Du Croz, A. Greenbaum, S. Hammarling, A. McKenney, S. Ostrouchov and D. C. Sorensen. *LAPACK Users' Guide Release 2.0.* SIAM, Philadelphia, 1995.
2. J.J. Dongarra, J. Du Croz, S. Hammarling, J. Waśniewski and A. Zemła. *LAPACK Working Note 101, A Proposal for a Fortran 90 Interface for LAPACK.* Report UNIC-95-9, UNI•C, Lyngby, Denmark, 1995.
3. M. Metcalf and J. Reid. *Fortran 90 Explained.* Oxford, New York, Tokyo, Oxford University Press, 1990.

ScaLAPACK Tutorial *

Jack Dongarra[1,2] and Antoine Petitet[1]

[1] Department of Computer Science, University of Tennessee, Knoxville,
TN 37996-1301, USA
[2] Mathematical Sciences Section, Oak Ridge National Laboratory, Oak Ridge,
TN 37831, USA

Abstract. This ScaLAPACK tutorial begins with a brief description of
the LAPACK library. The importance of block-partitioned algorithms
in reducing the frequency of data movement between different levels of
hierarchical memory is stressed. By relying on the Basic Linear Algebra
Subprograms (BLAS) it is possible to develop portable and efficient implementations of these algorithms across a wide range of architectures,
with emphasis on workstations, vector-processors and shared-memory
computers, as has been done in LAPACK.

The ScaLAPACK library, which is a distributed memory version of LA-
PACK is then presented. A key idea in our approach is the use of Basic
Linear Algebra Communication Subprograms (BLACS) as communication building blocks and the use of a distributed version of the BLAS,
the Parallel Basic Linear Algebra Subprograms (PBLAS) as computational building blocks. The BLACS and PBLAS features are in turn outlined and it is shown how these building blocks can be used to construct
higher-level algorithms, and hide many details of the parallelism from
the application developer. Performance results of ScaLAPACK routines
are presented validating the adoption of the block-cyclic decomposition
scheme as a way of distributing block-partitioned matrices yielding to
well balanced computations and scalable implementations.

Finally, future directions for the ScaLAPACK library are described and
alternative approaches to mathematical libraries are suggested that could
integrate ScaLAPACK into efficient and user-friendly distributed systems.

1 Introduction

Much of the work in developing linear algebra software for advanced architecture computers is motivated by the need to solve large problems on the fastest
computers available. In this tutorial, we focus on the development of standards

* This work was supported in part by the National Science Foundation Grant No.
ASC-9005933; by the Defense Advanced Research Projects Agency under contract DAAL03-91-C-0047, administered by the Army Research Office; by the Office of Scientific Computing, U.S. Department of Energy, under Contract DE-AC05-
84OR21400; and by the National Science Foundation Science and Technology Center
Cooperative Agreement No. CCR-8809615.

for use in linear algebra and the building blocks for a library and the aspects of algorithm design and parallel implementation.

The linear algebra community has long recognized the need for help in developing algorithms into software libraries, and several years ago, as a community effort, put together a *de facto* standard for identifying basic operations required in linear algebra algorithms and software. The hope was that the routines making up this standard, the Basic Linear Algebra Subprograms (BLAS), would be implemented on advanced-architecture computers by many manufacturers, making it possible to reap the portability benefits of having them efficiently implemented on a wide range of machines. This goal has been largely realized.

The key insight of our approach to designing linear algebra algorithms for advanced architecture computers is that the frequency with which data are moved between different levels of the memory hierarchy must be minimized in order to attain high performance. Thus, our main algorithmic approach for exploiting both vectorization and parallelism is the use of block-partitioned algorithms, particularly in conjunction with highly-tuned kernels for performing matrix-vector and matrix-matrix operations (BLAS). In general, block-partitioned algorithms require to move blocks, rather than vectors or scalars, so that the startup cost associated with the movement is greatly reduced because fewer messages are exchanged.

A second key idea is that the performance of an algorithm can be tuned by a user by varying the parameters that specify the data layout. On shared memory machines, this is controlled by the block size, while on distributed memory machines it is controlled by the block size and the configuration of the logical process mesh.

2 The Linear Algebra Package (LAPACK)

LAPACK [1] provides routines for solving systems of simultaneous linear equations, least-squares solutions of linear systems of equations, eigenvalue problems, and singular value problems. The associated matrix factorizations (LU, Cholesky, QR, SVD, Schur, generalized Schur) are also provided, as are related computations such as reordering of the Schur factorizations and estimating condition numbers. Dense and banded matrices are handled, but not general sparse matrices. In all areas, similar functionality is provided for real and complex matrices, in both single and double precision.

The original goal of the LAPACK project was to make the widely used EIS-PACK [15] and LINPACK libraries run efficiently on shared-memory vector and parallel processors. On these machines, LINPACK and EISPACK are inefficient because their memory access patterns disregard the multilayered memory hierarchies of the machines, thereby spending too much time moving data instead of doing useful floating-point operations. LAPACK addresses this problem by reorganizing the algorithms to use block matrix operations, such as matrix multiplication, in the innermost loops [6, 1]. These block operations can be optimized for each architecture to account for the memory hierarchy [5], and so provide a transportable way to achieve high efficiency on diverse modern machines.

LAPACK can be regarded as a successor to LINPACK and EISPACK. It has virtually all the capabilities of these two packages and much more besides. It improves on them in four main respects: speed, accuracy, robustness and functionality. While LINPACK and EISPACK are based on the vector operation kernels of the Level 1 BLAS, LAPACK was designed at the outset to exploit the matrix-matrix operation kernels of the Level 3 BLAS. Because of the coarse granularity of these operations, their use tends to promote high efficiency on many high-performance computers, particularly if specially coded implementations are provided by the manufacturer.

2.1 The BLAS as the Key to Portability and Efficiency

There are now three levels of BLAS:

Level 1 BLAS [2]: for vector operations, such as $y \leftarrow \alpha x + y$
Level 2 BLAS [3]: for matrix-vector operations, such as $y \leftarrow \alpha A x + \beta y$
Level 3 BLAS [4]: for matrix-matrix operations, such as $C \leftarrow \alpha AB + \beta C$.

Here, A, B and C are matrices, x and y are vectors, and α and β are scalars.

The Level 1 BLAS are used in LAPACK, but for convenience rather than for performance: they perform an insignificant fraction of the computation, and they cannot achieve high efficiency on most modern supercomputers.

The Level 2 BLAS can achieve near-peak performance on many vector processors, such as a CRAY Y-MP, or Convex C-2 machine. However, on other vector processors such as a CRAY-2, the performance of the Level 2 BLAS is limited by the rate of data movement between different levels of memory [11].

The Level 3 BLAS overcome this limitation. They perform $O(n^3)$ floating-point operations on $O(n^2)$ data, whereas the Level 2 BLAS perform only $O(n^2)$ operations on $O(n^2)$ data. The Level 3 BLAS also allow us to exploit parallelism in a way that is transparent to the software that calls them. While the Level 2 BLAS offer some scope for exploiting parallelism, greater scope is provided by the Level 3 BLAS, as Table 1 illustrates.

Table 1. Speed (Megaflops) of BLAS Operations on a CRAY Y-MP. All matrices are of order 500.

Number of processors:	1	2	4	8
Level 2: $y \leftarrow \alpha A x + \beta y$	311	611	1197	2285
Level 3: $C \leftarrow \alpha AB + \beta C$	312	623	1247	2425
Peak	333	666	1332	2664

2.2 A Block Partitioned Algorithm Example

We consider the Cholesky factorization algorithm, which factorizes a symmetric positive definite matrix as $A = U^T U$. To derive a block form of Cholesky factorization, we partition the matrices into blocks, in which the diagonal blocks of A and U are square, but of differing sizes. We assume that the first block has already been factored as $A_{00} = U_{00}^T U_{00}$, and that we now want to determine the second block column of U consisting of the blocks U_{01} and U_{11}.

$$
\begin{pmatrix} A_{00} & A_{01} & A_{02} \\ \cdot & A_{11} & A_{12} \\ \cdot & \cdot & A_{22} \end{pmatrix} = \begin{pmatrix} U_{00}^T & 0 & 0 \\ U_{01}^T & U_{11}^T & 0 \\ U_{02}^T & U_{12}^T & U_{22}^T \end{pmatrix} \begin{pmatrix} U_{00} & U_{01} & U_{02} \\ 0 & U_{11} & U_{12} \\ 0 & 0 & U_{22} \end{pmatrix}.
$$

Equating submatrices in the second block of columns, we obtain

$$
A_{01} = U_{00}^T U_{01}
$$
$$
A_{11} = U_{01}^T U_{01} + U_{11}^T U_{11}.
$$

Hence, since U_{00} has already been computed, we can compute U_{01} as the solution to the equation

$$
U_{00}^T U_{01} = A_{01}
$$

by a call to the Level 3 BLAS routine STRSM; and then we can compute U_{11} from

$$
U_{11}^T U_{11} = A_{11} - U_{01}^T U_{01}.
$$

This involves first updating the symmetric submatrix A_{11} by a call to the Level 3 BLAS routine SSYRK, and then computing its Cholesky factorization. Since Fortran 77 does not allow recursion, a separate routine must be called (using Level 2 BLAS rather than Level 3), named SPOTF2 in Figure 1. In this way, successive blocks of columns of U are computed. The LAPACK-style code for the block algorithm is shown in Figure 1.

```
do j = 0, n-1, nb
  jb = min( nb, n-j )
  call strsm( 'left', 'upper', 'transpose', 'non-unit', j, jb, one,
              a, lda, a(0,j), lda )
  call ssyrk( 'upper', 'transpose', jb, j, -one, a(0,j), lda, one,
              a(j,j), lda )
  call spotf2( 'upper', jb, a(j,j), lda, info )
  if( info .ne. 0 ) go to 20
end do
```

Fig. 1. The body of the "LAPACK-style" routine for block Cholesky factorization. In this code fragment, nb denotes the width of the blocks.

On a CRAY Y-MP, the use of Level 3 BLAS squeezes a little more performance out of one processor, but makes a large improvement when using all 8 processors. Table 2 summarizes the results.

Table 2. Speed (Megaflops) of Cholesky Factorization $A = U^T U$ for $n = 500$

CRAY Y-MP	1 processor	8 processors
j-variant: LINPACK Level 1 BLAS	72	72
j-variant: using Level 2 BLAS	251	378
j-variant: using Level 3 BLAS	287	1225

3 ScaLAPACK

The ScaLAPACK software library is extending the LAPACK library to run scalably on MIMD, distributed memory, concurrent computers [9]. For such machines the memory hierarchy includes the off-processor memory of other processors, in addition to the hierarchy of registers, cache, and local memory on each processor. Like LAPACK, the ScaLAPACK routines are based on block-partitioned algorithms in order to minimize the frequency of data movement between different levels of the memory hierarchy. The fundamental building blocks of the ScaLAPACK library are a set of Basic Linear Algebra Communication Subprograms (BLACS) [8] for communication tasks that arise frequently in parallel linear algebra computations, and the Parallel Basic Linear Algebra Subprograms (PBLAS), that are a distributed memory version of the sequential BLAS. In the ScaLAPACK routines, all interprocessor communication occurs within the PBLAS and the BLACS, so the source code of the top software layer of ScaLAPACK looks very similar to that of LAPACK.

3.1 Block Cyclic Data Distribution

The way in which a matrix is distributed over the processes has a major impact on the load balance and communication characteristics of the concurrent algorithm, and hence largely determines its performance and scalability. The block cyclic distribution provides a simple, yet general-purpose way of distributing a block-partitioned matrix on distributed memory concurrent computers. It has been incorporated in the High Performance Fortran standard [14].

The block cyclic data distribution is parameterized by the four numbers P_r, P_c, r, and c, where $P_r \times P_c$ is the process template and $r \times c$ is the block size.

Suppose first that we have M objects, indexed by an integer $0 \le m < M$, to map onto P processes, using block size r. The m-th item will be stored in the

i-th location of block b on process p, where

$$\langle p, b, i \rangle = \left\langle \left\lfloor \frac{m}{r} \right\rfloor \bmod P, \left\lfloor \frac{\left\lfloor \frac{m}{r} \right\rfloor}{P} \right\rfloor, m \bmod r \right\rangle .$$

In the special case where $r = 2^{\hat{r}}$ and $P = 2^{\hat{P}}$ are powers of two, this mapping is really just bit extraction, with i equal to the rightmost $\hat{r}$ bits of m, p equal to the next $\hat{P}$ bits of m, and b equal to the remaining leftmost bits of m. The distribution of a block-partitioned matrix can be regarded as the tensor product of two such mappings: one that distributes the rows of the matrix over P_r processes, and another that distributes the columns over P_c processes. That is, the matrix element indexed globally by (m, n) is stored in location

$$\langle (p, q), (b, d), (i, j) \rangle =$$

$$\left\langle (\left\lfloor \frac{m}{r} \right\rfloor \bmod P_r, \left\lfloor \frac{n}{c} \right\rfloor \bmod P_c), \left(\left\lfloor \frac{\left\lfloor \frac{m}{r} \right\rfloor}{P_r} \right\rfloor, \left\lfloor \frac{\left\lfloor \frac{n}{c} \right\rfloor}{P_c} \right\rfloor \right), (m \bmod r, n \bmod c) \right\rangle .$$

The nonscattered decomposition (or pure block distribution) is just the special case $r = \lceil M/P_r \rceil$ and $c = \lceil N/P_c \rceil$. Similarly a purely scattered decomposition (or two dimensional wrapped distribution) is the special case $r = c = 1$.

3.2   The Basic Linear Algebra Communication Subprograms

The **BLACS** (Basic Linear Algebra Communication Subprograms) [8] are a message passing library designed for linear algebra. The computational model consists of a one or two dimensional grid of processes, where each process stores matrices and vectors. The BLACS include synchronous send/receive routines to send a matrix or submatrix from one process to another, to broadcast submatrices to many processes, or to compute global reductions (sums, maxima and minima). There are also routines to set up, change, or query the process grid. Since several ScaLAPACK algorithms require broadcasts or reductions among different subsets of processes, the BLACS permit a process to be a member of several overlapping or disjoint process grids, each one labeled by a *context*. Some message passing systems, such as MPI [12], also include this context concept. The BLACS provide facilities for safe interoperation of system contexts and BLACS contexts.

3.3   PBLAS

In order to simplify the design of ScaLAPACK, and because the BLAS have proven to be very useful tools outside LAPACK, we chose to build a Parallel BLAS, or PBLAS [10], whose interface is as similar to the BLAS as possible. This decision has permitted the ScaLAPACK code to be quite similar, and sometimes nearly identical, to the analogous LAPACK code. Only one substantially new routine was added to the PBLAS, matrix transposition, since this is a complicated operation in a distributed memory environment [7].

We hope that the PBLAS will provide a distributed memory standard, just as the BLAS have provided a shared memory standard. This would simplify and encourage the development of high performance and portable parallel numerical software, as well as providing manufacturers with a small set of routines to be optimized. The acceptance of the PBLAS requires reasonable compromises among competing goals of functionality and simplicity.

The PBLAS operate on matrices distributed in a 2D block cyclic layout. Since such a data layout requires many parameters to fully describe the distributed matrix, we have chosen a more object-oriented approach, and encapsulated these parameters in an integer array called an *array descriptor*. An array descriptor includes

(1) the number of rows in the distributed matrix,
(2) the number of columns in the distributed matrix,
(3) the row block size (r in Section 3.1),
(4) the column block size (c in Section 3.1),
(5) the process row over which the first row of the matrix is distributed,
(6) the process column over which the first column of the matrix is distributed,
(7) the BLACS context (see Section 3.2), and
(8) the leading dimension of the local array storing the local blocks.

By using this descriptor, a call to a PBLAS routine is very similar to a call to the corresponding BLAS routine.

```
CALL DGEMM ( TRANSA, TRANSB, M, N, K, ALPHA,
             A( IA, JA ), LDA,
             B( IB, JB ), LDB, BETA,
             C( IC, JC ), LDC )


CALL PDGEMM( TRANSA, TRANSB, M, N, K, ALPHA,
             A, IA, JA, DESC_A,
             B, JB, DESC_B, BETA,
             C, IC, JC, DESC_C )
```

DGEMM computes C = BETA * C + ALPHA * op(A) * op(B), where op(A) is either A or its transpose depending on TRANSA, op(B) is similar, op(A) is M-by-K, and op(B) is K-by-N. PDGEMM is the same, with the exception of the way in which submatrices are specified. To pass the submatrix starting at A(IA,JA) to DGEMM, for example, the actual argument corresponding to the formal argument A would simply be A(IA,JA). PDGEMM, on the other hand, needs to understand the global storage scheme of A to extract the correct submatrix, so IA and JA must be passed in separately. DESC_A is the array descriptor for A. The parameters describing the matrix operands B and C are analogous to those describing A. In a truly object-oriented environment matrices and DESC_A would be the synonymous. However, this would require language support, and detract from portability.

The presence of a context associated with every distributed matrix provides the ability to have separate "universes" of message passing. The use of separate communication contexts by distinct libraries (or distinct library invocations) such as the PBLAS insulates communication internal to the library from external communication. When more than one descriptor array is present in the argument list of a routine in the PBLAS, it is required that the individual BLACS context entries must be equal. In other words, the PBLAS do not perform "intra-context" operations.

We have not included specialized routines to take advantage of packed storage schemes for symmetric, Hermitian, or triangular matrices, nor of compact storage schemes for banded matrices [10].

3.4 ScaLAPACK – Contents and Performance

Given the infrastructure described above, the ScaLAPACK version (PDGETRF) of the LU decomposition is nearly identical to its LAPACK version (DGETRF).

```
      SEQUENTIAL LU FACTORIZATION CODE

DO 20 J = 1, MIN( M, N ), NB
   JB = MIN( MIN( M, N )-J+1, NB )



   Factor diagonal and subdiagonal blocks and test for exact
   singularity.

   CALL DGETF2( M-J+1, JB, A( J, J ), LDA, IPIV( J ),
$              IINFO )

   Adjust INFO and the pivot indices.

   IF( INFO.EQ.0 .AND. IINFO.GT.0 ) INFO = IINFO + J - 1
   DO 10 I = J, MIN( M, J+JB-1 )
      IPIV( I ) = J - 1 + IPIV( I )
10 CONTINUE

   Apply interchanges to columns 1:J-1.

   CALL DLASWP( J-1, A, LDA, J, J+JB-1, IPIV, 1 )


   IF( J+JB.LE.N ) THEN

      Apply interchanges to columns J+JB:N.

      CALL DLASWP( N-J-JB+1, A( 1, J+JB ), LDA, J, J+JB-1,
$                 IPIV, 1 )

      Compute block row of U.

      CALL DTRSM( 'Left', 'Lower', 'No transpose', 'Unit',
$                JB, N-J-JB+1, ONE, A( J, J ), LDA,
$                A( J, J+JB ), LDA )
      IF( J+JB.LE.M ) THEN

         Update trailing submatrix.

         CALL DGEMM( 'No transpose', 'No transpose',
$                   M-J-JB+1, N-J-JB+1, JB, -ONE,
$                   A( J+JB, J ), LDA, A( J, J+JB ), LDA,
$                   ONE, A( J+JB, J+JB ), LDA )
      END IF
   END IF
20 CONTINUE
```

```
       PARALLEL LU FACTORIZATION CODE

DO 10 J = JA, JA+MIN(M,N)-1, DESCA( 4 )
   JB = MIN( MIN(M,N)-J+JA, DESCA( 4 ) )
   I = IA + J - JA

   Factor diagonal and subdiagonal blocks and test for exact
   singularity.

   CALL PDGETF2( M-J+JA, JB, A, I, J, DESCA, IPIV, IINFO )

   Adjust INFO and the pivot indices.

   IF( INFO.EQ.0 .AND. IINFO.GT.0 )
$     INFO = IINFO + J - JA




   Apply interchanges to columns JA:J-JA.

   CALL PDLASWP( 'Forward', 'Rows', J-JA, A, IA, JA, DESCA,
$               J, J+JB-1, IPIV )


   IF( J-JA+JB+1.LE.N ) THEN

      Apply interchanges to columns J+JB:JA+N-1.

      CALL PDLASWP( 'Forward', 'Rows', N-J-JB+JA, A, IA,
$                  J+JB, DESCA, J, J+JB-1, IPIV )

      Compute block row of U.

      CALL PDTRSM( 'Left', 'Lower', 'No transpose', 'Unit',
$                 JB, N-J-JB+JA, ONE, A, I, J, DESCA, A, I,
$                 J+JB, DESCA )
      IF( J-JA+JB+1.LE.M ) THEN

         Update trailing submatrix.

         CALL PDGEMM( 'No transpose', 'No transpose',
$                    M-J-JB+JA, N-J-JB+JA, JB, -ONE, A,
$                    I+JB, J, DESCA, A, I, J+JB, DESCA,
$                    ONE, A, I+JB, J+JB, DESCA )
      END IF
   END IF
10 CONTINUE
```

The Cholesky decompositions (PDPOTRF and DPOTRF) and QR decompositions (PDGEQRF and DGEQRF) are analogous.

In addition to the PBLAS, the ScaLAPACK library provides routines to solve linear systems of equations for dense general and positive definite matrices as well as the symmetric eigenproblem. The associated matrix factorizations (LU,

Cholesky, QR) and the matrix reductions to (upper) Hessenberg, tridiagonal and bidiagonal forms are also provided. Similar functionality is provided for real and complex matrices, in both single and double precision. Condition estimators, iterative refinement of solutions of linear systems of equations, least-squares problem solvers, nonsymmetric eigensolvers, as well as routines for banded systems will be added to the library in the very near future.

The ScaLAPACK codes runs efficiently on a wide range of distributed memory MIMD computers, such as the IBM SP-1 and SP-2, the Cray T3D, the Intel iPSCs, Delta, and Paragon. Extensive performance results can be found in [9]. On 8 wide nodes IBM SP-2 for example, a 13000×13000 LU factorization runs at 1.6 Gflop/s. A 2000×2000 LU factorization on the same machine reaches already 1.0 Gflop/s. These performance results correspond to a very efficient use of this machine. The ScaLAPACK library can also be used on a cluster of workstations on top of PVM [13]. Performance results on the CM-5, however, have been disapointing because of the difficulty of using the vector units in message passing programs.

4 Conclusions and Future Research Directions

Traditionally, large, general-purpose mathematical software libraries have required users to write their own programs that call library routines to solve specific subproblems that arise during a computation. Adapted to a shared-memory parallel environment, this conventional interface still offers some potential for hiding underlying complexity. For example, the LAPACK project incorporates parallelism in the Level 3 BLAS, where it is not directly visible to the user.

But when going from shared-memory systems to the more readily scalable distributed memory systems, the complexity of the distributed data structures required is more difficult to hide from the user. Not only must the problem decomposition and data layout be specified, but different phases of the user's problem may require transformations between different distributed data structures.

These deficiencies in the conventional user interface have prompted extensive discussion of alternative approaches for scalable parallel software libraries of the future. Possibilities include:

1. Traditional function library (i.e., minimum possible change to the status quo in going from serial to parallel environment). This will allow one to protect the programming investment that has been made.
2. Reactive servers on the network. A user would be able to send a computational problem to a server that was specialized in dealing with the problem. This fits well with the concepts of a networked, heterogeneous computing environment with various specialized hardware resources (or even the heterogeneous partitioning of a single homogeneous parallel machine).
3. General interactive environments like Matlab or Mathematica, perhaps with "expert" drivers (i.e., knowledge-based systems). With the growing popularity of the many integrated packages based on this idea, this approach would

provide an interactive, graphical interface for specifying and solving scientific problems. Both the algorithms and data structures are hidden from the user, because the package itself is responsible for storing and retrieving the problem data in an efficient, distributed manner. In a heterogeneous networked environment, such interfaces could provide seamless access to computational engines that would be invoked selectively for different parts of the user's computation according to which machine is most appropriate for a particular subproblem.

4. Domain-specific problem solving environments, such as those for structural analysis. Environments like Matlab and Mathematica have proven to be especially attractive for rapid prototyping of new algorithms and systems that may subsequently be implemented in a more customized manner for higher performance.

5. Reusable templates (i.e., users adapt "source code" to their particular applications). A template is a description of a general algorithm rather than the executable object code or the source code more commonly found in a conventional software library. Nevertheless, although templates are general descriptions of key data structures, they offer whatever degree of customization the user may desire.

Novel user interfaces that hide the complexity of scalable parallelism will require new concepts and mechanisms for representing scientific computational problems and for specifying how those problems relate to each other. Very high level languages and systems, perhaps graphically based, not only would facilitate the use of mathematical software from the user's point of view, but also would help to automate the determination of effective partitioning, mapping, granularity, data structures, etc. However, new concepts in problem specification and representation may also require new mathematical research on the analytic, algebraic, and topological properties of problems (e.g., existence and uniqueness).

We have already begun work on developing such templates for sparse matrix computations. Future work will focus on extending the use of templates to dense matrix computations.

We hope the insight we gained from our work will influence future developers of hardware, compilers and systems software so that they provide tools to facilitate development of high quality portable numerical software.

The EISPACK, LINPACK, LAPACK, BLACS and SCALAPACK linear algebra libraries are in the public domain. The software and documentation can be retrieved from *netlib* (http://www.netlib.org).

References

1. Anderson, E., Bai, Z., Bischof, C., Demmel, J., Dongarra, J., Du Croz, J., Greenbaum, A., Hammarling, S., McKenney, A., Ostrouchov, S., Sorensen, D.: LAPACK Users' Guide, Second Edition. SIAM, Philadelphia, PA, 1995.

2. Lawson, C., Hanson, R., Kincaid, D., Kincaid, D., Krogh, F.: Basic Linear Algebra Subprograms for Fortran Usage. ACM Transactions on Mathematical Software, 5:308–323, 1979.

3. Dongarra, J., Du Croz, J., Hammarling, S., Hanson, R.: Algorithm 656: An extended Set of Basic Linear Algebra Subprograms: Model Implementation and Test Programs. ACM Transactions on Mathematical Software, 14(1):18–32, 1988.

4. Dongarra, J., Du Croz, J., Duff, I., Hammarling, S.: A Set of Level 3 Basic Linear Algebra Subprograms. ACM Transactions on Mathematical Software, 16(1):1–17, 1990.

5. Anderson, E., Dongarra, J.: Results from the initial release of LAPACK. LAPACK working note 16, Computer Science Department, University of Tennessee, Knoxville, TN, 1989. institution = "Computer Science Department, University of

6. Anderson, E., Dongarra, J.: Evaluating block algorithm variants in LAPACK. LAPACK working note 19, Computer Science Department, University of Tennessee, Knoxville, TN, 1990.

7. Choi, J., Dongarra, J., Walker, D.: Parallel matrix transpose algorithms on distributed memory concurrent computers. In Proceedings of Fourth Symposium on the Frontiers of Massively Parallel Computation (McLean, Virginia), pages 245–252. IEEE Computer Society Press, Los Alamitos, California, 1993. (also LAPACK Working Note #65).

8. Dongarra, J., Whaley, R.C.: A User's Guide to the BLACS v1.0. Technical Report UT CS-95-281, LAPACK Working Note #94, University of Tennessee, 1995.

9. Choi, J., Demmel, J., Dhillon, I., Dongarra, J., Ostrouchov, S., Petitet, A., Stanley, K., Walker, D., Whaley, R.C.: ScaLAPACK: A Portable Linear Algebra Library for Distributed Memory Computers - Design Issues and Performance. Technical Report UT CS-95-283, LAPACK Working Note #95, University of Tennessee, 1995.

10. Choi, J., Dongarra, J., Ostrouchov, S., Petitet, A., Walker, D., Whaley, R.C.: A Proposal for a Set of Parallel Basic Linear Algebra Subprograms. Technical Report UT CS-95-292, LAPACK Working Note #100, University of Tennessee, 1995.

11. Dongarra, J., Duff, I., Sorensen, D., Van der Vorst, H.: Solving Linear Systems on Vector and Shared Memory Computers. SIAM Publications, Philadelphia, PA, 1991.

12. Message Passing Interface Forum. MPI: A Message Passing Interface Standard. International Journal of Supercomputer Applications and High Performance Computing, 8(3–4), 1994.

13. Geist, A., Beguelin, A., Dongarra, J., Jiang, W., Manchek, R., V. Sunderam, V.: PVM: Parallel Virtual Machine. A User's Guide and Tutorial for Networked Parallel Computing. The MIT Press, Cambridge, Massachusetts, 1994.

14. Koebel, C., Loveman, D., Schreiber, R., Steele, G., Zosel, M.: The High Performance Fortran Handbook. The MIT Press, Cambridge, Massachusetts, 1994.

15. Wilkinson, J., Reinsch, C.: Handbook for Automatic Computation: Volume II - Linear Algebra. Springer-Verlag, New York, 1971.

Highly Parallel Concentrated Heterogeneous Computing *

Ya.I. Fet and A.P. Vazhenin

Computing Center of SD RAS, 6 Lavrentiev Ave.,
Novosibirsk, 630090, Russia

Abstract. In this paper the possibilities are discussed of organizing heterogeneous computing in the so-called Combined Architecture systems consisting of a basic host subsystem (a massively parallel computer), and a set of high-performance specialized parallel coprocessors (hardware modules) executing the main workload. To optimize the choice of hardware modules, a classification of massive computations is suggested, based on the notion of processing types, which correspond to the character of data processing. A technique of parallel programming for the suggested concentrated heterogeneous systems is introduced, ensuring close matching of the tasks to the hardware modules.

1 Introduction

During the last decade, the most impressive gain in computer performance has been made by raising the *level of parallelism* and using of microprocessors with improved *physical characteristics*. Such important issue as the *specialization* of hardware was rarely addressed, though it presents a very significant source of further improvement of computer systems performance.

As a matter of fact, the most powerful contemporary supercomputers attain striking values of *peak performance*. However, of practical interest is the *real performance* reachable on important applications. Usually one has to make every effort in order to "embed" his algorithms and problems into the given system architecture.

The real problems are usually non-uniform. For different problems and various fragments of the same problem it is expedient to use features of different architectures. Utilization of a rigid architecture leads to a large gap between the peak and the real performance.

Recently, a wide interest was attracted by *Heterogeneous Computing* [1], which implies a *distributed* network system of several *commercially available* computers of diverse architectures. In such environment, the user is able to vary flexibly the style of programming, in accordance with the characteristics of his problems.

At the same time, the network approach in organizing heterogeneous computing seems to have a number of shortcomings (see, for instance, [2]):

---

* This research was partially supported by a Russian Foundation for Basic Research grant No.94-01-00574

1. Complete computers of serial production are used, each of them being not perfectly suited for the specific application problem.

2. Exploiting the networks, as well as the node supercomputers, involves large overheads.

3. The internode data transfer through the networks usually causes considerable delays.

In this paper a different approach to heterogeneous computing is suggested based on so-called *combined architecture* [3,4], in which different kinds of processing are executed by corresponding dedicated accelerators. Now, accelerators are widely used in computing systems for increasing their performance. However, nearly in all cases, the designers come merely to floating-point arithmetic processors, which are in essence *weakly* specialized.

In contrast with the known distributed heterogeneous systems, the combined architecture allows for building of Concentrated Heterogeneous Systems (CHSs).

The distinguishing characteristics of the present paper are as follows:

1. Use of a highly parallel SIMD computer as a basic (host) subsystem of combined architecture.

2. Broad application of *strongly* specialized accelerators (*hardware modules*) of different intentions and various hardware structures including parallel, pipeline, systolic, associative, etc.

3. Systematic approach to the selection of accelerators based on the classification of massive procedures according to so-called *processing types*.

4. Use of a definite technique of flexible heterogeneous programming ensuring the decomposition of a given application problem into processing types, with subsequent mapping of the designed algorithm onto the available set of hardware modules of a specific system.

5. Balanced interaction of various subsystem of the combined architecture.

In Section 2 the main features of the combined architecture are described. Section 3 is devoted to the issues of classification of massive parallel procedures. In Section 4 some examples of typical hardware modules are presented. The suggested technology of programming in CHSs is described in Section 5. In Section 6 some conclusions are made.

2 Combined Architecture

We call *combined architecture* a cooperation of a highly parallel host computer with a set of specialized processors. In this architecture, solving of any problem is considered as interaction of several processes, so that execution of each process is delegated to a specialized subsystem, most efficient in implementation of this process. The subsystems are controlled in such a way that their balanced operation might be ensured, and special complementing features of subsystems might be best exploited. For each subsystem a structure is chosen which best corresponds to the function it should perform.

At choosing the architecture of subsystems, the following considerations should be taken into account.

In the combined architecture (Figure 1), the main working load of the processing is delegated to the coprocessors. Hence, the requirements to the performance of the basic computer can be moderate. However, in order to ensure the effective interaction between the subsystems and the necessary data flows, this computer should have a sufficiently high degree of parallelism, a large capacity of the main memory, and an adequate software.

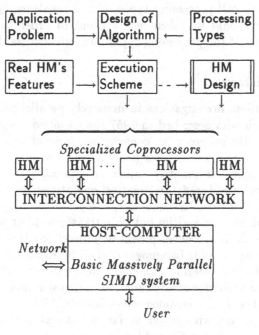

Figure 1

At the same time, in view of the general objectives of the system, extremely high demands should be made to the performance of each coprocessor. It means that special care is needed in selection of the structures of coprocessors. The most suitable architecture for the basic subsystem seems to be the fine-grained SIMD similar to DAP, or CM. These computers fit the above mentioned requirements (memory size, parallelism, etc.), while they are much cheaper of the other parallel systems because of using simple single-bit processing elements.

Looking at the present development of supercomputers, one can discover two important trends: increasing of the performance of processing elements, and extending of allowable programming styles. The combined architecture pertains to both these trends: the problem-oriented coprocessors of different structure provide a multiarchitectural environment, while the strong specialization and the massive parallelism of these processors ensure high performance.

3 Classification

The novelty of the present approach is that the specific type, or technique, of processing necessary for efficient execution of the most labor-intensive procedures involved in the implementation of a problem is used as a criterion for the selection of appropriate hardware architecture. As a rule, similar techniques are encountered as well in solving problems of other classes.

Analyzing the common numerical methods, algorithms, and programming languages, one can select a set of recurring general techniques, or styles of data treatment which we call processing types. It such analysis, specific features of existing hardware should be taken into account in order to make possible a reasonable mapping of the processing types onto efficient hardware modules.

3.1 A short history

One of the earliest investigations in massively parallel processing is due to Leonid Kantorovich who described in 1957 the so-called "large-block programming system" [5]. He proposed to consider as basic objects operated by the system ordered sets called *quantities* (such as vectors, matrices, etc.), a single number being the simplest quantity, called an *element*. Some special operations on quantities were introduced: *arithmetical* operations as extensions of usual arithmetic on any element of the quantity, and *geometrical* operations which do not change the values of quantities but only transform their structures.

A significant event in the development of parallel processing was the appearance of APL (A Programming Language) devised by Kennet Iverson [6]. In APL, the variables are *logical, integral, numerical, and arbitrary*. They can be *scalars* as well as rectangular *arrays* of any rank and dimensionality.

The following types of operations are defined in APL:

1. *Scalar* operations, with scalar variables both as arguments and results. The scalar operations are subdivided into *unary* and *binary*.

2. *Compound* operations, being an extension of scalar operations to arrays. Four kinds of extensions are provided: *component-wise processing, reduction, inner product*, and *outer product*.

3. *Mixed* operations, in which the ranks of the arguments and the results are different. These operations serve mostly for transforming the array structure.

Later on, some massive, large-block operations were further developed in such programming languages as PL-1, Algol-68, etc.

The development of new parallel systems stimulated more precise classification of massive operators, for better exploiting of the advantages of different architectures. One attempt to construct such a classification was made in [7], where five classes of operators were selected: *numerical component-wise processing, numerical reductive processing, logical component-wise processing, logical reductive processing*, and *data structure transformations*.

Recently, a valuable contribution to the classification of massively parallel operations has been made by Guy Blelloch [8]. In his book, a set of *primitive instructions* is discussed for the so-called "Scan Vector Model". Three classes of primitive instructions, relying to a considerable extent on the APL operations, are defined: *scalar, vector,* and *vector-scalar* instructions. The vector instructions are, in turn, divided into *element-wise, scan,* and *permutation* instructions.

It should be noted that Blelloch's primitives are virtually directed toward a definite (though ruther flexible) architecture of the Connection Machine. Following our approach, in selecting of the typical massive procedures to define corresponding processing types one should not limit himself by the features of a specific computer system, but rather examine a broad variety of existing and prospective hardware structures.

3.2 Classification of problems

Among the numerous problems requiring the power of supercomputers to be solved in adequate time, several prevailing classes can be chosen constituting the main part of the computer charge. These problems belong basically to the following two groups.

A. Numerical problems: Numerical approximation, Linear algebra, Ordinary differential equations, Equations of mathematical physics, Numerical simulation, Discrete transforms, Combinatorial problems, Error analysis, Computer graphics and geometry, Signal and image processing, Simulation of complicated objects.

B. Non-numerical problems: Searching and sorting, Symbolic processing, Text processing, Databases, Operational systems, Artificial intelligence (Production systems, Logical inference systems, Pattern and speech recognition, Genetic algorithms, Neurocomputing, Robotics, Computer-aided design).

Of course, this list of problems is incomplete, and needs further improving. Nevertheless, we suppose that the enumerated problems cover the needs of the most important applications. Hence, the analysis of the processing types involved in these problems may be ruther representative.

3.3 Processing types

Our goal here is to provide a classification of processing styles confronting them with the known hardware structures, in order to find a reasonable mapping: $PROCESSING\ TYPE \rightarrow HARDWARE\ MODULE$. Indeed, the variety of styles, or techniques involved in machine realization of different application problems is not too large.

We will describe the Processing Type (PT) as a three-tuple:

$$PT = \{\ A_1,\ A_2,\ T\ \},$$

where $A_1(A_2)$ corresponds to the data type of the first (second) argument, and T corresponds to the transformation to be executed upon these arguments.

In massive procedures, the terms A_1 and A_2 usually take the values "Vector (V)", "Scalar (S)", and "Binary (B)". Other possible types of arguments can be: "Array", "Set", "Relation", "Tree", etc. Examples of transformations T are: "Arithmetic (A)", "Logic (L)", "Permute (P)".

Table 1 shows the notations of the most common processing types. In this Table, some HMs are also shown suitable for implementation of different PTs. Of course, this list of basic processing types needs further extentions and corrections.

Table 1

A_1	A_2	T	PT	HM(s)
Vector	Vector	Arithmetic	**VVA**	Pipeline, PVM, PVA, Systolic
Vector	–	Arithmetic	**VoA**	Pipeline, PVM, PVA, Systolic
Vector	Vector	Logic	**VVL**	Set Intersection Processor
Vector	Scalar	Arithmetic	**VSA**	Pipeline
Vector	Scalar	Search	**VSS**	Associative Processor
Vector	Interval	Search	**VIS**	Associative Processor
Vector	Vector	Search	**VVS**	Set Intersection Processor
Vector	–	Order	**VoO**	Sorting Network, Systoic
Vector	–	maX	**VoX**	Extremum Selector
Vector	–	miN	**VoN**	Extremum Selector
Vector	Vector	Permute	**VVP**	Permutation Network
Vector	–	Permute	**VoP**	Permutation Network
Vector	–	Compress	**VoC**	Digital Compressor
Vector	–	Expand	**VoE**	Permutation Network
Vector	–	Logic	**VoL**	Associative Processor
Binary	Binary	Logic	**BBL**	Associative Processor
Binary	Binary	Permute	**BBP**	Permutation Network
Binary	–	Permute	**BoP**	Permutation Network

4 Specialized hardware modules

We have already mentioned that the combined architecture is an open system capable of including a variety of hardware modules of different architectures. Further investigations in the computational requirements set by problems, as well as the progress in hardware technology, will supplement the assortment of hardware modules with novel elaborate devices.

The goal of this Section is to give some examples of existing hardware modules.

Vector Pipeline Architecture. These devices are widely adopted in modern supercomputers (Cray, Convex, Fujitsu FACOM VP, and others), as well as in arithmetical accelerators (Weitek, Fujitsu μVP, etc.). The vector pipeline devices are appropriate for implementation of processing types VVA, VSA, VoA.

Systolic/Wavefront Arrays. These devices are locally connected networks of homogeneous cells, with a regular directed flow of data and results (see, for instance, [9]). Note, that a systematic method for design and optimization of systolic arrays has been developed in [10]. The systolic/wavefront processors fit well to implementation of processing types VVA, VoA, VoO, etc.

Permutation Networks. The permutation network (also called an interconnection network or a commutator) is now an indispensable component of any parallel computing system because of the necessity to provide fast data transfer between different nodes of the system. The permutation network may be considered as a specialized processor embedded into proper system. The most popular permutation networks are: the suffle/exchange network, the Ω-network, the Data Manipulator, the Flip network (see, for instance, [11]). Simple meshes are often used as permutation networks: the NEWS-grid (a rectangular four-neighbour mesh), and the X-grid (an eight-neighbour mesh). In the CM-5 system a novel efficient network was implemented called a *fat-tree* [12]. The permutarion networks realise the processing types VVP, VoP, VoE.

Sorting Devices. A variety of methods and tools of hardware sorting have been discussed in the literature (see, for instance, [13]). A remarkable achievement in this field was Batcher's *sorting network* [14], which is capable to implement the ordering of an N-element array with the time estimation of the order $\log_2^2 N$. Note also, that systolic arrays can be efficiently used for sorting. The sorting devices implement processing types VVP, VoO.

Associative Array Processors. Different Associative Array Processors (AAPs) were developed to implement non-numerical problems (database machines, information systems, etc.). An extensive literature is devoted to the architecture and applications of AAPs (see, for instance, [15]). The operation of AAP is based on the principle of content-addressable memory (CAM). As it was shown by various authors, the AAP can realize different algorithms and problems. However, they are most efficient for the processing types VVL, VSS, VIS, VoL, BBL.

Distributed Functional Structures. The CAM represents a strongly specialized processor oriented to a very important processing type VSS. Here, the algorithm of associative search is simulated in the course of signal flow along a specific distributed logical net of the CAM matrix. This kind of processing might be called *quasi-analogue simulation*. Different logical nets possessing such properties we call *Distributed Functional structures (DF-structures)* [16]. In [16] numerous examples of efficient hardware modules based on DF-structures are also described, for instance, Extremum Selector, Digital Compressor, Pipelined Vertical Adder (PVA), Pipelined Vertical Multiplier (PVM), Set Intersection Processor, etc.

5 Programming of CHS

5.1 Techique of problem solving

Prior to be executed in suggested CHS, a problem needs definite preparations taking into account the features of this system. The main phases of this work are shown in Figure 1. They are:

1. Design of a "coarse-grained" algorithm by means of decomposition of the given problem into separated procedures (steps), thinking of the PT classification (Table 1).

2. Justification of this algorithm by comparing the required PTs with the assortment of the available HMs of specific CHS.

3. Assigning of PTs to corresponding HMs. If there is no direct hardware support for some PT (absence of corresponding HM), this PT is realized either by combining operation of several existing HMs, or by the resources of the host.

4. Optimization of the execution scheme as a whole, and assembling of the final program taking into account the necessary consistency of data structures and dimensionalities in various procedures, as well as the synchronization of HMs operation.

Evidently, the CHS software should also provide for designing host programs responsible for the control and data exchange in the system.

In this Section the main principles are discussed of a language for the description of algorithms in CHS environment. Such a language should reflect the features of the basic fine-grained SIMD subsystem, as well as of its extension by the set of coprocessors. Some constructs of such language called VEPRAN were described in [17]. Here, we propose an extension of VEPRAN taking into account all requirements of the combined architecture and the classification of PTs.

5.2 A program model of CHS

The fine-grained SIMD architecture is notable for a huge number of simple single-bit processing elements (PEs) working synchronously under the control of a common program unit. Each PE has its own one-bit word local memory, executes bit-sequential data processing, and communicates with other PEs via interconnection network. The aggregate memory of the system can be considered as a bit matrix each row of which is connected with a corresponding PE. Then, the proper processing presents a sequence of bit operations on the binary vectors (slices) fetched/loaded from/in memory.

Three types of slices are usually distinguished in SIMD systems: **vertical** (columns), **word** (segments of a row), and **complex** (different from the first two). Most frequently are used the vertical slices, or simply **slices**. That is why we call such architectures Vertical Processing Systems [18]. A submatrix formed from adjacent vertical slices is called a **field**. In accordance with the type of processed data, the fields may contain numerical vectors, matrices, relational tables, etc. Each PE contains necessary logic circuits and several flip-flops to store the intermediate results of the bit operations. All PE flip-flops of the same name form specific bit slices considered as programming operational registers (PORs). The input and output buses of the HMs, included in the CHS, can be viewed in the same way. Thus, the extension of the basic SIMD system by co-processors, from the programmer's point of view, can be considered as enlarging of the number of PORs thus saving the style of SIMD computations.

5.3   VEPRAN-language

In VEPRAN language, suggested for the description of algorithms, not only the data types like **integer, index** (unsigned), **float, double, char, logical, structure**, but also their location in the system is used. So, the declaration **scalar** implies that data are placed in the control memory, while declarations **slice** (locate a slice) and **field** (locate a field) serve for the description of data placed in the parallel memory. The modes of data arranging within the fields of the parallel memory are declared by the following language construct:

place $< object >$ **in field** $< field[elements]$ **by coord** $< c_1, c_2, \cdots, c_n >$.

The following example:

    scalar index n;

    integer $A[n, n]$;

    field integer C;

    place $A[n, n]$ **in field** $C[n * n]$ **by coord** 1;

implies that the matrix A should be placed by columns in n^2 elements of an one-dimensional field C. The next instructions are an example of placement in the parallel memory of n rows of a relational table:

    scalar index n, l;

    structure $Person[n]\{$**char** $Name[8]$, **logical** Sex,

                              index Age, **char** $City[8]\}$;

    index $AgeData[l]$;

    field structure $PersonTable\{$**field char** $Name[8]$, **slice** Sex,

                              field index Age, **field char** $City[8]\}$;

    place $Person[n]$ **in field** $PersonTable[n]$ **by coord** 1;

In this case, the operator

    place $AgeData[l]$ **in field** $PersonTable.Age[n]$ **by coord** 1;

can be considered as placing of l values of the attribute Age in the subfield Age of the field $PersonTable$.

The operations on scalar data are indicated by a sign ":=". Parallel operations are indicated by "$\leftarrow$", identified with corresponding PT, and belong to the following modifications:

    $< dest > \leftarrow < \mathbf{PT} >$ - for all elements of field;

    $< dest > \leftarrow [slice] \leftarrow < \mathbf{PT} >$ - for elements unmasked by $slice$;

    $< dest[index] > \leftarrow < \mathbf{PT} >$ - for selective loading.

The control constructs in VEPRAN:

        do $< operations >$ **enddo**;

        if $(< condition >)$ **then** $< do - part >$ **else** $< do - part >$;

> **for** ($< operation >$, $< condition >$, $< operation >$) $< do - part >$;
> **while** ($< condition >$) $< do - part >$;
> **procedure** ($< formal\ parameters >$);
> **call** ($< real\ parameters >$);

are similar to those used in traditional languages. It should be noted only that the statement $< condition >$ may include parallel operations and be identified with PT.

5.4 Example CHS algorithms

Consider a well-known problem of matrix multiplication $\mathbf{C} = \mathbf{AB}$.

To solve this problem, n^2 inner products should be computed (where n is the order of the matrices). In our case, the inner product is a basic PT denoted **VVA** in Table 1. The detailed elaboration of the algorithm depends on the relationship between the order n and the level of system parallelism (that is the number of processing channels) m. If $m \geq n^2$, simultaneous computing is possible of n inner products, and the CHS algorithm takes the form shown in Figure 2.

```
procedure MulMatr1(A, B, C, n);
/* Algorithm 1: matrix multiplication for m ≥ n² */
    scalar index n;
    integer A[n, n], B[n, n], C[n, n];
    scalar index k;    /* the work index */
    integer field A, B, C, D;    /* define fields */
    slice S;    /* define work slice */
    place A[n, n] in field A[n * n] by coord 2;    /* locate A by rows */
    place B[n, n] in field B[n * n] by coord 1;    /* locate B by columns */
    S ← BBL[S, 0, AND];    /*clear slice of masks */
    for (k := 0; k ≤ n - 1; k := k + 1)    /* set masks*/
        do S[k * n + k] ← BBL[S, 1, OR]; enddo
    for (k := 1; k ≤ n; k := k + 1)
        do
            D ← VVA[A, B, *];    /* D ← A * B */
            D ← VoA[D, sum(n, n)];    /* compute n sums */
            C ← [S] ← VoP[D, none];    /* save n elements of C */
            B ← VoP[B, cshift(n * n, n)];    /* exchange columns of B */
            S ← BoP[S, cshift(n * n, n)];    /* exchange bits of mask S */
        enddo
end procedure
```

Figure 2. Matrix multiplication in CHS

¿From this example, it is easy to pick out the necessary PTs which can be implemented as specialized hardware modules.

As it was shown in [19], when using only the basic SIMD computer with an interconnection network of the Flip, or Hypercube type, the execution of the above algorithm requires $O_A(n \log_2 n)$ arithmetic operations and $O_B(n(s^2 + s \log_2 n))$ bit operations (where s is the wordlength of the data). This is because the procedures concerned with the PTs $\mathbf{VVA}[\cdots, \cdots, *]$ and $\mathbf{VoA}[\cdots, sum(\cdots, n)]$ are realized in the basic architecture in $O_B(s^2)$ and $O_B(s \log_2 n)$ correspondingly.

The use of strongly specialized HMs can sufficiently improve these estimations. Thus, in [17] was shown that application of coprocessors PVM and PVA (see Section 4) ensures execution of the PTs $\mathbf{VVA}$ and $\mathbf{VoA}$ in $O_B(s)$ and $O_B(s + \log_2 n)$ bit operations. Moreover, one can organize a programmed "chaining" of these two HMs, thus supporting the pipelining of computations. In this case, the described algorithm can be implemented in $O_B(n(s + \log_2 n))$ and $O_B(n^2(s + \log_2 n))$ bit operations, correspondingly.

6 Conclusion

In this paper a new approach is suggested to the organization of heterogeneous computing. This approach is based on the conception of combined architecture, which is a composition of a fine-grained SIMD computer with a set of high-performance strongly specialized processors.

The presence of a number of accelerators of various architectures at one site allows to organize heterogeneous computing within a single system. In contrast to the existing distributed heterogeneous systems, the proposed concentrated heterogeneous system does not need for high-bandwidth communications networks, and does not suffer from the delays arising in these networks at the data transfer. The accelerators (hardware modules) of the combined architecture are much cheaper than complete supercomputers and, at the same time, they can be better oriented to the necessary programming styles.

To optimize the choice of hardware modules, a classification of massive computations is suggested, based on the notion of processing types, which correspond to the character of data processing and are applicable for hardware implementation of different classes of problems. Several examples of efficient hardware modules of diverse architecture are listed.

A technique of parallel programming for the combined architecture is introduced based on the language VEPRAN extended by appropriate means to specify the processing types and to control the movement of parallel data in the given concentrated heterogeneous system. Using this language, one can analyze the algorithm of solution of any given problem and present it as a sequence of processing types. In course of the execution of such a program, each processing type is assigned to a definite hardware module (or a combination of some modules) taken from the set of real modules included in the system.

The combined architecture is considered as an open system which can be supplemented by additional hardware modules, according to the requirements of the user. The suggested approach is expected to help to design a family of

188

cost-effective supercomputers providing flexible programming, as well as high performance in a broad range of applications.

References

1. Khokhar A.A., et al. Heterogeneous computing: challenges and opportunities, *Computer*, Vol.26, No.6, pp.18-27, 1993.
2. Freund R.F. and Siegel H.J. Heterogeneous processing, *Computer*, Vol.26, No.6, pp.13-17, 1993.
3. Vazhenin A.P., Sedukhin S.G., Fet Ya.I. High-performance computing systems of combined architecture, *In: "Parallel Computing Technologies (PaCT-91)", Novosibirsk, Russia, 1991)*, Singapore, World Scientific, pp. 246-257, 1991.
4. Fet Ya.I. and Vazhenin A.P. Heterogeneous processing: a combined approach, *In: "Workshop on Parallel Scientific Computing (PARA'94-L)", 1994, Lingby, Denmark*, (LNCS, Vol.879), Berlin, Springer-Verlag, pp. 194-206, 1994.
5. Kantorovich L.V. On a system of mathematical symbols, convinient for electronic computer operations, *In: Dokl. Akad. Nauk SSSR*, Vol.113, pp.738-741, 1957. (In Russian).
6. Iverson K.E. *A Programming Language*, New York - London, Wiley, 1962.
7. Fet Ya.I. Hardware support of massive computations, *Optimization*, Novosibirsk, Inst. of Mathematics, Siberian Div. of the USSR Acad. Sci., No.22(39), pp.115-126, 1978. (In Russian).
8. Blelloch G.E. *Vector Models for Data-Parallel Computing*, Cambridge, Mass., MIT Press, 1990.
9. Kung H.T. Why systolic architectures?, *Computer*, Vol.15, No.1, pp.37-46, 1982.
10. Sedukhin S.G. and Sedukhin I.S. An interactive graphic CAD tool for the synthesis and analysis of VLSI systolic structures, *In: Parallel Computing Technologies (PaCT-93), Obninsk, Russia, 1993*, Moscow, ReSCo J.-S. Co., pp. 163-175, 1993.
11. Broomel G. and Heath J.R. Classification categories and historical development of circuit switching topologies, *ACM Computing Surveys*, Vol.15, No.2, pp.95-133, 1983.
12. Hillis W.D. and Tucker L.W. The CM-5 Connection Machine: a scalable supercomputer, *Comm. ACM*, Vol.36, No.11, pp.31-40, 1993.
13. Knuth D. *The Art of Computer Programming, Vol.3, Sorting and Searching*. New York, Addison-Wesley, 1973.
14. Batcher K.E. Sorting networks and their applications, *In: AFIPS Confer. Proc., 1968 SJCC*, Vol.32, pp.307-314, 1968.
15. Foster C.C. *Content Addressable Parallel Processors*, New York, Van Nostrand Reinhold, 1976.
16. Fet Ya.I. *Parallel Processing in Cellular Arrays*, Tounton, UK, Research Studies Press, 1995.
17. Vazhenin A.P. Hardware and algorithmic support of high-accuracy computations in vertical processing systems, *In: Parallel Computing Technologies (PaCT-93), Obninsk, Russia, 1993*, Moscow, ReSCo J.-S. Co., pp. 149-162, 1993.
18. Fet Ya.I. Vertical processing systems: a survey, *IEEE Micro*, Vol.15, No.1, pp.65-75, 1995.
19. Vazhenin A.P. Efficient high-accuracy computations in massively parallel systems, *In: "Workshop on Parallel Scientific Computing (PARA'94-L)", 1994, Lingby, Denmark*, (LNCS, Vol.879), Berlin, Springer-Verlag, pp. 505-519, 1994.

Adaptive Polynomial Preconditioning for the Conjugate Gradient Algorithm

Martyn R. Field

Hitachi Dublin Laboratory, O'Reilly Institute, Trinity College, Dublin, Ireland

Abstract. For the parallel conjugate gradient algorithm polynomial preconditioners are more suitable than the more common incomplete Cholesky preconditioner. In this paper we examine the Chebyshev polynomial preconditioner. This preconditioner is based on an interval which approximately contains the eigenvalues of the matrix. If we know the extreme eigenvalues of the matrix then the preconditioner based on this interval minimises the condition number of the preconditioned matrix. Unfortunately this does not minimise the number of conjugate gradient iterations. We propose an adaptive procedure to find the interval which gives optimal rate of convergence. We demonstrate the success of this adaptive procedure on three matrices from the Harwell-Boeing collection.

1   Introduction

In the numerical modelling of many scientific applications the most time consuming part of the algorithm is the solution of large systems of linear equations of the form

$$Ax = b,$$

where A is an $N \times N$ matrix and x and b are N-vectors. For a large number of these problems, e.g. FEM of static structural analysis problems, the matrix A is sparse, symmetric and positive definite. For these problems the most efficient and popular iterative solver is the conjugate gradient (CG) algorithm which was first proposed by Hestenes and Stiefel [6] and later popularised as an iterative method by Reid [9]. The performance of this algorithm can be greatly enhanced with the use of a preconditioner. The simplest preconditioners are basic iteration techniques such as Jacobi (diagonal scaling) and SSOR. One of the most widely used preconditioners is the incomplete Cholesky factorisation which was first proposed as a preconditioner by Meijerink and van der Vorst [8]. On sequential machines this has proven to be an excellent preconditioner but on a parallel machine it is difficult to implement efficiently. The solution of triangular systems which are the major part of applying the incomplete Cholesky preconditioner do not parallelise well.

In this paper we will consider an alternative group of preconditioners, namely, polynomial preconditioners. These use a polynomial in the matrix A to approximate the inverse of the matrix. Therefore they can be implemented using a series of matrix vector products. The advantage of this is that on a parallel machine the vector dot product operation is often the bottleneck whereas the matrix vector product can be efficiently carried out in parallel. Therefore the polynomial preconditioner reduces the total iteration time by reducing the number of iterations, and hence the number of vector products, at the cost of a few extra matrix vector products per iteration.

An outline of this paper is as follows. Firstly we describe the Chebyshev polynomial preconditioner. Next we discuss why we need an adaptive procedure to improve the convergence of one of these polynomial preconditioners and describe our new adaption procedure. We then test this adaption procedure on a number of test problems. Finally we draw some conclusions from this work and suggest some possible future improvements.

2 Polynomial Preconditioners

The idea of polynomial preconditioners is to find polynomial $C(A)$ which approximates the inverse of A. The simplest choice for $C(A)$ is based on the Neumann series

$$A^{-1} = (I - N)^{-1} = I + N + N^2 + N^3 + \ldots,$$

where $N = I - A$. This series converges if the spectral radius of N is less than one. The polynomial preconditioner is formed by truncating this series and has the advantage that it does not require the calculation of any coefficients. Unfortunately in general this is a poor preconditioner and for a given degree of polynomial there are much better choices.

Suppose the eigenvalues of A lie in the interval $[\lambda_1, \lambda_N]$ ($0 < \lambda_1 < \lambda_N$, since A is positive definite) then $C(\lambda)$ should approximate λ^{-1} on this interval. We define the 'best' polynomial to be the one which solves the problem

$$\min_{C(\lambda) \in \Pi_m} \|1 - C(\lambda)\lambda\|, \tag{1}$$

where Π_m is the set of polynomials of degree m and the norm is defined on the interval $[\lambda_1, \lambda_N]$.

2.1 Chebyshev Polynomial Preconditioner

The norm we consider is the uniform norm

$$\|f(\lambda)\| = \max_{\lambda \in [c,d]} |f(\lambda)|, \tag{2}$$

where $[c, d]$ is an approximation to the interval which contains the eigenvalues of A. The solution of (1) in this norm is a scaled and shifted Chebyshev polynomial

$$C_m(\lambda)\lambda = 1 - T_{m+1}\left(\frac{d + c - 2\lambda}{d - c}\right) \bigg/ T_{m+1}\left(\frac{d + c}{d - c}\right), \tag{3}$$

where $T_{m+1}(x)$ is the $(m + 1)$th Chebyshev polynomial of the first kind on the interval $[-1, 1]$. The Chebyshev polynomials have a number of nice properties which make them suitable for preconditioning. Firstly, they satisfy a three term recursion formula

$$T_{k+1}(x) = 2xT_k(x) - T_{k-1}(x),$$

where $T_0(x) = 1$ and $T_1(x) = x$. This means they can be applied efficiently. Secondly, the minimax property of the Chebyshev polynomials which states that the polynomial oscillates between -1 and 1, means that the preconditioned polynomial $P_{m+1}(\lambda) = C_m(\lambda)\lambda$ equioscillates about 1 on the interval $[c, d]$. Hence, Chebyshev polynomial preconditioning gives unbiased suppression of eigenvalues and is therefore well suited to matrices which have uniform distribution of eigenvalues as is typically the case when discretising PDE's. More specifically if the spectrum of A, $\sigma(A) \subset [\lambda_1, \lambda_N]$, lies entirely within the interval $[c, d]$ (i.e. $0 < c \le \lambda_1 < \lambda_N \le d$) then the spectrum after preconditioning, $\sigma(P_{m+1}(A))$, is uniformly distributed in the interval $[1 - \epsilon_{m+1}, 1 + \epsilon_{m+1}]$ where $\epsilon_{m+1} = 1/T_{m+1}\left(\frac{d+c}{d-c}\right)$. We can therefore give a bound on the condition number of the preconditioned matrix

$$\kappa(P_{m+1}(A)) \le \frac{1 + \epsilon_{m+1}}{1 - \epsilon_{m+1}}.$$

If $c = \lambda_1$, $d = \lambda_N$ and m is even then $P_{m+1}(\lambda_1) = 1 - \epsilon_{m+1}$ and $P_{m+1}(\lambda_N) = 1 + \epsilon_{m+1}$ so this bound is obtained.

3 Adaptive Polynomial Preconditioners

It has been observed [7], [10] that the conjugate gradient algorithm with the Chebyshev polynomial preconditioner converges in the fewest iterations when the polynomial is based on an interval $[c, d]$ which lies slightly inside the extreme eigenvalues of A (i.e. $\lambda_1 < c < d < \lambda_N$ where λ_1 and λ_N are the minimum and maximum eigenvalues of A). In general we will not know the extreme eigenvalues of A and even if we did there is no obvious way to calculate the optimal values for c and d. Therefore we wish to use an adaptive procedure to find c and d.

Firstly we note that the conjugate gradient algorithm is closely related to the Lanczos eigenvalue routine and therefore it is easy calculate successively better approximations to the extremal eigenvalues of the preconditioned matrix, $P_{m+1}(A)$. We know that the eigenvalues in the interval $[c, d]$ are mapped into the interval $[1 - \epsilon_{m+1}, 1 + \epsilon_{m+1}]$. Therefore, if $0 < \lambda_1 < c < d < \lambda_N$ and m is even then $P_{m+1}(\lambda_N) > 1 + \epsilon_{m+1}$ and $P_{m+1}(\lambda_1) < 1 - \epsilon_{m+1}$. If m is odd then both $P_{m+1}(\lambda_1)$ and $P_{m+1}(\lambda_N)$ are less than $1 - \epsilon_{m+1}$ so it is impossible to calculate estimates for λ_1 and λ_N. Therefore we take m to be even. If $\tilde{\mu}_1$ and $\tilde{\mu}_N$ are sufficiently accurate estimates to the minimum and maximum eigenvalues of the preconditioned matrix then we expect them to lie outside the interval $[1 - \epsilon_{m+1}, 1 + \epsilon_{m+1}]$ and consequently we can calculate estimates for λ_1 and λ_N. That is, if $\tilde{\mu}_1 < 1 - \epsilon_{m+1}$ then the estimate of the minimum eigenvalue of A is

$$\tilde{\lambda}_1 = \tfrac{1}{2}\left[(d + c) - (d - c)\cosh\left(\frac{1}{m+1}\cosh^{-1}\left(\frac{1 - \tilde{\mu}_1}{\epsilon_{m+1}}\right)\right)\right], \tag{4}$$

and if $\tilde{\mu}_N > 1 + \epsilon_{m+1}$ then the estimate of the maximum eigenvalue of A is

$$\tilde{\lambda}_N = \frac{1}{2}\left[(d+c) + (d-c)\cosh\left(\frac{1}{m+1}\cosh^{-1}\left(\frac{\tilde{\mu}_N - 1}{\epsilon_{m+1}}\right)\right)\right]. \qquad (5)$$

The simplest adaption strategy is to take $c = \tilde{\lambda}_1$ and $d = \tilde{\lambda}_N$ so the preconditioner converges to the exact Chebyshev polynomial preconditioner, this is similar to the procedure used in [1]. Although this choice minimises the condition number of the preconditioned matrix it does not necessarily give the minimum number of conjugate gradient iterations. The reason for this is that if, for example, $d = \lambda_N$ and the eigenvalues of A in the interval $[\lambda_1, c]$ are isolated then the conjugate gradient algorithm quickly picks these out and the minimum eigenvalue effectively becomes c. It has been found, [2], that although there is an optimal d less than λ_N it has a much smaller effect on the rate of convergence of the CG algorithm so we will assume that $d = \lambda_N$.

3.1 Adaption Based on Axelsson Analysis

Our adaption strategy is based on the convergence analysis of Axelsson [3]. We start by considering the effect of applying the Chebyshev polynomial preconditioner to a simple example. Suppose A is a matrix of size $N = 100$ and that the eigenvalues are evenly distributed between the minimum eigenvalue $\lambda_1 = 10^{-4}$ and the maximum eigenvalue $\lambda_N = 2$, i.e.

$$\lambda_i = \lambda_1 + \frac{i-1}{N-1}(\lambda_N - \lambda_1) \qquad \text{for} \qquad i = 1, \ldots, N.$$

Now suppose the Chebyshev polynomial preconditioner is of degree $m = 4$ and is based on the interval $c = 0.1$ and $d = 2$. Note that A has five eigenvalues less than c. In Fig. 1 we plot the eigenvalues of the preconditioned matrix. We can clearly see that the eigenvalues in the interval $[c, d]$ are mapped into the small interval, $[1 - \epsilon_{m+1}, 1 + \epsilon_{m+1}]$, about 1. The five eigenvalues of A outside this interval are mapped into the interval $[0, 1 - \epsilon_{m+1}]$ and as can be seen in Fig. 1 these eigenvalues are isolated. Increasing c reduces the size of the interval into which the majority of the eigenvalues of A are mapped while increasing the number of eigenvalues outside this interval. Reducing the size of the interval into which most of the eigenvalues are mapped reduces the effective condition number of the matrix in the CG algorithm once the isolated eigenvalues have been removed. Increasing the number of eigenvalues less than c increases the number of CG iterations to remove these eigenvalues. Also, if the number of eigenvalues outside the interval becomes too large then the CG algorithm will no longer treat them as isolated. The right combination of these two effects will give the minimum number of conjugate gradient iterations.

Let $\mathbf{e}_k = \mathbf{x} - \mathbf{x}_k$ be the error after k iterations of the CG algorithm and let $\|\mathbf{e}_k\|_A = \mathbf{e}_k^T A \mathbf{e}_k$ be the energy norm. Axelsson has shown that if A is a symmetric positive definite matrix with q small isolated eigenvalues

$$0 < \lambda_1 < \lambda_2 < \ldots < \lambda_q < c = \lambda_{q+1} \leq \ldots \leq \lambda_N = d$$

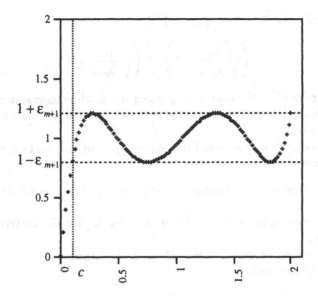

Fig. 1. Eigenvalues after applying Chebyshev polynomial preconditioner based on $[0.1, 2]$ where the eigenvalues are uniformly distributed between $\lambda_1 = 10^{-4}$ and $\lambda_N = 2$.

then the number of iterations required such that $\|e_k\|_A \leq \epsilon \|e_0\|_A$ is bounded by

$$k \leq \left(\log\left(\frac{2}{\epsilon}\right) + \sum_{i=1}^{q} \log\left(\frac{d}{\lambda_i}\right) \right) / \log\left(\sigma^{-1}\right) + q \tag{6}$$

where $\sigma = (\sqrt{\frac{d}{c}} - 1)/(\sqrt{\frac{d}{c}} - 1)$. This shows the balance between increasing c which decreases σ but increases q or decreasing c which decreases q but increases σ.

We wish to use this analysis in order to choose the lower endpoint of the interval, c, for the Chebyshev polynomial preconditioner such that it minimises this upper bound. Suppose after a few conjugate gradient iterations we have approximations $\tilde{\lambda}_1$ and $\tilde{\lambda}_N$ for the extremal eigenvalues of A. In order to calculate the bound (6) we also need estimates for the q smallest eigenvalues. Therefore, we make the assumption that the eigenvalues are uniformly distributed between $\tilde{\lambda}_1$ and $\tilde{\lambda}_N$, i.e. $\tilde{\lambda}_i = \tilde{\lambda}_1 + \frac{i-1}{N-1}(\tilde{\lambda}_N - \tilde{\lambda}_1)$. This assumption is a good approximation if the matrix A arises from a finite element or finite difference discretisation of a PDE. The eigenvalues of the preconditioned matrix are then $\tilde{\mu}_i = P_{m+1}(\tilde{\lambda}_i)$. Hence the upper bound on the number of iterations for a given c is

$$k^* = \left(\log\left(\frac{2}{\epsilon}\right) + \sum_{i=1}^{q} \log\left(\frac{\tilde{\mu}_N}{\tilde{\mu}_i}\right) \right) / \log\left(\sigma^{-1}\right) + q, \tag{7}$$

where q is such that $\tilde{\lambda}_q < c \le \tilde{\lambda}_{q+1}$ and

$$\sigma = \left(\sqrt{\frac{\tilde{\mu}_N}{\tilde{\mu}_{q+1}}} - 1 \right) \Bigg/ \left(\sqrt{\frac{\tilde{\mu}_N}{\tilde{\mu}_{q+1}}} + 1 \right).$$

We can minimise k^* with respect to q using a one dimensional nonlinear optimisation routine such as Brent's method [4] and this will give us a new estimate for the optimal c.

The adaption algorithm based on the analysis in this section can be summarise as follows:

1. Calculate eigenvalue estimates $\tilde{\mu}_1$ and $\tilde{\mu}_N$ for the preconditioned matrix, $P_{m+1}(A)$.
2. Calculate eigenvalue estimates $\tilde{\lambda}_1$ (4) and $\tilde{\lambda}_N$ (5) for the matrix A.
3. Set $d = \tilde{\lambda}_N$.
4. Find c which minimises (7) using Brent's algorithm.
5. Restart CG iteration.

The unknown in this adaption algorithm is the number of iterations between adaption steps and we will discuss this in the next section.

4   Numerical Experiments

We now look at how well these Chebyshev polynomial preconditioners perform in practice. The conjugate gradient algorithm with Chebyshev polynomial preconditioner and adaption procedures was implemented in Fortran 90 with double precision arithmetic. The numerical experiments were run on a on network of 4 HP 735/125 workstations. The convergence criteria for all problems is

$$(z_k^T r_k)^{\frac{1}{2}} \le \epsilon (z_0^T r_0)^{\frac{1}{2}}$$

where $r = b - Ax$ is the residual vector and $z = C_m(A)r$ is the preconditioned residual. The convergence tolerance ϵ is set to 10^{-6}. For all the problems we take the degree of the preconditioning polynomial to be $m = 4$.

4.1   Simple Example

We begin by showing how good the upper bound (7) is for a simple example. Consider a diagonal matrix of size $N = 100$ where the values on the diagonal, and hence the eigenvalues, are $\lambda_i = 10^{-4} + \frac{i-1}{N-1}(2 - 10^{-4})$ for $i = 1, \ldots, 100$. Therefore the eigenvalues are uniformly distributed between 10^{-4} and 2 and the condition number is 2×10^{-4}. In Fig. 2 we plot the number of iterations as we vary c. The lower curve is from the results of numerical experiments and the upper curve is the upper bound (7). Note that the upper bound gives a large over estimate in the number of iterations when c is small but at the minimum point on the curve the bound is very tight and more importantly the minimum of both curves is at the same point.

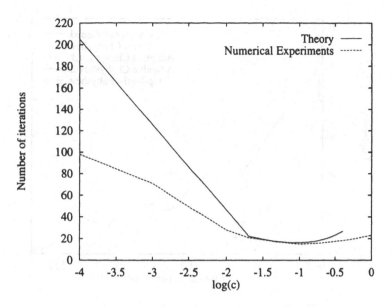

Fig. 2. Comparison of theoretical upper bound and numerical results.

4.2 Harwell-Boeing Test Matrices

We now consider how well we can find this optimal value of c using the two adaptive procedures in the previous section. We test the preconditioners on three matrices from the Harwell-Boeing collection of matrices, [5].

Before applying the Conjugate Gradient iteration we diagonally scale the matrices so that they have ones on the diagonal. In practice we have found that this dramaticly improves the results and in some cases the CG method fails to converge without this scaling. We compare the results for our adaptive preconditioner with four others, the unpreconditioned case, the exact Chebyshev polynomial ($c = \lambda_1$ and $d = \lambda_N$), the adaption scheme of Ashby et al. [2] and the optimal Chebyshev polynomial. The optimal choice of c and d is found by fixing d to be the maximum eigenvalue and testing a range of values for c to estimate the optimal value empirically.

We consider first the matrix BCS14 which has dimension 1806. After applying diagonal scaling the minimum and maximum eigenvalues are 4.8×10^{-4} and 3.34, respectively. These figures are typical for all three of these matrices. In Fig. 3 we plot the convergence histories for all five choices of preconditioner. Firstly, note that all the Chebyshev polynomial preconditioners give a significant reduction in the number of iterations required for convergence. Next note that the Chebyshev polynomial with the exact endpoints, hence the preconditioned matrix with minimum condition number, gives almost monotone convergence. The optimal Chebyshev polynomial preconditioner for this problem has $c = 0.05$ which is significantly larger than the minimum eigenvalue. This gives excellent results, with convergence reached in less than a quarter the number of iterations required in the unpreconditioned case. This is the best we can hope for our adaptive procedure to do.

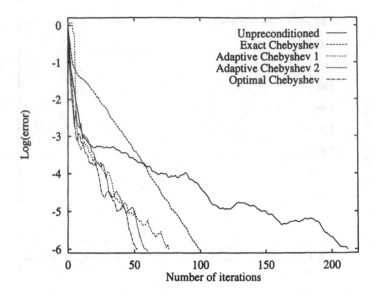

Fig. 3. Convergence histories for BCS14.

Next consider the adaptive procedure of Ashby et al. where the procedure is applied every 5 iterations which we found to be the optimal. Notice in Fig. 3 that after each change in c there is an increases in the error for the next iteration before starting to converge again. This effect does not seem to be too detrimental because if we apply the preconditioner every 10 iterations to reduce this effect then the total number of iterations increases to 84.

The adaption scheme based on Axelsson's analysis gets much closer to the optimal. For this scheme we found that it was best to apply it just twice because after the second application the estimates for c and d were close to optimal which explains why the convergence of this scheme closely matches that of the optimal choice. We choose to apply the scheme after 5 iterations and after 10 iterations. After 5 iterations we get $c = 0.284$ and $d = 2.564$ and after 10 iterations we get $c = 0.073$ and $d = 3.44$.

In Table 1 we also show the results for two larger matrices, BCS15 and BCS16, which have dimension 3948 and 4884, respectively. For BCS15 the differences between the preconditioners is much more pronounced. The exact Chebyshev polynomial only gives a one third improvement in the number of iterations over the unpreconditioned case. The optimal Chebyshev polynomial again reduces the number of iterations by over a factor of four. For this problem there is a large difference between the two adaption procedures. The results of the Ashby scheme are again for the case where the adaption is applied every 5 iterations though we can get a slight improvement if we instead apply it every 10

Table 1. Comparison of various preconditioners for a selection of Harwell-Boeing matrices.

Matrix	Preconditioner	Adaption Method	No. of iterations
BCS14	Unpreconditioned	None	212
(N=1806)	Chebyshev Polynomial	None	99
	Chebyshev Polynomial	Ashby et al.	76
	Chebyshev Polynomial	Axelsson analysis	60
	Optimal Chebyshev Polynomial	none	52
BCS15	Unpreconditioned	None	462
(N=3948)	Chebyshev Polynomial	None	303
	Chebyshev Polynomial	Ashby et al.	243
	Chebyshev Polynomial	Axelsson analysis	114
	Optimal Chebyshev Polynomial	none	105
BCS16	Unpreconditioned	None	149
(N=4884)	Chebyshev Polynomial	None	37
	Chebyshev Polynomial	Ashby et al.	42
	Chebyshev Polynomial	Axelsson analysis	43
	Optimal Chebyshev Polynomial	none	34

iterations. Our adaption scheme, applied after 5 and 10 iterations, again gives results which are close to the optimal.

Although the third test matrix, BCS16, is the largest, the unpreconditioned conjugate gradient algorithm converges in relatively few iterations. The optimal Chebyshev polynomial again gives the same improvement factor over the unpreconditioned case. For this problem though the exact Chebyshev polynomial converges in fewer iterations than the adaptive procedures. This is because the exact Chebyshev polynomial is already close to optimal and therefore the adaptive procedures will not be able to make up for the initial poor iterations before converging at the optimal rate. It should also be noted that for a general matrix the eigenvalues are unknown so we cannot use the exact Chebyshev polynomial preconditioner and therefore results of the adaptive preconditioners can only be compared with the unpreconditioned case where they show a large improvement.

5   Conclusions

We have shown that Chebyshev polynomial preconditioning can give excellent results if we have the optimal values for the endpoints of the interval. Unfortunately we cannot estimate these values before the conjugate gradient algorithm. We have shown though, that it is possible to estimate these values during the CG iteration and achieve a convergence rate which is close to that for the optimal values.

The adaption scheme of Ashby et al. which was based on the standard analysis of the convergence of the CG algorithm gives reasonable results. By using an adaption scheme based on a more complex analysis by Axelsson we were able to get good values for the endpoints after just two adaption stages and 10 iterations. Further work is required on when and how often to apply the adaption procedures. Testing on a large parallel computer is also required to establish the optimal degree of the preconditioning polynomial and to do comparisons with other preconditioners.

References

1. Ashby, S. F., Manteuffel, T. A., Otto, J. S.: Adaptive polynomial preconditioning for HPD linear systems. In R. Glowinski and A. Lichnewsky, editors, Proc. Ninth International Conference on Computing Methods in Applied Sciences and Engineering, pages 3—23, 1990.
2. Ashby, S. F., Manteuffel, T. A., Otto, J. S.: A comparison of adaptive Chebyshev and least squares polynomial preconditioning for Hermitian positive definite linear systems. SIAM Journal of Scientific and Statistical Computing 13 (1992) 1—29
3. Axelsson, O.: Iterative Solution Methods. Cambridge University Press, 1994.
4. Brent, R. P.: Algorithms for Minimisation without Derivatives, Chapter 5. Englewood Cliffs, NJ: Prentice-Hall, 1973.
5. Duff, I. S., Grimes, R. G., Lewis, J. G.: Sparse matrix test problems. ACM Transactions on Mathematical Software 15 (1989) 1—14
6. Hestenes, M. R., Stiefel, E.: Methods of conjugate gradients for solving linear systems. Journal of Research of the National Bureau of Standards 49 (1952) 409—435
7. Johnson, O. G., Micchelli, C. A., Paul, G.: Polynomial preconditioning for conjugate gradient calculations. SIAM Journal of Numerical Analysis 20 (1983) 362—376
8. Meijerink, J. A., van der Vorst, H. A.: An iterative solution method for linear systems of which the coefficient matrix is a symmetric M-matrix. Mathematics of Computation 31 (1977) 148—162
9. Reid, J. K.: On the method of conjugate gradients for the solution of large sparse systems of linear equations. In Proc. Conference on Large Sparse Sets of Linear Equations. Academic Press, New York, 1971.
10. Saad, Y.: Practical use of polynomial preconditionings for the conjugate gradient method. SIAM Journal of Scientific and Statistical Computing, 6 (1985) 865—881

The IBM Parallel Engineering and Scientific Subroutine Library

Salvatore Filippone *

IBM European Center for
Scientific and Engineering Computing
P.le Giulio Pastore 6,
I-00144, Roma, Italy

1 Introduction

The Parallel ESSL [3] product is a mathematical library specifically tuned for
the IBM RISC System/6000 Scalable POWERParallel Systems (SP) series of
machines; it allows programmers to access the computational power of the underlying hardware, with little development effort on the user side, scaling in
performance from a small number of processors up to the full machine configuration.

It is also designed to support existing standards in parallel computations,
such as the BLACS and ScaLAPACK, while adding new features of usability
and performance tuning.

In this paper we will review the underlying hardware and system software
features relevant to the library, then we will describe salient features of the
library itself, including different computational areas and usage hints and tips.

The Parallel ESSL has been developed by the IBM RISC System/6000 Div-
sion, Poughkeepsie, N.Y., the T.J. Watson Research Center, Yorktown Heights,
N.Y., and the European Center for Scientific and Engineering Computing, Rome,
Italy. The linear algebra part of Parallel ESSL is based on the ScaLAPACK [1]
public domain software, jointly developed by the University of Tennessee at
Knoxville, the Oak Ridge National Laboratory and the University of California,
Berkeley; ScaLAPACK is available from Prof. J. J. Dongarra, Computer Science
Department, University of Tennessee, Knoxville.

2 SP Hardware and Software

The SP series is a line of parallel machines based on the processor technology of
the IBM RISC System/6000 line of products [4]; it provides scalable performance
from small to very large configurations, with a flexible architecture, a consistent
software environment and integrated system management capabilities.

The current product is the SP2 machine; it consist of a collection of up to 128
computing nodes based on the POWER2 processor architecture. Parallel ESSL

---

* filippon@vnet.ibm.com

also supports computing nodes based on the POWER technology, such as those found in the SP1 machine.

Processing nodes for the SP2 are available in three models: the "Wide" nodes, the "Thin" nodes and the "Thin-2" nodes; salient machine characteristics are summarized in table 1.

	Wide	Thin	Thin-2
Clock	66.7 Mhz	66.7 Mhz	66.7 Mhz
Peak MFLOPS	266	266	266
Memory bus	256	64	128
Data cache	256 KB	64 KB	128 KB Level 1
			0-2 MB Level 2

Table 1. SP2 Node characteristics

Each POWER2 CPU has dual fixed-point units, dual floating point units and an instruction/branch unit, all capable of operating simultaneously; since each floating-point unit supports multiply-and-add operations the processor is capable of delivering 4 floating-point operations per clock cycle.

The processing nodes are assembled in frames; each frame can hold up to 8 wide nodes and up to 16 thin nodes; larger machine configuration are formed by connecting together up to 16 frames.

Nodes can communicate by means of different network interconnections; the main ones are:

- An internal Ethernet network, used for system purposes;
- A fast data communication adapter for the High Performance Switch, intended for parallel application support;
- A network connection to the external world; supported adapters include Ethernet, Token Ring and FDDI;

The system is managed by a single RISC System/6000 control workstation, connected to the frames constituting the machine by the Ethernet, the external network and one serial channel per frame dedicated to hardware monitoring.

The High Performance Switch provides a logical all-to-all communication connection, with a scalable structure supporting a point-to-point peak bidirectional bandwidth of 41 to 48 MB/sec, for a peak aggregate bandwidth of 5 GB/sec on 128 nodes; the hardware latency is 500 ns. The architecture of the switch is based on a multi-stage omega-network, and was developed at the T.J. Watson Research Center.

The machine is administered through the control workstation, which is in charge of managing the users, accessing and mounting external file systems, and handling software and hardware configuration files.

Each machine node runs a full operating system, the AIX/6000; in addition to the standard features, there is a set of tools for various tasks.

System administration software addresses such requirements as keeping a single, consistent system image throughout the machine and monitoring the hardware/software machine status; all administration operations are centralized on the control workstation.

Job management is handled through different tools, including the Resource Manager which keeps track of the configuration and reserves nodes for parallel execution, and the Partition Manager that defines subdivisions of the machine.

The parallel communication interfaces include the AIX Parallel Environment Message Passing Library, upon which Parallel ESSL is based; it provides basic point to point and collective communication functions. The communication interface can run in IP mode over both the High Performance Switch and the Ethernet, and in dedicated User Space mode on the High Performance Switch for best performance. The parallel environment includes a parallel debugger and a trace driven Program Visualization tool.

3   Parallel ESSL design features

The Parallel ESSL library was designed to fully exploit the computational capabilities of the hardware, while maintaining usability and compatibility with existing standards.

The parallel library is based on the serial ESSL/6000 library for local node computations; this ensures full exploitation of the computational capabilities of the POWER2 processing nodes. Techniques used to achieve full processor utilization include cache managements, efficient memory usage through data blocking and contiguous storage accesses, and minimized paging. Detailed information on tuning codes for the POWER2 processor may be found in [4].

3.1   Environment and communication interface

The user-level communication provided by PESSL is conformant to the BLACS standard [2], implemented on top of the AIX Parallel Environment Message Passing Library.

The BLACS interface provides primitives for setting up 2-dimensional logical process grids; therefore they provide a convenient way for the user to organize the parallel application structure. The user has the freedom of defining different logical grids and using them at the same time over the same physical processor set.

Communication is also handled via BLACS calls; there are both point-to-point and collective operations.

3.2   Usability features

The PESSL routines provide extensive parameter checking and error reporting; parameters are checked both for legality and for consistency across different processes.

Extensive documentation, including performance tips, also enhances the usability of the product.

4 Computational areas: Linear Algebra

The Linear Algebra part of Parallel ESSL is compatible with the public domain ScaLAPACK library [1]; therefore application programs using ScaLAPACK routines available in PESSL need only be linked with the new library.

Level 2 Parallel BLAS

Currently supported level 2 PBLAS are:

Routine	Operation
PDGEMV, PDSYMV	General and symmetric matrix-vector product
PDGER, PDSYR	General and symmetric matrix rank-1 update
PDSYR2	Symmetric matrix rank-2 update
PDTRMV	Triangular matrix-vector product
PDTRSV	Triangular system solve

Level 3 Parallel BLAS

Currently supported level 3 PBLAS are:

Routine	Operation
PDGEMM, PDSYMM, PDTRMM	General, symmetric and triangular matrix-matrix product
PDTRSM	Triangular system solution with multiple right-hand sides
PDSYRK, PDSYR2K	Symmetric rank k and rank $2k$ updates
PDTRAN	General matrix transpose

Linear Algebraic equations

Currently supported linear algebraic equation routines are:

Routine	Operation
PDGETRF and PDGETRS	General system solve $AX = B$, $A^T X = B$ via LU factorization
PDPOTRF and PDPOTRS	Symmetric positive definite system solve $AX = B$ via Cholesky factorization

Eigensystems and Singular Value Analysis

Currently supported eigensystems routines are:

Routine	Operation
PDSYEVX	Compute selected eigenvalues and eigenvectors of a symmetric matrix
PDSYTRD	Symmetric matrix reduction to tridiagonal form
PDGEHRD	General matrix reduction to Hessenberg form
PDGEBRD	General matrix reduction to bidiagonal form

5 Computational areas: Fourier Transforms

Parallel ESSL supports the following mixed-radix discrete Fourier transforms:

Routine	Operation
PDFCT2	Two-dimensional complex transform
PDRCFT2	Two-dimensional real to complex conjugate even transform
PDCRFT2	Two-dimensional complex conjugate even to real transform
PDFCT3	Three-dimensional complex transform
PDRCFT3	Three-dimensional real to complex conjugate even transform
PDCRFT3	Three-dimensional complex conjugate even to real transform

The Fourier transform routines assume a unidimensional logical processor arrangement.

6 Computational areas: Random Number generation

Parallel ESSL provides a uniform pseudo-random number generator based on the multiplicative congruential method:

$$s_i = (a(s_{i-1}) \bmod (m) = (a^i s_0) \bmod (m)$$

with s_0 the user-specified initial seed, and with parameters

$$a = 44485709377909.0$$
$$m = 2^{48}$$

7   Performance tuning considerations

Tuning a parallel application for best performance requires some action by the user in specifying a number of parameters.

The first and foremost tuning issue facing the user is the choice of the appropriate machine size; given a problem size there will be a definite number of processors guaranteeing the best trade-off between raw performance and utilization. Adding more processors may even, for small problems, cause a slow-down; as an extreme case consider solving a 100 × 100 linear system on 8 processors.

Taking as a reference linear algebra computations there are two main choices to be made when running the application:

1. Choice of the processor grid configuration;
2. Choice of the matrix distribution parameters.

As a general rule, processor grids closer to square are better suited to linear algebra computations. A notable exception to this rule is the LU factorization code; in this case partial pivoting imposes additional requirements onto the communication network, and therefore a grid $P \times Q$ with $P < Q$ is to be preferred.

The blocking size used to distribute the matrices also influences performance; many factors are involved in the optimal choice, including:

- Cache utilization;
- Communication and synchronization costs;
- Load balancing.

The Parallel ESSL Guide and Reference provides some guidance on these factors, with recommended values providing consistent performance across a wide range of cases.

8   Performance data

Obtaining best performance has been one of the main objectives of PESSL development; we show some performance data for selected routines.

In figure 1 we show performance in MFLOPS for the factor routine PDGETRF; for a given machine size grid configurations different from the ones shown here would give different performance results.

Table 2 lists performance data for the PDCFT2 complex two-dimensional Fourier transform; the data were obtained on wide nodes.

9   Conclusions

Parallel ESSL is a product fully committed to providing the best possible utilization of computational resources; it is also committed to providing support for computational standards, ease of use and superior support.

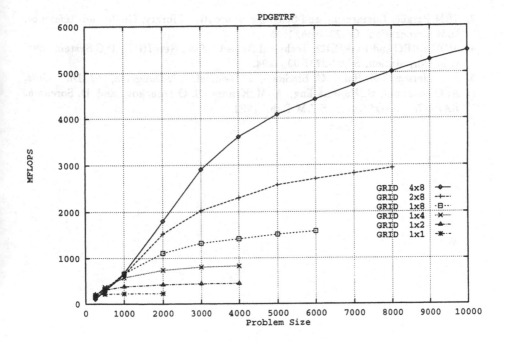

Fig. 1. PESSL PDGETRF performance

	Matrix order				
Nodes	1008	1024	2048	4096	5040
1	105	128			
2	133	154	202		
4	271	314	335		
8	503	586	650	695	
16	845	911	1003	1085	961

Table 2. Performance in MFLOPS for PDCFT2

References

1. Choi, J., Demmel, J., Dhillon, J., Dongarra, J., Ostrouchov, S., Petitet, A., Stanley, K., Walker, D. and Whalley, R. C. ScaLAPACK: A Portable Linear Algebra Library for Distributed Memory Computers LAPACK working note 95, available from http://www.netlib.org/
2. Dongarra, J., and Whaley, R. C. A User's Guide to the BLACS LAPACK working note 94, available from http://www.netlib.org/, 1995.

3. IBM Parallel Engineering and Scientific Subroutine Library. Guide and Reference. IBM Corporation, GC23-3836, 1995.
4. POWERPC and POWER2: Technical Aspects of the New IBM RISC System/6000 IBM Corporation, SA23-2737-00, 1994.
5. E. Anderson, Z. Bai, C. Bischof, J. Demmel, J. Dongarra, J. Du Croz, A. Greenbaum, S. Hammarling, A. McKenney, S. Ostrouchov, and D. Sorensen. *LAPACK Users' Guide*. SIAM Pub., 1992.

Some Preliminary Experiences with Sparse BLAS in Parallel Iterative Solvers

Salvatore Filippone[1] and Carlo Vittoli[2]

[1] filippon@vnet.ibm.com IBM ECSEC Rome, Italy
[2] vittoli@vnet.ibm.com IBM Cagliari, Italy

Abstract. A proposal has been put forward for standardization of sparse matrix operatons in [10]; we describe here some early experiments in adapting the proposed standard to parallel operations in distributed memory environments.

1 Introduction

Iterative solvers for linear systems are among the most important kernels in scientific and technical applications; therefore their implementation on high-end computing architectures is the subject of much activity in the scientific community. This paper explores some ideas that we believe useful in helping the prototyping and maintenance of these kernels.

Recent developments include devising efficient problem splitting techniques [14], identification of good data distributions for the inner kernels [4, 19], and identification of good preconditioning techniques [18].

One common problem encountered by researchers and practitioners is the mutual incompatibility of the data structures used on different machine architectures and/or different specific problems; the only successful standardization that we are aware of deals only with the format for exchanging data on external memory, commonly known as the Harwell-Boeing format [9].

There have been attempts at unifying the description of algorithms for sparse linear algebra; for instance, the template concept [3] provides a convenient framework for conveying the essential mathematical and computational details of an iterative method, without delving into the intricacies of any specific data structure.

Recently a proposal has been put forward to standardize the interface of the inner kernels of sparse iterative computations [10]: the sparse BLAS should enable researchers and developers to decouple the algorithmic and numerical issues from the data structure implementation details, allowing faster prototyping and easier reuse of software.

We describe here some early experiences in adapting the proposed serial standard interface to parallel implementations on distributed memory machines, such as the IBM 9076 SP2. The paper is organized as follows: in section 2 we give a brief overview of the proposed standard for serial computations; in section 3 we describe our prototype parallel implementation; in section 4 we give some results from numerical experiments, and we conclude with a look at future expansion directions.

2   User level sparse BLAS

Basic linear algebra operations have been successfully standardized for full matrices over the past 15 years, the last step being the development of the level 3 BLAS interface [8]; these are widely recognized as effective tools for exploiting computational capabilities of modern computer architectures, and most computer vendors support them in their computational libraries.

The same did not happen for operations performed on sparse matrices; the main reason is that most sparse linear systems arise from the solution of physical models with a very definite structure, and great care is usually exercised in exploiting this structure to match the problem with the available computational resources. While this approach has resulted in the successful development of applications capable of handling very large problems, up to millions of degrees of freedom, it has also created problems on the software engineering side, impairing maintainability and portability of programs, and also making prototyping a somewhat tedious task.

The aim of the sparse BLAS proposal presented in [10, 15] is to alleviate this situation for serial computation on modern computer architectures. Salient features are:

- The standard is targeting the essential kernels found in most popular iterative methods; these include the product of a sparse matrix times a dense matrix, the solution of a sparse triangular system of equations, and various data pre and post-processing routines;
- The sparse matrix data structure is specified in a general way, encompassing many different data structures currently in use. Sparse matrices are represented by six entities: a character string identifying the data structure, a character array containing additional structural information, a real and three integer arrays, whose interpretation details will vary according to the structure information; this structure bears a close relationship to the approach taken in ScaLAPACK [5] to describe matrices residing in a parallel machine;
- The inner layers of the kernel implementation take care of actually performing the various operations according to the storage scheme specified in the descriptors for the sparse matrix argument.

We end our discussion of the proposed standard by noting that the approach is somewhat limited by the choice of FORTRAN 77 as its implementation language; with the availability of some object-oriented support in Fortran 90 the same concepts could lead to cleaner codes.

3   A parallel iterative solver

Our aim in developing an iterative solver code is mainly to explore the extension of serial codes written in the sparse BLAS framework to a parallel setting with the minimal amount of changes. A number of consideration have to be included in the definition of a suitable data structure; recent works on these aspects include [4, 11, 19].

For our experiments we rely on a data distribution based on domain decomposition: each domain is assigned to a processor, inducing a row strip partition of the coeffcient matrix. Each local data structure therefore contains the coefficients of the local part of the matrix and the information about the boundary variables to be exchanged with neighbouring processors at each step to compute the matrix vector product. While our tests have been performed on a simple 2D domain with a tiled distribution, the scheme is general enough to accomodate different domain decomposition strategies.

The reference iterative algorithm is Bi-CGSTAB [20], because of its wide popularity and applicability; we use a local block ILU preconditioning; this choice is consistent with the main idea of the experiment, i.e. keeping at the barest minimum the number of changes to the serial code. The local ILU approach is appropriate in certain classes of problems; moreover other preconditioning strategies can be implemented within the same framework, such as those proposed in [13, 18], with little or no overhead in the preconditioning step, provided appropriate care is taken in the data setup phase.

The parallel implementation of the iterative algorithm requires communication among processor nodes in the matrix-vector product, in the preconditioning step and in the computation of scalar products.

For the matrix-vector product each processor needs to get the vector elements on the boundary of its own domain from the neighbouring processors. This communication step is local, and also subject to a surface to volume effect for matrices coming from the discretization of a PDE on a physical domain; therefore its effect can be kept under control provided the domain assigned to the processor is large enough, i.e. the global problem size is adequately fitting on the machine configuration. The communication can be effectively hidden inside the sparse BLAS call; the product

$$v \leftarrow Aq$$

would be coded as:

```
      CALL DCSMM(TRANSA,M,K,M,ONE,P1,FIDA,DESCRA,A,IA1,IA2,
     *           INFOA,P2,Q,LDB,ZERO,V,LDC,WORK,LWORK,IERRV)
```

where matrix A is represented through the arrays **FIDA, DESCRA, A, IA1, IA2** and **INFOA**; according to the layout of [10] information about the matrix structure is stored into the integer auxiliary arrays, extended to keep track of the boundary between domains.

The communication requirements of the preconditioning step depend on the exact choice of the preconditioner; for purely local ILU preconditioners no communication is required, whereas for overlapped preconditioner such as the ones in [13, 18] communication is required for the boundary values. In our case the step

$$\text{Solve} Lv = s; Us = v$$

is coded as

```
    CALL DCSSM(TRANSA,M,K,ONE,DIAGL,DV,P3,FIDL,DESCRL,L,IL1,IL2,
*             INFOL,P4,V,M,ZERO,S,M,WORK,LWORK,IERRV)
    CALL DCSSM(TRANSA,M,K,ONE,DIAGU,VDIAG,P5,FIDU,DESCRU,U,IU1,IU2,
*             INFOU,P6,S,M,ZERO,V,M,WORK,LWORK,IERRV)
```

The other source of communication requirements is the scalar product accumulation; this can be effectively hidden by providing a specialized routine to substitute for the level-1 BLAS DDOT. The main difference is that this is always a *global* communication step, synchronizing all participating processors.

Finally note that the best support for such a code would come from an HPF-like environment, provided it can overcome the limitations in the data distributions of HPF-1, viz. the requirement of regularity in the data distribution parameters.

The internal parallel data structure has been implemented as an extension of the Compressed Storage by Rows; however this is by no means essential to the parallelization of the code, and the same interaction among processors would work with different "local" data structures, fully optimized for the processor architecture, such as those described in [1].

We have imbedded the sparse BLAS calls in a self-contained iterative code; however they could also be used in a reverse communication iterative driver, at least in principle. This particular issue will be the subject of further investigation.

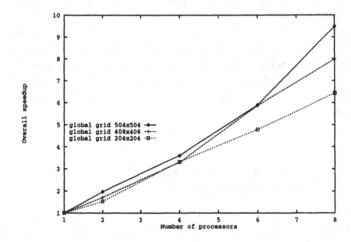

Fig. 1. Conjugate gradient test: fixed problem size, global speedup

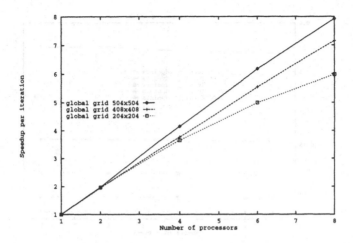

Fig. 2. Conjugate gradient test: fixed problem size, speedup per iteration

4 Numerical experiments

The reference test problem for our approach is a simple 2D finite-difference on a rectangular domain, mapped onto a rectangular grid of processors.

We begin by looking at the speedup for the different problem sizes with respect to the overall iterative process as shown in fig. 1; the apparently super-linear speedup in the 504×504 grid is due partly to fluctuations in the number of iterations to achieve convergence and partly to cache/memory allocation issues in the uniprocessor configuration taken as reference.

To evaluate the feasibility of the approach we also show the speedup per iteration of the same test cases in fig. 2; as expected the efficiency of parallelization will depend on the global problem size, i.e. small problem sizes will exhibit growing overhead. By isolating the behaviour in a single iteration we eliminate the dependence on the convergence properties of the particular preconditioner.

Another criterion, perhaps more realistic, is to evaluate the iteration speed keeping constant the number of grid points per node; this reflects the need of parallel processing to solve large problems saturating the capacity of a given machine. Results obtained on the SP2 up to 8 processors are shown in Fig. 3; the cost per iteration is approximately constant; this should be expected, because the communication for the matrix-vector product is subject to a surface-to-volume effect, and the global scalar product accumulation does not add too much on these machine configurations.

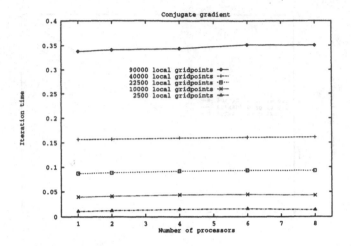

Fig. 3. Conjugate gradient test: fixed processor memory

5 Conclusions

In this paper we have examined some experiments on implementing parallel
iterative solvers for sparse matrices in the framework of a proposed sparse BLAS
standard; we believe that this is the right development environment for both
run-time efficiency and ease of development/maintenance.

Future development directions include exploring usage of other established
standards such as the BLACS, interfacing with high-level data-parallel languages, interoperability with domain decomposition tools.

References

1. Agarwal, R. C., Gustavson, F. G., and Zubair, M. A high performance algorithm
   using pre-processing for the sparse matrix-vector multiplication. *In Proceedings of
   Supercomputing '92*, Minneapolis, MN. Nov 16-20, 1992.
2. Amestoy, P. R., Daydé, M., and Duff, I. S. Use of Level 3 BLAS in the solution of
   full and sparse linear equations. *In High Performance Computing. Edited by J-L.
   Delhaye and E. Gelenbe.* North-Holland, 19–31, 1989.
3. Barret, R., Berry, M., CHan, T., Demmel, J., Donato, J., Dongarra, J., Eijkhout,
   V., Pozo, R., Romine, C. and van der Vorst, H. Templates for the Solution of
   Linear Systems SIAM, 1993.
4. Bercovier, M. and Schlesinger, A. Random methods for creating disjoint sets of
   elements for parallel sparse matrix-vector product Technical report R93024, Université Pierre et Marie Curie, Paris, 1993.
5. Choi, J., Demmel, J., Dhillon, J., Dongarra, J., Ostrouchov, S., Petitet, A., Stanley, K., Walker, D. and Whalley, R. C. ScaLAPACK: A Portable Linear Algebra

Library for Distributed Memory Computers LAPACK working note 95, available from http://www.netlib.org/

6. Dodson, D. S., Grimes, R. G., and Lewis, J. G. Sparse extensions to the Fortran Basic Linear Algebra Subprograms. *ACM Trans. Math. Softw.*, 17:253–263, 1991.

7. Dongarra, J. J., Du Croz, J., Duff, I. S., and Hammarling, S. A set of level 3 Basic Linear Algebra Subprograms. *ACM Trans. Math. Softw.*, 16:1–17, 1990.

8. Dongarra, J. J., Du Croz, J., Hammarling, S., and Hanson, R. J. An extended set of Fortran Basic Linear Algebra Subprograms. *ACM Trans. Math. Softw.*, 14:1-17, 1988.

9. Duff, I. S., Grimes, R. G., and Lewis, J. G. Sparse matrix test problems. *ACM Trans. Math. Softw.*, 15:1–14, 1989.

10. Duff, I., Marrone, M. and Radicati, G. A proposal for user level sparse BLAS Technical report TR/PA/92/85, CERFACS, Toulouse, 1992.

11. Eijkhout, V. Distributed sparse data structures for linear algebra operations Technical report CS-92-169, University of Tennessee, Knoxville, 1992.

12. Erhel, J. Sparse matrix multiplication on vector computers. *Int J High Speed Comput*, 2:101–116, 1990.

13. Filippone, S. Marrone, M. and Radicati di Brozolo, G. Parallel preconditioned conjugate-gradient type algorithms for general sparsity structures Intern. J. of Computer Math., Vol. 40, pp. 159–167, 1992.

14. Hendrickson, B. and Leland, R. An improved spectral graph partitioning algorithm for mapping parallel computations SIAM J. Sci. Computing, Vol. 16, No. 2, pp. 452–469, 1995.

15. Heroux, M. A proposal for a sparse BLAS toolkit. Technical Report TR/PA/92/90, CERFACS, Toulouse, France, 1992.

16. E. Anderson, Z. Bai, C. Bischof, J. Demmel, J. Dongarra, J. Du Croz, A. Greenbaum, S. Hammarling, A. McKenney, S. Ostrouchov, and D. Sorensen. *LAPACK Users' Guide*. SIAM Pub., 1992.

17. Lawson, C. L., Hanson, R. J., Kincaid, D. R., and Krogh, F. T. Basic linear algebra subprograms for Fortran usage. *ACM Trans. Math. Softw.*, 5:308–323, 1979.

18. Notay, Y. An efficient parallel discrete PDE solver, Technical report NM-IBM 94-01, Université Libre de Bruxelles, 1994.

19. Romero, L. F. and Zapata, E. L. Data distributions for sparse matrix vector multiplication Parallel Computing, Vol. 21, pp. 583–605, 1995.

20. Van der Vorst, H. A. BiCGSTAB: a fast and smoothly converging variant of BI-CG for the solution of nonsymmetric linear systems, SIAM J. Sci. Stat. Comput., Vol. 13, No. 2, pp. 631–644, 1992.

Load Balancing in a Network Flow Optimization Code

Enric Fontdecaba, Antonio González and Jesús Labarta

Universitat Politécnica de Catalunya,
Departament d'Arquitectura de Computadors, Modul D6,
08071 Barcelona, Spain

Abstract. This paper presents different schemes to parallelize a code that solves a Network Flow Optimization problem. These schemes have been implemented on a heterogeneous cluster of workstations with a multiprogrammed workload. The paper shows that for this situation, it is crucial to have a load balancing strategy that takes into account the computing power of each node as well as its load. Performance figures show that significant speed-ups are obtained when this load balancing strategy is followed.

Solving optimization problems usually requires a huge amount of calculations. In these cases, it may be desirable a parallel implementation in order to meet some given constraints in the execution time. One of these optimization problems that are only affordable with parallel machines is the hydroelectrical production planning, which can be formulated as a network flow optimization problem. This paper focuses on the parallel implementation of an algorithm to solve this problem. The target platform is a heterogeneous cluster of workstations with a multiprogrammed workload.

Load balancing and reduction of communication/synchronization costs are the key issues to attain a good performance on a parallel machine. For the type of hardware platform that we are considering, these objectives are not easy to achieve since the effective power of each node is different and variable. Besides, the computation cost of each part of the algorithm is not precisely known before execution.

In this paper, we present several load balancing strategies that can be used for this scenario. These strategies have been implemented and evaluated. The results presented are experimental results from real executions on a workstation cluster and also simulated results obtained with PARADIS [1], which is a tool for simulate and visualize the execution of parallel programs. The performance figures show that significant speed-ups are obtained when the load balancing strategy takes into account the characteristics of the application and the computing power of each node of the cluster as well as its load.

1 The Problem

In this section the algorithm to be parallelized is briefly described. For more details on it, the interested reader is referred to [2], [3].

1.1 The Network

The objective of this optimization code is to obtain the maximum benefit, measured as the ability to meet the user demand in each time interval, taking into account the bounding conditions which are the minimum and the maximum reservoir levels and flows between them. These conditions are needed to keep the rivers ecological flow, the human consumption and also the agricultural requirements. More precisely, we are interested in a long term planning. That introduces some non-deterministic parameters. It is not possible to know in advance the climate of a given month, neither to know precisely the consumption. These issues causes that it is not possible to know the outcoming and incoming flows of the reservoir network.

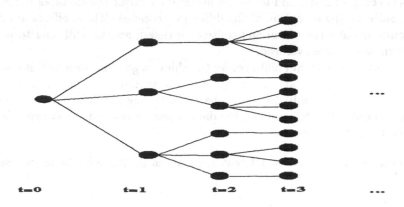

Fig. 1. Dependence graph of tasks

A possible approach to deal with this problem is to analyze the different potential scenarios and try to optimize the sum of the individual cost functions. A scenario can be seen as a set of possible inputs/outputs of the network along all the periods of time under study. In order to reduce the amount of computations and to obtain better solutions, all the results obtained for all the scenarios that are equal until a given point of time are forced to be equal. Notice that this restrictions may reduce the chances to exploit parallelism because it introduces some dependences among subproblems.

As it is shown in figure 1 the dependences among the different subproblems can be represented as a tree graph. Each node of the tree node corresponds to a small network flow optimization problem. In order to solve the whole problem it is necessary to traverse all the tree a certain number of iterations, solving for each node the corresponding subproblem. This subproblem is, in fact, another iterative problem.

In order to understand the load balancing strategies proposed in this paper, it is not necessary to have more knowledge about the problem. This is a very interesting point, which also means that all the results shown in this paper can be generalized to other problems. In particular, they can be applied to any problem that can be expressed as a set of tasks with a tree-like set of dependences among them.

2 Load Balancing Strategies

The objective of a parallelization of a code is to split all the work in chunks that can be executed in parallel by different processors. In order to achieve a good performance with a parallel program it is important to pay attention to two issues:

- One of these issues is the added work needed to do the communication between processors and to do the necessary control among tasks in order to coordinate the execution of the different processes. Those effects are called communication overhead and control overhead, and we will refer to both of them as runtime overhead.
- The other important point in order to achieve a good performance is load balancing. In order to take advantage of all the available computing resources, it is necessary to evenly distribute the tasks among the different processors. It is also important to minimize the data dependences and the synchronization points among processors.

In this work, we present different approaches to deal with these two issues.

2.1 Problems

Due to the application One of the problems to achieve a good load balancing of the parallel code is due to the fact that the computing time needed to solve each subproblem is not known in advance. As a consequence, it is not possible to partition the work obtaining a perfect load balancing. Nevertheless, this phenomenon is not so bad because an acceptable approximation of the computing cost of every problem is known. In addition, since each processor will compute a large number of subproblems (about 80 in our experiments), one can expect that the errors made in this approximation may be compensated. Besides, It has been observed that the time spent solving each subproblem becomes more homogeneous as the execution progesses. This is not surprising since this is an iterative solver and at each iteration the initial solution is closer to the final solution.

Due to the environment The most important cause for load imbalance reason is the due to computational environment. This code has been executed on a heterogeneous workstation cluster, with a multiprogrammed workload. The message passing library used is PVM.[4] [5] The main problem that arises in this kind of platform is derived from the fact that the machines and the communication network is shared with other users. It is not possible to do a perfectly balanced partitioning of the workload because it is not known the load of each machine at every moment.

We call effective power to the computing resource share from a machine that is expected to be available for our application. Of course, the effective power of a given machine will vary with the time, since the workload due to other users is not constant. It is important to remark the fact that the effective power variation of one machine has important effects on the other processors due to the synchronization points.

2.2 Standard Solutions

There are basically two groups of solutions in order to balance the load among processors: static load balancing and dynamic load balancing techniques. The main difference between them is that in the first one the task allocation does not change during execution while in the other it does.

Static load balancing In a static load balancing approach, the work is distributed at the beginning of the execution among all the available processors. This distribution is not changed in all the execution. This technique presents a minimum run time overhead and a optimum use of the memory, since it does not require to replicate data structures. Besides, it also minimizes the communications requirements and therefore the number of synchronization points. The latest advantage has a significant effect when the target machine is a workstation cluster, because one of the major drawbacks of this kind of computer is its low communication bandwidth.

Furthermore, a static load balancing approach does not compensate possible imbalancing that may occur during execution. In our case, there are two effects that may cause this kind of imbalancing: variations of effective power that occurs during the execution of the code and the fact that the computational cost of every subproblem is not precisely known in advance.

Summarizing, the main advantage of the static load balancing is its low run time overhead and the main drawback is its lack of adaptability.

Dynamic Load Balancing Due to the variable effective power of the each node of the cluster and the unknown computation requirements of each subproblem, a dynamic load balancing approach may be expected to achieve a better load balancing. With this technique the program can be adapted to different conditions at runtime. On the other hand, this technique increases the number of communication and synchronization operations and in additions, it needs to keep track of the execution in order to have all the calculations done in the correct order. Another drawback is its lack of scalability due to the fact that the control operations are centralized in one node.

2.3   Proposed Solutions

In this section, the different solutions that we have proposed to deal with the problems that exhibit the standard solutions described in the previous section. Four different load balancing schemes have been studied and three of them have been implemented.

Static Distribution The first approach consists in evenly distribute the tree of tasks among the nodes of the cluster. This distribution takes into account the computation cost of each task (an estimation) and tries to allocate neighbor tasks to the same node in order to reduce communication and synchronization overhead. This approach is in fact the static load balancing technique mentioned in the previous section. We use it as a reference point for the other schemes.

This code has been run on a heterogeneous cluster and also on a simulated dedicated homogeneous cluster showing a very different behavior in both cases. Whereas it performs quite well for a dedicated homogeneous cluster, its performance is rather low for a multiprogrammed heterogeneous cluster. This is in fact an expected result because of its lack of adaptability.

Adaptive Static Distribution One major drawback of the above technique is that it does not consider the fact that the cluster consists of nodes with a different effective power. The present approach overcomes this drawback by doing a completely static distribution of the tasks that takes into account the effective power of each node. This distribution is made only once at the beginning of each execution. When the estimation of the load of each node and the cost of each task is accurate this approach produces good results as it was expected, since the communication and synchronization requirements are minimal. Notice that for long executions the load of each node may substantially vary. In this cases, an adaptive dynamic approach may be more suitable.

Adaptive Dynamic Distribution This approach is an hybrid technique that combines static and dynamic load balancing of tasks. In this case, the tree of tasks is split into two parts. The first one (average size of 60-80 %) is statically distributed among the available processors taking into account the different effective power of each node. The effective power of a given node is computed as the product of its relative speed times the percentage of CPU available for our application. The second part is distributed on demand at run time by a process that is called the master. At each iteration, when a processor has finished its statically allocated work, it asks the master processor for more work. The tasks on the dynamic pool can be packed into sets containing several tasks in order to minimize the communication and synchronization overhead. The invariant set of data of the dynamically distributed work is replicated at the beginning of the execution in order to minimize the amount of communications required in the iterative process. Using this technique the amount of communication required is reduced at the expense of increasing the amount of memory used.

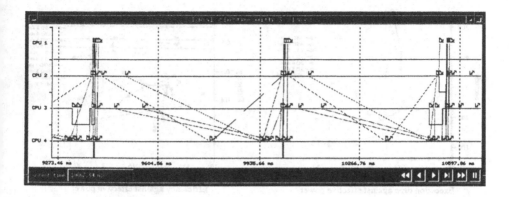

Fig. 2. A PARADIS Snapshot showing the behavior of an ideal homogeneous cluster

This algorithm is critically influenced by the communication bandwidth of the interconnection network, and also by the processor scheduling policy because it has a lot of critical communications and synchronization points. However, including a dynamic distributed part significantly increases the adaptability of the algorithm.

Readaptive Static Distribution A compromise between a static and a dynamic distribution of the workload may be to start with a adaptive static distribution and to redistribute it when a load imbalance situation is detected. We call this approach an Readaptive Static Distribution. This approach is attractive for long executions where the load conditions of each node vary significantly and then, the overhead introduced by redistributing the load at run time is justified.

There are several parameters that need to be set in order to achieve a good performance with this approach. It is necessary determine the time between load balancing measurements, which imbalance will cause a redistribution of the tree of tasks, how to allocate the static data in order to minimize the cost of redistribution, among others. This scheme has not been implemented because the lack of a suitable test set.

3 Results

Performance figures of the different schemes are presented in this section. Two type of performance figures are presented. One type corresponds to the measurements obtained with the PARADIS [1] simulation and visualization tool. The other set of figures corresponds to the real execution of the algorithm on a heterogeneous cluster of workstations, shared with other users. The tests has been done with 3 to 5 workstations with different computation power and load.

In figure 2, a snapshot of the simulated execution of the static distribution approach is shown for a dedicated homogeneous cluster. This results have been

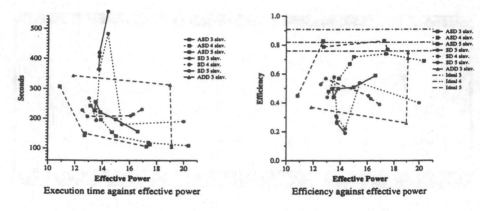

Fig. 3. Results from different machine configuration and load balancing techniques, (ASD, Adaptive Static Distribution; ADD Adaptive Dynamic Distribution; SD Static Distribution)

obtained with the PARADIS tool. It can be observed that all the processors are working most of the time, and also that the communication/synchronization operations do not degrade performance. A high efficiency can be easily obtained with the simplest load balancing technique since the computation power of every node of the cluster does not vary and it is the same for all of them.

Figure 3 shows the performance of the real execution on a multiprogrammed heterogeneous cluster. This graphs depict the execution time and efficiency of the different load balancing approaches against the effective power of the cluster. It is important to point out that the points of these graphs are not precisely place. In particular, the effective power of the cluster is measured only once at the beginning of the execution and this is the value shown in the X-axis. However, this parameter varies during the execution and in average, it could be somewhat different from the value obtain at the beginning. This may explain why sometimes with a higher effective power the execution time is higher. The reason may be that we had a higher effective power at the beginning but it decreased during execution.

In the efficiency graphs, it can be observed that the best result is obtained by a adaptive static distribution (ASD) with 3 slaves. Notice also that all the load balancing schemes suffer a significant degradation when the number of slaves grows. This behavior is due to the high cost of communication. Another interesting thing is the bad performance of the adaptive dynamic distribution (ADD) strategy with 3 slaves. The main reason is the high number of communications and synchronization points that are required by this scheme. As expected, the static distribution (SD) scheme has the worse performance due to the significant load imbalance that it introduces.

Figure 4 presents the curves obtained for another series of experiments, only with a three slave configuration. This experiments were done during the night, with a low traffic in the interconnection network, but with some batch processes

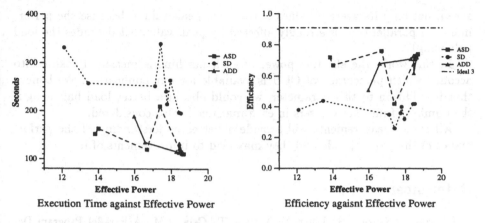

Execution Time against Effective Power Efficiency agaisnt Effective Power

Fig. 4. Results from different techniques with 3 slaves, (ASD, Adaptive Static Distribution; ADD Adaptive Dynamic Distribution; SD Static Distribution)

running. The figures show again a better performance for ASD and ADD algorithm than SD but now, ASD and ADD have a more similar performance. The main reason for that is the lower load of the interconnection network, and also the lower number of interactive users.

4 Conclusions

In this paper, we have presented several schemes to parallelize an optimization code. The main difference among them is the approach used to balance the load. The target machine is an heterogeneous cluster of workstations with a multiprogrammed workload.

The main conclusion of this work is that heterogeneous clusters can be successfully used for solving optimization problems in parallel (also problems that have a similar task structure). The problems arise when trying to use a normal departmental cluster of workstations which are usually made of different types of machines. These machines are usually loaded and, even worse, with many interactive users. A dynamic load balancing approach is not successful enough due to the communication overhead, so static approaches taking into account the load of each machine and having the possibility of redistributing the load are more effective.

There are a number of issues to be further analyzed. There are many factors that affect the performance of our application running on this kind of cluster. It would be beneficial to obtain figures of the communication speed of PVM in the target machine in order to take into account the overhead introduced by the communication/synchronization operations. Some partial experiments done in this work show a communication time of 2-4 millisecond per message, which is rather high. Another interesting effect to measure is the influence of a multiprogrammed workload especially on the communication cost. Another interesting thing to be evaluated is the influence of a NFS file system on the communication speed, not only its average values but also its peak values, because the performance of parallel code is strongly affected by peak values (it degrades the load balancing).

In this work, the effective power of each machine is measured taking into account which percentage of CPU is available for our application. We believe that in addition to this parameter, we could obtain a better load balancing if the number of interactive users in each machine is also considered.

All these measurements will provide a better understanding of the performance of the parallel code and they may lead to improvements of it.

References

1. Labarta, J., Girona, S., Pillet, V., Cortes, T., Cela, J.M.: A Parallel Program Developement Environment: CEPBA/UPC Report No. RR-95/02 (1995)
2. Ahuja, R., Magnanti, T., Orlin, J.: Network Flows, Theory, Algorithms and Applications: Prentice Hall, (1993)
3. Dembo, R.S.: Scenario optimization: Annals of Operation Research *30* (1991) 43-72.
4. Gueist, A., Beguelin, A., Dongarra, J., Jiang, W., Manchel, R., Sunderam, V.: PVM 3 User's Guide and Reference Manual: Oak Ridge National Laboratory, TM-12187 (May 1994)
5. Manchek, R.: Design and Implementation of PVM 3 M.S. Degree, Thesis, University of Tennessee (May 1994)

User-Level VSM Optimization and its Application*

Rupert W. Ford, Andrew P. Nisbet and J. Mark Bull

Centre for Novel Computing, Department of Computer Science,
The University of Manchester, U.K.

Abstract. This paper describes user-level optimisations for virtual shared memory (VSM) systems and demonstrates performance improvements for three scientific kernel codes written in Fortran-S and running on a 30 node prototype distributed memory architecture. These optimisations can be applied to all consistency models and directory schemes, whether in hardware or software, which employ an invalidation based protocol. The semantics of these optimisations are carefully stated. Currently these optimisations are performed by the programmer, but there is much scope for automating this process within a compiler.

1 Introduction

Virtual shared memory (VSM) systems provide the illusion of a shared address space on distributed memory architectures. A shared memory programming model is attractive because it is simple to program, thus speeding the implementation and porting of parallel programs, and enabling the parallelisation of complex adaptive programs which may be difficult to implement in message-passing. However, for problems with a regular communication pattern, VSM can fair significantly worse than message-passing. One of the main reasons for this performance difference is that a program's communication structure is explicitly coded in message-passing, whereas VSM implementations seldom provide facilities which allow this application-specific knowledge to be exploited.

Scalable VSM systems use a structure of directories (which typically record the location and the read/write access permissions of copies of data) and invalidation/update based protocols to maintain consistency. In the sequential consistency model with an invalidation protocol an attempt to write a new value is delayed until all remote copies are invalidated. Performance can be degraded both by the delay on the writing node, and by the resulting network traffic. A number of weak consistency models have been developed which offer the benefit of performance improvements of the order of 10% to 40% [11] over sequential consistency. However, weakening the consistency model increases the complexity of the programming model and complicates the coding of applications. Message-passing can be viewed as the ultimate weak consistency model where the programmer is entirely responsible for managing consistency.

* This work was funded by the U.K. Meteorological Office, the ESPRIT SODA and the ESPRIT APPARC projects.

VSM systems research has focussed on efficient software/hardware mechanisms for the implementation of sequential consistency, and on the development and implementation of weaker consistency models. The thrust of this paper is not a new consistency model, rather it is the use of application-specific knowledge to alleviate VSM performance problems. We describe the application of two user-level mechanisms; *local_invalidate (address)* and *local_exclusive (address, flag)*, which can be used in conjunction with knowledge of communication structure to reduce invalidation traffic and to remove synchronisation points whilst preserving program semantics. The work is applicable to any directory-based consistency model employing an invalidation based protocol.

In Section 2 of this paper we briefly discuss previous related work on VSM optimisation. We describe the prototype distributed memory architecture and the semantics and implementation of two user-level optimisations for VSM. Section 3 describes the applications which we use in our experiments. Section 4 describes the experimental method used to produce the results which are presented in Section 5. Conclusions and future work plans are discussed in Section 6.

2 VSM Policy

VSM Policy refers to the functionality and features provided by a VSM implementation. Policy is therefore determined by the underlying hardware/software used to support VSM on a computer platform. Most commercial platforms provide interfaces to sequentially consistent VSM, whereas research platforms may provide interfaces to multiple consistency models. Previous work on optimisations to VSM policy has investigated dynamically switching between consistency models [2] and between update/invalidation based protocols [12], based on static and dynamic analysis of a program's communication structure. Compiler technology that exploits such research by transforming sequentially consistent programs into programs which dynamically select the appropriate consistency model is very much in its infancy [10]. VSM optimisations which are similar to those described in this paper have been presented in [13, 8, 5]. Previous work has not presented detailed application studies, or has provided experimental results for implementations of VSM optimisations that are believed to be *unprotected*[2]. The work presented in this paper describes how protected implementations of *local_invalidate* and *local_exclusive* optimisations may be applied to applications which are representative of different communication structures. Such optimisations can be used to reduce invalidation traffic and to remove synchronisation points. Ongoing work (see Section 6) is attempting to automate the application of VSM optimisations within a research compiler infrastructure.

---

[2] Protection prevents an application from reading/writing an address outside of its own address space.

2.1 Target Architecture

The target architecture for the experiments is a 30 node EDS prototype [14] which was developed as a result of the European Declarative System (EDS) ESPRIT project EP2025. Each node has two 40MHz Cypress SPARCs with cache-coherent Memory Management Units (MMUs), external 64K direct-mapped unified caches and 64M RAM shared store. The Processing Unit executes user code, the operating system kernel and associated servers. The System Support Unit is used as a communications processor that handles the protocols for sending and receiving messages over the delta interconnection network having a link bandwidth of 20MByte/s. The prototype runs the EDS Machine EXecutive (EMEX) operating system. A supervisor level process known as the *mapper* provides a fixed-distributed scheme [9] for the management of VSM consistency on a 4K page basis through the manipulation of access permissions in MMU Page Table Entries (PTEs). In this scheme, write ownership may move but there is one fixed mapper which is responsible for dispatching ownership of particular virtually shared pages. The assignment of pages to mappers is based on the locus of creation of VSM. In our experiments, all VSM is created on the node on which an EMEX executable is initially loaded.

2.2 VSM Optimisations

Semantics: The *local_invalidate* and *local_exclusive* mechanisms currently assume that the data whose access permissions are to be manipulated are present in local memory. An error is returned if data is not present in local memory. The *local_invalidate(address)* mechanism revokes all access rights associated with an address such that a subsequent read or write access will cause a page fault that will in turn contact the VSM consistency management software (the mapper) in order to resolve the fault. The *local_exclusive(address,flag)* mechanism promotes the local access permissions from read to write, and additionally contacts the mapper in order to update the coherence directory to reflect a transition to single writer if the *flag* is set. Subsequent write accesses to the address will not cause a page fault. Store may be inconsistent with respect to coherence directory structure during a code section but will always leave the section in a consistent state.

EDS Target Implementation: The mechanisms are implemented as a single EMEX operating system call which manipulates access permissions in the appropriate PTEs and an additional mapper remote-procedure-call (RPC) which is used to update a coherence directory to reflect a transition to single writer semantics. Users are only able to manipulate the access permissions of user-created VSM addresses. Thus, protection is preserved as users cannot read/write any addresses outside of their own address space.

3 Applications

In this Section we describe the three kernel codes used in our experiments. These represent a simple explicit partial differential equation (PDE) solver, an iterative solver and PDE solvers with staggered grids and/or interpolation[3].

3.1 Simple PDE Solver

The code described in this Section implements a simple, two dimensional, five point stencil, finite difference method. The communication pattern is representative of the communication pattern in all stencil methods which form the the core of many scientific applications. This communication pattern can also be found in pde1 from the Genesis benchmark suite [6]. This method can be simply represented by three loop nests and two arrays. The first loop steps over iterations and encloses the other two loop nests. The second loop nest updates the new array using the values from the other array (OLD) which holds the values from the previous iteration. The third loop nest then copies the contents of the main array into the new array ready for the next iteration.

In single address space parallel machines the two inner loop nests can be simply parallelised. Parallelisers such as KAP [7] will do this automatically. A barrier is included after the second loop nest to ensure all nodes have read the correct values of OLD before they are updated in the third loop nest. A barrier is included after the third loop nest to ensure all values of OLD have been updated before they are read in the second loop (on the next iteration). Automatic parallelisers add barriers by default after parallel loops. For distributed memory parallel machines there must also be communication of parts of the array OLD. These overlapping areas are usually called halo regions.

Assuming the arrays are partitioned by columns, in the second loop nest a neighbouring node will take a copy of a node's data residing in the halo region. In the third loop nest this data will then be updated. In sequentially consistent VSM all remote copies must be invalidated before a new value is updated. Thus invalidation traffic is generated on the network and the node must stall, see Figure 1a. However when new values are written into the OLD array we know no other node needs to read them; this is ensured by barrier synchronisation. We also know where copies have been taken. We can therefore replace the default invalidation mechanism with a distributed invalidation mechanism. The node that reads the remote value is made responsible for locally invalidating its copy; this can be done at any point. Given this, the writing node can override the consistency management (which thinks there are valid remote copies) and locally set the tag value to exclusive and immediately proceed with the write. This cannot occur until all nodes which require a read copy have executed their read, so the *local_exclusive* must be placed after the first barrier and before the write. We therefore remove unnecessary invalidation traffic and node stalls, see Figure 1b. A similar approach is used in [13] when implementing their EM3D code using Tempest, a coherence interface, on Typhoon, a simulated VSM architecture.

---

[3] These codes are available by anonymous ftp at ftp.cs.man.ac.uk in /pub/cnc/vsm.

Fig. 1. Invalidation traffic on pages of OLD array.

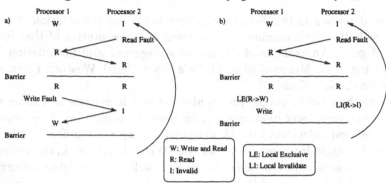

3.2 Iterative Solver

The code described in this Section implements a simple matrix iteration. The communication pattern represents the communication pattern found in iterative solvers such as the conjugate gradient method and SOR. Its communication pattern is similar to that found in the previous section, however, instead of a single remote reader we have P-1 remote readers, where P is the number of nodes used. This communication pattern can also be found in CG from the NAS parallel benchmarks [1] and in Solver from the Genesis benchmark suite [6].

This method can be simply represented by three loop nests, two vectors and one matrix. The first loop steps over iterations and encloses the other two loop nests. The second loop nest updates the new vector using the constant matrix and the values from the other vector (VOLD), which holds the values from the previous iteration. The third loop then copies the contents of the new vector into the vector VOLD ready for the next iteration.

In single address space parallel machines the two inner loop nests can be simply parallelised by partitioning the new vector. Parallelising software such as KAP [7] will do this automatically. A barrier is included after the second loop nest to ensure all nodes have read the correct values of VOLD before it is updated in the third loop. A barrier is included after the third loop to ensure all values of VOLD have been written before they are read in the second loop (in the next iteration). For distributed memory parallel machines there must also be communication of the array VOLD. In the second loop nest each node reads all of VOLD. In the third loop nest parts of VOLD are updated by separate nodes. In sequentially consistent VSM all remote copies must be invalidated before the new value is written, generating invalidation traffic on the network and causing the node to stall.

However, as in the example in the previous section, we know that when values in VOLD are updated no other node needs to read them; this is ensured by barrier synchronisation and we also know where copies are taken. In the same manner we can therefore replace the default invalidation mechanism with a distributed invalidation mechanism removing unnecessary invalidation traffic and node stalls.

3.3 Staggered Grid and/or Interpolation

The code described in this Section implements the explicit solution of a mixed derivative PDE. The communication pattern is representative of that found in staggered grids. An example of the use of a staggered grid formulation can be found in the U.K. Meteorological Office's Operational Weather Forecast and Climate Prediction Model [4].

This method can be simply represented by four loop nests and three arrays. The first loop steps over iterations and encloses the other three loop nests. The second loop nest calculates the *DX* derivative, using finite differences in the x-direction. The third loop nest calculates the *DXDY* derivative, in the y-direction. The fourth loop nest interpolates the results back to the original co-ordinates (see Figure 2a). In single address space parallel machines the three inner loop nests can be simply parallelised. Again parallelisers such as KAP [7] will do this automatically. In general a barrier is included after the fourth loop nest to ensure all nodes have updated the array (marked as +) before it is read in the second loop (on the next iteration). In fact, this is not needed in our implementation as we only partition on the outer dimension of the arrays, and therefore all reads of the array are local in the second loop, see Figure 2a. A barrier is included after the second loop nest to ensure all values of *DX* are updated before they are read in the third loop nest. A barrier is included after the third loop nest to ensure all values of *DXDY* are updated before they are read in the fourth loop nest. However, the shared values of *DX* and *DXDY* can in fact be calculated

Fig. 2. a) Original dependencies b) Dependencies with multiple writers

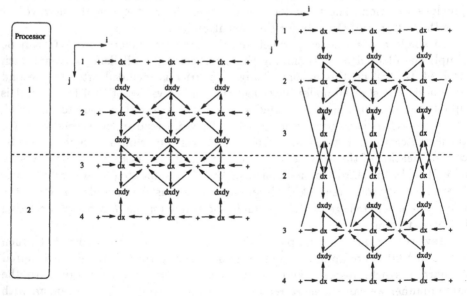

redundantly, see Figure 2b. This removes the need for the barriers after loop nests two and three but means a barrier must be added after the fourth loop nest. This further reduces the number of required barriers by one, although the amount of communication between nodes remains the same. This can be simply implemented in our VSM system by allowing multiple writers to the same address. All that is required is a small change in the indexing of the loops. When the redundant computation is complete we simply invalidate one of the writers. A similar approach is used in [10] to implement reductions, where multiple writers to the same address perform partial sums and a combine call is added after the loop. Note that the redundant computation could alternatively be performed with local temporaries, but this necessitates extra code restructuring.

4 Experimental Method

In VSM systems the unit of consistency between nodes varies in size from a few words with hardware support to whole pages in software only implementations. As the consistency unit is greater than a single word such systems can suffer from the problem of false sharing, where two or more nodes attempt to write to distinct elements which happen to be on the same coherency unit. This leads to a "ping pong" effect. In order to avoid these problems we chose our problem sizes, array alignment and the number of nodes so that false-sharing is eliminated.

In the second and third codes, to allow reasonable problem sizes, we expand the vectors so that they occupy one page per node, and access them with a non-unit stride. This is simply achieved by adding an inner dimension to the vectors of the appropriate size and indexing the vectors as VECTOR(1,INDEX) rather than VECTOR(INDEX). As we are interested in the communication and relative performance of the code, with and without our distributed invalidation mechanism, the effects on performance due to cache use and code generation are of no significance, but could be avoided by additional restructuring.

4.1 Execution Time Measurements

The experiments record the time in seconds (with micro-second resolution) taken to execute iterations 2 to N inclusive. The first iteration was not timed, in order to exclude any variable one-off overheads such as that caused by initial page faults on data/code segments. The PDE Solver (size 1024×512) and the Iterative Solver (size 1024×1024) were evaluated with $N = 51$ whilst the Staggered Grid codes of sizes 1024×1024, 512×512 and 256×256 was evaluated with $N = 21$. Experiments were performed on 2, 4, 8 and 16 nodes. Sequential Fortran codes with annotations were transformed by the Fortran-S [2] compiler into SPMD Fortran and then into "C" (using f2c) from which an EMEX loadable executable is produced. Sequential code was generated by manually removing all parallelisation annotations, and any unnecessary calls to runtime libraries in the "C" prior to generating an executable. Sequential timings were used to calculate the *Naive Ideal* lines in Figures 3, 4 and 5.

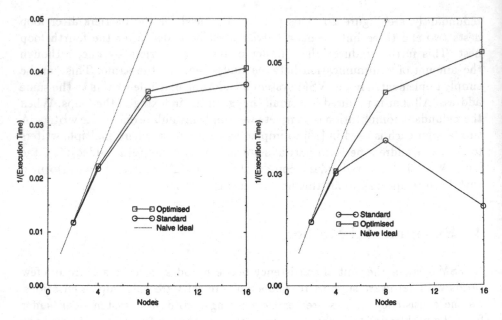

Fig. 3. Temporal Performance of (a) Simple PDE Solver 1024x512 and (b) Iterative Solver 1024x1024 Codes

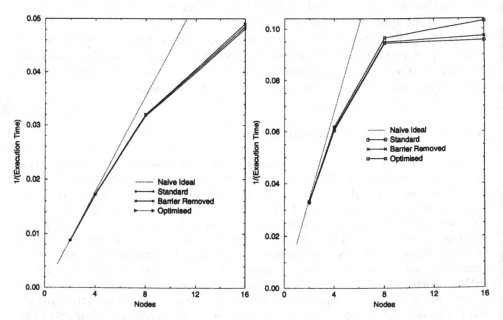

Fig. 4. Temporal Performance of (a) 1024x1024 Staggered Grid (b) 512x512 Staggered Grid Codes

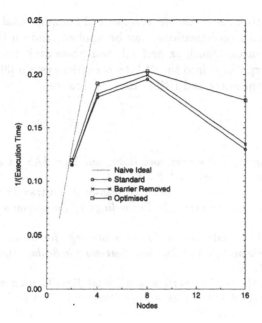

Fig. 5. Temporal Performance of 256x256 Staggered Grid Code

5 Results

Figures 3 to 5 show the temporal performance of the codes described in Section 3. The PDE Solver shows some improvement with optimisation applied but other overheads dominate execution time. The Iterative Solver shows a greater performance improvement over the unoptimised code, giving almost three times the temporal performance on 16 nodes. This is to be expected, as the benefit of the optimisation increases with the number of readers, which in turn increases with the number of nodes. The Staggered Grid codes give increasing relative benefit as the problem size is reduced, but barriers are not a significant factor of the total overheads in these experiments. Barrier overhead would become significant if the code were run on a sufficiently large number of nodes. On the EDS prototype performance improvements are made in spite of performing significant redundant computation.

6 Conclusions and Future Work

The major benefits of these VSM optimisations are when transitions from multiple readers to a single writer are repeatedly performed, especially when there are a large number of readers. The optimisations are flexible, powerful and applicable to all coherency schemes employing an invalidation-based protocol. The optimisations can be used to bring the performance of VSM applications closer to those of message-passing, where communication structure can be determined.

Future work is to focus on the development of compiler analysis to automatically determine when optimisations may be applied, and on the evaluation of other VSM optimisations (such as prefetch and poststore). Such analysis techniques will be incorporated into the MARS compiler system [3] for VSM under development at Manchester and IRISA Rennes, France.

References

1. D. Bailey, J. Barton, T. Lasinski, and H. Simon. The NAS Parallel Benchmarks. *NASA Technical Memorandum 103863*, 1993.
2. F. Bodin, L. Kervella, and T. Priol. Fortran-S: A Fortran Interface for Shared Virtual Memory Architecture. In *Proceedings of Supercomputing*. IEEE Press, November 1993.
3. F. Bodin and M.F.P. OBoyle. A Compiler Strategy for SVM. In *3rd Workshop on Languages, Compilers and Runtime Systems for Scalable Computing*. Kluwer Press, May 1995.
4. A. Dickinson et al. Implementation and Initial Results from a Parallel Version of the Meteorological Office Atmosphere Prediction Model. In *Coming of Age: Proceedings of the Sixth ECMWF Workshop on the use of Parallel Processors in Meteorology*. World Scientific, 1994.
5. B. Falsafi et al. Application-Specific Protocols for User-Level Shared Memory. In *Supercomputing 94*. IEEE Press.
6. A.J.G Hey. The GENESIS Distributed-Memory Benchmarks. *Parallel Computing 17(10-11)*, 1991.
7. Kuck and Associates Inc. Champaign Illinois. *KAP User's Guide*, 1988.
8. A.R. Lebeck and D.A. Wood. Dynamic Self-Invalidation: Reducing Coherence Overhead in Shared-Memory Multiprocessors. In *ISCA95*, pages 48–59, 1995.
9. K. Li and P. Hudak. Memory Coherence in Shared Virtual Memory Systems. *ACM Transactions on Computer Systems*, 7(4):321–359, 1989.
10. R. Mirchandaney, S. Hirandani, and A. Sethi. Improving the Performance of DSM Systems via Compiler Involvement. In *Proceedings of Supercomputing*, 1994.
11. D. Mosberger. Memory Consistency Models. *ACM SIGOPs Review*, 27(1), 1993.
12. F. Mounes-Toussi and D.J. Lilja. The Potential of Compile-Time Analysis to Adapt the Cache Coherence Enforcement Strategy to the Data Sharing Characteristics. *IEEE Transactions on Parallel and Distributed Systems*, 6(5), May 1995.
13. S.K. Reinhardt, J.R. Larus, and D.A. Wood. Tempest and Typhoon: User-level Shared Memory. In *Proceedings of the 21st Annual Iternational Symposium on Computer Architecture*, 1994.
14. C.J. Skelton et al. EDS a Parallel Computer System for Advanced Information Processing. In *Parallel Architectures and Languages Europe, PARLE92*, pages 3–18, 1992.

Benchmarking the Cache Memory Effect *

Vladimir Getov

School of Computer Science and Information Systems Engineering
University of Westminster, 115 New Cavendish St., London W1M 8JS, U.K.

Abstract. A new performance model of the memory hierarchy is first
introduced, which describes all possible scenarios for the calculation process, including the important case when the cache memory is bypassed. A
detailed study of each scenario is then given along with the derivation of
corresponding formulae. In these formulae the cache load time associated
with the penalty which must be paid to transfer data between the main
memory and the cache is also taken into account. A two-parameter linear
model for performance characterisation of cache memory effect is introduced. The double-performance parameter, n_2, is defined to describe the
performance degradation for problem sizes that do not fit into the cache
memory. This parameter determines the problem size required to preserve twice the asymptotic performance. Excellent agreement is shown
between the estimated performance figures and several benchmark measurements on iPSC/860.

1 Introduction

The increased architectural complexity of the new parallel machines, when compared with traditional vector supercomputers, make the problem of reliable performance prediction exceptionally problematic. In particular, inefficient cache
utilization leads to a significant performance penalty on massively-parallel computers as the performance losses in a single node must be multiplied by the
total number of nodes in the system. The challenge is to create parallel programs that can take advantage of the fastest possible memory access but this
is extremely difficult to achieve as the cache memory is usually programmer-
transparent from a high-level language. Nevertheless, there are implicit methods
for tuning the efficiency of cache memory through the implementation of the
so-called block algorithms [1]. The particular block size used in block algorithms
is machine-dependent, and thus becomes a tuning parameter. The diversity of
massively-parallel computers, however, can not be adequately captured by a
single blocking parameter. We need a limited but sufficient set of parameters
to estimate the performance of specific parallel computers with reasonable accuracy. Therefore, performance characterisation of the cache memory effect is

---

* The results presented here were previously published in an extended version of this
  article which appeared in *Supercomputer 63*, vol XI (5), 1995. They are re-produced
  with kind permission of ASFRA BV, The Netherlands.

crucially important for achieving the highest possible overall performance on current massively-parallel computers.

For vector supercomputers without a cache memory, Hockney's ($r_\infty,n_{\frac{1}{2}}$) characterisation proved to be a valuable abstraction for estimating applications' performance [2]. The performance of vector computers is characterised by their pipelines, which have a significant start-up time, t_0. The share of this start-up time in the total execution time varies depending on the lengths of the vectors involved. The execution time, t, for a vector instruction of length n is approximated by a straight line crossing the t-axis above the origin of the coordinate system (1). In a vector pipeline the maximum performance occurs for infinitely long vectors and is designated by r_∞. One can tell how close the performance will be to the asymptote by comparing the vector length with the second parameter $n_{\frac{1}{2}}$. If the vector length equals $n_{\frac{1}{2}}$, the performance will be half of the asymptotic maximum of r_∞. Therefore $n_{\frac{1}{2}}$ is known as the half-performance parameter.

$$t = \frac{n + n_{\frac{1}{2}}}{r_\infty} \qquad\qquad r = \frac{n}{t} = \frac{r_\infty}{1 + \frac{n_{\frac{1}{2}}}{n}} \qquad (1)$$

2 Cache memory characterisation

2.1 Performance model

Most often within the various memory hierarchy models that have been developed so far, the ALU and registers are at the bottom, followed by the cache, the main memory, the interconnection network, and finally the magnetic disks and tapes at the top. Toward the top of the hierarchy, memory is larger and cheaper, but slower. Since operations such as multiplication and addition must be done at the ALU/registers level, data has to move down through the various memory layers to the bottom to be processed, and then up again to be stored. Clearly, codes must be designed to minimizes the memory traffic in the hierarchy and to reuse data with care in order to run as close as possible to the sustained calculation rate instead of running at memory access speed.

In general, none of the existing memory hierarchy models cover adequately the notable case when the cache is bypassed and the data are transferred directly between the processor and the main memory. In fact, sometimes this is really the best option, as there is no guarantee that the efficiency of the cache memory will always be very high. It depends on the locality of reference, which is a key aspect of performance on DM systems with a hierarchical memory model for each node. Indeed, there are many algorithms and programs which lack this property. In such cases the cache miss rate will be high and the cache may actually hurt the execution time. Therefore, some state-of-the-art computers provide a way to bypass the cache memory, and use direct access from the processor to the main memory instead.

The implementation of the above feature requires a direct data path within the hardware, between the CPU and the main memory. Secondly, there must be corresponding assembly instructions which allow caching to be inhibited by software, if needed. Thirdly, the compiler should implement an appropriate strategy to decide whether to use or to bypass the cache. For instance, the i860 microprocessor has two forms of the load instruction: `ld.x` loads a word into a register and updates the data cache if the line containing that word is not currently in cache; `pfld.x`, however, is a pipelined load instruction that loads data from main memory directly into a register bypassing the cache. There is only one store instruction (`st.x`), as the i860 does not update the cache on store misses. A caching strategy implemented within the compiler helps to decide which type of instructions to use when generating the executable code. The behaviour of such parallel systems is more complex and does not conform to the simple linear pipeline performance model described above. This phenomenon has been observed many times in the GENESIS benchmarks measurements. Similar results have also been reported by other researchers [3, 4, 5].

Let us consider a performance model which deals only with those layers from the memory hierarchy that are directly involved in calculations; namely the CPU, the cache and the main memory (Fig. 1). They are interconnected with each other by data paths, including a direct path from the CPU to the main memory. The data path between the cache and the main memory is characterised by the cache load time, $t^{(l)}$, associated with the penalty which must be paid to transfer data between the cache and the main memory. It is also important to consider both the fast cache calculation rate, $r_\infty^{(1)}$, and the slow main memory access rate, $r_\infty^{(2)}$. Furthermore, our basic notation includes the number of memory elements accessible through the cache memory, $n^{(1)}$; the number of memory elements accessible directly from the main memory, $n^{(2)}$; and the cache size, n_{cb}. Note that $n = n^{(1)} + n^{(2)}$. We also define the number of cache loads, $q^{(l)}$, as the greatest integer less than or equal to the ratio n/n_{cb}, $q^{(l)} = \left\lfloor \dfrac{n}{n_{cb}} \right\rfloor$.

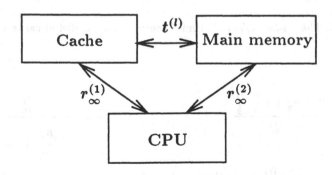

Fig. 1. Performance model

Here, we assume that the entire problem fits freely in the main memory for the whole execution period of a given program. Therefore, the external magnetic memory is excluded from the performance model for simplicity. We also assume a separate characterisation of the communication performance for the message-passing programming model [6]. In general, there are three possible scenarios for the calculation process within this model.

2.2 All-through-cache scenario

In this case the entire memory access is performed through the cache ($n^{(2)} = 0; n = n^{(1)}$), and therefore we call this option the *all-through-cache* scenario. Most existing memory hierarchy models only describe this scenario, where the execution time is a step function of the number of memory elements (Fig. 2). Using the ($r_\infty, n_{\frac{1}{2}}$) notation, the execution time can be represented as a sum of three terms (2). The first term is the total number of memory elements, divided by the fast cache calculation rate[2]. This is actually the time component spent in real calculations. The second is the initial startup time which takes into account the pipeline effect, if any, in the same formula. The third term is the cache load time multiplied by the floor function, $q^{(l)}$. One has to pay this time overhead, which is proportional to the total number of memory elements required, in order to benefit from the fast cache calculation rate, if $n > n_{cb}$.

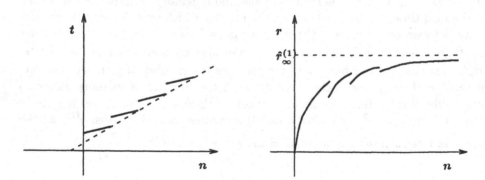

Fig. 2. The *all-through-cache* timing/performance model of cache memory.

$$t = \frac{n + n_{\frac{1}{2}} + q^{(l)} n_{\frac{1}{2}}^{(l)}}{r_\infty^{(1)}} \qquad\qquad r = \frac{n}{t} = \frac{r_\infty^{(1)}}{1 + \dfrac{n_{\frac{1}{2}}}{n} + \dfrac{q^{(l)} n_{\frac{1}{2}}^{(l)}}{n}} \qquad (2)$$

The parameter $n_{\frac{1}{2}}^{(l)}$ shows the number of memory elements which would have been involved in useful calculations at the fast cache calculation rate during

---

[2] Throughout this section we assume that the number of floating point operations equals the number of memory elements, which is true if the computational intensity, $f = 1$. In the case of a real code, the number of memory elements in all analytic expressions should be replaced by the corresponding flop count.

the period of the cache load time. This parameter is a constant which is added to the aggregate $n_{\frac{1}{2}}$ each time a jump in the timing model occurs. Therefore, for the *all-through-cache* scenario the aggregate half-performance parameter is a variable which depends on the total number of memory elements required. If $n \to \infty$, we may neglect the effect of $n_{\frac{1}{2}}$ and the floor properties of $q^{(l)}$. In this case $r \to \hat{r}_\infty^{(1)}$, where

$$q_{n \to \infty}^{(l)} = \frac{n}{n_{cb}} \qquad\qquad \hat{r}_\infty^{(1)} = \frac{r_\infty^{(1)}}{1 + \dfrac{n_{\frac{1}{2}}^{(l)}}{n_{cb}}} \qquad\qquad (3)$$

Thus, the aggregate asymptotic performance, $\hat{r}_\infty^{(1)}$, for the *all-through-cache* scenario is lower than the fast cache calculation rate, $r_\infty^{(1)}$.

2.3 *All-out-of-cache* scenario

There are two options within this scenario. In the first one, the cache memory is completely bypassed and the model follows the behaviour of pipeline or scalar computations as described in section 2.2 ($n^{(1)} = 0; n = n^{(2)}$). The reason for choosing this option may simply be the lack of a suitable programming section to fit into the cache. Alternatively, it may be because of the confidence that switching off the cache will result in better performance for sufficiently large problems, which may well be true in quite some cases [4]. Whatever the reason, this option shows a complete absence of suitability between the specific cache memory architecture and the application code under consideration.

In the second option, the cache memory is out of use ($n^{(1)} = 0; n = n^{(2)}$) only for $n > n_{cb}$. This may be due to a loop-carried dependence within the code, which makes it difficult for the compiler to exploit the cache (for example by strip-mining), if the problem is too large to be held entirely in it [7]. If, however, the code can fit into the cache for $n \leq n_{cb}$, the model follows the *all-through-cache* behaviour and the performance asymptotically tends to $r_\infty^{(1)}$ (Fig. 3). The graph for $n = n_{cb}$ is not continuous; the performance drops down substantially to $r_\infty^{(2)}$ and remains at this value for any $n > n_{cb}$.

$$t = \begin{cases} \dfrac{n}{r_\infty^{(1)}} + t_0^{(1)}, & \text{if } n \leq n_{cb} \\[2ex] \dfrac{n}{r_\infty^{(2)}}, & \text{if } n > n_{cb} \end{cases} \qquad\qquad r = \begin{cases} \dfrac{r_\infty^{(1)}}{1 + \dfrac{n_{\frac{1}{2}}^{(1)}}{n}}, & \text{if } n \leq n_{cb} \\[3ex] r_\infty^{(2)}, & \text{if } n > n_{cb} \end{cases} \qquad (4)$$

2.4 *Mixed codes* scenario

In reality even small code fragments written in high-level languages or vendor-supplied library routines produce a mixture of both the *all-through-cache* and

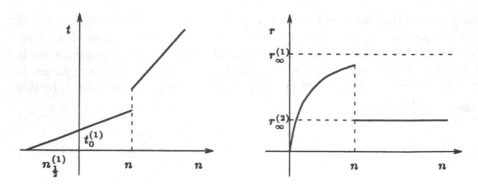

Fig. 3. The combined *all-through-cache/all-out-of-cache* timing/performance model of cache memory.

the *all-out-of-cache* behaviours. Therefore, a consideration of the *mixed codes* scenario is necessary, where the execution requires memory elements accessible from both the cache and the main memory ($n^{(1)} > 0; n^{(2)} > 0$). This scenario follows the *all-through-cache* behaviour if the code can fit into the cache ($n \leq n_{cb}$). If, however, the number of memory elements is bigger than n_{cb}, the total execution time is a sum of two components - the time for the *through-cache* calculations, $t^{(1)}$, and the time for the *out-of-cache* calculations, $t^{(2)}$. Thus, the execution time for the *mixed codes* scenario is:

$$t = \begin{cases} \dfrac{n^{(1)}}{r_\infty^{(1)}} + t_0^{(1)}, & \text{if } n \leq n_{cb} \\[3mm] \dfrac{n^{(1)}}{r_\infty^{(1)}} + \dfrac{n^{(2)}}{r_\infty^{(2)}} + t_0^{(1)} + t_0^{(2)} + q^{(l)}t^{(l)}, & \text{if } n > n_{cb} \end{cases} \tag{5}$$

If $n \to \infty$, we may neglect the effect of the startup time components, $t_0^{(1)}$ and $t_0^{(2)}$. In this case $t \to t_\infty$, where

$$t_\infty = \frac{n^{(1)}}{r_\infty^{(1)}} + \frac{n^{(2)}}{r_\infty^{(2)}} + \frac{n^{(1)}}{n_{cb}} \frac{n_{\frac{1}{2}}^{(l)}}{r_\infty^{(1)}} = \frac{n^{(1)}}{r_\infty^{(1)}} \left(1 + \frac{n_{\frac{1}{2}}^{(l)}}{r_\infty^{(1)}} \right) + \frac{n^{(2)}}{r_\infty^{(2)}} = \frac{n}{r_\infty}. \tag{6}$$

After substituting with the formula for $\hat{r}_\infty^{(1)}$ (3) in the last equality of (6) we can derive the aggregate asymptotic performance, r_∞, for the *mixed codes* scenario which is the weighted harmonic mean

$$r_\infty = \frac{n}{\dfrac{n^{(1)}}{\hat{r}_\infty^{(1)}} + \dfrac{n^{(2)}}{r_\infty^{(2)}}}. \tag{7}$$

The performance formula for application codes following this scenario is:

$$r = \begin{cases} \dfrac{r_\infty^{(1)}}{1 + \dfrac{n_{\frac{1}{2}}^{(1)}}{n}}, & \text{if } n \leq n_{cb} \\[3ex] \dfrac{r_\infty}{1 + \dfrac{r_\infty}{n}\left(t_0^{(1)} + t_0^{(2)}\right)}, & \text{if } n > n_{cb} \end{cases} \tag{8}$$

Similarly to the other two scenarios, a pipeline effect is observed if the problem can fit into cache. Here, the $(r_\infty, n_{\frac{1}{2}})$ performance characterisation is perfectly appropriate, with the two parameters being $r_\infty^{(1)}$ and $n_{\frac{1}{2}}^{(1)}$. This model applies until the transition point at $n = n_{cb}$, where the *mixed codes* scenario does not have continuous properties. If the problem is too large to be held entirely in the cache, this scenario enters a completely different behaviour which is a superposition of the *through-cache* and the *out-of-cache* models. Indeed, in this region the performance, r, is a decreasing function which tends towards r_∞ according to the weighted harmonic mean (7). The reason for this is the slow *out-of-cache* calculations which now coincide with the fast *through-cache* calculations. Of course, the exact execution time also depends on the problem size and the other parameters as shown in formula (5). None of the existing linear models can adequately describe the above phenomenon which is specific to the cache memories. Therefore, a new timing/performance model is necessary (Fig. 4).

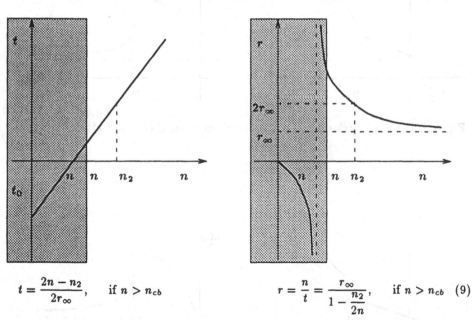

$$t = \frac{2n - n_2}{2r_\infty}, \quad \text{if } n > n_{cb} \qquad\qquad r = \frac{n}{t} = \frac{r_\infty}{1 - \dfrac{n_2}{2n}}, \quad \text{if } n > n_{cb} \tag{9}$$

Fig. 4. The two-parameter (r_∞, n_2) timing/performance model of the cache memory effect. The graphs and the corresponding formulae (9) satisfy the requirements of the model only for the region outside the shadow.

The execution time is modelled by a straight line crossing the t-axis beneath the origin of the coordinate system. The negative startup time, t_0, shows the exact time that is saved when using the cache memory. This is the guaranteed minimum time interval which has to be added to the actual execution time if the cache memory is switched off by the compiler or it does not exist at all. The performance graph is a reciprocal function like in the pipeline effect model. For the cache memory effect, however, the requirements of the model are satisfied only in part of the decreasing region of the curve. Therefore, in order to provide a computer user with an appropriate problem-related parameter for the decreasing performance curve, we need to define the new *double-performance parameter*, n_2. The value of n_2 is twice the intercept of the vertical asymptote (n_a) with the n-axis. By contrast with the half-performance parameter $n_{\frac{1}{2}}$, this parameter determines the memory elements that are required in order to preserve twice the asymptotic memory access rate, $r_\infty^{(2)}$. After a little rearrangement the timing/performance formulae (5) and (8) can be rewritten in terms of the new double-performance parameter, $n_2^{(2)}$.

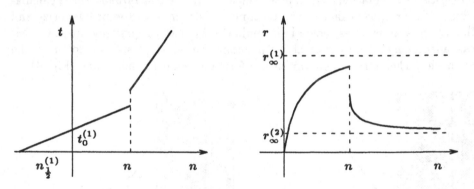

Fig. 5. The timing/performance model for mixed codes if the ratio $n^{(1)}/n^{(2)} = const.$

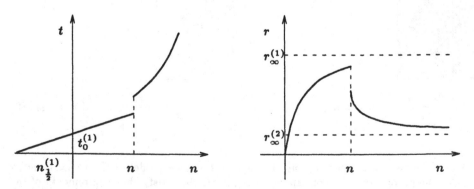

Fig. 6. The timing/performance model for mixed codes if the ratio $n^{(1)}/n^{(2)} = var.$

$$t = \begin{cases} \dfrac{n^{(1)}}{r_\infty^{(1)}} + \dfrac{n_{\frac{1}{2}}^{(1)}}{r_\infty^{(1)}}, & \text{if } n \leq n_{cb} \\[3ex] \dfrac{n}{r_\infty^{(1)}} + \dfrac{n_{\frac{1}{2}}^{(1)}}{r_\infty^{(1)}} - \dfrac{n_2^{(2)}}{2r_\infty^{(2)}}, & \text{if } n > n_{cb} \end{cases} \tag{10}$$

$$r = \begin{cases} \dfrac{r_\infty^{(1)}}{1 + \dfrac{n_{\frac{1}{2}}^{(1)}}{n}}, & \text{if } n \leq n_{cb} \\[5ex] \dfrac{r_\infty}{1 + \dfrac{r_\infty}{n}\left(\dfrac{n_{\frac{1}{2}}^{(1)}}{r_\infty^{(1)}} - \dfrac{n_2^{(2)}}{2r_\infty^{(2)}}\right)}, & \text{if } n > n_{cb} \end{cases} \tag{11}$$

In both equations r_∞ is the weighted harmonic mean as defined by formula (7). Its actual value depends on the ratio of $n^{(1)}$ to $n^{(2)}$ as follows:

- $n^{(1)}/n^{(2)} = const$ – this means that the aggregate asymptotic performance is also a constant and the timing graph has linear properties (Fig. 5);
- $n^{(1)}/n^{(2)} = var$ – in this case the weighted harmonic mean (7) is a function of this ratio and the model of the execution time is a non-linear one (Fig. 6).

3 Examples on iPSC/860

The formulae introduced in the previous section may be rewritten as functions of the problem size, N, for a given parallel kernel or application. As mentioned earlier, in this case we should use the number of floating point operations within the mathematical expressions instead of the number of memory elements. Here, one can also define the block size, N_{cb}. This is the maximum problem size which fits into cache. All other parameters can be substituted by their measured or estimated values for the target machine. The examples given below are based on the FFT1 benchmark from the GENESIS suite [8, 9]. This benchmark follows the classical vectorised implementation of the 1-dimensional FFT algorithm [2].

The *all-through-cache* example makes use of the FFTCT1 subroutine which calculates part of a single butterfly layer within the message-passing version of the benchmark. It is a single loop which does or does not fit into cache, depending on the problem size and the block size for the target machine. The final choice of the scenario for the calculation process depends also on the compiler options. If the compiler options are -O4 -Mvect, then the execution of FFTCT1 follows the *all-through-cache* scenario (Fig. 7). The number of flops for this subroutine is $5N$, which after substitution gives the following timing/performance formulae:

$$T(N) = \frac{5N}{r_\infty} + t_0 + q^{(l)}t^{(l)} \qquad\qquad R_B(N) = \frac{r_\infty}{1 + \dfrac{n_{\frac{1}{2}} + q^{(l)}n_{\frac{1}{2}}^{(l)}}{5N}} \tag{12}$$

As it was shown in the previous section the asymptotic performance for the *all-through-cache* scenario depends on the number of cache loads required. For the first tooth of the sawtooth performance graph its value is 7.5 Mflop/s. The penalty of the first cache load allows an increase in the asymptotic performance to 8.6 Mflop/s. The influence of each subsequent cache load is smaller and smaller and the aggregate asymptotic performance is 9.1 Mflop/s.

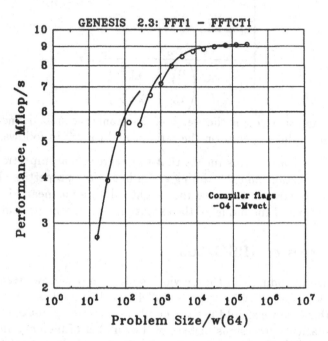

Fig. 7. *All-through-cache* example: Benchmark performance, $R_B(N)$, as a function of the problem size, N. The plot symbols are measured values, and the lines are the predicted values from the performance formula (12).

If, however, the same FFTCT1 subroutine is compiled using only -04 compiler optimisation, the execution follows the *all-out-of-cache* scenario (Fig. 8). In this experiment, the flop count is the same as in the previous example and after substitution in (4) one can obtain the following timing/performance formulae:

$$T(N) = \begin{cases} \dfrac{5N}{r_\infty^{(1)}} + t_0^{(1)}, & \text{if } N \le N_{cb} \\[2ex] \dfrac{5N}{r_\infty^{(2)}}, & \text{if } N > N_{cb} \end{cases} \qquad R_B(N) = \begin{cases} \dfrac{r_\infty^{(1)}}{1 + \dfrac{n_{\frac{1}{2}}^{(1)}}{5N}}, & \text{if } N \le N_{cb} \\[3ex] r_\infty^{(2)}, & \text{if } N > N_{cb} \end{cases} \qquad (13)$$

The asymptotic performance for small problem sizes ($N \le N_{cb}$) is 11.2 Mflop/s but directly after reaching 10 Mflop/s it drops down as the problem size becomes too big for the cache memory on iPSC/860. Here, the performance results enter a transitional area, which is characterised with a rapid performance degradation. For large enough problems the cache memory is bypassed and the performance drops to the constant value of 2.8 Mflop/s constrained by the slow main memory access rate.

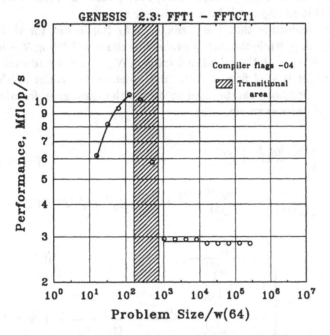

Fig. 8. *All-out-of-cache* example: Benchmark performance, $R_B(N)$, as a function of the problem size, N. The plot symbols are measured values, and the lines are the predicted values from the performance formula (13).

The *mixed codes* example makes use of the butterfly subroutine from the single node version of the 1-dimensional FFT benchmark. The performance analysis is based on the nominal flop-count of $5N \log N$ real operations. It is convenient to consider the butterfly calculation as a two-dimensional matrix of $\log N$ layers with $5N$ floating point operations in each of them. The layers represent the outer loop of the calculations, while each layer contains a variable number of inner loops. This number doubles from one layer to the next, beginning with one inner loop in the first layer. Thus, the length of the inner loops varies and some of them may or may not fit into cache.

The calculation of the butterfly subroutine requires two vectors, equal to the problem size in length. Taking into account that the cache size on iPSC/860 is 8 Kbyte and that the FFT1 benchmark is implemented using the double precision data type[3], one can calculate the block size for this experiment, $N_{cb} = 512$ word(64). The *through-cache* flop count can be derived using the fact that the bottom $\log N_{cb}$ layers of our calculation matrix can be handled entirely through the cache memory, irrespective of the chosen problem size. This is because the length of the inner loops in this part of the matrix is always less then the size of the cache memory. Note, however, that the amount of calculation work to be carried through cache still varies as the layer's length depends on the problem size. This makes the ratio *through-cache/out-of-cache* a variable and the timing function of this example will be a non-linear one.

One can determine that the *through-cache* flop count for this experiment is $5N \log N_{cb}$ flop, while the *out-of-cache* flop count is $5N(\log N - \log N_{cb})$ flop. After substituting with the calculated value of N_{cb}, one can find that the *through-cache* flop count is $45N$ flop, and the *out-of-cache* flop count is $5N(\log N - 9)$ flop. These expressions can be used to derive the timing/performance formulae for the *mixed codes* example:

$$T(N) = \begin{cases} \dfrac{5N \log N}{r_\infty^{(1)}} + t_0^{(1)}, & \text{if } N \le N_{cb} \\[3mm] \dfrac{45N}{r_\infty^{(1)}} + \dfrac{5N(\log N - 9)}{r_\infty^{(2)}} + t_0^{(1)} + t_0^{(2)} + q^{(l)}t^{(l)}, & \text{if } N > N_{cb} \end{cases} \tag{14}$$

$$R_B(N) = \begin{cases} \dfrac{r_\infty^{(1)}}{1 + \dfrac{n_{\frac{1}{2}}^{(1)}}{5N \log N}}, & \text{if } N \le N_{cb} \\[6mm] \dfrac{r_\infty}{1 + \dfrac{n_{\frac{1}{2}}^{(1)}}{5N \log N}\dfrac{r_\infty}{r_\infty^{(1)}} - \dfrac{n_2^{(2)}}{5N \log N}\dfrac{r_\infty}{2r_\infty^{(2)}}}, & \text{if } N > N_{cb} \end{cases} \tag{15}$$

The asymptotic performance for small problem sizes ($N \le N_{cb}$) is 9.5 Mflop/s and again it drops down rapidly as the problem size becomes too big for the cache memory (Fig. 9). After the transitional area, however, the performance results enter the region of the weighted harmonic mean. Here the asymptotic performance is a decreasing function which depends on the ratio *through-cache/out-of-cache* with an extreme lowest value of 2.8 Mflop/s.

4 Conclusions

As can be seen from the experimental results presented in this paper, none of the existing cache memory scenarios can guarantee the highest possible performance across different problem sizes, not to mention different algorithms and their implementations in a high-level programming language. Therefore, a great deal

---

[3] One double precision word is equal to 8 bytes or 64 bits.

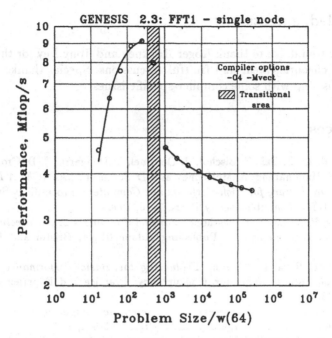

Fig. 9. *Mixed codes* example: Benchmark performance, $R_B(N)$, as a function of the problem size, N. The plot symbols are measured values, and the lines are the predicted values from the performance formula (15).

of knowledge and sensitivity is necessary in order to make the right decision and avoid the misfortune of selecting the scenario which delivers the worst possible performance. This decision is very often taken during the compilation, depending on a number of factors including the compiler switches.

The above conclusion is particularly important for the interpretation of benchmark results and also assists in the understanding of both good and bad performance. The hierarchical approach adopted by both GENESIS [10] and PARK-BENCH [11] benchmarks helps to achieve this, because performance parameters measured by low-level benchmarks can, in principle, be used in timing formulae to predict and thereby understand the performance of the higher-level benchmarks. Whilst this process has not been taken very far to date, it is an important requirement to define a small but sufficient set of key performance parameters which not only cover all major aspects of state-of-the-art parallel architectures, but are also related to the corresponding properties of the parallel applications. The double-performance parameter, n_2, as defined in this paper, is a strong candidate for inclusion into this set. From this point of view the characterisation of the cache memory effect is a considerable step towards achieving this goal.

The performance analysis presented in this paper is easily extendable for computer systems with multiple-level cache designs. Similar performance characterisation is applicable to the upper layers of the memory hierarchy, and in particular to the external memory.

Acknowledgements

The author would like to thank Roger Hockney and Tony Hey for their support, continuous encouragement and fruitful discussions. Special thanks go to Doug Miles for his help with some compiler-specific issues.

References

1. Anderson, E., Z. Bai, C. Bischof, J. Demmel, J. Dongarra, J. DuCroz, A. Greenbaum, S. Hammarling, A. McKenney and D. Sorensen, *LAPACK: A Portable Linear Algebra Library for High-Performance Computers*, Proceedings Supercomputing'90, IEEE Computer Society Press, 1-10, 1990.
2. Hockney, R.W. and C.R. Jesshope, *Parallel Computers 2: Architecture, Programming and Algorithms*, IOP Publishing/Adam Hilger, Bristol and Philadelphia, 1988.
3. Berrendorf, R. and J. Helin, *Evaluating the Basic Performance of the Intel iPSC/860 Parallel Computer*, Concurrency: Practice and Experience, 4(3), 223-240, 1992.
4. Bailey, D.H., *RISC Microprocessors and Scientific Computing*, Proceedings Supercomputing'93, IEEE Computer Society Press, 645-654, 1993.
5. Schönauer, W. and H. Häfner, *Explaining the Gap between Theoretical Peak Performance and Real Performance for Supercomputer Architectures*, Scientific Programming, 3(2), 157-168, 1994.
6. Getov, V.S. and R.W. Hockney, *Comparative Performance Analysis of Uniformly Distributed Applications*, Proceedings of Euromicro Workshop on Parallel and Distributed Processing, IEEE Computer Society Press, 259-262, January 1993.
7. Miles, D., *Beyond Vector Processing: Parallel Processing on the CRAY APP*, Proceedings COMPCON Spring'93, IEEE Computer Society Press, 321-328, February, 1993.
8. Getov, V.S., *1-Dimensional Parallel FFT Benchmark on SUPRENUM*, in: Etiemble, D. and J.-C. Syre (Eds.), PARLE'92, Parallel Architectures and Languages Europe, Lecture Notes in Computer Science, 605, Springer Verlag, 163-174, 1992.
9. Hey, T., R. Hockney, V. Getov, I. Wolton, J. Merlin and J. Allwright, *The GENESIS Distributed Memory Benchmarks. Part 2: COMMS1, TRANS1, FFT1 and QCD2 Benchmarks on the SUPRENUM and iPSC/860 Computers*, Concurrency: Practice and Experience, 1995 (to appear).
10. Hey, A.J.G., *The GENESIS Distributed Memory Benchmarks*, Parallel Computing, 17(10 - 11), 1275-1283, 1991.
11. PARKBENCH Committee (assembled by R. Hockney and M. Berry, with contributions from D. Bailey, M. Berry, J. Dongarra, V. Getov, T. Haupt, T. Hey, R. Hockney and D. Walker), *PARKBENCH Report - 1: Public International Benchmarks for Parallel Computers*, Scientific Programming, 3(2), 101-146, 1994.

Efficient Jacobi Algorithms on Multicomputers

Domingo Giménez[1], Vicente Hernández[2*] and Antonio M. Vidal[2*]

[1] Departamento de Informática y Sistemas. Univ de Murcia. Aptdo 4021. 30001 Murcia. Spain. (domingo@dif.um.es)
[2] Departamento de Sistemas Informáticos y Computación. Univ Politécnica de Valencia. Aptdo 22012. 46071 Valencia. Spain. (cpvhg@dsic.upv.es, cpavm@dsic.upv.es)

Abstract. In this paper, we study the parallelization of the Jacobi method to solve the symmetric eigenvalue problem on distributed-memory multiprocessors. To obtain a theoretical efficiency of 100% when solving this problem, it is necessary to exploit the symmetry of the matrix. The only previous algorithm we know exploiting the symmetry on multicomputers is that in [10], but that algorithm uses a storage scheme appropriate for a logical ring of processors, thus having a low scalability. In this paper we show how matrix symmetry can be exploited on a logical mesh of processors obtaining a higher scalability than that obtained with the algorithm in [10]. Algorithms for ring and mesh logical topologies are compared experimentally on the PARSYS SN-1040 and iPSC/860 multicomputers.

1 Introduction

The symmetric eigenvalue problem appears in many applications. In some applications in computational chemistry [3, 6] it is necessary to solve the eigenvalue problem obtaining all the eigenvalues of a symmetric, dense and real matrix of the size of some hundred or a few thousand. In this case it could be interesting to solve the problem by efficient Jacobi methods on multicomputers.

The basic idea behind Jacobi-like methods for finding the eigenvalues and eigenvectors of a real symmetric matrix is to construct a matrix sequence $\{A_l\}$ by means of

$$A_{l+1} = Q_l A_l Q_l^t , \ l = 1, 2, \ldots \tag{1}$$

where $A_1 = A$, and Q_l is a Jacobi rotation in the (p, q) plane with $0 \leq p, q \leq n-1$. Under certain conditions sequence $\{A_l\}$ converges to a diagonal matrix D,

$$D = Q_k Q_{k-1} \cdots Q_2 Q_1 A Q_1^t Q_2^t \cdots Q_{k-1}^t Q_k^t \tag{2}$$

whose diagonal elements are then the eigenvalues of A.

Each product $Q_l A_l Q_l^t$ represents a similarity transformation that annihilates a pair of nondiagonal elements, a_{ij} and a_{ji}, of matrix A_l. Matrix Q_l coincides

* Partially supported by ESPRIT III Basic Research Programm of the EC under contract No.9072 (Project GEPPCOM) and partially supported by Generalitat Valenciana Project GV-1076/93

with the identity matrix except in elements $q_{ii} = c$, $q_{ij} = s$, $q_{ji} = -s$, and $q_{jj} = c$, where $c = \cos\ \theta$, $s = \sin\ \theta$, and [5]

$$\tan\ 2\theta = \frac{2a_{ij}}{a_{ii} - a_{jj}}\ . \tag{3}$$

A cyclic Jacobi method works by making successive sweeps until some convergence criterion is fulfilled, normally until $off(A) < bound$. A sweep consists of successively nullifying the $n(n - 1)/2$ nondiagonal elements in the upper-triangular part of the matrix (and the corresponding symmetrical part). To perform a sweep it is necessary to order the pairs of indices (i, j), $0 \leq i < j \leq n - 1$, grouping them in $n - 1$ disjointed sets, each one formed by $n/2$ pairs, then we can nullify the nondiagonal elements in $n - 1$ iterations. Some of these orderings are intended to simplify parallelization of the method. To design efficient parallel Jacobi algorithms it is necessary to use a cyclic ordering with little movement of indices to pass from a set of pairs of indices to the next set of pairs, because this movement implies a data movement between processors. It is also necessary to combine the cyclic ordering with the topology we are using. In our implementations we have used the odd-even [8] and Eberlein [2] orderings with the ring topology, and the Round-Robin [1] ordering with a mesh topology.

When we apply the sequential Jacobi method to symmetric matrices we can exploit the symmetry by updating only the upper-triangular part of the matrix. In this case the cost of the method per sweep is $3n^3$ flops. It is not so easy to exploit the symmetry in parallel Jacobi methods. There are some papers where the symmetry is exploited on shared-memory multiprocessors [5] and systolic arrays [1]. Even the more recent works on Jacobi methods on distributed-memory multiprocessors applied to the Symmetric Eigenvalue Problem [7, 9] do not exploit the symmetry of the matrix, and the theoretical efficiency has an upper bound of 50%. The only previous algorithm we know exploiting the symmetry on a distributed-memory multicomputer is that presented in [10]. This algorithm has been designed for a logical ring, thus having the problem of a low scalability.

In section 2 two storage schemes, allowing us to exploit the symmetry on a logical ring, are presented. One of these storage schemes is that used in [10], and with the other storage scheme a better use of BLAS 1 can be achieved, due to a more compact use of the memory. In section 3 a parallel algorithm exploiting the symmetry on a logical mesh is proposed. In section 4 this algorithm is compared experimentally with those in section 2 on the PARSYS SN-1040 and iPSC/860 multicomputers. The experimental results confirm that the algorithm for mesh topology is more scalable than those for a ring topology.

2   Parallel Algorithms on a Logical Ring

To exploit the symmetry of the matrix on Jacobi methods on multicomputers it is necessary to obtain storage schemes storing symmetric elements in the same processor, and to combine the storage scheme with adequate orderings and topologies.

In [10] the *block Hankel-wrapped* storage scheme (figure 1.a) is combined with the odd-even ordering to obtain a parallel Jacobi method on a ring of processors.

For the purpose of comparing the proposed algorithm for a logical mesh, with the previous algorithm, we have implemented the algorithm using the *block Hankel-wrapped* storage scheme, and another algorithm also adequate for ring topology but with a more compact storage scheme that allows a better use of BLAS 1. We call the storage scheme used in this second algorithm a storage scheme by *frames* (figure 1.b) and it has been combined with the Eberlein ordering.

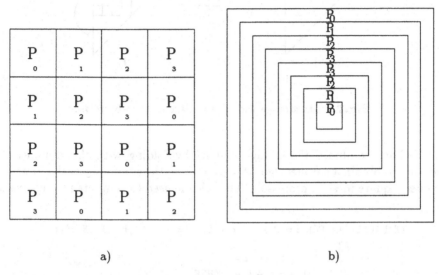

a)   b)

Fig. 1. Storage schemes to exploit the symmetry on a logical ring of processors: a) *block Hankel-wrapped* storage scheme, b) storage scheme by *frames*.

These two algorithms for ring topology have a theoretical efficiency of 100%, but they present a low scalability.

3 A Parallel Algorithm on a Logical Mesh

In this section we show a storage scheme to exploit the symmetry of the matrix on a mesh of processors. Suppose $p = r^2$ processors are available and they are connected in a mesh, and the columns and rows of processors are numbered from 0 to $r - 1$, and the processors designed by P_{ij}.

To distribute the matrix in the multiprocessors system we divide the upper-triangular part of matrix A in three submatrices of size $\frac{n}{2} \times \frac{n}{2}$ (figure 2.a), distribute the square submatrix upon the mesh (figure 2.b), and fold the upper-triangular matrices over the processors in the way shown in figure 2.c). So, we call this storage scheme by *folding*. If matrix A is divided in blocks A_{ij} with

$i = 0, 1, \ldots, 2r-1$, $j = i, i+1, \ldots, 2r-1$, the data assignation is done in the following way:

to $P_{i,r-1-i}$, with $i = 0, 1, \ldots, r-1$, blocks A_{ii}, $A_{2r-1-i,2r-1-i}$ and $A_{i,2r-1-i}$,

to P_{ij}, with $i+j < r-1$, blocks $A_{i,j+r}$ and $A_{i,r-1-j}$,

and to P_{ij}, with $i+j > r-1$, blocks $A_{i,j+r}$ and $A_{2r-1-i,j+r}$.

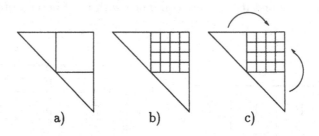

a)   b)   c)

Fig. 2. Data storage with the storage scheme by *folding*.

We have combined the storage scheme by folding with the Round-Robin ordering to obtain a parallel algorithm exploiting the symmetry of the matrix on an open, square mesh of processors. The scheme used to compute the eigenvalues is:

```
IN PARALLEL: FOR i = 0,1,...,r-1; j = 0,1,...,r-1 IN P_ij:
      REPEAT
            FOR k = 1,2,...,n-1
                  IF i = r-1-j THEN
                        Compute Rotations
                  ENDIF
                  Transfer Parameters
                  Update Matrix
                  Transfer Columns
                  Transfer Rows
            ENDFOR
            Compute off(A)
      UNTIL off(A) < bound
```

Here is a brief explanation of each one of the procedures in the algorithm:

Only processors in the main antidiagonal of the mesh ($i = r-1-j$) compute rotations, because they contain the blocks in the main diagonal of the matrix.

The transference of parameters is not a Broadcast. The parameters obtained from the block A_{ii} must be transferred to processors containing blocks A_{ik}, with $i+1 \leq k \leq 2r-1$, and A_{ki}, with $0 \leq k \leq i-1$, so they must be transferred in the column and row of processors where the block is. But not all the communications are necessary, because parameters of blocks A_{ii}, $i < r$, must be transferred horizontally to all the processors in the row, but vertically only to the north; and parameters of blocks A_{ii}, $r \leq i$, must be transferred vertically to all the processors in the column, but horizontally only to the east. In figure 3 the data transference from processor P_{11} with $p = 9$ is shown. In a) the movement of data in the matrix is represented, and in b) the corresponding movement in the mesh is represented.

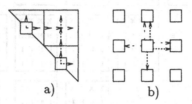

Fig. 3. Transference of parameters: a) transference in the matrix, b) transference in the mesh.

In **Update Matrix** each processor updates the part of the matrix it contains. So, this work is balanced.

In **Compute** $off(A)$ each processor computes the part of $off(A)$ corresponding to elements it contains. After this, the partial results are accumulated in P_{00} and the final result is broadcasted from P_{00} to the other processors in the system.

To explain procedures **Transfer Columns** and **Transfer Rows** we use an example with $n = 24$ and $r = 3$.

After **Transfer Columns** and **Transfer Rows** the upper-triangular part of A must be modified. We will see how these data are modified in **Transfer Rows**, and then we will deduce what data must be modified by **Transfer Columns**.

Figure 4 shows the transference of rows. With I, S, N, E and W we mark data that must be transferred. I indicates an internal transference, and S, N, E and W indicate external tranferences to the south, north, east and west, respectively. Each X represents an element that must be modified in the corresponding movement of data, but which does not participate in the movement.

We need also to store some blocks 2×2 in the lower-triangular part of A. These blocks appear in the figure with a letter, and they are stored in the processor where their symmetric blocks are stored (we want to exploit the symmetry).

In **Transfer Columns** we need to transfer data modifying those in figure 4. Figure 5 shows how to do this tranference.

In each step, each processor in the main antidiagonal computes $\frac{n}{2\sqrt{p}}$ rotations. So, the cost of **Compute Rotations** per sweep is

$$\frac{11n^2}{2\sqrt{p}} \quad flops. \tag{4}$$

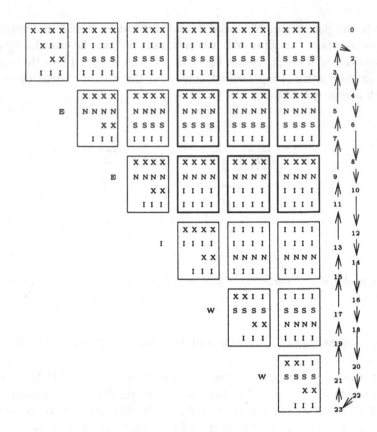

Fig. 4. Transference of rows with the storage scheme by *folding* and the Round-Robin ordering.

In **Update Matrix** each processor not in the main antidiagonal updates $\frac{n^2}{8p}$ blocks 2×2, and each processor in the main antidiagonal updates $\frac{n^2}{8p} + \frac{n}{4\sqrt{p}}$ blocks 2×2. So, the cost per sweep of **Update Matrix** is

$$\frac{3n^3}{p} + \frac{6n^2}{\sqrt{p}} - \frac{3n^2}{p} \quad flops. \tag{5}$$

And the arithmetic cost per sweep is

$$\frac{3n^3}{p} + \frac{23n^2}{2\sqrt{p}} - \frac{3n^2}{p} \quad flops. \tag{6}$$

If β is the start-up time and τ the word-sending time, in each step the transference of rotations has a cost of $\sqrt{p}\beta + n\tau$, the transference of columns expends a time of $6\beta + \left(\frac{4n}{\sqrt{p}} + 4\right)\tau$, and the transference of rows expends a time of

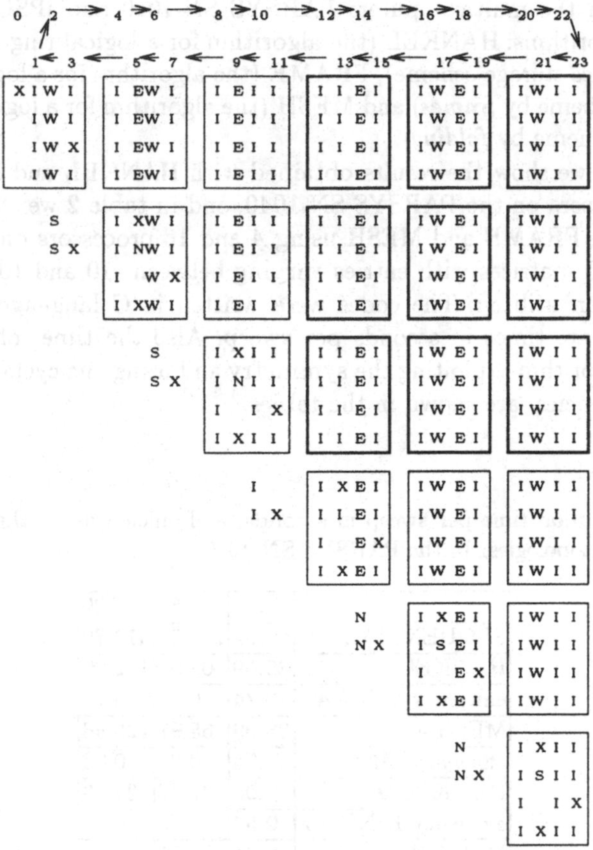

```
   0  2 → 4 →6 → 8 →10 →12→14 →16→18 →20→22
   1← 3 ← 5← 7 ← 9← 11← 13←15← 17←19← 21←23

 X I W I    I EW I    I EI I    I I EI    I W EI    I W I I
 I W I      I EW I    I EI I    I I EI    I W EI    I W I I
 I W X      I EW I    I EI I    I I EI    I W EI    I W I I
 I W I      I EW I    I EI I    I I EI    I W EI    I W I I

    S       I XW I    I EI I    I I EI    I W EI    I W I I
    S X     I NW I    I EI I    I I EI    I W EI    I W I I
            I   W X   I EI I    I I EI    I W EI    I W I I
            I XW I    I EI I    I I EI    I W EI    I W I I

               S      I X I I   I I EI    I W EI    I W I I
               S X    I N I I   I I EI    I W EI    I W I I
                      I     I X I I EI    I W EI    I W I I
                      I X I I   I I EI    I W EI    I W I I

                         I      I X EI    I W EI    I W I I
                         I X    I I EI    I W EI    I W I I
                                I   E X   I W EI    I W I I
                                I X EI    I W EI    I W I I

                                   N      I X EI    I W I I
                                   N X    I S EI    I W I I
                                          I   E X   I W I I
                                          I X EI    I W I I

                                             N      I X I I
                                             N X    I S I I
                                                    I   I X
                                                    I X I I
```

Fig. 5. Transference of columns with the storage scheme by *folding* and the Round-Robin ordering.

$6\beta + \left(\frac{4n}{\sqrt{p}} + 2\right)\tau$. Hence, the communication cost per sweep can be approximated by

$$n\left(12 + \sqrt{p}\right)\beta + n^2\left(1 + \frac{8}{\sqrt{p}}\right)\tau. \tag{7}$$

Because the cost per sweep of a sequential algorithm exploiting the symmetry is $3n^3$, the theoretical efficiency of the parallel algorithm is 100%.

4 Experimental Results

In order to compare our proposed algorithm for a logical mesh of processors, with the algorithms for a logical ring in section 2, we have implemented the algorithms on the multicomputers PARSYS SN-1040 and iPSC/860. We will call these algorithms: HANKEL (the algorithm for a logical ring using the *block Hankel-wrapped* storage scheme), FRAME (the algorithm for a logical ring using the storage scheme by *frames*) and MESH (the algorithm for a logical mesh using the storage scheme by *folding*).

In table 1 we show the results obtained with HANKEL and MESH using 4 and 16 processors on the PARSYS SN-1040, and in table 2 we show the results obtained with FRAME and MESH using 4 and 16 processors on an iPSC/860. Dense random matrices with entries varying between -10 and 10, and different sizes have been utilized. The codes were written in C language and the time in the tables are times in seconds per sweep. Also the times obtained with a sequential algorithm exploiting the symmetry and using the cyclic-by-row order, and the efficiencies, are shown in the tables.

Table 1. Execution time per sweep in seconds, and efficiencies of the algorithms for ring and mesh topologies. In the PARSYS SN-1040.

	192	256	320
SEQUENTIAL	90.52	214.52	417.79
HANKEL 4	30.59	69.69	132.80
efficiency RING 4	0.74	0.77	0.79
MESH 4	28.99	65.92	125.56
efficiency MESH 4	0.78	0.81	0.83
HANKEL 16	9.98	21.37	39.23
efficiency RING 16	0.57	0.63	0.67
MESH 16	8.64	18.75	34.62
efficiency MESH 16	0.65	0.72	0.75

The theoretical efficiency is 100%, but the experimental efficiency is still far from this limit when the matrix size is not very great, and this is due mainly to the broadcast, the contention time in the synchronizations of **Transfer Rows** and **Transfer Columns**, and the overhead due to the irregular distribution of data in the multiprocessor. This difference is bigger in the iPSC/860 than in the PARSYS SN-1040, and this is due to the fact that in the implementations for the iPSC/860 BLAS 1 has been used, reducing the execution time of the arithmetic part of the programs, and also to a greater cost of the communications in relation to the arithmetic operations in the iPSC/860.

The algorithm on mesh topology presents better performances (when the number of processors increases) than the algorithms on ring topology, and this is due to a lower cost of communications and to the possible overlapping of arithmetic and communication work on the algorithm for mesh.

To analize the scalability of the algorithms, in table 3 the efficiencies are compared on the iPSC/860. The table shows the efficiencies obtained with the

Table 2. Execution time per sweep in seconds, and efficiencies of the algorithms for ring and mesh topologies. In an iPSC/860.

	256	384	512
SEQUENTIAL	9.36	35.35	88.19
FRAME 4	4.59	11.30	22.85
efficiency RING 4	0.51	0.78	0.96
MESH 4	4.77	13.01	28.68
efficiency MESH 4	0.49	0.68	0.77
FRAME 16	4.67	8.59	14.99
efficiency RING 16	0.13	0.26	0.37
MESH 16	2.77	6.00	11.23
efficiency MESH 16	0.21	0.37	0.47

algorithms FRAME and MESH with 4 and 16 processors and with different matrix sizes, maintaining the number of elements per processor constant when the number of processors changes. The efficiency decreases in both algorithms when the number of processors increases, but with the topology of ring a big reduction in the efficiency is obtained, and with the topology of mesh this reduction is considerably lower. So a better experimental scalability is obtained with the algorithm we propose.

Table 3. Comparison of the scalability of the algorithms FRAME and MESH, comparing the efficiency when the number of processors increases maintaining the number of elements per processor. On the iPSC/860.

processors	matrix size	FRAME	MESH
4	128	0.18	0.24
16	256	0.13	0.21
4	192	0.33	0.40
16	384	0.26	0.37
4	256	0.51	0.49
16	512	0.37	0.47

5   Conclusions

From the analysis of the previous results we can conclude that it is possible to exploit the symmetry of the matrix when computing eigenvalues of symmetric matrices on a logical mesh of processors. The algorithm presented has been compared experimentally with the only previous algorithm we know exploiting the symmetry in the symmetric eigenvalue problem on multicomputers [10], and with another algorithm for a logical ring allowing a better use of BLAS 1. The algorithm for mesh presents a better experimental scalability, solving the problem of the previous algorithm. In addition, these storage schemes can be combined with schemes to work by blocks to obtain more efficient algorithms [4].

Acknowledgments. This work was performed in part on the 32 node Intel iPSC/860 operated by the University of Texas Center of High Performance Computing. We thank Professor Robert van de Geijn of the Department of Computer Sciences of the University of Texas for allowing and facilitating the use of this equipment.

References

1. R. P. Brent and F. T. Luk. A systolic architecture for almost linear-time solution of the symmetric eigenvalue problem. Technical Report TR-CS-82-10, Department of Computer Science, Australian National University, Camberra, August 1982.
2. P. J. Eberlein and H. Park. Efficient implementation of Jacobi algorithms and Jacobi sets on distributed memory architectures. *Journal of Parallel and Distributed Computing*, 8:358–366, 1990.
3. A. Edelman. Large dense linear algebra in 1993: The parallel computing influence. *The International Journal of Supercomputer Applications*, 7(2):113–128, 1993.
4. D. Giménez, V. Hernández, R. van de Geijn and A. M. Vidal. A jacobi method by blocks to solve the symmetric eigenvalue problem on a mesh of processors. ILAS Conference, 1994.
5. G. H. Golub and C. F. Van Loan. *Matrix Computations*. The Johns Hopkins University Press, 1989.
6. I. N. Levine. *Molecular Spectroscopy*. John Wiley and Sons, 1975.
7. M. Pourzandi and B. Tourancheau. A Parallel Performance Study of Jacobi-like Eigenvalue Solution. Technical report, March 1994.
8. G. H. Stewart. A Jacobi-like algorithm for computing the Schur decomposition of a nonhermitian matrix. *SIAM J. Sci. Stat. Comput.*, 4:853–864, 1985.
9. V. Strumpen and P. Arbenz. Improving Scalability by Communication Latency Hiding. In D. H. Bailey, P. E. Bjørstad, J. R. Gilbert, M. V. Mascagni, R. S. Schreiber, H. D. Simon, V. J. Torczon and L. T. Watson, editor, *Proceedings of the Seventh SIAM Conference on Parallel Processing for Scientific Computing*, pages 778–779. SIAM, 1995.
10. R. A. van de Geijn. Storage schemes for Parallel Eigenvalue Algorithms. In G. H. Golub and P. Van Dooren, editor, *Numerical Linear Algebra. Digital Signal Processing and Parallel Algorithms*, volume 70 of *NATO ASI Series*. Springer-Verlag, 1991.

Front Tracking: A Parallelized Approach for Internal Boundaries and Interfaces *

James Glimm[1], **   John Grove[1],   Xiao Lin Li[2],   Robin Young[1],

Yanni Zeng[1],   Qiang Zhang[1]

[1] The University at Stony Brook, Stony Brook, NY 11794-3600
[2] Department of Mathematics, Indiana University Purdue University, Indianpolis

Abstract. Internal boundaries and interfaces are an important part of many fluid and solid modeling problems. Front Tracking contains a general interface framework, closely related to the non-manifold geometry used in CAD solid modeling packages, to support numerical simulation of such fluid problems. It can thus be considered to be a systematic application of the ideas of computational geometry to computational fluid dynamics. It is based on the principle that finite differences behave best when applied to differentiable functions, and that weak derivatives of nondifferentiable functions can be replaced by regularized expressions such as jump conditions. Front Tracking offers superior resolution for fluid problems with important discontinuities and interfaces, and in some cases, it has provided the unique method to obtain correct answers. Here we present Computer Science issues which have contributed to the success of Front Tracking: software design and organization – modularity, data structures and data hiding.

1   Introduction

Front tracking is a numerical method which assigns explicit computational degrees of freedom to surfaces of discontinuity. Additional, grid based degrees of freedom, represent continuously varying solution values. This method is ideal for solutions in which discontinuities are an important feature, and especially where their accurate computation is difficult by other methods. Phase transition boundaries, deposition and etching fronts, material boundaries, slip surfaces, shear bands, and shock wavesprovide important examples of such problems. Richtmyer [25] first proposed this method. Moretti [22] [23] [24] used it for high quality aerodynamic computations.

* Supported by the Applied Mathematics Subprogram of the U.S. Department of Energy DE–FG02-90ER25084.
** Also supported by the Army Research Office, grant DAAL03-92-G-0185 and through the Mathematical Sciences Institute of Cornell University under subcontract to the University at Stony Brook, ARO contract number DAAL03-91-C-0027, and the National Science Foundation, grant DMS-9201581.

A systematic development of front tracking in two dimensions has been carried out by the authors and coworkers [12], [7], [13], [14], [11], [5], [17]. See [27], [26], [20], [6], [1], [21] for other approaches to front tracking in two dimensions. Special purpose front tracking codes have been also developed, as well as a number of one dimensional front tracking codes.

We draw two conclusions from this body of work. First, front tracking is feasible for geometrically complex fronts. Examples include complex shock and wave front interaction problems, with bifurcations, i.e. changes of wave front topology, as occurs after interaction, or crossing of one wave (tracked discontinuity) by another. The second conclusion is that the front tracking solutions are often (a) better and (b) obtained on significantly coarser grids [16], [19], [18], [2], [5], [3], [4].

In Fig. 1, we show the interaction of a shock wave with a randomly perturbed planar contact discontinuity in two dimensions, representing a density discontinuity layer between two gases. The original computation and two levels of zoomed enlargement are displayed in the three panels. The most enlarged panel occupies only a few mesh blocks, and shows a highly resolved and complex set of incident, reflected, and transmitted waves.

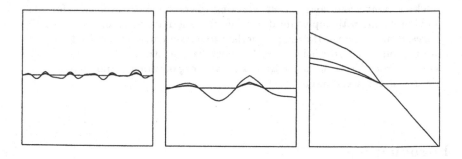

Fig. 1. Tracked waves to represent the passage of a shock wave through a perturbed planar interface separating two compressible fluids of different densities. The three frames show successive zoom enlargments of a single timestep in the interaction.

2 Modularity and Data Structures

Modern programming languages and modular organization have been an integral part of our methodology for many years, and are essential for the work presented here. Due to the growing interest in object oriented programming and modular algorithm design, we include a discussion of this aspect of our methods. The discussion is organized by increasing levels of specificity:

1. Utilities and software tools.
2. Geometrical and topological structures (*e.g.* grids). Only the independent variables of the problem are available to the routines defined at this level.
3. General equation and problem types and solution methods: hyperbolic, parabolic, elliptic (*e.g.* Godunov, conjugate gradient, finite elements, interpolation). The dependent variables of the problem are available in symbolic form only, as addresses of storage locations. As such, routines at this level can copy (bitwise) and allocate storage for dependent variables and pass them as arguments of functions.
4. Physical laws: compressible gas dynamics, elastic-plastic flow, etc. (*e.g.* Riemann solvers). The full set of dependent and independent variables defined in the problem are available to the routines defined at this level.
5. Material specification (*e.g.* equation of state, constitutive laws). The equations of physics close at this level, with all nonlinear function fully defined in their functional form. Material constants would normally not appear in the source code, but exist in separate tables, to be read upon run time initialization, or as initialization parameters. If the constituitive response functions are give in tabular form, these tables are read at run time as input data for level 3 interpolation functions available in the source code.

Modular code is organized into libraries, each carrying a specificity level (1 to 5), which governs the scope and sharing of data and variables. Public data and structures defined at level n are usually available to all higher specificity levels $m \geq n$, but not to lower specificity levels.

2.1 Interface

We first discuss the interface library, a level 2 library which describes the geometry and topology of piecewise smooth manifolds with piecewise smooth boundaries, embedded in R^3. Boundary operators, to map from a manifold to its boundary, and to the manifolds which it bounds, are included in this library. We begin with a description of the main data structures (whose names are in capital letters) and their interrelationships. The library compiles and runs independently of other libraries with equal or higher specificity. At a continuum level, an INTERFACE in R^2 [15] is a collection of non-intersecting oriented CURVEs, meeting only at endpoints (called NODEs), and thus dividing the R^2 into distinct connected components. We designate as COMPONENT, some labeling scheme, *i.e.* equivalence class, for components. Thus several components may constitute one COMPONENT. Each SIDE of each CURVE is labeled by the COMPONENT adjacent to that SIDE. Each NODE has a list of its incoming and outgoing CURVEs. The same concepts apply in R^3, with the addition of an oriented SURFACE as a new structure. The SURFACEs divide R^3, so that each SIDE of a SURFACE is labeled by a COMPONENT. The SURFACEs are bounded by and may meet along CURVEs. There is no consistent orientation for the INTERFACE as a whole. SURFACEs and the CURVEs which bound them do not have consistent orientation. The boundary of a SURFACE may consist of several CURVEs, and each CURVE may bound several SURFACEs (or none at all). In cases where the components on the two sides of the SURFACE are the same, SIDE is useful as a local generalization of COMPONENT. The discretized version of the INTERFACE has this same structure with a piecewise linear description built from doubly linked lists of simplices of the appropriate dimensions. In Fig. 2, we illustrate the geometric data structures used for the front tracking method in three dimensions.

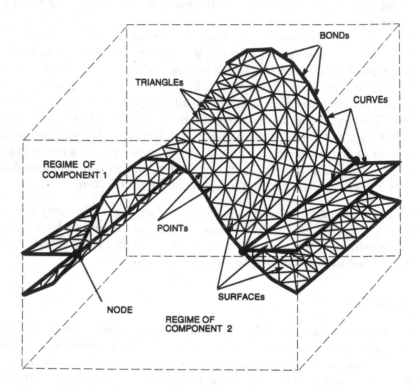

Fig. 2. An illustration of the geometric data structures used for the front tracking method in three dimensions.

An INTERFACE, in the C language, is a data structure with arrays of pointers to SURFACEs, CURVEs, and NODEs. The SURFACEs, CURVEs, and NODEs are also data structures. They contain arrays of pointers to their bounding and co-bounding objects, *i.e.* arrays of pointers to CURVEs, for the SURFACEs, etc. The SURFACEs and CURVEs also contain reference to the first and last simplices (TRIANGLEs and BONDs) in the linked list which defines them.

The TRIANGLEs and BONDs are data structures defined in terms of POINTs and neighbors (adjacent TRIANGLEs or BONDs). For computational efficiency, they contain additional information, namely, length for BONDs, area and positive unit normal for TRIANGLE. A POINT is also a data structure, with data to represent its coordinate description.

To achieve modularity between dimensions, the coordinates are represented as pointer to an array of dim $= 1$, 2 or 3 floating point variables. Here dim is a run time variable associated with a specific INTERFACE. Since one of the applications of the interface library is to support the efficient interpolation of piecewise smooth functions, it would be beneficial to allow arbitrary positive integer values for the dimension dim. In this case, the interface is a generalization of the notion of a simplicial complex [10] of dimension dim $- 1$ imbedded in R^{dim}. Multiple INTERFACEs and their uses, even within a single computation, prevent dim from being a globally defined variable.

Each of the elementary objects in an INTERFACE (including the INTERFACE itself) has support routines, for initializing, copying, printing, reading of printed format, and modifying. These routines are exported, and publically available to the rest of the code. In addition, the INTERFACE has hidden, or private, data and support functions.

The interface library supports its own storage allocation scheme. Storage allocation is a level 1 module, built as an extension of the Unix routine malloc. As a hidden variable, a linked list of all active interfaces is maintained, and for each a separate instance of the level 1 storage allocation scheme is maintained. Storage for an interface is allocated in blocks of a designated size. These are used as needed for the allocation of SURFACEs, etc. In this way, the storage is (more nearly) contiguous in physical memory, and paging inefficiencies for access of computationally related variables are minimized. Deleted objects are not deallocated; rather the knowledge of their addresses is eliminated, so that they are deaddressed. The reason for this choice is that available deaddressed space is highly fragmented. Much of the data consists of pointers, so that compression by recopy of data is incorrect. Upon deletion of the entire INTERFACE, all of its storage is deallocated and returned to the system. The combination of copy INTERFACE to get a new INTERFACE and delete (old) INTERFACE will free deaddressed storage, reset pointers correctly and accomplish compression. Storage allocation is a private aspect of the publicly callable functions to initialize an INTERFACE object.

It is frequently necessary to determine the topology associated with an INTERFACE. INTERFACEs are required to be non-self intersecting (so that SURFACEs can meet only along their bounding CURVEs, etc.). After each timestep in the dynamical evolution, it is necessary to check the propagated INTERFACE for intersections. If intersections arise, signaling a bifurcation of the topology, a call will be made to a physics-specific routine, to modify and reconnect the INTERFACE, with the possible introduction of additional reflected waves, as required by the physics. A second topological requirement is to determine the COMPONENT of a given location in space. These requirements lead to a private INTERFACE data structure of hashed lists of hypersurface simplices (BONDs or TRIANGLEs) stored according to their intersection with each mesh cell.

The intersection routine checks all pairs of hypersurface simplices for intersections, and returns a list of intersecting hypersurfaces and their intersection locations. In three dimensions, the intersections are organized into CURVEs, while in two dimensions, they are isolated NODEs. By use of this hashed list, intersections are tested only for pairs meeting a common mesh block. Since the local density of hypersurface elements is normally bounded, the $O(n^2)$ intersection computation is reduced to $O(n)$ in complexity for typical problems.

2.2 Front

An INTERFACE, with generic physics dependent information, is called a FRONT. The resulting algorithms define a level 3 library. At level 3, the POINT acquires new structure: physical STATEs, associated with each SIDE of the hypersurface on which the POINT is located. At level 3, a STATE is the address of allocated storage of known size. It can be passed as a function argument, and its contents can be copied (bitwise), but it cannot be otherwise accessed. For co-dimension two POINTs, such as where multiple SURFACEs meet along a CURVE, there are two STATEs (one for each hypersurface SIDE) for each such hypersurface; this storage is associated with the hypersurface, rather than with the POINT. The hypersurfaces also acquire new structure: a wave type, which designates physics specific information about the type of front. At level 4 specificity, these wave types can be read fully, but at level 3, only generic wave types defining boundary conditions (NEUMANN, DIRICHLET, PERIODIC) can be read.

Passing to specificity level 4, meaning is attached to the (floating point and integer) data a STATE contains. Thus the simplest idea of a STATE for three dimensional compressible fluid flow would be five floating point numbers. It is convenient to have the equation of state addressible through the STATE itself, for application to multicomponent or multiphase flow problems. Thus the address of the equation of state data base is added to the STATE structure, at specificity level 5.

Apart from support routines for front-associated data structures, the main operations performed within the front library are (a) remeshing of the FRONT, (b) propagation routines for both regular (co-dimension one) and irregular (co-dimension two or higher) POINTs (§3), and (c) untangle of self-intersecting FRONTs with only scalar degrees of freedom (§3). A FRONT with scalar degrees of freedom is one whose physical quantities transform as a scalar (e.g. a material boundary or contact discontinuity). Such fronts generally have degenerate light cones and simple wave interaction laws. In contrast, vector fronts, such as shock waves, interact so as to create both reflected and transmitted waves, even in the simplest cases. The FRONT data structure contains parameters to control these operations (a) – (c), including function pointers to level 4 physics functions for propagation details.

2.3 Interpolation

Three other libraries complete the level 3 code for conservation laws. A hyperbolic library is concerned with states and propagation at regular grid points. A driver library contains the main program initialization, time loop control, and i/o capabilities. Another level 3 library, discussed here, supports interpolation, based on state data from the front and hyperbolic libraries.

The ability to interpolate piecewise smooth functions with arbitrary discontinuities across interfaces is of considerable independent interest, and has been developed as an isolated capability. It is used to support equation of state tables, with phase transitions as the source of discontinuity [8], [9].

To ensure the integrity of the interpolation process, only state values from a single COMPONENT can be interpolated. State data is stored at grid cell centers. For interpolation, we consider the dual grid, with states stored at grid cell corners. For a regular (dual grid) cell, $i.e.$ one which does not intersect the front, bilinear interpolation gives interpolated state values. For an irregular cell, in two dimensions, we introduce a triangulation which respects the interface, and which uses only the points for which the states are already known: the dual grid corners and the (one-sided) front points.

Three dimensional simplicial cells (tetrahedra) are constructed on each side of the interface within a rectangular grid block. We use the Delaunay triangulation algorithm to achieve an optimal aspect ratio for the tetrahedra, so that the interpolation is based on the most reasonable simplices.

The triangulated three dimensional cells near the interface and the intersection between the interface and the rectangular grid lines provide a dynamical coupling between the front and the interior fluid components. The former provides a robust way to calculate the fluid states for the propagation of front and the latter seperates different fluid components and interior numerical solvers so that no numerical diffusion will be cross the interface.

3   Mathematical Theory and Modeling

The solution of conservation laws, of the form

$$U_t + \boldsymbol{\nabla} \cdot \mathbf{F}(U) = \text{Source}$$

are supported by the general framework discussed in §2. The nature of the propagation is governed by the codimension of the point. The codimension is the codimension of the maximal space-time manifold containing the point locally, on which the solution is smooth. In simple examples, the codimension equals the number of (simple jump) discontinuities present simultaneously in the solution at a given point in space and time. For time dependent problems in dim space dimensions, $0 \leq$ codimension $\leq \dim + 1$.

The codimension 1 points are located on the front, but are otherwise regular. They are points of simple jump discontinuity. They lie on the interior of a surface in 3 spatial dimensions and on the interior of a curve in 2 dimensions. The propagation of the front coordinates and states is performed in a single step. Operator splitting, in a rotated coordinate system, allows a separate propagation step in the directions normal to and tangential to the front. The propagation of a discontinuity, in the direction normal to the discontinuity surface, is given by the solution of a Riemann problem. This is the one dimensional Cauchy problem,

Fig. 3. The triangulated interface between two fluids of differing density, subject to a gravitational (Rayleigh-Taylor) instability. This computation shows four bubbles of light fluid rising. The computational mesh is $10 \times 10 \times 20$. Periodic boundary conditions are imposed in the x and y directions. The fluid interface is patched with a buffer zone at the periodic boundary and at the parallelization subdomain boundaries.

with idealized initial conditions consisting of a single jump discontinuity. An extensive body of theory has been developed in recent years for the analysis of such Riemann solutions.

4 Applications and Results

The study of instabilities of an interface separating distinct fluids has been one of the important applications of front tracking. Acceleration, either impulsive (as in the Richtmyer-Meshkov instability) or steady (as in the Rayleigh-Taylor instability) provide examples. We consider here a gravitational acceleration force. The mixing zone is defined as the portion of the z-axis from the lowest to the highest interface position. The size and growth rate of the mixing zone is a matter of considerable interest. Consistent results comparing laboratory experiments, front tracking simulations and simple theoretical models (based on the renormalization group) have been obtained for the mixing rate. In Fig. 3 we present a simulation of an unstable interface in three dimensions.

References

1. Bell , J. B., Colella , P., Welcome, M. L.: Conservative front-tracking for inviscid compressible flow. UCRL-JC-105251, preprint (1991)
2. Boston, B., Glimm, J., Grove, J. W., Holmes, R., Zhang, Q.: Multiscale Structure for Hyperbolic Waves. Report No. SUNYSB-AMS-93-18, State Univ. of New York

at Stony Brook (1993) In: Proceedings of the International Conference on Nonlinear Evolution Partial Differential Equations, Beijing, P.R. China 1993

3. Boston, B., Grove, J. W., Henderson, L. F., Holmes, R., Sharp, D. H., Yang, Y., Zhang, Q.: Shock Induced Surface Instabilities and Nonlinear Wave Interactions. Report No. SUNYSB-AMS-93-20 (1993), State Univ. of New York at Stony Brook In: Proceedings of Eleventh Army Conference on Applied Mathematics and Computing

4. Boston, B., Grove, J. W., Holmes, R.: Front Tracking Simulations of Shock Refractions and Shock Induced Mixing. Report No. SUNYSB-AMS-93-19 (1993), State Univ. of New York at Stony Brook. In: Proceedings of the 19th International Symposium on Shock Waves

5. Chen, Y., Deng, Y., Glimm, J., Li, G., Sharp, D. H., Zhang, Q.: A Renormalization Group Scaling Analysis For Compressible Two-Phase Flow. Phys. Fluids A **5** (1993) 2929-2937

6. Chern, I-L., Colella, P.: A Conservative Front Tracking Method for Hyperbolic Conservation Laws. LLNL Rep. No. UCRL-97200 (1987)

7. Chern, I-L., Glimm, J., McBryan, O., Plohr, B., Yaniv, S.: Front Tracking for Gas Dynamics. J. Comput. Phys. **62** (1986) 83-110

8. Coulter, L., Grove, J. W.: The Application of Piecewise Smooth Bivariate Interpolation to Multiphase Tabular Equation of States. Report No. SUNYSB-AMS-92-11 (1992) University at Stony Brook

9. Lisa Osterman Coulter: Piecewise Smooth Interpolation and the Efficient Solution of Riemann Problems with Phase Transitions. Ph.D. Thesis New York Univ. 1991

10. Eilenberg, S., Steenrod, N.: Foundations of Algebraic Topology. Princeton University Press, Princeton, 1952

11. Glimm, J., Grove, J., Lindquist, W. B., McBryan, O., Tryggvason, G.: The Bifurcation of Tracked Scalar Waves. SIAM J. Sci. Stat. Comput. **9** (1988) 61-79

12. Glimm, J., Isaacson, E., Marchesin, D., McBryan, O.: Front Tracking for Hyperbolic Systems: Adv. Appl. Math. **2** (1981) 91-119

13. Glimm, J., Klingenberg, C., McBryan, O., Plohr, B., Sharp, D., Yaniv, S.: Front Tracking and Two Dimensional Riemann Problems. Adv. Appl. Math. **6** (1985) 259-290

14. J. Glimm W. B. Lindquist O. McBryan L. Padmanabhan A Front Tracking Reservoir Simulator, Five-Spot Validation Studies and the Water Coning Problem. In: Frontiers in Applied Mathematics. SIAM, Philadelphia, PA, **1** (1983) 107

15. Glimm, J., McBryan, O.: A Computational Model for Interfaces. Adv. Appl. Math. **6** (1985) 422-435

16. Grove, J., Holmes, R., Sharp, D. H., Yang, Y., Zhang, Q.: Quantitative Theory of Richtmyer-Meshkov Instability. Phys. Rev. Lett. **71** (1993) 3473-3476

17. Grove, J. W.: Applications of Front Tracking to the Simulation of Shock Refractions and Unstable Mixing. J. Appl. Num. Math. **14** (1994) 213-237

18. Grove, J. W., Yang, Y., Zhang, Q., Sharp, D. H., Glimm, J., Boston, B., Holmes,R.: The Application of Front Tracking to the Simulation of Shock Refractions and Shock Accelerated Interface Mixing. In: Proceedings of the 4th International Workshop on the Physics of Compressible Turbulent Mixing Cambridge Univ., Cambridge (1993), Report No. SUNYSB-AMS-93-21 State Univ. of New York at Stony Brook

19. Holmes, R., Grove, J. W., Sharp, D. H., Numerical Investigation of Richtmyer-Meshkov Instability Using Front Tracking. J. Fluid Mech. (To Appear 1995)

20. LeVeque, R. J., Shyue, K.-M.: Two-dimensional front tracking based on high resolution wave propagation methods: submitted to J. Comput. Phys.

21. Mao, D.-K.: A treatment of discontinuities for finite difference methods in the two-dimensional case. J. Comp. Phys. **104** (1993) 377–397
22. Moretti, G.: Thoughts and Afterthoughts About Shock Computations. Rep. No. PIBAL-72-37, Polytechnic Institute of Brooklyn, 1972
23. Moretti, G.: Computations of Flows with Shocks. Ann Rev Fluid Mech, **19** (1987), 313–337
24. Moretti, G., Grossman, B., Marconi, F.: A Complete Numerical Technique for the Calculation of Three Dimensional Inviscid Supersonic Flow. American Institute for Aeronautics and Astronautics, Rep. No. 72-192, (1972)
25. Richtmyer, R., Morton, K.: Difference Methods for Initial Value Problems. Interscience, New York, 1967
26. Zhu, Y.-L., Chen, B.-M., Wu, X.-H., Xu, Q.-S.: Some New Developments of the Singularity-Separating Difference Method. Lecture Notes in Physics, Springer-Verlag, Heidelberg **170** (1982)
27. Zhu, Y.-L., Chen, B.-M.: A Numerical Method with High Accuracy for Calculating the Interactions between Discontinuities in Three Independent Variables. Scientia Sinica **23** (1980)

Program Generation Techniques for the Development and Maintenance of Numerical Weather Forecast Grid Models

Victor V. Goldman[1] and Gerard Cats[2]

[1] Dept. of Computer Science, University of Twente,
PB 217, 7500 AE Enschede, The Netherlands
[2] Royal Netherlands Meteorological Institute,
PB 201, 3730 AE De Bilt, The Netherlands

Abstract. This article presents computer-algebra based techniques for the automatic generation and maintenance of numerical codes based on finite difference approximations. The various generation phases – specification, discretization, implementation and translation – as well as their respective knowledge bases, are discussed and specific attention is given to data mappings in the implementation phase and to high-performance language extensions in the Fortran translation phase. The generation of Fortran source for the dynamics part of a limited area weather forecasting grid-point model is discussed and is illustrated by showing the production of a few variants of the surface-pressure tendency code using the present prototype. Finally, we indicate briefly how adjoints can be obtained using the present methodology.

1 Introduction

A common problem facing efficient maintenance of large scientific codes is the necessity of recoding for porting to new parallel architectures. Although one can nowadays appeal to a variety of tools to assist in such a task, most of them can usually be characterized as source-transforming packages which either compile directly or produce new source appropriate for the target architecture and compiler. Converters such as CMAX [16] fall into the latter category. Experience shows however [19], that most tools do not perform satisfactorily without additional manual intervention or additional automated preprocessing [6], especially for porting from vector to MPP architectures. Although the standardization of data parallel languages such as HPF is expected to significantly improve portability, many proposed automatic data layout and distribution tools assume coded Fortran source as starting point. In the present context, a difficulty with automatic source transformation is that important knowledge which is used in developing the code is not available to the automatic tool at the coded program stage. This knowledge is often useful for attaining an acceptable level of optimality. The problem is usually compounded when extensibility (e.g. amending or extending the algorithm) is added to the portability issue.

In this paper we adopt a program generation methodology to address such maintenance issues. The generation process, which is partially rule-driven, starts from a compact equational specification. After a number of transformations which include discretization and data mapping, the generator produces source code: Fortran 77 or 90, optionally with some high-performance constructs. It also generates code using memory mappings suitable for vector as well as for MPP architectures. The programming platform of the generation is the Lisp-based computer algebra system REDUCE [7] and the translation of intermediate code makes use of the REDUCE-based GENTRAN [4] package for Fortran 77 and GENTRAN 90 [2], its Fortran 90 extension. To demonstrate the potential of the present methodology, it was applied to the equations that are solved by the HIRLAM [3] model. This is a numerical weather forecasting model that is in operational use in most of the meteorological institutes of the countries that participate in the HIRLAM project.

Because of the time constraints on weather forecast production it is essential that the HIRLAM model code is efficient on the available hardware. Therefore models like this pose a real challenge on automatic code generation systems with respect to code optimality. Currently, however, we accept that the generation of optimal code is aided by the manual specification of a set of heuristic rules to the generator. Our research is more directed towards the generation of (Fortran) codes for a range of models from a basic set of equations: The model proper, its linearized version and the adjoint of the latter. A reason to choose HIRLAM in particular is that currently within the HIRLAM project group a task force is operating to develop a linearized version of the HIRLAM model and its adjoint. For sake of brevity, in this paper we will restrict ourselves to the generation of the equation for surface pressure tendency as solved by HIRLAM and its linearization and adjoint. Within this research, we will assume that the model equations are given in discretized form. The derivation of the discretized equations from the continuous forms is a task for numerical experts where other constraints than the model equations (e.g. conservation of energy) ([1]) play a role. Here we do not attempt to build an expert system to automate that process.

The HIRLAM forecast model and the discretized pressure tendency in particular are explained in Sect. 2. In Sect. 3 the structure of the prototype generator is described. The procedure for generating surface-pressure tendency code is presented in the Sect. 4. Finally in Sect. 5, we briefly describe automatic generation adjoints for data assimilation using the present methodology.

2 The Discretized Forecast Model

Overview. The main components of the HIRLAM forecasting system consist of a data assimilation scheme, to construct the initial state of the atmosphere from observations, and the forecast model, to integrate the atmospheric equations of

---

[3] The HIRLAM model was developed by the HIRLAM project group, a cooperative project of Denmark, Finland, Iceland, Ireland, The Netherlands, Norway and Sweden.

motions in time. The current formulations of these components allow efficient implementation of the forecast model and its adjoint on massively parallel systems, but not of the data assimilation scheme. This is one of the reasons that currently efforts are directed towards reformulation of the latter scheme into one based on variational techniques using the adjoint equation of the forecast model ([15]). Therefore, it is not a serious restriction that we will limit ourselves here to the forecast model and its adjoint.

The forecast model consists essentially of two main parts. The first, called 'dynamics', solves the primitive equations of motion, e.g., conservation of mass, momentum, and energy, by time and space discretization techniques. The second, called 'physics', describes the effect of sub-grid scale processes on the discretized model variables. It is a characteristic of the division into physics and dynamics that the former do not contain horizontal exchange of information: Physics are formulated column-wise, all horizontal 'communications' take place within the dynamics part. In general, the dynamics form the challenging part for code generation systems because physics are embarrassingly parallel under horizontal domain decomposition (which is the usual vectorization and parallelization strategy, [3]). For this study we selected from the dynamics the equation describing conservation of mass, because it is representative for most dynamical equations, yet conceptually simple. A class of equations we do not address here consists of those implying 'global communications', like Helmholtz equations or Fourier transforms. The following is thus restricted to the grid-point version of the HIRLAM model.

The Surface-Pressure Tendency Equation. In its vertically integrated form, conservation of mass reads:

$$\frac{\partial p_s}{\partial t} = -\int_0^1 \nabla \cdot \left(\mathbf{v}_h \frac{\partial p}{\partial \eta} \right) d\eta \tag{1}$$

In here, p_s is surface pressure, t time, $\nabla\cdot$ the divergence operator, p pressure, and $\mathbf{v}_h$ the horizontal wind. The vertical coordinate is η ([13]). Equation (1) is the pressure tendency equation.

The vertical discretization is performed by defining the pressure at NLEV $+ 1$ so-called half-levels by two sets of numbers, A and B:

$$p_{k+1/2} = A_{k+1/2} + B_{k+1/2} p_s(x,y) \; ; \; k = 0,\ldots,\text{NLEV} \tag{2}$$

Wind components u and v are given on NLEV 'full-levels', at which the pressure is the arithmetic average of the pressures at the two neighboring half-levels. The surface is half-level NLEV $+ 1/2$, so $A_{\text{NLEV}+1/2} = 0$ and $B_{\text{NLEV}+1/2} = 1$.

The horizontal layout of the grid points is the Arakawa-C grid ([1]) where the wind components u and v are given half a grid distance to the East and North, resp., with respect to surface pressure p_s. (Fig. 1). The surface pressure tendency equation (1) can be discretized in a variety of ways. In combination with suitable discretizations of the other model equations, the following form satisfies the additional constraints like conservation of energy:

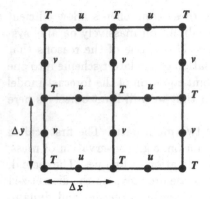

Fig. 1. The Arakawa-C grid. The figure shows the position of the grid points of wind components (u and v) with respect to the main grid points (T), carrying all other prognostic model variables

Define:

$$U_k = \overline{\Delta p_k}^x u_k \; ; \; V_k = \overline{\Delta p_k}^y v_k \tag{3}$$

The overbar denotes the operator to obtain the value at a grid point where there is no directly available value; simple arithmetic averaging is sufficiently accurate (second order in Δx). Further symbols to be used are: a: earth radius; h_x and h_y: map factors (to describe e.g. convergence of meridians towards the poles); Δ: vertical difference; δ: horizontal difference. Then:

$$\frac{\partial p_s}{\partial t} = - \sum_{j=1}^{\text{NLEV}} \frac{1}{a h_x h_y} \{\delta_x(h_y U_j) + \delta_y(h_x V_j)\}$$

$$= -\frac{1}{a h_x h_y} \left\{ \delta_x \left(h_y \sum_{j=1}^{\text{NLEV}} U_j \right) + \delta_y \left(h_x \sum_{j=1}^{\text{NLEV}} V_j \right) \right\} \tag{4}$$

Equations (3) and (4) are the equations we will consider in the sequel.

3   The Generation Process

Overview. Program generation, or more broadly, *software synthesis* [12], is a branch of software engineering which concerns itself with the automation of program writing. In a knowledge-driven approach, coded knowledge is used by the generator to transform a compact specification into a ready to compile and execute program. An important aspect of knowledge-based software synthesis is that once represented and coded, the knowledge can be easily reused, amended, or extended, thus facilitating maintenance. In the field of computational science, there is a great deal of research being done on the many issues in this area: from computational intelligence, knowledge bases, and environments (see e.g.

[8, 18, 10]), to expression manipulation, optimization and translation (see e.g. [9, 4, 2]). Many software generators [17, 10, 18] make use of computer algebra systems, especially to manipulate mathematical expressions. The present generator is embedded in the computer algebra system REDUCE [7]. The overall architecture of the generator is depicted in Fig. 2. The specifications contain computational *scripts* which delineate concisely the computational trajectory. High-level optimization heuristics are also part of scripts, but routine low-level optimizations are implicitly left to the compiler.

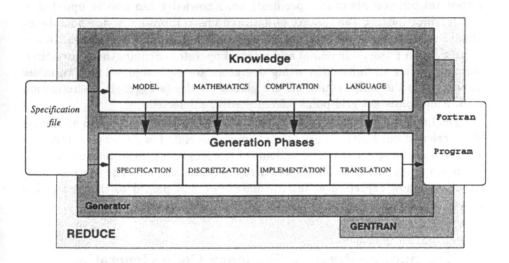

Fig. 2. General architecture of the generator

Knowledge Sources. On an abstract level there are essentially four knowledge sources: *model knowledge, mathematical knowledge, computational implementation knowledge,* and *target language knowledge.* A great deal of this knowledge is implemented in the form of equations and rules. When invoked by an object, a set of REDUCE rules will recursively perform substitutions very much like rewrite systems. The knowledge categories are not fully independent of each other: knowledge in one category is sometimes expressed in terms of knowledge of another.

Model knowledge includes model equations such as those discussed in Sect. 2 and forms the basis for constructing *scripts.* Model discretization also belongs to this category. Mathematical knowledge embodies the necessary mathematical machinery. Besides implicit knowledge contained in the computer algebra system, it includes definitions of finite difference operators and approximations for differential and integral operators. Computational implementation knowledge is used in transforming the discretized equations into a computation on a computational grid. Information concerning program variables, their memory mappings and shapes is also contained here. Finally, translation knowledge embodies target language syntax and target architecture knowledge, the former being mostly contained in the GENTRAN [4] and GENTRAN 90 [2] packages.

Generation Phases. Generation proceeds through *specification, approximation-discretization, implementation* and *translation* phases. They are executed procedurally, drawing from the declarative knowledge sources which were summarized above. Besides scripts, variable declarations and data mapping directives form important components of the specifications. Knowledge can also be updated in the specifications. In the context of data parallel computing, script boundaries should correspond as much as possible to program phase[4] boundaries. In the discretization phase, differential and integral operators acquire their discretized form and, after all finite differencing operators are evaluated, all field variables appearing in the script equations are placed on the (staggered) discretization grid. An inconsistent grid-point reference will be detected.

The implementation phase transforms the discretized scripts into an imperative calculation. Indexed variables are now associated with arrays; staggered variables get integer array indices. Some of the program transformations include index mappings, array reduction and array declarations, etc. The abstract computation is now expressed as intermediate code and passed to the translation phase. The latter is performed by the GENTRAN packages.

4   The Surface-Pressure Tendency Code Generation

Specification and Scripts. We illustrate below a portion of the specification for the pressure tendency code. It is written in the syntax of REDUCE and its associated support language Rlisp. Certain entries such as grid definition and associated field variables declarations are not shown. Another script defining temporary variables to improve efficiency through common subexpression elimination, exchanging divisions for multiplications, as well as other heuristics is also omitted. Equation (2) for p and equation (3) for U and V are represented through the assignments of the generator variables p_r, uaux_r and vaux_r respectively. The operator df is the REDUCE symbolic differentiation operator. The discretization rules determine its finite difference approximation. The operator centr_av is a central average operator, and div2_op represents the horizontal divergence expressed in curvilinear coordinates.

---

[4] The word *phase* is overloaded in this article. When the generator itself is discussed, the conventional meaning of the word is intended. When referring to generated programs, as is done here, the word takes its more recently acquired technical designation: a program segment with specific array reference characteristics, (see e.g. [11]). Confusion of this type is nearly unavoidable when discussing programs which generate programs. Other program attributes such as *implementation* suffer also to some extent from the same ambiguity.

```
% ------------------------------------------------------------------
%                      Pressure Tendency Script
% ------------------------------------------------------------------
p_r:=A+B*ps                                                       $
    da_eq:=      d_a=df(a,eta)                          *    d_eta$
    db_eq:=      d_b=df(b,eta)                          *    d_eta$
    dp_eq:=      d_p=df(p_r,eta)                        *    d_eta$
    dp_eq:=      (dp_eq where
                    heuristics_rules0({da_eq,db_eq}))             $

uaux_r:=u*centr_av(d_p,x)                                 *2/d_eta$
vaux_r:=v*centr_av(d_p,y)                                 *2/d_eta$


    u_aux_eq:=  u_aux=numeric_integral(uaux_r,eta,0,1)           $
    v_aux_eq:=  v_aux=numeric_integral(vaux_r,eta,0,1)           $


    repla_unknown!*:='((hx.hxv) (hy.hyu))                        $

eq_4:=         ddt_ps= -div2_op(u_aux,v_aux)              /   2$
ps_tendency_script:={da_eq,db_eq,dp_eq,u_aux_eq,v_aux_eq,eq_4}   $
% ------------------------------------------------------------------
```

Embodied in the REDUCE program variable ps_tendency_script (last program line), is the actual definition of the script used to generate the computation of (4). It is a concise list of equations (also defined in the specification) which delineate the steps of the computation on a high level. It was obtained from the second equality in (4). The list of equations could be modified, e.g. to represent the first equality in (4), should that lead to more efficient implementations on available hardware architectures. The present breakdown corresponds roughly to the one in the current (Fortran 77) reference version of the HIRLAM model, chosen because it leads to a more efficient code for vector architectures than the first equality. A posteriori it became apparent that a small number of factorizations would further improve efficiency. They appear in the script flushed to the right for the sake of clarity.

Data Mappings. The storage association of the field variables is determined by specifying index transformation functions. The corresponding inverse transformations are also needed because one is dealing with the mapping of functions and not of index values. A mapping of the horizontal index set (m, n) onto a column-major storage index set (ii) can be specified by the REDUCE expression:

```
vec_map := { {ii= m-low(m)+1+(n-low(n))*npts(m)},
               {n=floor((ii-1)/npts(m))+low(n),
                m=(ii-1)-floor((ii-1)/npts(m))*npts(m)+low(m)} };
```

The second and third lines represent the inverse map. Presently a mapping is applied globally to all field quantities, but future implementations will enable assigning a specific mapping to a specific variable as well as to a specific program phase.

Translation. Omitting the automatically generated array declarations, as well as temporary variables code, the program segment generated from the above script and mapping is shown below. The loop ranges were obtained via an iter-

```
        DO 25012 KK=1,MLEV                      DO 25019 II=1,MLAT*MLON-MLON
        ZDAK(KK)=AHYB(KK+1)-AHYB(KK)            V_AUX(II)=0.0
25012 CONTINUE                            25019 CONTINUE
        DO 25013 KK=1,MLEV                      DO 25020 KK=1,MLEV
        ZDBK(KK)=BHYB(KK+1)-BHYB(KK)            DO 25021 II=1,MLAT*MLON-MLON
25013 CONTINUE                                 V_AUX(II)=PVZ(II,KK)*ZDPK(II+MLON,KK)+
        DO 25014 KK=1,MLEV                    . PVZ(II,KK)*ZDPK(II,KK)+V_AUX(II)
        DO 25015 II=1,MLAT*MLON            25021 CONTINUE
        ZDPK(II,KK)=PPSZ(II)*ZDBK(KK)+ZDAK(KK) 25020 CONTINUE
25015 CONTINUE                                 DO 25022 II=MLON+2,MLAT*MLON-MLON-1
25014 CONTINUE                                 PDPSDT(II)=ZRDLOH*HYU(II-1)*U_AUX(II-1)
        DO 25016 II=1,MLAT*MLON-1            . *ZRHXHY(II)-(ZRDLOH*HYU(II)*U_AUX(II)*
        U_AUX(II)=0.0                         . ZRHXHY(II))+ZRDLAH*HXV(II-MLON)*V_AUX(
25016 CONTINUE                               . II-MLON)*ZRHXHY(II)-(ZRDLAH*HXV(II)*
        DO 25017 KK=1,MLEV                    . V_AUX(II)*ZRHXHY(II))
        DO 25018 II=1,MLAT*MLON-1         25022 CONTINUE
        U_AUX(II)=PUZ(II,KK)*ZDPK(II+1,KK)+PUZ(
      . II,KK)*ZDPK(II,KK)+U_AUX(II)
25018 CONTINUE
25017 CONTINUE
```

ated DEFINE/USE consistency scheme as a way of treating boundaries. Two iterations were needed. Certain optimizing transformations such as factorizations, loop-fusion, and code motion will improve the code further. Implementation of such transformations is in progress.

Presently, maintenance of forecast models involves porting code with two-dimensional arrays for vector machines to fully three-dimensional arrays for MPP architectures. By specifying a three-dimensional array storage scheme instead of the one above, a similar code is generated with the correct array references and three-level loop nesting. We show instead below the elemental assignment version with **eoshift** and **sum** operators in Fortran 90 syntax using GENTRAN 90.

```
      real,dimension(mlev+1)::ahyb,bhyb,zdak,zdbk
      real,dimension(mlon,mlat)::ppsz,pdpsdt,zhx,zhy,hxv,hyu,u_aux,v_aux&
   & ,zrhxhy,hxhy,rhxu,rhyv
      real,dimension(mlon,mlat,mlev+1)::zpkp,zdpk,puz,pvz,ke

      zdak=eoshift(ahyb,1,1)-ahyb
      zdbk=eoshift(bhyb,1,1)-bhyb
      zdpk=ppsz*zdbk+zdak
      u_aux=sum(eoshift(zdpk,1,1)*puz+puz*zdpk,3)
      v_aux=sum(eoshift(zdpk,2,1)*pvz+pvz*zdpk,3)
      pdpsdt=zrdloh*eoshift(hyu,1,-1)*eoshift(u_aux,1,-1)*zrhxhy-zrdloh*&
   & hyu*u_aux*zrhxhy+zrdlah*eoshift(hxv,2,-1)*eoshift(v_aux,2,-1)*    &
   & zrhxhy-zrdlah*hxv*v_aux*zrhxhy
```

The array declarations as well as the DIM and SHIFT parameters of eoshift were inferred automatically using special generic procedures. We have not yet studied in detail the efficiency of the code in this form. However its compactness makes it more amenable to automatic common subexpression elimination and other transformations, obviating the manual breakdown of the scripts discussed earlier, as shown by our preliminary work using the SCOPE [9] package.

5 Adjoint Code Generation

The maintenance of large model codes is usually accompanied by the maintenance of the adjoint model. The adjoint has been traditionally hand-coded but more recently tools have been constructed for its automatic generation through the use of *automatic differentiation* techniques (see e.g. [5] for an overview). Like many automatic tools, automatic adjoint generators take (hand-)coded source as their starting point. As already implied in the introduction, source code often contains programming constructs which are more relevant to the target language and hardware architecture. These usually mask model and mathematical knowledge which is needed to perform what is in principle a mathematical transformation. The generation techniques presented here allow us to couple the generation of the adjoint model to the generation of the forward model with relative ease. Within the framework shown in Fig. 2, the adjoint generation is initiated at the specification/discretization level.

In adjoint modeling it is the adjoint $\mathcal{D}^*$ of the linearized form of the forward model D which is needed. The action of $\mathcal{D}^*$ on a vector a in the range space of D can be expressed in terms of the inner product (a, Du), as:

$$\mathcal{D}^* a = \frac{\delta(a, Du)}{\delta u} , \tag{5}$$

as discussed by Thacker [14] in the context of automatic differentiation. The adjoint code can thus be generated by differentiating the appropriately discretized inner product. In terms of specification scripts, we introduce the variable adj_ddt_ps adjoint to the surface pressure tendency. Its contribution to the inner product for the dynamics will be the horizontal sum of adj_ddt_ps*ddt_ps. By prepending the script of the forward model one obtains a full script for the inner product. An adjoint script adj_script is then generated by performing differentiation on inner_prod_script in reverse mode with respect to the model variables u, v, and ps. The scripts for the inner product as well as for the adjoint are shown below. The latter was automatically generated from the former, and has the recursive structure typical of reverse mode differentiation. The last three equations in the script represent the required adjoints of ps, u and v respectively. Fortran code generation follows by going through the same transformations as for the forward model. Details of this work, which is in progress, will be presented elsewhere.

```
inner_prod_script:=                  adj_script :=
  {ps_tendency_script,                 {adj_a1001=df(inner_prod,ddt_ps'),
                                        adj_b1001=mat_mul(adj_a1001,df(ddt_ps,u_aux')),
  inner_prod=                           adj_b1002=mat_mul(adj_a1001,df(ddt_ps,v_aux')),
    discr_integral(discr_integral(      adj_c1001=mat_mul(adj_b1001,df(u_aux,d_p')) +
    adj_ddt_ps*ddt_ps,                            mat_mul(adj_b1002,df(v_aux,d_p')),
    x,0,rmaxlon),y,0,rmaxlat)}$         adj_c1002=mat_mul(adj_b1001,df(u_aux,u')),
                                        adj_c1003=mat_mul(adj_b1002,df(v_aux,v')),
                                        adj_d1001=mat_mul(adj_c1001,df(d_p,ps')),
mod_vars!*:='( ps u v)$                 adj_d1002=mat_mul(adj_c1002,df(u,u')),
mod_range_vars!*:='( ddt_ps)$           adj_d1003=mat_mul(adj_c1003,df(v,v'))}        $
```

6 Conclusions

We have presented program generation techniques for large scale numerical models. We have concentrated more on procedural components than on high-level inference in order to be able to assess the functionality needed to be able to obtain the necessary leverage for realistic maintenance and development issues. In spirit, our work is mostly related to that of Ref. [17] and, to some extent, to that of Ref. [10] where the architecture has much in common with ours. That work however has a much more developed inference system and more formalized specifications. Specifications in the present work are less constrained so as to allow us to deal more directly with the issues mentioned in the article. Future work includes generating other parts of the forecast model, and implementing additional code transformations and more sophisticated data layout schemes.

7 Acknowledgments

The authors have benefited from discussions with Lex Wolters, Robert van Engelen, Paul ten Brummelhuis, and Hans van Hulzen. We are grateful to André Koopal for providing LaTeX forms of the HIRLAM equations.

References

1. Arakawa, A., Lamb, V. R.: A potential enstrophy and energy conserving scheme for the shallow water equations. Mon. Wea. Rev.**109** (1981) 18–36.
2. Borst, W. N., Goldman, V.V., van Hulzen, J. A.: GENTRAN90: A REDUCE package for the generation of FORTRAN 90 code. (ISSAC '94),Proceedings Int. Symp. on Symbolic and Algebraic Computation (1994) 45–51. ACM Press, New York.
3. Cats, G., Middelkoop, H., Streefland, D., Swierstra, D.: In: The dawn of massively parallel processing in meteorology, Springer Verlag, Berlin (1990) 47–75.
4. Gates, B. L., Dewar, M. C.: GENTRAN User's Manual - REDUCE VERSION, (1991)
5. Griewank, A., Corliss, G.F., eds.: *Automatic Differentiation of Algorithms.* SIAM, Philadelphia, (1991).

6. Hammond, S. W., Loft, R. D., Dennis, J. M., Sato, R. K.: A data parallel implementation of the NCAR Community Climate Model (CCM2). In *Proc. Seventh SIAM Conference on Parallel Processing for Scientific Computing*, D.H. Bailey, P.E. Bjørstad, J.R. Gilbert, M.V. Mascagni, R.S. Schreiber, H.D. Simon, V.J. Torczon and L.T. Watson, (eds.), (1995) 125–130. SIAM, Philadelphia.
7. Hearn, A. C.: REDUCE 3.5 Manual (1993), The Rand Corporation, Santa Monica.
8. Houstis, E. N., Rice, J., R., Vichnevetsky, R., (Eds.): Proceedings of the Third International Conference on Expert Systems for Numerical Computing, West-Lafayette,May 1993. Math. Comput. Simulation **36** (1994) 269–520
9. van Hulzen, J. A.,: SCOPE 1.5, a Source-Code Optimization Package for REDUCE 3.5 – User Manual. Memorandum INF-94-17, Department of Computer Science, University of Twente, (1994).
10. Kant, E., Yau, A. S-H., Liska, R., Steinberg, S.: A problem solving environment for generating certifiably correct programs to solve evolution equations. Preprint (1995). Department of Mathematics and Statistics, University of New Mexico, Albuquerque.
11. Kremer, U., Mellor-Crummey, J., Kennedy, K., Carle, A.: Automatic data layout for distributed-memory machines in the D programming environment. Technical Report CRPC-TR93298-S, Center for Research on Parallel Computation, (1993).
12. Setliff, D., Kant, E., Cain, T.: Practical Software Synthesis. IEEE Software **10** (1993) 6–10
13. Simmons, A., Burridge, D. M.: An energy and angular-momentum conserving vertical finite-difference scheme and hybrid vertical coordinates. Mon. Wea. Rev.**109** (1981) 758–766.
14. Thacker, W. C.: Automatic differentiation from an oceanographer's perspective. In [5], pp. 191–201.
15. Thépaut, J.N., Courtier, P.: Four-dimensional variational data assimilation using the adjoin of a multilevel primitive-equation model. Quart. J. Roy. Meteor. Soc. **117** (1991) 1225-1254.
16. Thinking Machines Corporation: *Using the CMAX converter.* Manual version 2.0 (1994).
17. Wang, P. S.: FINGER: A Symbolic System for Automatic Generation of Numerical Programs in Finite Element Analysis, J. Symb. Comp. **2** (1986) 305–316
18. Weerawarana, S., Houstis, E. N., Rice, J. R.: An interactive symbolic-numeric interface to Parallel ELLPACK for building general PDE Solvers. In B.R. Donald, D. Kapur, and J. L. Mundy, editors: *Symbolic and Numerical Computation for Artificial Intelligence*, Academic Press (1992) 303–321.
19. Wolters, L., Cats, G., Gustafsson, N.: Limited area numerical weather forecasting on a massively parallel computer. *Proceedings of the 8th ACM International Conference on Supercomputing*, July 11-15 1994, Manchester, England, ACM press, (1994),289-296.

High Performance Computational Chemistry: NWChem and Fully Distributed Parallel Applications

M.F. Guest[1], E. Aprà[2], D.E. Bernholdt[2], H.A. Früchtl[2], R.J. Harrison[2],
R.A. Kendall[2], R.A. Kutteh[2], X. Long[2], J.B. Nicholas[2], J.A. Nichols[2],
H.L. Taylor[2], and A.T. Wong[2], and G.I. Fann[3], R.J. Littlefield[3] and
J. Nieplocha[3]

[1] Theory and Computational Science Dept, CCLRC Daresbury Laboratory,
Daresbury, Warrington WA4 4AD, Cheshire, UK
[2] High Performance Computational Chemistry Group, Environmental Molecular
Sciences Laboratory, Pacific Northwest Laboratory, PO Box 999, Mail Stop K1-90,
Richland, WA. 99352, USA
[3] High Performance Computing Software Support, Environmental Molecular Sciences
Laboratory, Pacific Northwest Laboratory, PO Box 999, Mail Stop K1-90, Richland,
WA. 99352, USA

Abstract. We describe the development, implementation and performance of the NWChem computational chemistry package. Targeting both
present and future generations of massively parallel processors (MPP),
this software features a variety of efficient parallel algorithms for a broad
range of methods commonly used in computational chemistry. The emphasis throughout is on scalability and the distribution, as opposed to
the replication, of key data structures. To facilitate such capabilities,
we describe a shared non-uniform access memory model which simplifies
parallel programming while at the same time providing for portability
across both distributed- and shared-memory machines. As an example
of the capabilities of NWChem, we outline the development and performance of a highly efficient and scalable algorithm for conducting self-
consistent field Density Functional calculations on molecular systems. A
performance analysis of this module in calculations on the zeolite fragment $Si_{28}O_{67}H_{30}$ (1673 basis functions) is presented for representative
MPP systems, the IBM-SP2, Kendall Square Research KSR-2, the Intel
Paragon and Cray T3D. Finally we consider, based on these figures, the
scale of performance likely to be achieved on the next generation of MPP
systems.

1 Introduction

Computational chemistry covers a wide spectrum of activities ranging from
quantum mechanical calculations of the electronic structure of molecules, to classical mechanical simulations of the dynamical properties of many-atom systems,
to the mapping of both structure-activity relationships and reaction synthesis
steps. Although chemical theory and insight play important roles in the work of

computational chemists, the prediction of physical observables is almost invariably bounded by the available computer capacity. Massively parallel computers, with hundreds to thousands of processors (MPPs), promise to significantly outpace conventional supercomputers in both capacity and price-performance. Indeed the mission of the High Performance Computational Chemistry (HPCC) group at the Pacific Northwest Laboratory is to develop molecular modeling software applications that realize this promise, providing 10-100 times more computing capability than has been available with more traditional hardware. Such improvements will enable substantial scientific and commercial gains by increasing the number and complexity of chemical systems that can be studied.

While increases in raw computing power alone will greatly expand the range of problems that can be treated by theoretical chemistry methods, a significant investment in new algorithms is needed to fully exploit the potential of present and future generations of MPPs. Merely porting presently available software to these parallel computers does not provide the efficiency required to exploit their full potential. Most existing parallel applications show a significant deterioration in performance as greater numbers of processors are used. In some cases, the efficiency is so poor that the use of additional processors decreases, rather than increases, the performance. Thus, new algorithms must be developed that exhibit parallel scalability (i.e., show a near linear increase in performance with the number of processors). Although perfect scalability is very difficult to achieve, we have demonstrated our ability to approach this level of performance with our new self-consistent field (SCF) codes (see below).

In addition to the creation of new algorithms for computational chemistry on parallel processors, we are also creating the high-level data and control structures needed to make parallel programs easier to write, maintain, and extend. Our efforts have culminated in a package called NWChem (for Northwest Chemistry). The package includes a broad range of functionality, and will continue to grow for some time yet. Algorithms for Hartree-Fock self-consistent field calculations [1, 2], a four-index transformation [3], several forms of second-order perturbation theory [3-5], density functional theory, and molecular dynamics with classical, quantum mechanical, or mixed force-fields are already in the code, and others such as multiconfiguration SCF (MCSCF) and higher-level correlated methods are under development. A wide range of properties are also available or under development: gradients, Hessians, electrical multipoles, NMR shielding tensors, etc.

This paper is organized in two sections. First, we focus on the design philosophy, structure, and tools which make NWChem an effective environment for the development of computational chemistry applications. Although our focus has been on the efficient use of parallel computers, almost all of the ideas and tools we present here are generally applicable. Second, we consider the computational requirements for molecular computations as a function of the level of theory and required accuracy, and provide, by way of example, an outline of the development and performance of a highly efficient and scalable algorithm for conducting self-consistent field Density Functional Theory (DFT) studies on molecular systems. Based on calculations of the zeolite fragment $Si_{28}O_{67}H_{30}$ (with 1673 basis functions), we present a performance analysis of this module on representative MPP systems e.g., the IBM-SP2, Kendall Square Research KSR-2, the Intel Paragon and Cray T3D. We consider, using these figures, the scale of performance likely to be achieved on the next generation of MPP systems.

2 Design Philosophy

The philosophy and structure of NWChem has developed in response to our experiences with other computational chemistry packages, and a number of constraints imposed on us.

Our initial experiments led us to conclude that while it was possible to modify existing software to implement parallel algorithms on a "proof of principle" basis, the effort involved in turning them into fully scalable (distributed rather than replicated data) codes was tantamount to writing the codes over from scratch. At the same time, as in most software development efforts, we have limitations on the personnel and time available to the development effort. Consequently, we must have a system which is easy to program, and in particular in which it is easy to prototype new parallel algorithms.

For a large software package, maintenance costs (in terms of human effort) can also be significant. In particular, the current state of parallel computing is that useful software standards are just beginning to emerge. Recognizing that we will probably have to develop some of our own software tools, we wish to maintain a path that will allow us, as far as possible, to transition from locally developed parallel programming tools to vendor-supported, standard-based implementations.

Finally, there is the issue of portability. In parallel computing, at present, the need for portability presents particular problems. There are a number of architectures in use, which can be broadly categorized as shared- or distributed-memory according to whether each processor can address all or only a portion of the machine's memory directly. In addition, an increasing number of machines combine these models by using small clusters of shared-memory CPUs as nodes in a distributed-memory machine. Portability in this case means finding a programming model which can be used effectively across this variety of platforms.

In order to meet all these requirements, we have adopted ideas from the object-oriented (OO) style of software development [6], which has come into wide use throughout the commercial software community. The basic principles of this methodology are abstraction, encapsulation, modularity, and hierarchy. Abstraction, in this case, is the separation of the problem to be solved from the process chosen to solve it. This facilitates the introduction of better methods as soon as they are available. In most complex systems, abstraction can be carried out at many levels, resulting in a hierarchy in which it may be possible to relate many components and develop further abstractions. An example of this in NWChem is the parallel programming model implemented in the Global Array toolkit (see below), which generalizes ideas common to the distributed-memory, shared-memory, and cluster architectures. In a complex system, it is also useful to encapsulate data structures or other objects so that they can only be manipulated in carefully controlled and well-defined ways. This helps to eliminate the common problem of large software systems of unexpected interactions between components that were thought to be independent. Modularity, the use of small packages of well-defined functionality which can be reused easily, is a familiar concept to most scientific programmers.

While there are OO programming languages, such as C++ [7], which allow the formal aspects of the methodology to be taken straight into the actual code, we have chosen a compromise approach in the development of NWChem. Object-oriented ideas are used at the design stage, while the implementation is a mixture of Fortran and C; languages which are much more familiar to current computational chemists than C++. Our experience is that although it may take some extra time, a thorough, careful design phase is probably the most important aspect of the OO methodology, and this effort is quickly recouped in easier implementation and fewer instances of modification of code after implementation. A design based on OO principles can be implemented in a non-OO programming language with only a small amount of discipline on the programmer's part.

3 Software Development Tools

We have selected or developed a number of low-level tools to facilitate the development of our parallel chemistry applications. Although some tools have been designed specifically to meet the needs of computational chemistry applications, these tools are generally applicable.

3.1 Message Passing

TCGMSG [8] is a toolkit for writing portable parallel programs using a message passing model. It has distinguished itself from other toolkits in its simplicity, high performance and robustness. The limited functionality provided includes point-to-point communication, global operations and a simple load-balancing facility, all designed with chemical applications in mind. This toolkit is available in the public domain (from ftp.tcg.anl.gov), and is distributed with a set of example chemical applications. The package has been ported to a wide range of workstations, heterogeneous networks of computers, and true parallel computers. The message-passing interface (MPI) standard [9] will soon obsolete the TCGMSG interface. However, TCGMSG has superior robustness and performance when compared with current portable MPI implementations. In addition, the simple TCGMSG interface is readily implemented using MPI.

3.2 Memory Allocation

One of the key problems on high performance parallel supercomputers is managing the memory utilization on each node. Since the Fortran77 standard does not include a dynamic memory allocation scheme, many computational chemistry and other Fortran-based applications have developed simple schemes to emulate dynamically allocated memory. These approaches typically place a substantial burden on the programmer to correctly handle the sizing and alignment of different data types, which may vary from one platform to another. In our experience, this is one of the most error-prone parts of working on many computational chemistry codes. In response to this problem, we have developed a library, called MA, that comprises a dynamic memory allocator for use by C, Fortran, or mixed-language applications. MA provides both heap and stack memory management disciplines, debugging and verification support (e.g., detecting memory leaks), usage statistics, and quantitative memory availability information. It is also "type safe", which means that unlike most Fortran dynamic allocation schemes, objects allocated by MA are of a specific Fortran data type which is recognized by the compiler. C applications can benefit from using MA instead of the ordinary malloc() and free() routines because of the extra features MA provides. Once Fortran90 is widely available, library tools such as MA may not be required but it is currently essential for memory management in complex applications, and offers many features not provided by the Fortran90 standard which are extremely useful during the code development process.

3.3 Global Arrays

Another important consideration in using MPPs is how data are stored. So-called replicated-data schemes require that a copy of each data item in the program be stored on each processor, so that the size of the problem that can be handled is limited by the memory of a single processor. In distributed-data applications each processor holds only a part of the total data; in such cases, the problem size is limited only by the total memory of the machine, allowing much larger problems to be treated. Our efforts focus on distributed-data applications. They span virtually the whole range of computational chemistry methods, with significant efforts in Hartree- Fock and correlated techniques, multi-configuration self-consistent field methods, density functional theory, semi-empirical methods, and classical Monte Carlo and molecular dynamics.

No emerging standards for parallel programming languages (notably just High Performance Fortran (HPF-1)) provide extensive support for multiple instruction multiple data (MIMD) programming [10]. The only truly portable MIMD programming model is message passing, for which a standard interface has been recently proposed [11]. It is, however, very difficult to develop applications with fully distributed data structures using the message-passing model [12, 13]. The shared-memory programming model offers increased flexibility and programming ease but is less portable and provides less control over the interprocessor transfer cost. What is needed is support for one-sided asynchronous access to data structures (here limited to one- and two-dimensional arrays) in the spirit of shared memory. With some effort this can be done portably [14]; in return for this investment, we gain a much easier programming environment, speeding code development and improving extensibility and maintainability. We also gain a significant performance enhancement from increased asynchrony of execution of processes [15]. Message passing forces processes to cooperate (e.g., by responding to requests for a particular datum). Inevitably, this involves waiting for a collaborating process to reach the same point in the algorithm, which is only partially reduced by the use of complex buffering and asynchronous communication strategies. With a one-sided communication mechanism, where each process can access what it needs without explicit participation of another process, all processes can operate independently. This approach eliminates unnecessary synchronization and naturally leads to interleaving of computation and communication. Most programs contain multiple algorithms, some of which may naturally be task-parallel (e.g., Fock matrix construction), and others that may be efficiently and compactly expressed as data-parallel operations (e.g., evaluating the trace of a matrix product). Both types of parallelism must be efficiently supported. Consideration of the requirements of the self-consistent field algorithm discussed later in this paper, the parallel COLUMBUS configuration interaction program [16], second order many-body perturbation theory [5] and parallel coupled-cluster methods [17] led to the design and implementation of the Global Array (GA) toolkit [14] to support one-sided asynchronous access to globally-addressable distributed one- and two-dimensional arrays.

The GA toolkit provides an efficient and portable "shared-memory" programming interface for distributed-memory computers. Each process in a MIMD parallel program can asynchronously access logical blocks of physically distributed matrices, without need for explicit co-operation by other processes. Unlike other shared-memory environments, the GA model exposes the programmer to the non-uniform memory access (NUMA) timing characteristics of the parallel computers and acknowledges that access to remote data is slower than to local data. From the user perspective, a global array can be used as if it were stored in shared memory, except that explicit library calls are required to access it. Details of actual data distribution and addressing are encapsulated in the global array objects. Matrices are physically distributed blockwise, either regularly or as the Cartesian product of irregular distributions on each axis. The information on the actual data distribution can be obtained and exploited whenever data locality is important. Each process is assumed to have fast access to some "local" portion of each distributed matrix, and slower access to the remaining "remote" portion. A very efficient, direct access to the local data is supported. Remote data can be accessed through operations like "get", "put" or "accumulate" (floating point sum-reduction) that involves copying the globally accessible data to/from process-private buffer space. Processes can communicate with each other by creating and accessing GA distributed matrices, and also (if desired) by conventional message-passing. The toolkit provides operations for the creation, and destruction of distributed arrays, for the synchronization of all processes, and for inquiries about arrays and their distribution. In addition, primitive operations: such as get, put, accumulate, atomic read and increment, gather and scatter, and direct access to the local portion of an array are supported. A number of BLAS-like data-parallel operations have been developed on top of these primitives and are also included in the toolkit.

Additional functionality is provided through a variety of third party libraries made available by using the GA primitives to perform the necessary data rearrangement. These include standard and generalized real symmetric eigensolvers (PeIGS, see below), and linear equation solvers (SCALAPACK) [18, 19, 20]. The $O(N^2)$ cost of data rearrangement is observed to be negligible in comparison to that of $O(N^3)$ linear-algebra operations. These libraries may internally use any form of parallelism appropriate to the host computer system, such as co-operative message passing or shared memory.

The GA interface has been designed in the light of emerging standards. In particular HPF-1 and subsequent revisions will certainly provide the basis for future standards definition of distributed arrays in Fortran. A long term goal must be to migrate to full language support, and to eliminate as much as possible the practice of parallel programming through subroutine libraries. The basic functionality described above (create, fetch, store, accumulate, gather, scatter, data-parallel operations) may be expressed as single statements using Fortran90 array notation and the data-distribution directives of HPF. However, HPF currently precludes the use of such operations on shared data in MIMD parallel code. There is reason to believe that future versions of the HPF standard will rectify this problem, as well as provide for irregular distributions, which we have found lead to a significant increase in performance in computational chemistry applications.

3.4   Distributed Matrices and Linear Algebra

Many electronic structure computations are formulated in terms of dense or nearly dense matrices of size roughly N by N, where N is the number of basis functions. Two distinct classes of operations are performed on these matrices: random access to small blocks, for the purpose of constructing matrices as a function of many one- and two-electron integrals; and linear algebra operations on the entire matrix, such as eigensolving, Cholesky decomposition, linear system solution, inversion, and matrix-matrix multiplication. Both types of operations must work on distributed matrices if the resulting application is to be truly scalable.

A particular focus of our distributed linear algebra work has been the development of a scalable, fully parallel eigensolver whose numerical properties satisfy the needs of the chemistry applications. This package, called PeIGS, solves dense real symmetric standard (Ax=lx) and generalized (Ax=lBx) eigenproblems. The numerical method used by PeIGS is multisection for eigenvalues [21, 22] and repeated inverse iteration and orthogonalization for eigenvectors [23]. Accuracy and orthogonality are similar to LAPACK's DSPGV and DSPEV [24]. Unlike other parallel inverse iteration eigensolvers in the current literature, PeIGS guarantees orthogonality of eigenvectors even for arbitrarily large clusters that span processors.

PeIGS is both fast and scalable – on a single processor it is competitive with LAPACK, and parallel efficiency remains high even for large processor counts.

For example, in one of our SCF applications, the standard eigenproblem Ax=lx was solved for all eigenpairs of a 2053 by 2053 matrix. This computation required only 110 seconds on 150 processors of an Intel Paragon computer, a time-to-solution estimated as 87 times faster than LAPACK for the same problem on 1 processor (The full 2053 by 2053 problem is too large to fit on a single processor, so the 1-processor time is extrapolated from smaller problems and confirmed from larger processor counts).

Internally, PeIGS uses a conventional message passing programming model and column-distributed matrices. However, it is more commonly accessed through an interface provided by the GA toolkit. The necessary data reorganization is handled by the interface, and is very fast compared to the $O(N^3/P)$ linear algebra times.

3.5 Run Time Data Base

The run time data base (RTDB) is the parameter and information repository for all application modules comprising NWChem. This is similar in spirit to the GAMESS [25, 26, 27, 28] dumpfile or GAUSSIAN [29] checkpoint file. An input parsing module(s) stores the input and each application module obtains the appropriate parameters for execution and communicates to other modules via the RTDB. The RTDB library is accessible from both Fortran and C, is based on simple UNIX data base tools, and is widely portable. The storage is based on a string to data (of a single type) mapping, with complex data structures stored as separate data blocks of the appropriate type. There is also contextual levels built into this tool; for example, the gradient convergence threshold or print level would be a parameter applicable to various modules and would thus have a different value for each method/module. The appropriate context can be set to arbitrate between the possible values available.

4 Chemistry-Related Software Tools

The software tools outlined above are examples of libraries with a standard application program interface (API). The designing of a standard API for each functionality that is required by many of our codes promotes code reuse, decreases the effort to develop new applications and also hides complexity, which further decreases development and maintenance costs. There are many chemistry-specific components that can be similarly defined, and with proper design and implementation we may realize similar benefits.

Primary examples of this idea in NWChem are the integral, basis set and geometry "objects". While not "objects" in the strict sense of an OO language, these are small modules which encapsulate specific data and provide access to it through a well-defined API. The geometry object specifies the physical makeup of the chemical system by defining atomic centers, their respective position in space and nuclear charge as well as an associated name for the center. It also includes a possible applied electric field, the symmetry (point group or periodic) and other characteristics of the system. The basis set object handles the details of the Gaussian (and eventually other types of) basis sets. The combination of a basis set object and a geometry object provides all information required by an ab initio calculation which is not specific to a particular electronic state. The APIs to these modules provide high-level inquiries about the number of centers and basis functions, such as many chemistry applications require, as well as lower-level information about the details of the basis set or coordinates, which are normally used only by the integral package and similar modules.

The integral API is a layer between the actual integral package and the application module, and allows the application programmer to essentially ignore the details of how the integrals are computed. This also facilitates the incorporation of different integral technologies into the applications that use the code; currently, the API requests integrals of various types based on computing shell blocks of integrals, with two different integral packages used for production.

The explicit separation of these objects has greatly simplified the development of several of the chemistry modules in NWChem. For example, density functional methods, and those using the so-called "resolution of the identity" integral approximations [30, 31] generally involve the use of more than one basis set. The basis set and integral modules are designed to allow more than one basis set to be used at a time, so implementation of these methods was straightforward.

This design may be applied at all levels within the code. Modules that would normally be considered stand-alone high-level chemistry applications may be invoked by other such modules. For instance, our ab initio molecular dynamics module invokes the self-consistent field and self-consistent field gradient modules as simple components through a standard API.

5 Chemistry Applications on MPPs

A wide range of computational chemistry methods have been implemented in NWChem, and more are planned, representing the core functionality required of any general-purpose computational chemistry package. In accord with our mission, the novel aspects of their implementation in NWChem are the parallel algorithms we employ, along with other ideas for reducing the scaling of the resource requirements of calculations as a function of their size.

Current MPP technology typically delivers Livermore FORTRAN Kernel (LFK) maximum megaflops per second (MFLOPS) ratings of 100-200, interprocessor communication bandwidths of 30-150 megabytes (Mb) per second, and point-to-point latencies of 20-60 microseconds (ms). Figure 1 illustrates the relative computing power required for molecular computations at four different levels of theory (and accuracy): configuration interaction, Hartree-Fock, density functional, and molecular dynamics. Each of these levels of theory formally scale from N^2 to N^6 with respect to computational requirements. Standard screening techniques which in essence ignore interactions for atoms far apart can bring these scalings down significantly. Each level of theory also exhibits differing requirements for interprocessor communication (bandwidth and latency) and we will provide a brief description of this requirement in the next section. It is our impression that the next generation MPP will have LFK maximum rates of 300-400 MFLOPS, interprocessor communication bandwidths of 100-300 Mb per second, and point-to-point latencies of 10-30 ms. This type of MPP (with several hundred nodes) will provide us with the technology to address important problems in our support of the environmental remediation efforts in progress at the Environmental Molecular Sciences Laboratory (EMSL).

Complete descriptions of all of the algorithms employed in NWChem is not possible in the space available here, and are being published separately. In order to demonstrate how the structure and tools we have described above are used by a high-level application, we will sketch the operation of the density functional theory module of NWChem. We will then mention briefly some of the other high-level functionality which is already in place or under development for the package.

5.1 Density Functional Theory - Self Consistent Field Module

An essential core functionality in an electronic structure program suite is the direct self consistent field (SCF) module. It is increasingly evident that the application of direct SCF to large molecules is best performed on MPPs due to the enormous computational requirements. However, targeting systems in excess of 1,000 atoms and 10,000 basis functions requires a re-examination of the conventional algorithm and assumed memory capacities. The majority of previous parallel implementations use a replicated-data approach [10] which is limited in scope since the size of these arrays will eventually exhaust the available single processor memory.

Our implementation of the parallel direct DFT (as in the Hartree-Fock module) distributes these arrays across the aggregate memory using the GA tools. This ensures that the size of systems treated scales with the size of the MPP and is not constrained by single processor memory capacities.

The software we have developed is a MPP implementation of the Hohenberg-Kohn-Sham formalism [32, 33] of DFT. This method yields results similar to those from correlated ab initio methods, at substantially reduced cost. It assumes a charge density and approximations are made for the Hamiltonian (the exchange correlation functional); in contrast with traditional ab initio molecular orbital methods that assumes an exact Hamiltonian and chooses approximations to the wavefunction [34]. The Gaussian basis DFT method in NWChem breaks the Hamiltonian down into the same basic one-electron and two-electron components as traditional Hartree-Fock methods, with the two-electron component further reduced to a Coulomb term and an exchange-correlation term. The treatment of the former can be accomplished in a fashion identical to that used in traditional SCF methods, from a fitted expression similar to that found in RI-SCF [31], or from the commonly used Dunlap fit [35, 36]. DFT is really distinguished from other traditional methods, however, by the treatment of the exchange-correlation term. This term is typically integrated numerically on a grid, or fit to a Gaussian basis and subsequently integrated analytically.

It is instructive to elaborate on the computation/communication requirements of these time-consuming components. The computationally dominant step, the construction of the Fock matrix, is readily parallelized as the integrals can be computed concurrently. A strip-mined approach is used where the integral contributions to small blocks of the Fock matrix are computed locally and accumulated asynchronously into the distributed matrix. By choosing blocking over atoms, the falloff in interaction between distant atoms can be exploited, while simultaneously satisfying local memory constraints. The three time-consuming steps contributing to the construction of the Fock matrix are: fit of the charge density (FitCD), calculation of the Coulomb potential (VC), and evaluation of the exchange-correlation potential (VXC). The fit of the charge density and the calculation of the Coulomb potential both typically consume similar amounts of floating point cycles; primarily evaluating three-center two-electron integrals at a rate of about 300-400 flops each. Very few communications are required beyond a shared counter and global array accumulate; these components are easily parallelized and display high efficiencies. The computation of the exchange-correlation potential requires far fewer flops (evaluating gaussian functions numerically on a grid, as opposed to analytical integrations) but approximately the same communication efforts. This results in a higher communication/computation ratio then the preceding two components. Figure 2 shows the speedup on a variety of MPPS, using up to 128 processors to build the Fock matrix. Near linear speedups are obtained with all of the platforms; the total time to solution is approximately the same for three of the four MPPs (for different reasons).

Data in Table 1 further contrasts the architectural differences in todays MPPs. For example, the high efficiencies for the VXC construction on the KSR-2 and T3D are a direct reflection of the hardware/software shared memory. The degradation of efficiency in the construction of VC on the SP2 is a direct reflection of the processor power and the need to increase the granularity of work; this is a simple modification and has already been implemented.

5.2 Other Application Modules

NWChem is designed to be a full-featured computational chemistry package serving a broad user community. As such, we have developed, or will be developing, a wide range of application modules for the package. The Hartree-Fock (HF) module differs slightly from the DFT module described above in that it is based on a quadratically convergent SCF (QCSCF) approach [37]. The SCF equations are recast as a non-linear minimization which bypasses the diagonalization step. This scheme consists of only data parallel operations and matrix multiplications which guarantees high efficiency on parallel machines. Perhaps more significantly, QCSCF is amenable to several performance enhancements that are not possible in conventional approaches. For instance, orbital-Hessian vector products may be computed approximately which significantly reduces the

Table 1. Time in wall clock seconds for the construction of the primary components of the density functional theory Fock matrix: Dunlap fit of the charge density, FitCD; construction of the Coulomb Potential, VCoul; construction of the exchange-correlation potential, VXC; and the total time

Machine	Nodes	FitCD	VCoul	VXC	Total
IBM SP2	32	1667	801	2330	4798
	64	708	571	1430	2709
	96	445	425	1040	1910
	128	370	338	832	1540
Intel Paragon	52	1675	1889	2895	6459
	128	774	830	1611	3215
KSR-2	16	3785	4252	3775	11812
	32	1875	2138	1870	5883
	48	1247	1439	1298	3984
	64	929	1066	1007	3002
	128*	461	538	513	1512
Cray T3D	32	2144	2447	1391	5982
	64	1039	1212	736	2987
	128	519	619	417	1555
	256	260	305	242	807

computation expense with no effect on the final accuracy. Gradients are available for both HF and DFT codes, with analytic Hessians and other properties under development. A parallel multiconfiguration self-consistent field (MCSCF) module is under development as well. In addition to efficient parallel scaling of algorithms as a route to large-scale simulations, we are also interested in alternative formulations of methods which may allow for more efficient calculations on large systems. Therefore, besides the traditional direct second-order many-body perturbation theory (MP2) code, we are exploring methods which will allow us to take advantage of the locality of interactions in large systems [5, 38, 39, 40], as well as the use of integral approximations such as the so-called "resolution of the identity" approximation [30-31], which we have implemented at both the SCF [41] and MP2 [4] levels. NWChem is also capable of molecular dynamics calculations using classical, quantum mechanical, or mixed Hamiltonians. Quantum mechanical forces may be obtained from any method in NWChem for which analytical gradients are implemented as well as from a number of third-party packages which can be easily interfaced with the NWChem structure.

6 Conclusions

We have described work in progress to develop the molecular modeling software application, NWChem, for both present and future generations of MPPs. The emphasis throughout has been on scalability and the distribution, as opposed to the replication, of key data structures. An important part of this effort has been the careful design and implementation of a system of supporting libraries and modules using ideas from OO methodology. Having a well thought out, easily understood toolkit to work with has allowed us to concentrate on our primary scientific task, the development of efficient parallel algorithms for a wide range of methods in computational chemistry. The DFT module, which we have described in some detail, is just one example of the high level of performance which can be achieved through this approach. A performance analysis of calculations on the zeolite fragment $Si_{28}O_{67}H_{30}$ on present generation MPP systems suggests that our goal of scalability, and hence the ability to treat more complex species in routine fashion, is well in hand. With the next generation of MPP systems likely to provide an improvement by a factor of three in the key hardware attributes e.g., CPU, communication latencies and bandwidth, we are confident that our on-going software development strategy will lead to both the type of calculations described above being performed interactively, a far cry from the present situation on conventional supercomputers, coupled with the ability to handle larger and more complex species. Important problems in biotechnology, pharmaceuticals and materials await just such advances. As stressed at the outset, however, raw computing power alone will not be sufficient to achieve these goals. Continuing theoretical and computational innovation will be the real key in realising the full potential of the teraFLOPs computers scheduled for the mid-late 1990's.

Acknowledgements The Pacific Northwest Laboratory is a multi-program laboratory operated by the Battelle Memorial Institute for the U.S. Department of Energy under contract DE-AC06-76RLO-1830. This work was performed under the auspices of the Office of Scientific Computing and under the Office of Health and Environmental Research, which funds the Environmental and Molecular Sciences Laboratory Project, D-384.

References

1. Foster, I.T., Tilson, J.L., Wagner, A.F., Shepard, R., Bernholdt, D.E., Harrison, R.J., Kendall, R.A., Littlefield, R.J., Wong, A.T.: High Performance Computational Chemistry: (I) Scalable Fock Matrix Construction Algorithms. J. Computat. Chem. (1995) in press
2. Harrison, R.J., Guest, M.F., Kendall, R.A., Bernholdt, D.E., Wong, A.T., Stave, M., Anchell, J.L., Hess, A.C., Littlefield, R.J., Fann, G.I., Nieplocha, J., Thomas, G.S., Elwood, D., Tilson, J., Shepard, R.L., Wagner, A.F., Foster, I.T., Lusk, E., Stevens, R.: High Performance Computational Chemistry:(II) A Scalable SCF Program. J. Computat. Chem. (1995) in press
3. Wong, A.T., Harrison, R.J., Rendell, A.P.: Parallel Direct Four Index Transformation, manuscript in preparation
4. Bernholdt, D.E., Harrison, R.J.: Large-Scale Correlated Electronic Structure Calculations: the RI-MP2 Method on Parallel Computers, manuscript in preparation

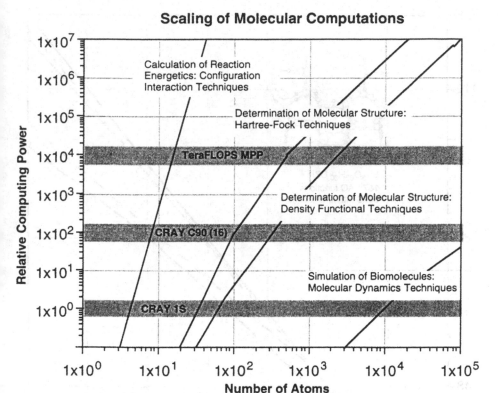

Fig. 1. The relative computing power required for molecular computations at four levels of theory. The formal scaling for configuration interaction, Hartree-Fock, density functional, and molecular dynamics is: N^6, N^4, N^3, and N^2, respectively. Standard screening techniques which in essence ignore interactions for atoms far apart can bring these scalings down significantly.

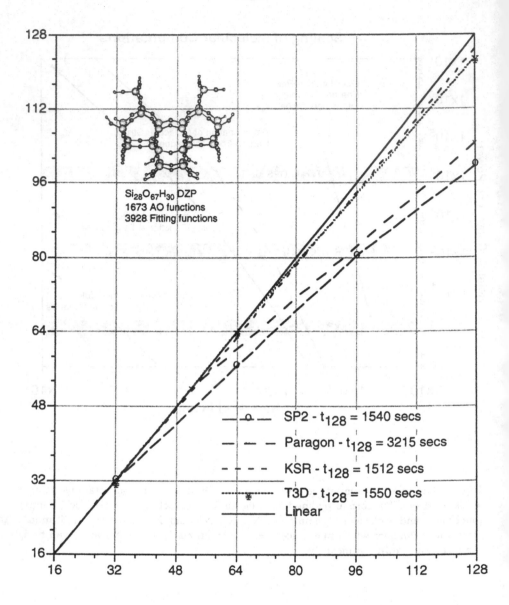

Fig. 2. Parallel scaling of the NWChem Fock matrix construction for the zeolite fragment $Si_{28}O_{67}H_{30}$ using density functional theory (in the local density approximation) on a variety of massively parallel systems.

5. Bernholdt, D.E., Harrison, R.J.: Orbital Invariant Second-Order Many-Body Perturbation Theory on Parallel Computers. An Approach for Large Molecules, J. Chem. Phys. (1995) in press
6. Booch, G.: Object-Oriented Analysis and Design, 2nd edition, (Benjamin/Cummings Publishing Co., Inc., 1994)
7. Stroustrup, B.: The C++ Programming Language, 2nd edition (Addison-Wesley Publishing Company, 1991)
8. Harrison, R.J.: Int. J. Quant. Chem. **40** (1991) 337–347
9. MPI, University of Tennessee, MPI: A Message-Passing Interface Standard (1994)
10. Kendall, R.A., Harrison, R.J., Littlefield, R.J., Guest, M.F.: Reviews in Computational Chemistry, K. B. Lipkowitz, D. B. Boyd, Eds. (VCH Publishers, Inc.., New York, 1994).
11. The MPI Forum: A Message Passing Interface, Supercomputing '93 (IEEE computer Society Press, Los Alamitos, California, Portland, OR, 1993), pp. 878-883
12. Colvin, M.E., Janssen, C.L., Whiteside, R.A., Tong, C.H.: Theoretica Chimica Acta **84** (1993) 301–314
13. Furlani, T.R., King, H.F.: J. Comp. Chem. (1995) in press
14. Nieplocha, J., Harrison, R.J., Littlefield, R.J.: Global Arrays; A Portable Shared Memory Programming Model for Distributed Memory Computers, Supercomputing '94 (IEEE Computer Society Press, Washington, D.C., 1994)
15. Arango, M., Berndt, D., Carriero, N., Gelernter, D., Gilmore, D.: Supercomputer Review (1990) **3(10)**
16. Schuler, M., Kovar, T., Lischka, H., Shepard, R., Harrison, R.J.: Theoretica Chimica Acta **84** (1993) 489–509
17. Rendell, A.P., Guest, M.F., Kendall, R.A.: J. Comp. Chem. **14** (1993) 1429–1439
18. Choi, J., Dongarra, J.J., Ostrouchov, L.S., Petitet, A.P., Walker, D.W., Whaley, R.C.: The Design and Implementation of the SCALAPACK LU, QR, and Cholesky Factorization Routines, Oak Ridge National Laboratory, Oak Ridge, TN, report ORNL/TM-12470 (LAPACK Working Note 80), September 1994. Available by anonymous FTP from ftp.netlib.org
19. Choi, J., Dongarra, J.J., Walker, D.W.: The Design of A Parallel Dense Linear Algebra Software Library: Reduction to Hessenberg, Tridiagonal, and Bidiagonal Form, Oak Ridge National Laboratory, Oak Ridge, TN, report ORNL/TM-12472 (LAPACK Working Note 92), January 1995. Available by anonymous FTP from ftp.netlib.org
20. Choi, J., Dongarra, J.J., Ostrouchov, L.S,, Petitet, A.P., Whaley, R.C., Demmel, J., Dhillon, I., Stanley, K.: LAPACK Working Note ??: Installation Guide for ScaLAPACK, Department of Computer Science, University of Tennessee, Knoxville, TN, February 28, 1995. Available by anonymous FTP from ftp.netlib.org
21. Jessup, E.R.: Ph. D. Thesis (1989) Yale University
22. Lo, S.S., Phillipps, B., Sameh, A.: SIAM J. Sci. Stat. Comput. **8(2)** (1987)
23. Fann, G., Littlefield, R.J.: Parallel Inverse Iteration with Reorthogonalization, Sixth SIAM Conference on Parallel Processing for Scientific Computing (SIAM, 1993), 409–413
24. Anderson, E., Bai, Z., Bischof, C., Demmel, J., Dongarra, J.J., Du Croz, J., Greenbaum, A., Hammerling, S., McKenney, A., Ostrouchov, S., Sorensen, D.: LAPACK User's Guide (SIAM, 1992)
25. Dupuis, M., Spangler, D., Wendolowski, J.J., University of California, Berkeley, NRCC Software Catalog (1980)
26. Schmidt, M.W., et al., QCPE Bull. **7** (1987) 115

27. Guest, M.F., Harrison, R.J., van Lenthe, J.H., van Corler, L.C.H.: Theo. Chim. Acta **71** (1987) 117
28. Guest, M.F., et al., Computing for Science Ltd, SERC Daresbury Laboratory, Daresbury, Warrington WA4 4AD, UK
29. Frisch, M., et al., Gaussian, Inc., 4415 Fifth Avenue, Pittsburgh, PA 15213, USA, (1992)
30. Feyereisen, M., Fitzgerald, G., Komornicki, A.: Chem. Phys. Lett. **208** (1993) 359–363
31. Vahtras, O., Almlof, J., Feyereisen, M.: Chem. Phys. Lett. **213** (1993) 514
32. Hohenberg, P., Kohn, W.: Phys. Rev. B. **136** (1964) 864–871
33. Kohn, W., Sham, L.J.: Phys. Rev. A. **140** (1965) 1133–1138
34. Wimmer, E.: in Density Functional Methods in Chemistry, Labanowski, J.K., Andzelm, J.W., Eds. (Springer-Verlag, 1991) 7–31
35. Dunlap, B.I., Connolly, J.W.D., Sabin, J.R.: J. Chem. Phys. **71** (1979) 3396
36. Dunlap, B.I., Connolly, J.W.D., Sabin, J.R., J. Chem. Phys. **71** (1979) 4993
37. Bacskay, G.B.: Chem. Phys. **61** (182) 385
38. Pulay, P.: Chem. Phys. Lett. **100** (1983) 151–154
39. Pulay, P., Saebo, S.: Theor. Chim. Acta **69** (1986) 357–368
40. Saebo, S., Pulay, P.: Annu. Rev. Phys. Chem. **44** (1993) 213–236
41. Fruechtl, H.A., Kendall, R.A.: A Scalable Implementation of the RI-SCF algorithm, manuscript in preparation

Parallel Ab-initio Molecular Dynamics

B. Hammer[1], and Ole H. Nielsen[1,2]

[1] Center for Atomic-scale Materials Physics (CAMP), Physics Dept., Technical University of Denmark, Bldg. 307, DK-2800 Lyngby, Denmark,
[2] UNI•C, Technical University of Denmark, Bldg. 304, DK-2800 Lyngby, Denmark.

Abstract. The Car-Parrinello ab-initio molecular dynamics method is heavily used in studies of the properties of materials, molecules etc. Our Car-Parrinello code, which is being continuously developed at CAMP, runs on several computer architectures. A parallel version of the program has been developed at CAMP based on message passing, currently using the PVM library. The parallel algorithm is based upon dividing the "special k-points" among processors. The number of processors used is typically 6-10. The code was run at the UNI•C 40–node SP2 with the IBM PVMe enhanced PVM message passing library. Satisfactory speedup of the parallel code as a function of the number of processors is achieved, the speedup being bound by the SP2 communications bandwidth.

1 Car-Parrinello ab-initio molecular dynamics

The term *ab-initio* molecular dynamics is used to refer to a class of methods for studying the dynamical motion of atoms, where a huge amount of computational work is spent in solving, as exactly as is required, the entire quantum mechanical electronic structure problem. When the electronic wavefunctions are reliably known, it will be possible to derive the forces on the atomic nuclei using the Hellmann-Feynman theorem[1]. The forces may then be used to move the atoms, as in standard molecular dynamics.

The most widely used theory for studying the quantum mechanical electronic structure problem of solids and larger molecular systems is the density-functional theory of Hohenberg and Kohn[2] in the *local-density approximation*[2] (LDA). The selfconsistent Schrödinger equation (or more precisely, the Kohn-Sham equations[2]) for single-electron states is solved for the solid-state or molecular system, usually in a finite basis-set of analytical functions. The electronic ground state and its total energy is thus obtained. One widely used basis set is "plane waves", or simply the Fourier components of the numerical wavefunction with a kinetic energy less than some cutoff value. Such basis sets can only be used reliably for atomic potentials whose bound states aren't too localized, and hence plane waves are almost always used in conjunction with *pseudo-potentials*[3] that effectively represent the atomic cores as relatively smooth static effective potentials in which the valence electrons are treated.

Car and Parrinello's method[4] is based upon the LDA, and uses pseudopotentials and plane wave basis sets, but they added the concept of updating iteratively the electronic wavefunctions simultaneously with the motion of atomic nuclei (electron and nucleus dynamics are coupled). This is implemented in a standard molecular dynamics paradigm, associating dynamical degrees of freedom with each electronic Fourier component (with a small but finite mass). The efficiency of this iteration scheme has opened up not only for the mentioned pseudopotential based molecular dynamics studies, but also for static calculations for far larger systems than had previously been accessible. Part of this improvement is due to the fact that some terms of the Kohn-Sham Hamiltonian can be efficiently represented in real-space, other terms in Fourier space, and that Fast Fourier Transforms (FFT) can be used to quickly transform from one representation to the other.

Since the original paper by Car and Parrinello[4], a number of modifications[5, 6] have been presented that improve significantly on the efficiency of the iterative solution of the Kohn-Sham equations. The modifications include the introduction of the conjugate gradients method[5, 6, 7] and a direct minimization of the total energy[6].

The present work is based upon the solution of the Kohn-Sham equations using the conjugate gradients method. We use Gillan's all-bands minimization method[7] for simultaneously updating all eigenstates, which is important when treating metallic systems with a Fermi-surface.

The Car-Parrinello code (written in Fortran-77) employed by us has been used for a number of years, and has been optimized for vector supercomputers and workstations. On a single CPU of a Cray C90 the code performs at about 350-400 MFLOPS (out of 952 MFLOPS peak), mainly bound by the performance of Cray's complex 3D FFT library routine. On a single node of a Fujitsu VPP-500/32 at the JRCAT computer center in Tsukuba, Japan the code achieves about 500 MFLOPS (out of 1600 MFLOPS peak).

2   A parallel Car-Parrinello algorithm

Given the virtually unlimited need for computational resources required for studying large systems using *ab-initio* molecular dynamics, it is obvious that parallel supercomputers must be employed as vehicles for performing larger calculations. Parallel Car-Parrinello implementations were pioneered by Joannopoulos et al.[8] and Payne et al.[9], and by now several parallel codes are being used[10, 11].

The parallelization approaches for the Car-Parrinello algorithm focus on the distribution of several types of data[10]: 1) the Fourier components making up the plane wave eigenstates of the system, and 2) the individual eigenstates ("bands") for each k-point of the calculation. The first approach requires the availability of an efficiently parallelized 3D complex FFT, whereas the second one does the entire FFT in local memory, but needs to communicate for eigenvector orthogonalization.

The selfconsistent iterations require that a sum over the electronic states in the system's Brillouin-zone be carried out. For very large systems, and especially for semiconductors with fully occupied bands, it may be a good approximation to use only a single k-point (usually $k = 0$) in the Brillouin-zone, and this is done in several of the current implementations.

However, it is our goal to treat systems that consist only of a few dozen atoms, and which typically consist of transition metals with partially occupied d-electron states. This requires rather large plane wave basis sets, as well as a detailed integration over the k-points in the Brillouin-zone, in order to define reliably the band occupation numbers of states near the metal's Fermi surface. Hence we typically need to use about a dozen k-points.

With such a number of k-points, and given that many parallel supercomputers consist of relatively few processors with rather much memory (128 MB or more), it becomes attractive to pursue a parallelization strategy based upon farming out k-points to processors in the parallel supercomputer. Since traditional electronic structure algorithms have always contained a serial loop over k-points, each iteration being in principle independent of other iterations, this is a much simpler task than the other two approaches referred to above. This approach is not any better than the other approaches, except that it is well suited for the problems that we are studying, and it could eventually be combined with the other approaches in an ultimate parallel code.

The parallelization over k-points is in principle straightforward. If we for simplicity assume that our problem contains N k-points and we have N processors available to perform the task, each processor will contain only the wavefunctions (one vector of Fourier components for each of the bands) for its own single k-point. A very significant memory saving results from each processor only having to allocate memory for its own k-point's wavefunctions (in general, $1/N$-th of the k-points). The wavefunction memory size is usually the limiting factor in Car-Parrinello calculations. A number of tasks are not inherently parallel: a) Input and output of wavefunctions and other data from a standard data file, b) accumulation of k-point-dependent quantities such as charge densities, eigenvalues, etc., c) the calculation of total energy and forces, and the update of atomic positions, and d) analysis of data depending only upon the charge density. These tasks should be done by a "master" task.

We have chosen to implement parallelization over k-points by modest modifications of our existing serial Fortran-77 code. Using conditional compilation, the code may be translated into a master task, a slave task, or simply a non-parallel task to be used on serial and vector machines. The master-slave communication is implemented by message-passing calls (send/receives and broadcasts).

The k-point parallelization is not as trivial as it might seem at first sight. Even though each iteration of the serial loop over k-points is algorithmically independent of other iterations, the wavefunction data actually depend crucially on the result of previous iterations, when the standard Car-Parrinello iterative update of wavefunctions takes place. When the k-points are updated in parallel with the same initial potential, one may experience slowly converging or even unstable calculations for systems with a significant density-of-states at the Fermi level.

One can understand this behavior by considering the screening effects that take place during an update of the electronic wavefunction: The electrons will tend to screen out any non-selfconsistency in the potential. When a standard algorithm loops serially through k-points, the first k-point will screen the potential as well as possible, the second one will screen the remainder, and so on, leading to an iterative improvement in the selfconsistency. However, when all k-points are calculated from the same initial potential and in parallel, they will all try to screen the same parts of the potential, leading to an over-screening that gets worse with increasing numbers of processors, possibly giving rise to an instability.

Obviously, some kind of "damping" of the over-screening is needed in order to achieve a stable algorithm. We have selected the following approach: Each k-point contains a number of electronic bands (eigenstates), which are updated band by band using the conjugate gradients method. The screening of the potential takes place through the Coulomb and LDA exchange-correlation potentials derived from the total charge density. We construct the total charge density and the screening potentials after the update of each band (performed for all k-points in parallel): When all processors have updated band no. 1, the charge density and potentials are updated before proceeding to band no. 2, etc. This damping turns out to very effectively stabilize the selfconsistency process, even for difficult systems such as Ni with many states near the Fermi level.

The algorithmic difference may be summarized by the following pseudo-code. The standard serial algorithm is:

```
DO k-point = 1, No_of_points
  DO band = 1, No_of bands
    Update wavefunction(k-point,band)
    Calculate new charge density and screening potentials
  END DO
END DO
```

whereas our parallel algorithm is:

```
DO band = 1, No_of bands
  DO (in parallel) k-point = 1, No_of_points
    Update wavefunction(k-point,band)
  END DO
  Calculate new charge density and screening potentials
END DO
```

It is understood that an outer loop controls the iterations towards a self-consistent solution. The conjugate gradient algoritm actually requires the calculation of an intermediate "trial" step in the wavefunction update, so that the work inside the outer loop is actually twice that indicated in the pseudo-code. In addition, the subspace rotations (not shown here) also require updates of the charge density.

3 Results on IBM SP2

The parallel algorithm described above is implemented in our code using the Parallel Virtual Machine (PVM) message-passing library, specifically the IBM PVMe implementation on the IBM SP2. At the time this work was carried out, PVMe was the most efficient message-passing library available from IBM, but we envisage other libraries to be substituted easily for PVM with time, or when porting to other parallel supercomputers such as the Fujitsu VPP-500.

In order to show the performance of the k-point-parallel algorithm, we choose a problem similar to a typical production problem of current interest to us. We perform fully selfconsistent calculations of a slab of a NiAl crystal with a (110) surface and a vacuum region, with 4 atoms in the unit cell. The plane wave energy cutoff was 50 Ry. We choose a Brillouin-zone integration with $N_k = 24$ k-points so that an even distribution of k-points over processors means that the job can be run on $N_{proc} = 1, 2, 3, 4, 6, 8, 12$ and 24 processors, respectively. At each k-point $N_{bands} = 18$ electronic bands were calculated, and the charge density array had a size of (16,24,96), or $N_{CD} = 0.295$ Mbytes. Only a single conjugate gradient step ($N_{CG} = 1$) is used. The starting point was chosen to be a selfconsistent NiAl system, where one of the Al atoms was subsequently moved so that a new selfconsistent solution must be found.

In our parallel algorithm, after each band has been updated by the slaves, the charge density array needs to be summed up from slaves to the master task, and subsequently broadcast to all slaves. Since the charge density array is typically 0.5 to 5 Mbytes to be communicated in a single message, our algorithm requires the parallel computer to have a high communication bandwidth and preferably support for global sum operations. Communications latency is unimportant, at the level provided on the IBM SP2.

Using the IBM PVMe optimized PVM version 3.2 library, a few minor differences between PVMe and PVM 3.3 are easily coded around. Unfortunately, the present PVMe release (1.3.1) lacks some crucial functions of PVM 3.3: The **pvm_psend** pack-and-send optimization, and the **pvm_reduce** group reduction operation, which could be implemented as a binary tree operation similar to the reduction operations available on the Connection Machines. Both would be very important for optimizing the accumulation of the charge density array from all slave tasks onto the master task; fortunately, they are included in a forthcoming release of PVMe.

The estimated number of bytes exchanged between the master and all of the slaves per iteration is $2N_{CD}(N_{proc} - 1)(2N_{bands}N_{CG} + 2)N_k/N_{proc}$, or $538 \times (N_{proc} - 1)/N_{proc}$ Mbytes total for the present problem, ignoring the communication of smaller data arrays. Since the IBM PVMe library doesn't implement reduction operations, the charge density has to be accumulated sequentially from the slaves. IBM doesn't document whether the PVMe broadcast operation is done sequentially or using a binary tree, so we assume that the data is sent sequentially to each of the slaves. If we take the maximum communication bandwidth of an SP2 node with TB2 adapters to be $B = 35$ Mbytes/sec, we have an estimate of the communication time as $538 \times (N_{proc} - 1)/(N_{proc}B)$ seconds per iteration.

The runs were done at the UNI•C 40–node SP2, where for practical reasons we limited ourselves to up to 12 processors. The timings for a single selfconsistent iteration over all k-points is shown in Table 1, including the speedup relative to a single processor with no message-passing. Since the number of subspace rotations[7] varies depending on the wavefunction data, the calculation of k-points may take different amounts of time so that some processors will finish before others. The resulting load-imbalance is typically of the order of 10%. We show the average iteration timings in the table.

Number of processors	Iteration time (sec)	Speedup
1	361	1.00
2	221	1.64
3	157	2.30
4	119	3.04
6	87	4.13
8	74	4.85
12	56	6.48

Table 1. Time for a single selfconsistent iteration, and the speedup as a function of the number of processors.

The speedup data is displayed in Fig. 1 together with the "ideal" speedup assuming an infinitely fast communication subsystem. More realistically, we include the above estimate of the communication time in the parallel speedup for the two cases of $B = 20$ and $B = 35$ Mbytes/sec, respectively. We see that the general shape of the theoretical estimates agree well with the measured speedups, and that a value of the order of $B = 20$ Mbytes/sec for the IBM SP2 communication bandwidth seems to fit our data well. This agrees well with other measurements[12] of the IBM SP2's performance using PVMe.

From the above discussion it is evident that any algorithmic changes which would reduce the number of times that the charge density needs to be communicated, while maintaining a stable algorithm, would be most useful. We intend to pursue such a line of investigation. Better message passing performance could be achieved by optimizing carefully the operations that are used to communicate, mainly, the electronic charge density. Efficient implementations of the global summation as well as the broadcast of the charge density (for example, using a "butterfly"-like communication pattern) should be made available in the message passing library, or could be hand-coded into our application if unavailable in the library.

4   Conclusions

A Car-Parrinello ab-initio molecular dynamics method used hitherto on workstations and vector-computers has been parallelized using a master-slave model

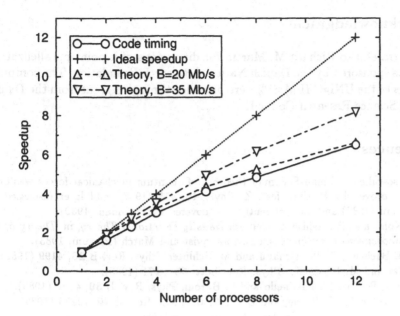

Figure 1. Speedup of the parallel code relative to the CPU time on a single processor (cf. Table 1). Besides the measured code timings, the linear "ideal" speedup is shown, along with the model estimate of the message passing time (discussed in the text) for the two bandwidths of 20 and 35 Mbytes/sec.

with message-passing. A fairly simple parallel algorithm based on farming a modest number of k-points out to slave processors has been used, complementary to other parallel Car-Parrinello algorithms. The memory savings are significant in our algorithm, since each processor only holds the wavefunction array for a single (or a few) k-points.

We find that k-point-parallel algorithms are non-trivial because of the changed convergence properties, owing to changes in the way the potential is screened. Updating the charge density after each band has been treated (in parallel) makes the algorithm stable.

The speedups measured for a test problem show satisfactory results for up to about 6 processors (depending on the problem at hand), which is more than adequate for our large-scale production jobs.

The timings obtained on an IBM SP2 show that the present parallel algorithm is bound by the communication bandwidth between the master processor and its slaves. Two options are identified to alleviate this bottleneck: 1) the investigation of less communication intensive algorithms, and 2) the efficient implementation of global reduction and broadcast operations within the message passing library.

5 Acknowledgments

We are grateful to Richard M. Martin for discussions of k-point parallelization. CAMP is sponsored by the Danish National Research Foundation. The computer resources of the UNI•C IBM SP2 were provided through a grant from the Danish Natural Science Research Council.

References

1. The so-called Hellman-Feynman theorem of quantum mechanical forces was originally proven by P. Ehrenfest, Z. Phys. **45**, 455 (1927), and later discussed by Hellman (1937) and independently rediscovered by Feynman (1939).
2. W. Kohn and P. Vashishta, *General Density Functional Theory*, in *Theory of the Inhomogeneous Electron Gas*, eds. Lundqvist and March (Plenum, 1983).
3. G. B. Bachelet, D. R. Hamann and M. Schlüter, Phys. Rev. B **26**, 4199 (1982).
4. R. Car and M. Parrinello, Phys. Rev. Lett. **55**, 2471 (1985).
5. I. Stich, R. Car, M. Parrinello and S. Baroni, Phys. Rev. B **39**, 4997 (1989).
6. M. P. Teter, M. C. Payne, and D. C. Allan, Phys. Rev. B **40**, 12255 (1989).
7. M. J. Gillan, J. Phys.: Condens. Matter **1**, 689 (1989).
8. K. D. Brommer, M. Needels, B. E. Larson, and J. D. Joannopoulos, Comput. Phys. **7**, 350 (1992).
9. I. Stich, M. C. Payne, R. D. King-Smith, J. S. Lin and L. J. Clarke, Phys. Rev. Lett. **68**, 1359 (1992).
10. J. Wiggs and H. Jónsson, Comput. Phys. Commun. **87**, 319 (1995), and their refs. 13-17.
11. T. Yamasaki, in these proceedings.
12. Oak Ridge performance measurements available on World Wide Web at the <URL:http://www.epm.ornl.gov/pvm/perf-graph.html>.

Dynamic Domain Decomposition and Load Balancing for Parallel Simulations of Long-Chained Molecules

David F. Hegarty, M. Tahar Kechadi and K.A. Dawson

Advanced Computational Research Group,
Centre for Soft Condensed Matter
University College Dublin,
Belfield, Dublin 4, Ireland

Abstract. It is clear that the use of parallelism for scientific computing will increase in the future as it is the only method available to tackle large scale problems. However, employing parallel methods to produce viable scientific tools brings about its own inherent problems which are not present in sequential implementations. Primary among these are problems in selecting the parallelisation method to be used, and inefficiencies brought about by load imbalance. In this paper we look at both these issues when applied to simulations of long-chained molecules. We introduce a decomposition scheme, dynamic domain decomposition (DDD), which parallelises a molecular simulation using geometrical information, and when combined with a migration function leads to reduced communications. We present a load balancing algorithm, positional scan load balancing (PSLB), which equalises processor work load to increase the efficiency, and present some experimental results.

1 Introduction

The calculation of equilibrium and dynamical properties of polymers, proteins and DNA is a major application of computer simulation. The methods used for the simulation of these long-chained molecules can be broadly classified into two categories: Monte-Carlo methods [17] which randomly sample the space of legal molecular conformations, and molecular dynamics methods [9] which update a molecular conformation according to particular equations of motion.

Parallel methods for these types of simulations are clearly of interest, since their size and complexity are such as to make sequential simulations of systems approaching biologically plausible size impossible. In common with many other real world simulations, e.g. dynamic fluid flow, weather modelling, molecular simulations are well modelled using the data parallel paradigm combined with some data domain decomposition scheme, where the data domain is broken into a number of sub-domains, each with similar properties and structure, which are processed concurrently.

Constructing a simulation using the data parallel paradigm involves first specifying the data decomposition, which can be viewed as assigning each sub-domain to a virtual processor, and second specifying a mapping from virtual to physical processor. Each of these parts is very important to the overall efficiency of the simulation. If the data decomposition is incorrectly specified, large load imbalance can result, while in implementing the mapping from virtual to physical processor, communications play a vital role. Decomposition locality is an important property that the mapping should usually try to conserve, since neighbouring sub-domains usually communicate more often with each other than with more distant ones.

Since these problems are dynamic it is not possible to predict their evolution. This evolution leads to load imbalance among the physical processors, since the amount of work each virtual processor has to do changes over time. Dynamic load balancing is then an integral part of any efficient simulation.

Section 2 presents domain decomposition methods, with Sect. 2.1 covering basic domain decomposition and its problems, and Sect. 2.2 presenting dynamic domain decomposition, a decomposition scheme to take evolving data dependencies into account. We present load balancing algorithms for these types of problems in Sect. 3, outlining scan directed load balancing in Sect. 3.1 and presenting positional scan load balancing in Sect. 3.2. Section 4 presents some of our experimental results of a massively parallel implementation and we give our conclusions in Sect. 5.

1.1 Model

We model a homo-polymer using a freely-jointed bead-spring model [1]. A bead corresponds to a group of monomers that form a rigid group. All beads interact with pairwise Van der Waals forces, which are modelled by the Lennard-Jones potential, given for any two beads i, j by $V_{i,j} = \frac{A}{r_{ij}^{12}} - \frac{B}{r_{ij}^6}$, where A and B are constants, and r_{ij} is the distance between the two beads i and j. This potential is repulsive at very close range, which gives each bead an excluded volume, and attractive at longer ranges, falling quickly towards zero. Nearest neighbouring beads along the chain are connected by springs, and so there is an additional harmonic interaction between them, with the potential given by $I_{i,j} = -\frac{1}{2}kr_{ij}^2$.

The Lennard-Jones potential falls off with distance in the sixth power, i.e. quite rapidly. This means that the interaction between two beads can be neglected if they are separated by more than some cutoff distance. Therefore each bead varies in the amount of interactions it must calculate, depending on how close or how distant it is from other beads compared to the cutoff distance.

In the sections that follow we consider a general model where each processor is assigned a number of *work units*, or simply *units*, with the size of each unit termed its *activity*.

1.2 Review

There has been a large resurgence of interest in domain decomposition methods for solving various problems in the past number of years due to their suitability for parallel implementation [6, 7, 14]. These schemes can be distinguished by whether they use the system's geometrical layout or the topology of the system when performing the decomposition. The latter include methods based on graph partitioning (many based on the heuristic method given in [13]) and spectral partitioning methods [20].

Because of the unpredictable behaviour of the majority of applications, including biomolecular simulations [18], their parallelisation using domain decomposition leads to an unequal distribution of work across the processors. To resolve the problem several load-balancing techniques have been proposed for SIMD [2, 11] and MIMD [10, 16, 19, 21, 22]. Their performance depends on several factors in addition to the nature of the application: the interconnection network, the speeds of the processors and communications, the number of processors and the size of the problem [15].

In general load balancing techniques are composed of three important phases: 1) the triggering phase where the load balancing procedure is invoked to determine whether the system is imbalanced or not, 2) the comparison phase where the overhead induced by the redistribution is compared to the work remaining to be done, and 3) redistributing the work among all the processors. The techniques in the literature differ in the importance they attribute to each of these phases, as well as in the method used to carry them out. The load balancing technique presented in this paper is characterised by having a static phase 1, assuming phase 2 always true, and carrying out the important phase 3 using parallel prefixes which allow us to calculate the work flow for each processor.

2 Domain Decomposition Techniques

2.1 Basic Domain Decomposition

The basic domain decomposition method applied to long-chained molecule simulations would assign a length of $\frac{n}{k}$ consecutive units of a molecule with n units to each of k processors, as shown in Fig. 1.

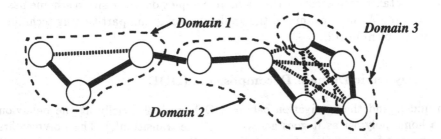

Fig. 1. The basic domain decomposition method. Heavy solid lines indicate the fixed topology of the molecule, while dotted lines indicate time-dependent data dependencies.

However, this scheme does not take a number of important points into account, including:

1. Geometrical arrangement
2. Activity of work units
3. The parallel architecture

Systems of long-chained molecules can be characterised by two different relationships: the fixed topology of the system, and the data-dependencies, which evolve over time. As mentioned in Sect. 1.2, domain decomposition techniques can be classified into those which decompose using the fixed topology, and those which use geometrical information. The basic domain decomposition method falls squarely into the former category, and thus cannot account for the dynamical nature of these systems, whereby any part of the chain can at some stage come into contact with any other part. Methods like this basic one can lead to large amounts of communication between domains, as can be seen with domains 2 and 3 in Fig. 1. To address this problem we introduce dynamic domain decomposition (DDD) in Sect. 2.2.

The second problem, that of load imbalance, occurs in most decomposition schemes because of the dynamical nature of the tasks involved. Again this can be seen from Fig. 1 where domains 2 and 3 each have many more interactions to calculate than domain 1. We discuss this problem in Sect. 3.

A parallel architecture can generally be classified as a set of processing elements connected by a communications network. Most domain decomposition methods are suitable for systems with identical processor-nodes in respect of their processing capabilities and speeds, and so do not take differences in underlying architecture into account. The first aspect of the parallel architecture, whereby the powers of the processing elements can vary, can be tackled by a simple modification to the assignment of work units to processors to take the power into account. Let $\pi_1, \cdots, \pi_k$ be the powers of the processors. The total power Π is given by $\Pi = \sum_{i=1}^{k} \pi_i$. Then let γ_i be the normalised relative power of processor i, with $\gamma_i = \pi_i/\Pi$. We then assign $\lceil \gamma_i n \rceil$ work units to the ith processor. This means that each processor is given the correct fraction of units to look after according to its processing power. If the interconnection network is also to be taken into account, then a more complex decomposition scheme has to be specified. We have examined the application of graph partitioning techniques to this problem in [12].

2.2   Dynamic Domain Decomposition (DDD)

To understand the motivation behind DDD, we look briefly at the behaviour of a homo-polymer as it goes across a collapse transition[5]. The polymer first undergoes a fast transition from a fully open Flory coil to a structure with locally collapsed sections strung out along the chain, with a much longer kinetic regime following where these locally collapsed sections migrate along the chain to form a single globule.

The difficulty in defining a decomposition mapping from work units to processors in these types of problems lies in the time dependence of the data dependencies. It is not possible to predict which units will need to communicate with which other units. DDD takes a given mapping and defines a migration function which allows us to move from one mapping to another as the problem evolves in order to reduce the communication between domains.

DDD decomposes the work unit domain into regions based on the geometrical arrangement of the long-chained molecule. Since we have a cutoff distance beyond which interactions are ignored, we can group the work units together according to their geometric cohesiveness, so that units need only communicate with other units that are within their domain. As such, it falls into the second category of decomposition schemes, since it takes geometrical information into account when performing the decomposition. However, due to the chain-connected structure of the molecule, topological information also plays a part in this decomposition since nearest neighbours along the chain will in general also be among the geometrical neighbours.

We introduce a migration function to decide whether one or more work units should migrate from one domain (processor) to another, leading to a reduction in the transfer of information between those domains. The migration function has the effect of organising the work units within any one domain in such a way that those units which have more data dependencies with units in neighbouring domains form the border of the domain, while those with less dependencies on neighbouring domains form the second layer within the domain, and so on. In this way membranes are formed around the border of each domain, with the migration function deciding to migrate units across membranes into neighbouring regions (see Fig. 2). A more formal description of the migration function is detailed in [12].

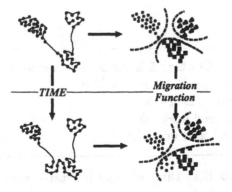

Fig. 2. Membranes defined by decomposition and migration function

An important property the migration function should have is to preserve the decomposition locality. This is so that any decomposition will conserve the connectivity of the chain molecule.

The migration function therefore takes as parameters the data dependencies of the work units, the definition of neighbourhood (i.e. the cutoff distance), and the current mapping (membrane positions). The decomposition scheme together with the migration function do of course lead to load imbalanced decompositions, and must therefore be combined with a load balancing technique to achieve an efficient simulation. We discuss this in the next section.

3   Load Balancing

Domain decomposition techniques usually lead to a load imbalanced system, simply because of the evolving nature of the tasks. Dynamic load balancing techniques are therefore necessary to increase the efficiency of any parallel implementation. To address the load imbalance issues raised in previous sections, we look at some techniques based on prefix computations, or scans.

Definition 1. A scan $(\oplus, V)$ on a vector $V = (V_1, \cdots, V_n)$ with the associative operator $\oplus$ gives as a result the vector of partial results $(I_\oplus, V_1, V_1 \oplus V_2, \cdots, V_1 \oplus \cdots V_{n-1})$ where $I_\oplus$ is the identity for $\oplus$.

This operation can be carried out in $O(\log n)$ time [4]. Load balancing techniques based on this operation are interesting because they preserve locality, i.e. given a definition of neighbourhood then tasks which are neighbours before the load balancing step will be neighbours afterwards as well.

3.1   Scan Directed Load Balancing

Biagioni[2] presented the scan directed load balancing (SDLB) algorithm which we give in Fig. 3. The main benefit of this algorithm is that it preserves decomposition locality.

---

1. Calculate $S = (+, L)$ where L is the activity vector, $L = \{L_0, \cdots, L_{k-1}\}$
2. Calculate the average work $\mu = \frac{L_{k-1} + S_{k-1}}{k}$
3. Calculate the flow $F_i = i\mu - S_i$
4. WHILE $F_i > 0.5$ send work left
5. WHILE $F_i < -0.5$ send work right

Fig. 3. Scan Directed Load Balancing algorithm

The algorithm first performs a scan of the activity of each processor, from which it calculates the *flow*, which is the difference between the processor index multiplied by the average work and the value for the scan in that processor. The absolute value of the flow in any particular processor represents the activity that must be sent out of that processor. After executing the algorithm, the work load on each processor is exactly equalised (to within one work unit). The algorithm is particularly suitable for a machine with a SIMD architecture.

3.2 Positional Scan Load Balancing

We propose an algorithm called Positional Scan Load Balancing (PSLB). We note that in SDLB each unit of work is only communicated one step at a time. In PSLB the work is communicated directly to the destination processor, and to the correct position within that processor to maintain the decomposition locality. This makes it a more suitable algorithm than SDLB for MIMD systems.

The algorithm is given in Fig. 4 for equipotent processors. The first step is to calculate a scan S of the activity vector L as before, and the resulting average activity μ. Now, each work unit w_j^i in the jth position on processor i can be assigned a work unit index, given by $I_j^i = j + \sum_{\ell=0}^{i-1} L_\ell$ where L_i is the total activity of processor i. Then the destination processor for that work unit is calculated as $\lfloor \frac{I_j^i}{\mu} \rfloor$, and the position within that processor as $I_j^i \bmod \mu$.

---

1. Calculate $S = (+, L)$ where L is the activity vector, $L = \{L_0, \cdots, L_{k-1}\}$
2. Calculate the average work $\mu = \frac{L_{k-1}+S_{k-1}}{k}$
3. For each work unit w_j^i with index j on processor i, calculate the destination processor $k_j^i = \lfloor \frac{i+S_i}{\mu} \rfloor$ and the position within the processor $l_j^i = (j + S_i) \bmod \mu$

Fig. 4. Positional Scan Load Balancing algorithm for equipotent processors

Figure 5 gives the algorithm for varipotent processors. In this case we need to pre-calculate the λ_i, which result from a plus-scan of the relative normalised processor powers, i.e. $\lambda = (+, \gamma)$, where $\gamma = \{\gamma_0, \cdots, \gamma_{k-1}\}$ is calculated as in Sect. 2.1. Then we proceed as before, except that we now calculate the destination processor for a work unit w_j^i as the smallest $\ell \geq 0$ such that $\lambda_\ell > \frac{I_j^i}{W}$, where W is the total activity on all processors. The position within that processor is given by $I_j^i \gamma_\ell - \lambda_\ell \gamma_\ell W$.

The scan operation for determining the activity partial sums can be carried out in $O(\log n)$ time. Determining the destination processor and destination index for both the equipotent and varipotent cases can be done in constant time, so the algorithm as a whole has a time complexity function of $O(\log n)$.

---

1. Calculate $S = (+, L)$ where L is the activity vector, $L = \{L_0, \cdots, L_{k-1}\}$
2. Calculate the total work $W = S_{k-1} + L_{k-1}$
3. For each work unit w_j^i with index j on processor i, calculate the destination processor k_j^i as the first ℓ such that $\lambda_\ell > \frac{i+S_i}{W}$ and the position within the processor $l_j^i = (j + S_i)\gamma_{k_j^i} - \lambda_{k_j^i}\gamma_{k_j^i}W$

Fig. 5. Positional Scan Load Balancing algorithm for varipotent processors

4   Experimental Results

We implemented a simulation of the homo-polymer model combined with scan directed load balancing on a MasPar MP-1 massively parallel SIMD processor array [3, 8]. We conducted our simulations for 128 polymers executing simultaneously each with 128 beads. These numbers of simulation runs are necessary when gathering statistics about polymer properties.

Figure 6(a) shows the time for calculations with and without 1-dimensional scan directed load balancing. The time is the total time taken for calculation of the Lennard-Jones potentials summed over all processors. We note an improvement from approximately 35% total processor-usage to approximately 65% usage. A 2-dimensional load balancing (since the MasPar is a 2-dimensional mesh) would result in 100% usage. This graph shows phase 3 in operation as described in Sect. 1.2, i.e the redistribution of work to reduce imbalance.

Figure 6(b) gives the gain obtained by load balancing when all overhead is taken into account for differing amounts of work. Small amounts of work result in negative gain since the algorithm communication overhead outweighs the time reduction obtained, with the crossover into positive gain occurring at around 7 seconds, or 15 times the work necessary for calculating the potentials for one work unit in the simple case of a short homo-polymer. This time scale is easily justified for the calculation of more complex potentials (the Lennard-Jones potential requires only a small number of operations), and for larger molecules. The information in this graph is used for phase 3 of Sect. 1.2, to tell when the gain from load balancing outweighs the overhead.

5   Conclusions

In this paper we focussed on two of the most important issues that arise when parallelising an application, the decomposition of the application into concurrent tasks, and attacking the load imbalances that arise.

To tackle the first problem we argue that for this case a scheme which takes only topological information into account leads to large communication overhead. The decomposition method DDD is suitable for a coarse grained parallelisation. The number of locally collapsed regions for the homo-polymer case is around

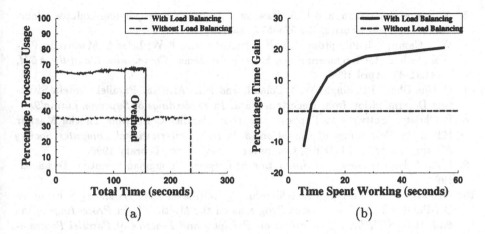

(a)　　　　　　　　　　　　　(b)

Fig. 6. (a) shows the time for calculating Lennard-Jones potentials with and without scan directed load balancing. (b) shows the gain obtained from load balancing including all communication overhead for varying amounts of work

$n/7$ rapidly declining to around $n/30$ [5]. These figures are reasonable for parallelisation of a long chain simulation onto workstations. Much work is being directed towards obtaining similar kinetic information for more complex long-chained molecules, such as proteins and DNA. We will further examine DDD and the migration function, as well as another decomposition scheme, Unstructured Domain Decomposition, which is suitable for systems with irregular topologies.

The second major problem is addressed using scan directed load balancing, a method suitable for SIMD architectures. SDLB preserves decomposition locality, which is important for maintaining the chain connectivity, and is adapted for massively parallel SIMD machines. Positional Scan Load Balancing is suitable for MIMD architectures. However PSLB suffers from a problem with communication conflict. We are examining the algorithm in more detail, including its complexity and scalability, and shall present a more detailed analysis. We shall also consider a more detailed experimental analysis of SDLB on massively parallel machines.

References

1. A. Baumgartner. Statics and dynamics of the freely jointed polymer chain with lennard-jones interaction. *Journal of Chemical Physics*, 72(2):871–879, Jan. 1980.
2. E.S. Biagioni. *Scan Directed Load Balancing.* PhD thesis, University of North Carolina at Chapel Hill, 1991.
3. T. Blank. The MasPar MP-1 Architecture. In *Proceedings of CompCon, 35th IEEE Computer Society International Meeting*, pages 20–24. IEEE, Computer Society Press, February 1990.
4. G.E. Blelloch. Scans as primitive parallel operations. *IEEE Transactions on Computers*, 38, November 1989.

5. A. Byrne, P. Kiernan, and K.A. Dawson. Kinetics of homopolymer collapse. *Journal of Chemical Physics*, 102:573–577, January 1995.

6. W.J. Camp, S.J. Plimpton, B.A. Hendrickson, and R.W. Leland. Massivelly Parallel Methods for Engineering and Science Problems. *Communication of the ACM*, 37(4):31–41, April 1994.

7. C. Che Chen, J.P. Singh, W.B. Poland, and R.B. Altman. Parallel Protein Structure Determination from Uncertain Data. In *Proceedings of Supercomputing'94*.

8. P. Christy. Software to Support Massively Parallel Computing on the MasPar MP-1. In *Proceedings of CompCon, 35th IEEE International Computer Society Meeting*, pages 29–33. IEEE, Computer Society Press, Febrary 1990.

9. E.A. Colbourn. *Computer Simulation of Polymers*. Longman Scientific Technical, 1994.

10. M. Furuichi, K. Taki, and N. Ichiyoshi. A Multi-level Load Balancing Scheme for OR-Parallel Exhaustive Search Programs on the Multi-PSI. In *Proceedings of the 2nd ACM SIGPLAN Symposium on Priciples and Practice of Parallel Programming*, pages 50–59, 1990.

11. G. Karypis and V. Kumar. Unstructured Tree Search on SIMD Parallel Computers. Technical Report 92-21, Department of Computer Science, University of Minnesota Minneapolis, MN 55455, April 1992.

12. T. Kechadi, A. Moskalenko, D.F. Hegarty, and K.A. Dawson. On the Optimal Solution for Connected Graph partitioning. Technical Report TR-9504, Advanced Computational Research Group, University College Dublin, Ireland, July 1995.

13. B.W. Kernighan and S. Lin. An Efficient Heuristic Procedure for Partitioning Graphs. *Bell Systems Technical Journal*, 49:291–306, February 1970.

14. D. E. Keyes and W. D. Gropp. A Comparison of Domain Decomposition Techniques for Elliptic Partial Differential Equations and their Parallel Implementation. *SIAM Journal of Scientific and Statistical Computing*, 8:166–202, March 1987.

15. V. Kumar and A. Gupta. Analyzing Scalability of Parallel Algorithms and Architectures. In *Proceedings of the International Conference on Supercomputing, Germany*, 1991.

16. M.H. Willebeek-Le Mair and A.P. Reeves. Strategies for Dynamic Load Balancing on Highly Parallel Computers. *IEEE Transactions on Parallel and Distributed Systems*, 4(9), September 1993.

17. J. Mazur and F.L. McCrackin. Monte Carlo Studies of Configurational and Thermodynamic Properties of Self-Interacting Linear Polymer Chains. *The Journal of Chemical Physics*, 49:648–665, July 1968.

18. B. Ostrovsky, M.A. Smith, and Y. Bar-Yam. Applications of Parallel Computing to Biological Problems. *Annual Reviews of Biophysics and Biomolecular Structure*, 1995.

19. S. Patil and P. Banerjee. A Parallel Branch and Bound Algorithm for Test Generation. *IEEE Transactions on Computer Aided Design*, 9(9), March 1990.

20. A. Pothen, H. Simon, and K.P. Liou. Partitioning Sparse Matrices with eigenvectors of graphs. *SIAM J. Mat. Anal. Appl.*, 11:430–452, 1990.

21. W. Shu and L.V. kale. A Dynamic Scheduling Strategy for the Chare-kernel System. In *Proceedings of Supercomputing'89*, pages 389–398, 1989.

22. B.W. Wah and Y.W. Eva Ma. MANIP - A Multicomputer Architecture for Solving Combinatorial Extremum-Search Problems. *IEEE Transactions on Computers*, C-33(5), May 1984.

Concurrency in Feature Analysis

Hilary J. Holz and Murray H. Loew

Department of Electrical Engineering and Computer Science
The George Washington University
Washington, DC 20052 USA

Abstract. Classifier–independent feature analysis is a classic problem whose solution exhibits exponential growth. In previous research, we developed a new approach to classifier–independent feature analysis based on *relative feature importance* (rfi), a metric for the relative usefulness of the feature in the optimal subset. Because finding the optimal subset requires exhaustive search, we have also developed an estimator for rfi. The estimator uses adaptive techniques to reduce the computational load. The implementation of both algorithms, direct calculation of rfi and the estimator, on a Connection Machine (CM–5) in CM Fortran is described in this paper. Direct calculation of rfi lends itself naturally to implementation in CM Fortran because the computationally intensive components of the algorithm involve manipulation of large arrays. The adaptive nature of the estimator, however, makes implementing it on the CM–5 more challenging and less efficient.

1 Introduction

The goal of classifier–independent feature analysis is to determine the usefulness of features for classification given only an initial set of features and a set of correctly–classified examples. Examples of classifier–independent feature analysis problems are wavelength selection for multi–component analysis and the selection of tests for medical diagnosis. Those problems are classifier–independent because the desired result is an analysis of how much class information the features possess (*e.g.*, how useful each test is in making a diagnosis) without introducing classifier-specific assumptions.

Classifier–independent feature analysis is based on the *non–parametric discriminatory power (npdp)* of the features. Conceptually, npdp is the potential of a feature to induce separation between classes, based only on the structure of the data. Ranking features based on npdp requires first finding the optimal subset of the features, because independent rankings cannot be used to choose feature sets [1]. Finding the optimal subset requires exhaustive search, as any non–exhaustive technique can do arbitrarily poorly in the general case [2]. In previous research, we developed a new approach to classifier–independent feature analysis based on relative feature importance (rfi), a metric for non–

This research partially supported by a grant from the NRL/ARPA Connection Machine Facility, an effort to rapidly prototype Massively Parallel Processing.

parametric discriminatory power [3]. Because finding the optimal subset requires exhaustive search, we have also developed an estimator for rfi.

Direct calculation of the metric is preferable when possible, while efficient estimation is necessary for large problems. Both the algorithm for direct calculation of rfi and the estimator have been implemented on a connection machine (CM–5) in CM Fortran. Due to the differences in the algorithms, direct calculation of rfi utilizes the concurrency of the CM–5 more efficiently than the estimator.

2 Computational complexity of relative feature importance

Non–parametric discriminatory power can be measured by *relative feature importance (rfi)*. Because feature rankings are dependent on the subset, rfi is positive only within the optimal subset. Features not in the optimal subset are assigned a rfi of zero. Rfi can thus be decomposed into two subproblems: identifying the optimal subset, and ranking the features within the optimal subset. A flowchart for calculating rfi is given in Figure 1.

Finding the optimal subset requires exhaustive search, an algorithm with $O(2^f)$, where f is the number of features. The order of magnitude of the exhaustive search is, therefore, the dominant factor in the algorithm for direct calculation of rfi.

Rfi measures separation between classes using the ratio of the between–class scatter to the within–class scatter of the learning sample. Scatter, a measure of variability, is calculated using non–parametric density estimates. A k–nearest-neighbor (kNN) estimate of the first moment of the class–conditional joint feature distributions is made for each data point. Thus the first major computational challenge within each subset is to find the distances between all points, and sort them in ascending order. In certain cases, *e.g.*, when the distance measure is not a metric, the distance between two points can be asymmetric, increasing the computational load.

The limitations imposed by the original representation chosen for the features are minimized through *feature extraction*. The extracted features are *optimal* in that they maximize separation between the class–conditional joint feature distributions as measured by the scatter matrices. Because it ranks the original features on the basis of the optimal extracted features, rfi measures the classification *potential* of the features.

Rfi extracts new features from the originals via non–parametric discriminant analysis [4]. Non–parametric discriminant analysis is an eigenvalue decomposition technique based on the non–parametric scatter matrices. Thus eigenvalue decomposition must be done for each subset. Feature extraction is used both in finding the optimal subset and ranking the features.

The total potential for separation offered by each subset is measured using either the k–nearest–neighbor classification error estimate, or a non–parametric version of the trace. The kNN error estimate, which asymptotically minimizes classifier–specific assumptions[1], requires a second set of distance calculations made in the rotated feature space.

Once the optimal subset is determined, the contribution of each original feature to the separation potential in that subset is estimated. WAWS, the algorithm used to estimate the contributions of the original features, is not computationally intensive.

3 Implementation platform

Both algorithms, direct calculation of RFI and GENFIE, have been implemented on the CM–5 at the Connection Machine Facility of the Naval Research Laboratory. The CM–5 is an array processor with multiple communications paths and a sophisticated masking capability. The fastest communication facility is a grid network (called NEWS) which connects each node to neighbors in an n-dimensional grid. Slower than NEWS, but still quite fast is a broadcast capability between the front-end and each node. Slowest, but most flexible, is a general communications router with full connectivity, which allows many processors to send data to many other processors simultaneously [5].

The CM–5 at NRL has 256 superSPARC nodes, each of which has 4 vector processors with 128 megabytes of local memory. Multiple partitions can be specified on a single CM–5. The algorithms in this research are designed for a partition of size 256, the size of the partition on NRL's CM–5 designated for production runs.

The algorithms were coded in CM Fortran, which is based on FORTRAN 77, but extended with routines from Fortran 90, CM specific routines, and routines from early versions of the Fortran 90 standard which were later eliminated. CM Fortran is an array processing language with extensive options for controlling and manipulating array sections. Array section manipulation commands such as WHERE and FORALL are implemented in CM Fortran, as well as boolean masking capabilities. Masking is available for most functions, and specifies on which elements of an array a function operates.

The CM Software Scientific Library (CMSSL) was used extensively in coding both algorithms. The CMSSL offers optimized implementations of numerical routines, most of which involve sparse or dense matrix manipulation. Many CMSSL routines allow embedded arrays, thus enabling multiple copies of arrays to be operated on in parallel. Thus, a program could find the transpose of fifteen matrices in parallel using the gen_matrix_transpose routine with fifteen matrices embedded in a three–dimensional array. Use of CMSSL routines is particularly important because of the difference in speed between the communications schemes on the CM; the CMSSL routines make optimal use of the communications paths.

---

[1] Asymptotically, the NN error is bounded below by the Bayes error and above by twice the Bayes error.

4 Calculating rfi

Calculating rfi directly lends itself naturally to implementation in CM Fortran. The computationally–intensive components of the rfi algorithm are: calculating all subsets, finding the k–nearest neighbors, and solving the eigensystems. These components all require parallel independent operations on large homogeneous arrays, at which CM Fortran excels. Figure 2 illustrates the data flow used to calculate rfi on the CM–5.

All subsets are processed in parallel by replicating the data and masking out the appropriate (unused) features for each subset. The computational gain is limited by the number of nodes in the CM but, theoretically, a $O(2^f)$ algorithm is calculated in parallel. Data transfer between subsets is non–existent, since the data are spread out in a single step before the subsets are considered. Masks are also established at this time, and the subsets are compressed (the s features used in a subset are placed in the first s slices of the array) due to the requirements of the CMSSL routines used later.

The most intensive calculation for each subset –– finding the k–nearest–neighbors –– is also spread across the CM. The all–to–all rotation routines optimized to the CM are not appropriate since only the k smallest distances need to be retained (where k is the number of nearest neighbors), and since the all–to–all rotations do not sort the distances, or provide information concerning which samples yielded which distances. Sorting is done by linear insertion sort since k is generally quite small in comparison to n, the number of samples, (k is generally less than 10, while n frequently numbers in the hundreds or thousands).

Thus the distance calculations are reduced from $O(n^2)$ to $O(n)$. Note that the sample size necessary for statistical significance is a function of problem difficulty, which in turn is influenced by the number of features; thus, n is not independent of f. While the reduction introduced by parallelizing the distance calculations theoretically is negligible next to calculating all subsets in parallel, sample sizes in the thousands are sometimes required to solve problems with three or four features [3]. The distance calculations themselves are matrix calculations, and use CMSSL routines.

The eigenvalue decompositions used in extracting the optimal features for each subset are executed in parallel. The scatter matrices are symmetric, so the CMSSL Jacobi eigensystems routines are used. Since eigensystem solutions are computationally demanding, this represents significant savings in compute time.

Since direct calculation of rfi exhibits exponential growth, the main benefit of implementing the algorithm on the CM–5 is to increase the number of problems which can be solved directly. Two benefits accrue from direct calculation of rfi. First, direct calculation yields a more accurate answer than estimation, particularly in that it is guaranteed to find the optimal subset[2]. Second, a larger set of problems can be used to measure the accuracy of estimation techniques for rfi.

Direct calculation of rfi is a problem well matched to the architecture of the CM–5, and the structures of CM Fortran. The dominant computational factor,

---

[2] Any non–exhaustive search technique can do arbitrarily poorly in the general case.

the exponential growth resulting from the need to consider all possible subsets, is easily handled by replicating the sample arrays and masking out the unused features. The computational load of the distance calculations and the eigenvalue decompositions can be greatly reduced using routines from the CMSSL.

5 The GENFIE estimator

GENFIE is a genetic algorithm/neural network hybrid estimator for rfi. The estimator uses adaptive techniques to address two computational issues which dominate direct calculation of rfi: the computational requirements to find the optimal subset, and the difficulty in calculating the kNN density estimates. Figure 3 illustrates the GENFIE algorithm.

A genetic algorithm (GA) explores a *population* of possible solutions in parallel rather than exploring a single solution at a time, as has been done in traditional search strategies. The solutions are typically encoded in binary strings. Within the population, information is shared between individual solutions through simulated genetic reproduction, crossover, and mutation, on the basis of their performance, or *fitness*. By searching multiple solutions at the same time, a GA essentially decomposes the search for the optimal subset into smaller problems. Sharing information between the solutions allows the GA to be robust to dependencies between subsets. GAs are robust to noisy fitness functions, thus have no requirement that the subset fitness criterion (in this case, separability) be monotonic. Genetic algorithms have previously successfully been used for classifier–specific subset selection and for classifier–specific feature ranking [6, 7].

While GAs typically search only a small portion of a problem's solution space, they can still visit thousands of solutions. GAs are inherently concurrent in that at each generation, a population of solutions is considered independently of one another. Population sizes range between fifty and several hundred, depending on the size and complexity of the problem. GENFIE estimates the ranks of the features directly, through gray–scale representation, therefore each feature requires more than one bit. Thus eight features requires at least eight distinct non–zero ranks, or four bits per feature, yielding a string length of thirty–two. Problems with hundreds of features can necessitate string lengths of a thousand or more. More notably, rather than having a solution space of $O(2^f)$ as in direct calculation of rfi, the search space is expanded to $O(2^{f*\ln(f)})$. However, only a small portion of that space is actually visited.

The hypotheses are tested using a neural network which does unsupervised clustering. Two neural networks have been implemented to date: a simple self–organizing feature map (sofm), and Dystal, a neural classifier which does unsupervised clustering [8, 9, 10]. Thus each population member requires that a net be trained and tested to determine the fitness of that set of ranks.

Neural network clustering techniques such as the sofm and Dystal perform well with small sample sizes, and have greatly reduced computational requirements when compared to kNN classifiers [10]. The clustering in these techniques is unsupervised in that neither the number of clusters nor an initial location for the clusters is required. Thus a much smaller sample size is used. In addition, it is no longer necessary to compute the distance to all known samples to classify a new element, but rather just to compute the distance to a small number of cluster centers as represented in the net.

6   Calculating GENFIE

GENFIE has been implemented on the CM–5. However, the implementation requires far more use of masking, and less use of the CMSSL routines. Figure 4 illustrates the data flow used in the CM–5 implementation of GENFIE.

All population members are tested in parallel on the CM–5. However, this in itself does not yield a particularly significant gain. Recall that typical population sizes are 50 to 100 for problems with eight features, ranging up to populations of several hundred for problems with hundreds of features. Compared to calculated thousands of distances in parallel, calculating 50 fitness estimates in parallel is not particularly significant.

However, each node of each sofm is also calculated in parallel. Since sofms typically reach a thousand nodes, this represents a greater gain. However, Dystal tends to be smaller, and grows nodes as needed. In order to process all nodes in parallel for Dystal, an extremely complex mask structure is required. To be truly efficient, a maximum number of nodes for each run should be selected, allowing for efficient replication. When nodes are added, the masks need to be recalculated. Thus the fluid nature of Dystal makes adaptation to the CM–5 more challenging than the more fixed structure of the sofm.

A set of distance calculations still must be done for the samples used in GENFIE. The distance between each test sample and the nodes in the net must be calculated in order to estimate classification accuracy. These distance calculations can be done in parallel, for all the test samples. However, the corresponding distance calculations for the training samples must be done one training sample at a time, since the net learns from each training sample. While the distances to all the nodes in the net can be calculated at once, each training sample must be calculated independently.

While GENFIE can be implemented on the CM–5, the very adaptive nature of the GENFIE estimator lessens the efficiency of the implementation. The adaptation in GENFIE serves to decrease the computational load algorithmically. However, the cost of the adaptation is an increase in the computational load due to a loss of real concurrency. As problem size increases, the algorithmic gains outweigh the concurrency losses.

7   Conclusions

For small problems, the differences between the implementations of rfi and GENFIE are more dramatic. However, for small problems, direct calculation of rfi is preferable in any case. A problem with five features, two classes, difficult distributions (five clusters per class), and 5000 sample points takes about 30 minutes of processing time on the CM–5 to solve using rfi. The same problem can be solved by the estimator using 500 sample points, but requires approximately the same 30 minutes. For smaller problems, direct calculation of rfi is faster than the estimator. For larger problems, the estimator is faster.

Implementing both the direct calculation of rfi and the estimator has resulted in significant gains in processing time. The architecture of the CM–5 and the structure of CM Fortran lend themselves more easily to the direct calculation of rfi. The estimator required extremely complex masks, and far more data transfer between nodes, since the algorithm adapted to the data, rather than analyzing the data.

References

1. Cover, T.M.: The Best Two Independent Measurements Are Not the Two Best. IEEE Trans. Sys, Man, Cyber, SMC-4 116-117.
2. Van Campenhout, J.M.: The Arbitrary Relation Between Probability of Error and Measurement Subset. Journal of the American Statistical Association, No. 367, 75 104-109.
3. Holz, H.J., Loew, M.H.: Relative Feature Importance: A classifier-independent approach to feature selection. Pattern Recognition in Practice IV NY: Elsevier Science Publishers (1994) 473-487
4. Fukunaga, K., Mantock, J.M.: Nonparametric discriminant analysis. Transactions of the IEEE on Pattern Analysis and Machine Intelligence PAMI-5 (1983) 671-678.
5. CM Fortran Programming Guide. Cambridge, MA: Thinking Machines Corporation (1993).
6. Kelly, J.D., Davis, L.: Hybridizing the Genetic Algorithm and the K Nearest Neighbors Classification Algorithm. Proceeding of the 4th International Conference on Genetic Algorithms Belew and Booker, eds., Morgan Kaufman, (1991) 377-383.
7. Punch, W.F., et. al.: Further Research on Feature Selection and Classification Using Genetic Algorithms. Proceedings of the 5th International Conference on Genetic Algorithms S. Forrest, ed., Morgan Kaufman (1993) 557-564.
8. Kohonen, T.: Self-Organization and Associative Memory, 2nd Edition. Springer-Verlag, London (1988).
9. Huang, W.Y., Lippman, R.P.: Neural Net and Traditional Classifiers. Neural Info. Proc. Syst., NY: American Institute of Physics (1988) 387-396.
10. Blackwell, K.T., Vogl, T.P., Hyman, S.D., Barbour, G.S., Alkon, D.L.: A new approach to hand-written character recognition. Pattern Recognition 25 (1992) 655-665.

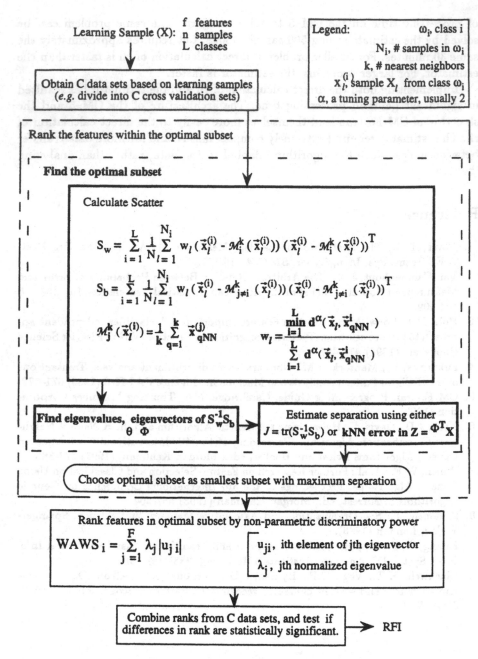

Figure 1: A flowchart for RFI. Steps in boldface are computationally intensive.

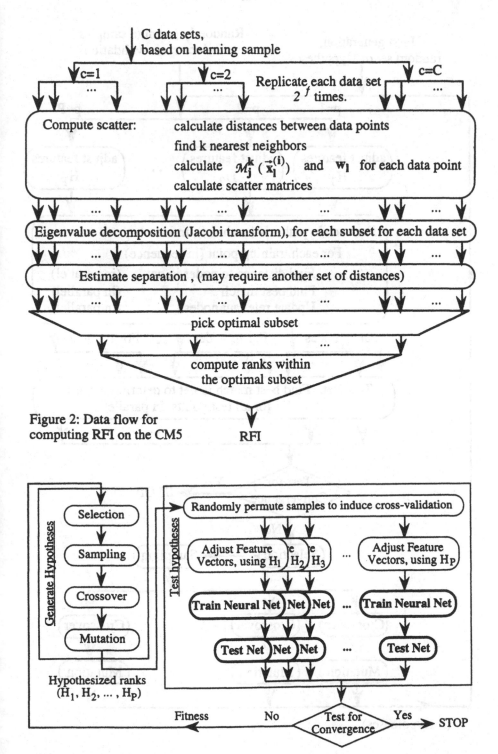

Figure 2: Data flow for computing RFI on the CM5

Figure 3: GENFIE flowchart; the steps in boldface are computationally intensive. The first generation uses a random set of hypotheses. (P: # elements in population)

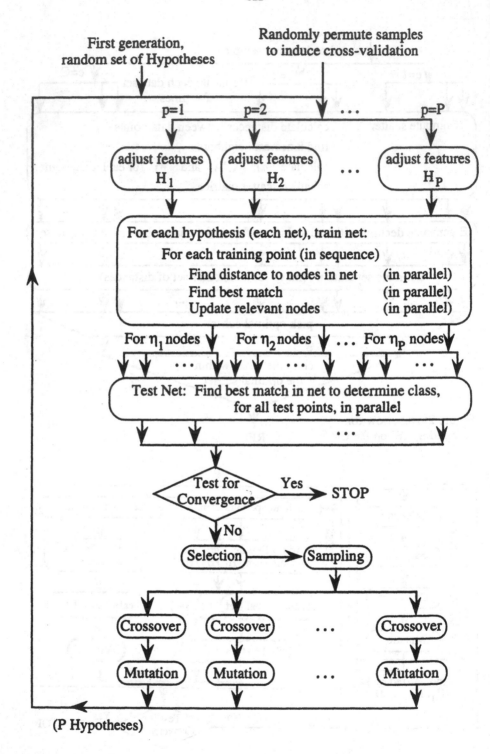

Figure 4: Data Flow for GENFIE on the CM5.

A Parallel Iterative Solver for Almost Block-Diagonal Linear Systems

D. Hu[1], F. Lou[2] and A. Sameh[1]

[1] Department of Computer Science, University of Minnesota, Minneapolis***
[2] Fidelity Investments, Boston, Massachusetts

1 Introduction

Many applications in science and engineering give rise to large sparse linear system of equations. An important class of such applications is that which involves the numerical handling of partial differential equations via finite difference or finite element discretization. In many cases, proper ordering of grid points, or proper permutations of the rows and columns of the linear system, yields a sparse matrix in which the nonzero elements are contained in a band of width much smaller than the number of equations. In such situations, the matrix can be organized in the block-tridiagonal, or almost block-diagonal form.

In this paper we present an iterative scheme for solving general linear systems of this form, which is suitable for parallel computers. First, we illustrate our approach by applying the scheme to diagonally dominant systems in which any diagonal block is nonsingular. The effectiveness of the scheme is demonstrated using the Generalized Minimal Residual (GMRES) algorithm, [7]. This parallel scheme is also generalized to handle general systems in which any diagonal block may be singular. This parallel scheme has its roots in the direct solver introduced in [10] for tridiagonal systems, and those in [2], [4], and [9] for block-tridiagonal systems.

Fig-1 gives an example of the systems being considered in this paper, in which the system is partitioned into four diagonal blocks, with three small upper and lower off-diagonal blocks. This example can be generalized easily into a matrix A with p diagonal blocks A_1, A_2, ..., A_p, and $(p-1)$ connecting off diagonal blocks V_i and W_i's. For simplicity we assume that all the A_i's are of the same size n_l, where n_l divides n, the dimension of A, and the V_i's and W_i's are each of size m, where $m \ll n_l$. In addition, we assume that the A_i's are sparse and of no particular structure.

2 A Parallel Iterative Scheme - Diagonal Dominance

Consider the linear system

$$Ax = f, \tag{1}$$

*** Research supported by NSF grants No. CCR-9396332 and CDA-9414015, and ARPA/NIST grant No. USDOC/60NANB4D1615.

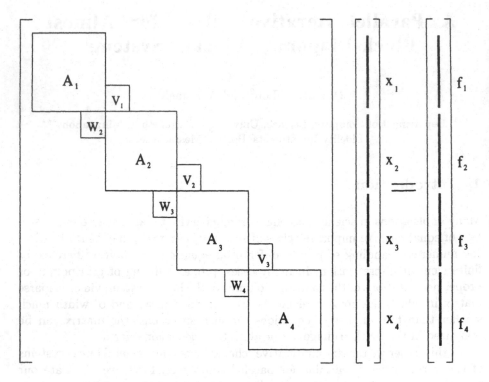

Fig. 1. Almost Block-diagonal Linear System $Ax = f$

where A is as described above, with each A_i being nonsingular. The idea behind this iterative 'SPIKE' algorithm, is the transformation of (1) into a "reduced" form as the follows. According to the partitioning of A, x and f can be partitioned as

$$x = \begin{pmatrix} x_1 \\ x_2 \\ \cdots \\ x_p \end{pmatrix}, \qquad f = \begin{pmatrix} f_1 \\ f_2 \\ \cdots \\ f_p \end{pmatrix}.$$

Fig-1 also shows the partitioning of x and f for $p = 4$.

Since the A_i's are nonsingular, we can define B_i, C_i such that

$$A_i B_i = \begin{pmatrix} W_i \\ O_{(n-m) \times m} \end{pmatrix}, \qquad A_i C_i = \begin{pmatrix} O_{(n-m) \times m} \\ V_i \end{pmatrix} \tag{2}$$

respectively. Also, let h_i be the solution of the linear system $A_i h_i = f_i$. Considering the block diagonal matrix

$$D = diagonal\{A_1, A_2, \ldots, A_p\},$$

as an outer preconditioner, we get the preconditioned system

$$D^{-1} Ax = h \tag{3}$$

where $h = D^{-1}f$. This preconditioned system is of the form, for $p = 4$,

$$\begin{pmatrix} I & \tilde{C}_1 & & \\ \tilde{B}_2 & I & \tilde{C}_2 & \\ & \tilde{B}_3 & I & \tilde{C}_3 \\ & & \tilde{B}_4 & I \end{pmatrix} \begin{pmatrix} x_1 \\ x_2 \\ x_3 \\ x_4 \end{pmatrix} = \begin{pmatrix} h_1 \\ h_2 \\ h_3 \\ h_4 \end{pmatrix} \tag{4}$$

where the I's are identity matrices of dimension n_l and

$$\tilde{B}_i = (O, B_i), \tilde{C}_i = (C_i, O)$$

with B_i and C_i's being dense $n_l \times m$ matrices. In order to obtain a reduced system, we need to partition the "spikes" B_i and C_i, as well as x_i and h_i. Let $B_t^{(i)}$ and $B_b^{(i)}$ be the top and bottom m rows of B_i, respectively. Similarly, we can define $C_t^{(i)}$ and $C_b^{(i)}$, $x_t^{(i)}$ and $x_b^{(i)}$, and $h_t^{(i)}$ and $h_b^{(i)}$. In Fig-2, we illustrate

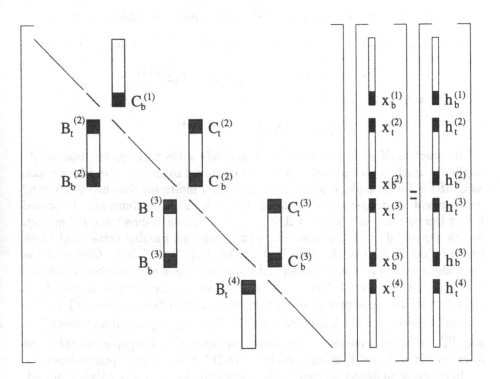

Fig. 2. Forming a Reduced System from $D^{-1}Ax = h$

how these partitions are formed. It can be observed that those small portions, described above and which appear in the shaded areas, actually form an independent subsystem of equations which only involves the $x_t^{(i)}$ and $x_b^{(i)}$'s. In general, we can prove that for a permutation matrix P, we have

$$P(D^{-1}A)P^t = \begin{pmatrix} A_r & O \\ T & I \end{pmatrix}; Px = \begin{pmatrix} x_r \\ y \end{pmatrix}; Ph = \begin{pmatrix} h_r \\ s \end{pmatrix} \tag{5}$$

and,

$$A_r x_r = h_r \tag{6}$$

is called the reduced system. For the system shown in Fig-2, $p = 4$, the reduced system has the following structure:

$$\begin{pmatrix} I_m & C_b^{(1)} & & & & \\ B_t^{(2)} & I_m & O_m & C_t^{(2)} & & \\ B_b^{(2)} & O_m & I_m & C_b^{(2)} & & \\ & & B_t^{(3)} & I_m & O_m & C_t^{(3)} \\ & & B_b^{(3)} & O_m & I_m & C_b^{(3)} \\ & & & & B_t^{(4)} & I_m \end{pmatrix} \begin{pmatrix} x_b^{(1)} \\ x_t^{(2)} \\ x_b^{(2)} \\ x_t^{(3)} \\ x_b^{(3)} \\ x_t^{(4)} \end{pmatrix} = \begin{pmatrix} h_b^{(1)} \\ h_t^{(2)} \\ h_b^{(2)} \\ h_t^{(3)} \\ h_b^{(3)} \\ h_t^{(4)} \end{pmatrix} \tag{7}$$

which is of order $6m$. In general, the reduced system is of order $n_r = 2m(p-1)$.

Solving the reduced system yields x_r, and the solution x of (1) may be retrieved by:

$$x_1 = h_1 - C_1 x_t^{(2)}, \quad x_p = h_p - B_p x_b^{(p-1)}$$

and for $i = 2, \ldots, p-1$:

$$x_i = h_i - B_i x_b^{(i-1)} - C_i x_t^{(i+1)}.$$

In practice, if n_l is still large, it is not advisable to form the matrix A_r explicitly. Rather, the reduced system can be solved by an iterative scheme such as GMRES, for example, where the matrix-vector multiplication involves solving independent linear systems of the form $A_i u_i = b_i$. These systems may be solved by an iterative scheme (e.g. GMRES), again or using a direct algorithm such as the sequential block Gaussian elimination scheme, parallel variants of block cyclic reduction ([3]) or the SPIKE algorithm ([2], [9] and [10]). Choice of the algorithms used in this case will depend upon the degree of parallelism available. If direct solvers are adopted in this inner matrix-vector multiplication, and if the the original linear system is diagonally dominant, the reduced system (6) may be preconditioned by its block-diagonal part (i.e. by dropping the submatrices $C_t^{(i)}$'s and $B_b^{(i)}$'s). In the numerical experiments we present in this paper, we solve the linear systems $A_i u_i = b_i$ using GMRES with ILUT as an inner preconditioner.

In applying an iterative scheme, whenever a portion of B_i's or C_i's is required, the iterative solver is called to solve linear systems of order n_l, each involving A_i. For example, in order to compute $w := C_t^{(2)} v$, where w and v are vectors of size m, we can first compute $r := \begin{pmatrix} O_{(n_l - m) \times m} \\ V_2 \end{pmatrix} v$ and then solve the linear system $A_2 w = r$. Therefore the matrix-vector product operation $y_r = A_r x_r$ can be carried out in parallel via solving several independent linear systems.

Once the reduced system (6) is solved by an iterative scheme such as GMRES, the right-hand side is updated and the remaining components of x are obtained again by solving independent linear systems involving the diagonal blocks A_i.

3 Comparison with the classical preconditioned GMRES

Note that the reduced system (6) should have the same convergence property as that of the preconditioned system $D^{-1}A$. Since its size has been largely reduced, and since it can be solved in parallel, applying an iterative solver such as GMRES should yield great advantage in both solution time and storage requirements.

In our numerical experiments, we compare our parallel GMRES scheme, denoted by **pGMRES**, with the classic GMRES scheme, denoted by **sGMRES**, preconditioned via incomplete LU transformation(ILUT, [1] and [8]).

The linear system A is obtained through the finite difference discretization of the 3-dimensional model convection-diffusion equation:

$$-\Delta u + 2c_1 u_x + 2c_2 u_y + 2c_3 u_z - c_4 u = F(x, y, z) \text{ in } \Omega,$$
$$u = 0 \quad \text{on } \partial\Omega.$$

Here c_1 is a function of x, c_2 a function of y, c_3 is a function of z and c_4 is a constant. We discretize the Laplacian by the standard seven-point stencil and the first order derivatives by centered finite differences. The mesh sizes are chosen such that the matrix A has a size $n = 16384$ and can be partitioned with $p = 4$. Each of the A_i's has size $n_l = 4096$ and the off-diagonal blocks W_i's and V_i's are each of size $m = 64$.

Our tests are conducted on an SGI Challenge workstation with 4CPUs. In applying **pGMRES** we use all 4CPUs, and GMRES subroutines are used in both solving the reduced system and the systems involving A_i. When solving these systems in parallel, ILUT of each A_i is used as a preconditioner.

For the sake of comparison, **sGMRES** is performed on a single processor, with ILUT of the whole matrix A as a preconditioner.

We keep the fill-in, denoted by fl, in ILUT of **sGMRES** the same as that of **pGMRES**. For **sGMRES** the test parameters are k the Krylov subspace size; and for **pGMRES** the parameters are k_r, the Krylov subspace for solving the reduced system, k_l the Krylov subspace size for solving the independent systems involving the A_i's. Thus the working space required by **sGMRES** is $(2fl + k)n$, and that required by **pGMRES** is less than $(2fl + k_l)n + k_r n_r$. Note that when $n_r \ll n$, the term $k_r n_r$ can be ignored.

First we choose $fl = 50$. The following table shows the timing and working spaces required for both schemes with different choices of parameters. An asterisk, '*', indicates a case in which convergence is not reached.

Note that in the above example, when solving the reduced system, a GMRES iteration with a Krylov subspace of size 40 completes in one step. Since in this case the reduced system has a rather small size, $n_r = 384$ we fix the Krylov subspace size to be 40.

sGMRES			pGMRES			
k	time(s)	space	k_r	k_l	time(s)	space
80	114	180 n	40	25	330	125 n
40	789	140 n	40	15	365	115 n
25	*	125 n				

Table 1. fill-in: 50, problem size: $n = 16384$, tolerance: 1.0e-10

From Table-1, we can tell that when the same level of fill-in is applied to both schemes, **sGMRES** requires a minimal working space which is much larger than that of **pGMRES**. And also, when the Krylov subspace size k of **sGMRES** is close to its 'break-down' point, the performance is much worse than that of **pGMRES** with smaller working space required.

This trend appears more clearly in the data shown in Table-2 for a level of fill-in 35.

sGMRES			pGMRES			
k	time(s)	space	k_r	k_l	time(s)	space
100	280	175 n	40	50	501	120 n
50	*	120 n	40	25	553	90 n

Table 2. fill-in: 35, problem size: $n = 16384$, tolerance: 1.0e-10

Reducing the fill-in to 25, results in total failure of **sGMRES** even for a Krylov space size of 200. **pGMRES**, however, is successful. (see Table-3).

sGMRES			pGMRES			
k	time(s)	space	k_r	k_l	time(s)	space
200	*	225 n	40	50	1083	100 n
			40	25	1051	75 n

Table 3. fill-in: 25, problem size: $n = 16384$, tolerance: 1.0e-10

4 A Parallel Iterative Scheme - General Case

If the nonsingularity of the diagonal blocks A_i, in (1), cannot be guaranteed, the above algorithm is not applicable. Following an earlier approach for symmetric positive definite systems [6] the almost block-diagonal system (1) can be partitioned into overlapping diagonal blocks. For the sake of illustration, we consider the case in which the general matrix A in (1) is partitioned only into two overlapping diagonal blocks. Let

$$Ax = f \tag{8}$$

where

$$A = \begin{pmatrix} A_{11} & A_{12} & \\ A_{21} & A_{22} & A_{23} \\ & A_{32} & A_{33} \end{pmatrix}$$

in which A_{22} is of order $k << n$, and the leading submatrix

$$\begin{pmatrix} A_{11} & A_{12} \\ A_{21} & A_{22} \end{pmatrix}$$

is singular. The matrix A can be represented as two overlapping diagonal blocks A_1 and A_2 each of order $(n+k)/2$, as shown in Fig-3(a). Where

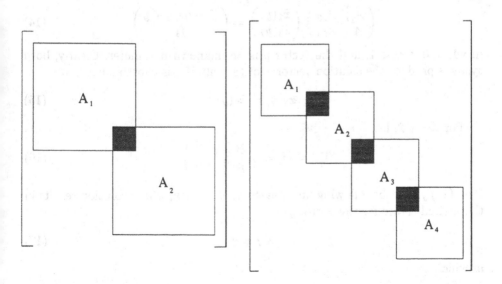

Fig. 3. (a): A with 2 overlapping Blocks; (b): A with 4 overlapping Blocks

$$A_1 = \begin{pmatrix} A_{11} & A_{12} \\ A_{21} & A_{22}^{(1)} \end{pmatrix} \tag{9}$$

and

$$A_2 = \begin{pmatrix} A_{11}^{(2)} & A_{13} \\ A_{32} & A_{33} \end{pmatrix} \tag{10}$$

where,

$$A_{22}^{(1)} = \alpha A_{22} + \beta I, \tag{11}$$
$$A_{11}^{(2)} = (1-\alpha)A_{22} - \beta I$$

with $0 < \alpha < 1$, and β chosen so as to assure nonsingularity of A_1 and A_2. It can be shown, proof omitted in this short communication, that A_1^{-1} and A_2^{-1} exist provided that β is chosen to assure diagonal dominance of $A_{22}^{(1)}$ and $A_{11}^{(2)}$. Now, if the original system (8) is written as

$$\begin{pmatrix} A_{11} & A_{12} & O \\ A_{21} & A_{22} & A_{23} \\ O & A_{32} & A_{33} \end{pmatrix} \begin{pmatrix} x_1 \\ x_2 \\ x_3 \end{pmatrix} = \begin{pmatrix} f_1 \\ f_2 \\ f_3 \end{pmatrix}, \tag{12}$$

it can be split into two systems,

$$\begin{pmatrix} A_{11} & A_{12} \\ A_{21} & A_{22}^{(1)} \end{pmatrix} \begin{pmatrix} x_1(y) \\ x_2(y) \end{pmatrix} = \begin{pmatrix} f_1 \\ \gamma f_2 - y \end{pmatrix} \tag{13}$$

and,

$$\begin{pmatrix} A_{11}^{(2)} & A_{23} \\ A_{32} & A_{33} \end{pmatrix} \begin{pmatrix} \tilde{x}_2(y) \\ x_3(y) \end{pmatrix} = \begin{pmatrix} (1-\gamma)f_2 + y \\ f_3 \end{pmatrix} \tag{14}$$

in which $0 < \gamma < 1$, and the vector y is an unknown parameter. Clearly, both systems produce the solution vector x of (8) only if y is chosen such that

$$x_2(y) = \tilde{x}_2(y). \tag{15}$$

For $j = 1, 2$, let,

$$A_j^{-1} = B_j = \begin{pmatrix} B_{11}^{(j)} & B_{12}^{(j)} \\ B_{21}^{(j)} & B_{22}^{(j)} \end{pmatrix}, \tag{16}$$

then (15) yields the following linear system, of order k, whose solution results in the desired choice for the vector y,

$$My = d \tag{17}$$

in which

$$M = B_{11}^{(2)} + B_{22}^{(1)} \tag{18}$$
$$d = B_{21}^{(1)} f_1 + [\gamma B_{22}^{(1)} - (1-\gamma)B_{11}^{(2)}]f_2 - B_{12}^{(2)} f_3.$$

Obviously, neither M nor d is available explicitly; rather, d is obtained by,

$$d = x_2(0) - \tilde{x}_2(0), \tag{19}$$

i.e. as the difference of portions of the solution vector of two independent linear systems with y set to be zero vector. The matrix M can be shown to be nonsingular, by examining the structure of the nonsingular matrix

$$G = \begin{pmatrix} A_1^{-1} & O \\ O & I_\nu \end{pmatrix} A \begin{pmatrix} I_\nu & O \\ O & A_2^{-1} \end{pmatrix} \qquad (20)$$

where $\nu = (n - k)/2$. Here, G is of the form

$$G = \begin{pmatrix} I & B_{12}^{(1)} & O \\ O & M & B_{12}^{(2)} \\ O & O & I \end{pmatrix} \qquad (21)$$

The reduced system (17) can be solved using an iterative scheme such as GMRES. Note that even though M is not available explicitly, matrix-vector multiplications involving M can be carried out as the follows,

$$\begin{aligned} q &= Mp \qquad\qquad (22) \\ &= d - [x_2(p) - \tilde{x}_2(p)]. \end{aligned}$$

In other words, the main computational step in solving the reduced system (17) via an iterative scheme is solving the independent systems (13) and (14), respectively, with different parameters y. Thus, it is essential that either effective factorization be available for the coefficient matrices for direct solution of (13) and (14), or effective preconditioners be available if iterative solvers are adopted. Once y converges, x_2 is taken as the average of $x_2(y)$ and $\tilde{x}_2(y)$.

The above approach can be generalized to allow a large number of diagonal overlapped blocks, and have a large number of smaller independent systems, e.g. the matrix with 4 overlapping diagonal blocks shown in Fig-3(b), in which case, the implicit matrix M is block tridiagonal,

$$M = \begin{pmatrix} * & * & O \\ * & * & * \\ O & * & * \end{pmatrix}.$$

Our preliminary numerical tests on Matlab show that this scheme works successfully for ill-conditioned linear systems with dominant skew-symmetric parts which render other iterative schemes ineffective . Our tests also indicate that, in many cases, the summation of the inverses of the small overlapping blocks yields a good approximation to the matrix M in (17), e.g. for the case of two overlapping blocks in Fig-3(a),

$$(A_{11}^{(2)})^{-1} + (A_{22}^{(1)})^{-1} \approx B_{11}^{(2)} + B_{22}^{(1)} = M,$$

which leads to a convenient preconditioner for the reduced system.

References

1. D'Azevedo, E., Forsyth, F. and Tang, W.: Towards a cost effective ILU preconditioner with high level fill. BIT **31** (1992) 442–463
2. Berry, M., Sameh, A.: Multiprocessor Schemes for Solving Block Tridiagonal Linear
   Systems. Int'l J. Supercomputer Applic. **2** (1988) 37–57
3. Bondeli, S.: Divide and conquer: A parallel algorithm for the solution of a tridiagonal
   linear system of equations. Parallel Computing **17** (1991) 419–434
4. Dongarra, J., Sameh, A.: On Some Parallel Banded System Solvers. Parallel Computing **1** (1984) 223–235
5. Lawrie, D., Sameh, A.: The Computation and Communication Complexity of a
   Parallel Banded System Solver. ACM Trans. Math. Software **10** (1984) 175–186
6. Lou, F., Sameh, A.: A Parallel Spliting Method for Solving Linear Systems TR.
   AHPCRC, University of Minnesota **125** (1992)
7. Saad, Y., Schultz, M.: GMRES: A Generalized Minimal Residual Algorithm for
   Solving Nonsymmetric Linear Systems. SIAM J. Scientific and Stat. Comp. **7** (1986)
   856–869
8. Saad, Y.: SPARSKIT: A basic tool kit for sparse matrix computations. TR. Research
   Inst. for Adv. Computer Sci., NASA Ames Research Center, Moffet Field, CA **90-20**
   (1990)
9. Sameh, A.: On Two Numerical Algorithms for Multiprocessors. Proc. of NATO Adv.
   Res. Workshop on High-Speed Comp. (Series F: Computer and Systems Sciences)
   7 (1983) 311–328
10. Sameh, A., Kuck, D.: On Stable Parallel Linear System Solvers. J. ACM **25** (1978)
    81–91

Distributed General Matrix Multiply and Add for a 2D Mesh Processor Network

Bo Kågström and Mikael Rännar

Department of Computing Science, University of Umeå, S-901 87 Umeå, Sweden.
Email addresses: bokg@cs.umu.se and mr@cs.umu.se.

Abstract. A distributed algorithm with the same functionality as the single-processor level 3 BLAS operation GEMM, i.e., general matrix multiply and add, is presented. With the same functionality we mean the ability to perform GEMM operations on arbitrary subarrays of the matrices involved. The logical network is a 2D square mesh with torus connectivity. The matrices involved are distributed with non-scattered blocked data distribution. The algorithm consists of two main parts, alignment and data movement of subarrays involved in the operation and a distributed blocked matrix multiplication algorithm on (sub)matrices using only a square submesh. Our general approach makes it possible to perform GEMM operations on non-overlapping submeshes simultaneously.

1 Introduction

The level 3 Basic Linear Algebra Subprograms (BLAS) [6] were developed to exploit the memory hierarchy of modern advanced architectures, including vector and RISC-based processors and parallel computers. With optimized level 3 BLAS for different machines it is possible to design portable high performance codes for many matrix computational problems (e.g. the LAPACK library [1]). The "most basic" operation in level 3 BLAS is the general matrix multiply and add (GEMM) operation

$$C \leftarrow \alpha \mathrm{op}(A)\mathrm{op}(B) + \beta C,$$

where $\mathrm{op}(X)$ denotes the matrix X or its transpose operation X^T, C is $m \times n$, $\mathrm{op}(A)$ is $m \times k$ and $\mathrm{op}(B)$ is $k \times n$, and α, β are scalars.

Our objective is to design and implement a general GEMM routine for scalable distributed memory machines (DMM) with the same functionality as the single processor level 3 BLAS matrix multiply and add operation. With the same functionality we mean the ability to perform GEMM operations on arbitrary subarrays of the matrices involved. Moreover, all necessary communication in the GEMM operation should be performed implicitly (and hidden for the user).

A lot of work concerning distributed GEMM on a 2-dimensional (2D) processor mesh has been published [4, 10, 14]. The most successful approaches (with respect to performance and scalability) distribute the data blockwise over the processor grid. This can be done in several ways (pure-block, block scattered (or block cyclic), virtual 2D torus map etc). When considering a distributed

GEMM in isolation pure-block matrix mapping is optimal [3] at least concerning communication overhead. Common for these approaches is that they only can perform block matrix multiplications on subarrays that conform with the block distribution of the arrays on the processor mesh.

A library version should be portable between different architectures. This is solved by using a set of communication primitives available on several platforms, e.g. PICL [9] and PVM [8]. By using optimized uniprocessor GEMM as the computational kernel and having a well-defined interface, which makes it easy to use your favourite distributed matrix multiplication algorithm, we accomplish high performance on a wide variety of DMM architectures.

Section 2 introduces the logical processor topology and the blocked data decomposition used. In Section 3 we present the algorithm for the distributed general GEMM operation (implemented in Fortran 77 as routine ddgemm). Section 4 presents the interface to ddgemm. In Section 5 we show some measured performance results from two DMM systems. Finally, Section 6 gives a brief summary of our work.

2 Processor Topology and Data Decomposition

Logical processor network. The processor network should be a square 2D mesh with torus connectivity, or at least able to embed that topology. It must also be possible to embed square subtorus networks, which in this context means a square 2D submesh with torus connectivity. The processor nodes are referenced by their row and column indices in the 2D mesh and the (i, j) node is denoted P_{ij}. The size of the complete square 2D mesh is DIM × DIM. Similarly, the size of a square 2D submesh is SubDIM × SubDIM.

Matrix partitioning and mapping. The matrices A, B, and C are distributed blockwise over the 2D mesh with equal-sized, possibly rectangular, blocks. When distributing an $m \times n$ matrix A over a $p \times p$ processor mesh we partition the matrix in p row blocks and p column blocks such that

$$
A = \begin{bmatrix} A_{11} & \cdots & A_{1p} \\ \vdots & \ddots & \vdots \\ A_{p1} & \cdots & A_{pp} \end{bmatrix},
$$

where block A_{ij} has dimension $m_i \times n_j$. For all blocks, except possibly the last block row and block column, $m_i = \lceil m/p \rceil$ and $n_j = \lceil n/p \rceil$. If p is not an even multiple of m and n, then the blocks in the last block row and the last block column will have smaller dimensions. Initially, node P_{ij} holds the blocks A_{ij}, B_{ij} and C_{ij}.

With arbitrary subarrays we mean any consecutive rows and columns of the full matrix, no matter which block boundaries we have due to the data distribution over the processor mesh. Since we allow arbitrary subarrays that cross the block partitioning, different processors may store blocks of subarrays of different sizes, which has to be dealt with separately. Notice, when it is clear from the context, A_{ij} may refer to a block of a subarray A.

3   Algorithm for a Distributed General GEMM

We present an algorithm for computing $C = \alpha \text{op}(A)\text{op}(B) + \beta C$, where $\text{op}(X) = X$ or X^T, on any subarrays of the matrices A, B and C with a single call to a subroutine ddgemm, without having to think about on which nodes different parts of the subarrays are stored. The acronym stands for distributed, double precision real, general matrix, multiply and add, i.e., ddgemm implements a distributed general GEMM operation.

A novelty of ddgemm is that it only uses a (square) 2D submesh large enough to hold the different subarrays. This makes it possible to have several distributed GEMM operations going on in the complete 2D mesh, which, e.g. is of interest in a GEMM-based approach [11, 12]. As in the single node dgemm, C is the only subarray changed, and since copies of subarrays A and B are used in data transfers between nodes it is possible to use one of the subarrays A and B as "C", e.g. $A = A + B \cdot A$.

The distributed GEMM algorithm consists of the following four major steps:

1. Determine the size (SubDIM) and the placement of the subtorus to be used in the distributed block algorithm.
2. Adjust subarrays A and B such that the local GEMM operations of the distributed algorithm are well-defined.
3. Move subarrays A and B into subtorus using the complete network.
4. Choose your favourite distributed algorithm for a general matrix multiply and add operation.

These four steps are captured in separate subroutines that are briefly described below. For more details see [13].

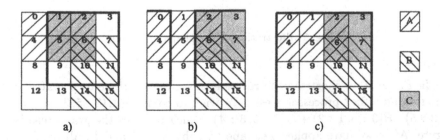

a)   b)   c)

Fig. 1. Illustration of the expand operation. (Subtorus inside the thick lines.)

expand: The routine sets SubDIM to the largest number of processors storing rows and columns of the subarrays involved in the GEMM operation. For example, in Figure 1 the subarrays A and B are stored on a 2×3 and 3×2 mesh, respectively, so SubDIM $= 3$. The starting processor of the subtorus is set to be the processor storing the $(1,1)$ block of subarray C. In Figure 1 we show how the submesh can be selected. All three cases have the same sizes of the subarrays involved in the GEMM operation, but in a) and b) with different starting indices of subarray C. In Figure 1 c) the whole processor mesh is selected as the subtorus. Case c) illustrates that **expand** can be changed to utilize facts about the processor network, e.g. that you always want to use the complete 2D mesh.

reshape: The routine adjusts a subarray so that given the subarray C's placement and size, all local multiplications in the distributed multiplication algorithm are well-defined. The reshape operation is performed on both A and B in the GEMM operation, while the subarray C is never touched. Each call performs zero, one or two shift operations (send/receive some rows/columns to/from a nearest neighbour) such that the new *local* starting row of subarray A is the same as the local starting row of subarray C and the new local starting column of subarray B is the same as the local starting column of subarray C.

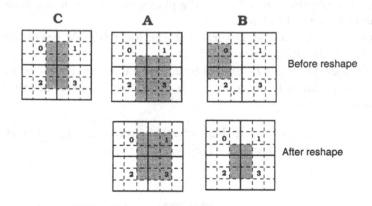

Fig. 2. Illustration of a reshape operation.

In Figure 2 we illustrate the reshape operation on three 6×6 matrices distributed on a 2×2 processor mesh. We wish to compute $C(2:5,3:4) = A(3:6,3:5) \cdot B(3:5,1:2) + C(2:5,3:4)$, which involves the gray areas in the figure. After we have applied **reshape** to subarrays A and B, all local matrix multiplications in the distributed algorithm are legal. For instance, node 0 will now multiply a 2×1 matrix with a 1×1, and add the result to a 2×1 matrix, which is a well-defined GEMM operation.

If the subarrays A and B do not overlap then the alignment of the subarrays are done in parallel, but nodes that store parts of both subarrays will have to do the reshape operation sequentially.

move: Since we only use the subtorus to perform the distributed matrix multiplication we need to move the subarrays A and B inside the subtorus, which is accomplished with the routine **move**. The $(1,1)$ blocks are moved to the node storing the $(1,1)$ block of subarray C, and so on for all blocks in the subarrays. The subarray C is never moved, because the result of the operation is stored in C, and we would have to move C back again. In some situations it could be more efficient to move C too.

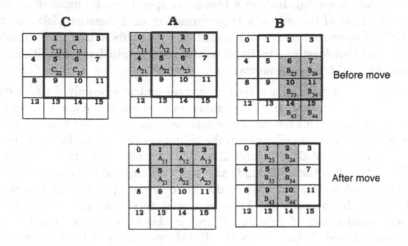

Fig. 3. Illustration of the move operation.

The shortest way is chosen with respect to the torus connectivity. The subarray is moved vertically first, and then horizontally, with only nearest neighbour communication. If the subarrays A and B originally have some nonoverlapping parts, then some of the sends and receives are done in parallel. In the end of the moves, e.g. inside the subtorus, the communication must be done serially. In Figure 3 we show an example of how the distribution may look like before and after the calls to **move**. After the move operation, all nodes outside of the subtorus return from **ddgemm** and are free to do other work (continue in the node program).

distblockalg: The routine performs the actual GEMM operation, and the "input" to the routine is a square subtorus (represented by a starting node and size), sizes of the local blocks of the subarrays, neighbours, and row and column indices of the node in the subtorus. Given all this information it is possible to implement different distributed algorithms for computing $C = \alpha \mathrm{op}(A)\mathrm{op}(B) + \beta C$, utilizing an optimized single node GEMM routine. We have implemented blocked variants of two algorithms [7], [2, 5].

The Block Torus Matrix Multiply (BTMM) algorithm is a modified blocked variant of Cannon's algorithm where each node stores a (rectangular) block of each matrix A, B, and C [2, 5]. Initially, A is skewed rowwise and B is skewed columnwise such that P_{ij} holds the blocks A_{ik} and B_{kj}, where $k = (i + j - 2)$ mod SubDIM $+ 1$. From now on the communication between nodes is minimized since A and B need to be shifted circularly only one position (east and south, respectively) for each of the SubDIM iterations of the block matrix multiplication.

In the Broadcast Multiply and Roll (BMR) algorithm [7] no initial skewing of the matrices is needed. Instead a broadcast operation of a block of subarray A along the rows of the submesh is performed in each iteration, followed by a local GEMM operation and a roll operation that shifts subarray B circularly one row. The broadcast can be implemented using a pipelined algorithm, with only nearest neighbour communication.

The original algorithms are used on square matrices, evenly distributed on the square processor mesh. Here the subarrays need not be square, and the local blocks of subarrays can even be of size zero on some nodes. We handle the zero matrix problem by letting all nodes in the subtorus take part in the algorithms as if they all had parts of all subarrays. If a node does not have a block of subarray A or B to send then the node sends an empty message, and the receiving node then knows that it does not have to do the local multiplication in this step of the algorithm. In the next step however it may receive a full block. It is only the nodes storing part of subarray C and in this step have nonempty blocks of subarrays A and B that do local GEMM operations. All other nodes just take part in data movements. For example, in Figure 3, nodes 1, 2, 5, and 6 do local multiplications, and all nodes inside the thick line (defining the subtorus) take part in data movement operations, e.g. row-broadcast of A in the BMR algorithm.

Communication overhead. The routines reshape, move and distblockalg contribute to the communication overhead of ddgemm. Notice that there is no communication at all in the routine expand. In reshape and move the data movements do not only depend on the block sizes, but also on starting indices of the subarrays involved in the GEMM operation and whether BLOCK is .TRUE. or .FALSE. (see Section 4). For example, if BLOCK = .TRUE., reshape is not called. Otherwise, reshape contributes with communication overhead for each subarray that has to be adjusted in row and/or column directions. An analysis of the communication overhead of reshape and move is presented in [13]. The communication cost of distblockalg is of course dependent on which distributed algorithm that is used. For example, the total communication of the BMR algorithm involves SubDim^2 send and receive operations of full blocks.

4   Interface to ddgemm

The calling sequence of ddgemm is extended with some parameters that define the data layout and the subarrays involved in the operation.

```
ddgemm(TransA, TransB, m, n, k, alpha, A, LDA, B, LDB, beta, C, LDC,
       Ablksz, Bblksz, Cblksz, BLOCK, StartA, StartB, StartC, TmpM,
       SIZE, Info)
```

The arguments of ddgemm on the first line are the same as for the level 3 BLAS dgemm routine [6] (see argument BLOCK for the exceptions on m, n, and k). The additional arguments are described below, where all except Info are input arguments. Examples of the use of ddgemm can be found in [13].

- Xblksz, integer(2) Row and column block sizes of matrix X.
- BLOCK, logical Set to .TRUE. if call made with block indices, otherwise .FALSE.. Normally, BLOCK is set to .FALSE. only if any of the subarrays involved in the distributed GEMM operation cross the block partitioning boundaries.
  The six parameters m, n, k, StartA, StartB, StartC are all controlled by BLOCK, e.g. if BLOCK is .TRUE. and $m = n = 1$, then the size of subarray C is Cblksz(1) $\times$ Cblksz(2), but if BLOCK is .FALSE. then the size of subarray C is 1×1.
- StartX, integer(2) Starting row and column indices of subarray X. The values of StartX is controlled by BLOCK. If BLOCK = .TRUE., StartX holds global block indices of subarray X. If BLOCK = .FALSE., StartX holds global elementwise indices of subarray X.
- TmpM, double precision(*) Workspace of size SIZE.
- SIZE, integer The size of the workspace in double precision words. At least large enough to hold copies of the local blocks of A and B, plus workspace for the larger of the two, i.e., if $Asize = $ Ablksz(1) $\cdot$ Ablksz(2) and $Bsize = $ Bblksz(1) $\cdot$ Bblksz(2) then SIZE must be at least $Asize + Bsize + \max(Asize, Bsize)$.
- Info, integer Diagnostic argument that indicates the success or failure of the computation.

A template of the node program for ddgemm can be found in [13]. To be able to have several distributed GEMM operations going on simultaneously in the 2D mesh we have to distinguish messages from different calls to ddgemm. We use an internal counter that is updated every time a call is made. To keep this information correct we have to impose that all nodes make a call to ddgemm. Some of these nodes may only perform data movements, and some perform both data movements and arithmetic computations, while some nodes may not be involved in any part of the operation and they can return immediately from ddgemm, and may, e.g. take part in another distributed GEMM operation.

5 Some Performance Results

We have tested ddgemm on two DMM computers, Intel iPSC/2 (Umeå) and Intel Paragon (Bergen). Here we show results from two computational experiments. First we let the subarrays start on the same block index and vary the sizes and "shapes" of the subarrays A, B, and C. In the second experiment we fix the sizes of the subarrays and the placement of the subarrays A and C, while varying the starting index of subarray B.

We measure the parallel speedup, S_p and the efficiencies, E_{p_c} and E_p as

$$ S_p = \frac{\text{time of optimized single node dgemm}}{\text{time of ddgemm on } p \text{ nodes}} \quad , \quad E_{p_c} = \frac{S_p}{p_c} \quad , \quad E_p = \frac{S_p}{p} \quad , \quad (1) $$

where p_c is the number of nodes storing a block of subarray C and p is the number of nodes in the submesh (SubDIM · SubDIM).

Maximal parallel speedup. Since only nodes storing parts of C are involved in the arithmetic computations of the distributed GEMM operation the maximum value of S_p is limited to p_c. For example, in Figure 3 we are using 9 nodes in the submesh, but there are only 4 nodes storing subarray C. This explains why we are measuring E_{p_c}. Notice that $E_p = p_c/p \cdot E_{p_c} \leq E_{p_c}$, with equality if $p_c = p$.

Experimental speedup. We have tested ddgemm on a number of different shaped subarrays, e.g. A square, B and C rectangular. Some of the results for the Intel iPSC/2 and the Intel Paragon are presented in tables 1 and 2. The GEMM operation is done on 16 nodes (a 4×4 mesh). In both cases we have used the largest matrix possible that fit into the memory of one processor of the parallel machine, 192×192 on the iPSC/2 and 920×920 on the Paragon. At least one of m, n or k is as large as possible, which means that the dimension of the torus involved in the multiplication is 4, i.e., the whole 4×4 processor mesh is always used.

Since all subarrays start on the same global index, the reshape and move operations do not need to be performed. Therefore this experiment is a test on how close to the maximal parallel speedup p_c we can reach. The first column of tables 1 and 2 displays the sizes of the involved subarrays. The second and third columns display the corresponding times (measured in milliseconds (ms)) for the optimized single node dgemm and ddgemm executing on 16 nodes, respectively.

We can see from Table 1 that E_{p_c} increases with k, on the iPSC/2. If $k = 192$ the efficiency stays around 98%, but if $k = 48$, it drops to about 87%. This degradation is correlated to the number of blocks in "k-direction", which is equal to the number of local GEMM operations done by each node storing a block of subarray C. The more local multiplications, the better performance, because more communication can be hidden during arithmetic work. For example, if $k = 192$ all C-nodes make 4 local calls to dgemm, but if $k = 48$, there is only 1 call. In both cases the number of send and receive operations are the same.

Table 1. Performance results on Intel iPSC/2, using 16 nodes (4 × 4 mesh)

Matrix sizes			dgemm	ddgemm	S_p	p_c	E_{p_c}
m	n	k	1 node	16 nodes			(%)
48	48	192	2833	2847	0.995	1	100
48	192	192	11332	2887	3.93	4	98
48	192	96	5448	1485	3.67	4	92
192	48	48	2688	769	3.50	4	87
192	192	48	10745	786	13.67	16	85
96	192	48	5365	767	6.99	8	87
192	48	192	11238	2880	3.90	4	98
192	192	192	44956	2930	15.34	16	96
96	192	96	10974	1488	7.38	8	92
96	48	192	5632	2864	1.97	2	98
96	96	192	11261	2885	3.90	4	98
48	96	192	5665	2858	1.98	2	99
48	192	48	2723	766	3.55	4	89

The results for the Intel Paragon are similar, as shown in Table 2, but with lower efficiencies which is mainly due to the higher communication/computation ratio of the Intel Paragon. The efficiency should increase if we use larger subarrays and the same proportions between m, n and k.

Timing with move. To see how **ddgemm** behaves if the subarrays do not start on the same node, we have done some test runs where the subarrays A and C are fixed (they always start in the same place), but subarray B moves around to different starting points. The submesh is 4 × 4 and has its upper left corner on the node that stores the $(1, 1)$ block of subarray C. All multiplications are done on full blocks. The experiment is "worst case" in the sense that the subarrays A, B and C are non-overlapping and the subarrays A and B (after position 4) are outside the submesh (see Figure 4).

Table 3 shows the results for Intel Paragon. The first column in the table gives references to the numbers in Figure 4, where we illustrate how the subarray B is placed and moved around in the processor mesh. All matrices are 3680 × 3680, and the multiplications are done on subarrays with $m = k = 1840, n = 460$.

In this test we have used the same dimensions as in example number 7 in the experimental speedup test. The difference is that now we have an 8 × 8 processor mesh, instead of a 4 × 4 mesh, and larger global matrices (but the local sizes are the same). The GEMM operation is still done in a 4 × 4 submesh, starting at node P_{54}.

As we can see in Table 3 the difference is only a few percent even if the distance from B to the submesh increases. Sometimes the execution time even becomes smaller as the distance increases, e.g. between positions 7 and 8 in

Table 2. Performance results on Intel Paragon, using 16 nodes (4 × 4 mesh).

Matrix sizes			dgemm	ddgemm	S_p	p_c	E_{p_c}
m	n	k	1 node	16 nodes			(%)
230	230	920	2231	2601	0.86	1	86
230	920	920	8575	2697	3.18	4	79
230	920	460	4378	1546	2.83	4	71
920	230	230	2275	936	2.43	4	61
920	920	230	8730	936	9.33	16	58
460	920	230	4461	847	5.27	8	66
920	230	920	8717	2590	3.37	4	84
920	920	920	34187	2737	12.49	16	78
460	920	460	8751	1538	5.69	8	71
460	230	920	4448	2661	1.67	2	84
460	460	920	8740	2664	3.28	4	82
230	460	920	4343	2694	1.61	2	81
230	920	230	2252	895	2.52	4	63

Table 3. Times with move on Paragon, using 64 nodes (logically a 8 × 8 mesh).

No.	Time (ms)	No.	Time (ms)	No.	Time (ms)
1	20091	7	20762	13	20812
2	20250	8	20289	14	20726
3	20251	9	20260	15	20597
4	20561	10	20580	16	20842
5	20386	11	20642	17	20851
6	20728	12	20979		

Table 3. The reason is that most of the move of B can be done in parallel with the move of A, as there are different nodes involved. If we compare the results in Table 3 with the "ideal" case (18266 ms), when the subarrays A, B and C start on the same processor, using 16 nodes, then it costs 10–15 % of the time to use the move operation. We somehow have to move the subarrays around and ddgemm does it implicitly.

6 Conclusions

We have developed a distributed general matrix multiply and add routine ddgemm in Fortran 77 for double precision real data that is portable, easy to use and can operate on subarrays as well as on full matrices. Instead of having to explicitly redistribute the subarrays A, B, and C, the user just specify indices that specify the subarray operation in almost the same way as in an ordinary single processor

343

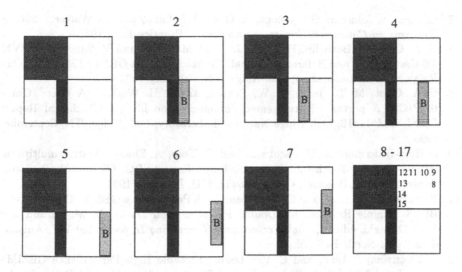

Fig. 4. Subarrays A, B and C on a 8×8 processor mesh, where the placement of B is varied to 17 positions with different starting indices. Each square (e.g. the area A) is a 4×4 mesh.

dgemm routine. Another novelty of **ddgemm** is that it only uses a square 2D submesh large enough to store the different subarrays, which makes it possible to perform several GEMM operations (or other computations) on non-overlapping submeshes simultaneously. Computational experiments show that our approach is effective.

References

1. E. Anderson, Z. Bai, C. Bischof, J. Demmel, J. Dongarra, J. Du Croz, A. Greenbaum, S. Hammarling, A. McKenney, S. Ostrouchov, and D. Sorensen. *LAPACK Users' Guide*. Society for Industrial and Applied Mathematics, Philadelphia, 1992.
2. V. Cherkassky and R. Smith. Efficient mapping and implementation of matrix algorithms on a hypercube. *Journal of Supercomputing*, 2(1):7–27, 1988.
3. J. Choi, J. J. Dongarra, and D. W. Walker. Level 3 BLAS for distributed memory concurrent computers. In *CNRS–NSF Workshop on Environments and Tools for Parallel Scientific Computing (Saint Hilaire du Touvet, France, September 7–8, 1992)*. Elsevier Science Publishers, 1992.
4. J. Choi, J. J. Dongarra, and D. W. Walker. PUMMA: Parallel Universal Matrix Multiplication Algorithms on distributed memory concurrent computers. Technical Report ORNL/TM–12252, Oak Ridge National Laboratory, Oak Ridge, TN, April 1993.
5. E. Dekel, D. Nassimi, and S. Sahni. Parallel matrix and graph algorithms. *SIAM Journal of Computing*, 10(4):657–675, November 1981.
6. J. Dongarra, J. Du Croz, I. Duff, and S. Hammarling. A set of level 3 basic linear algebra subprograms. *ACM Trans. Math. Software*, 18(1):1–17, 1990.

7. G. Fox, M. Johnson, G. Lyzenga, S. Otto, J. Salmon, and D. Walker. *Solving Problems on Concurrent Processors*, volume 1. Prentice–Hall, 1988.

8. G. A. Geist, A. Beguelin, Dongarra J. J., R. Manchek, and V. Sunderam. PVM 3.0 User's Guide and Reference Manual. Technical Report ORNL/TM–12187, Oak Ridge National Laboratory, Oak Ridge, TN, February 1993.

9. G. A. Geist, M. T. Heath, B. W. Peyton, and P. H. Worley. A Users' Guide to PICL: A portable instrumented communication library. Technical Report ORNL/TM–11616, Oak Ridge National Laboratory, Oak Ridge, TN, September 1990.

10. S. Huss-Lederman, E. M. Jacobson, and G. Tsao, A. Zhang. Matrix multiplication on the Intel Touchstone Delta. Technical Report SRC–TR–93–101 (Revised), Supercomputing Research Center, Bowie, MD, February 1994.

11. B. Kågström, P. Ling, and C. Van Loan. High Performance GEMM–Based Level 3 BLAS: Sample Routines for Double Precision Real Data. In M. Durand and F. El Dabaghi, editors, *High Performance Computing II*, pages 269–281, Amsterdam, 1991. North–Holland.

12. B. Kågström, P. Ling, and C. Van Loan. Portable High Performance GEMM–Based Level 3 BLAS. In Richard F. et al Sincovec, editor, *Parallel Processing for Scientific Computing*, pages 339–346, Philadelphia, 1993. SIAM Publications.

13. M. Rännar. A Distributed, Portable and General GEMM Operation for a 2D Mesh Processor Network. Report UMINF–95.xx, Department of Computing Science, Umeå University, S–901 87 Umeå, Sweden, 1995.

14. R. van de Geijn and J. Watts. SUMMA: Scalable universal matrix multiplication algorithm. Technical Report UT CS-95-286, LAPACK Working Note # 96, University of Tennessee, 1995.

Distributed and Parallel Computing of Short-Range Molecular Dynamics

J. Kitowski

[1] Institute of Computer Science, AGH, al. Mickiewicza 30, 30-059 Cracow, Poland
[2] Academic Computer Centre CYFRONET, ul. Nawojki 11, Cracow, Poland
email: kito@uci.agh.edu.pl

Abstract. In this paper an overview of timing results obtained for short-range 12/6 Lennard-Jones molecular dynamics is presented. The computational box adopted in this study is a long cylinder thus the metod is called the *Pipe Method*. For parallel computing two methods were applied: domain decomposition and use of virtual shared memory. Fixed and scalable problem sizes were applied.

Different kinds of computers were used: networks of computers with message passing programming environments, a set of T800 transputers running Parasoft Express and Helios operating system, iPSC/860 and Intel Paragon with NX-2 environment, CM5 with CM Fortran, and Convex Exemplar SPP1000/XA with 16 processing nodes (using ConvexPVM and virtual global shared memory paradigm).

1 Introduction

For the last couple of years parallelism has got significant interest from the community using numerical intensive computing as a tool for getting insight at origins of base phenomena. Interest on simple and relatively primitive parallel computers at the beginning of that period became complete soon with efforts toward distributed computing using network of computers, which, in turn came back to use of parallel computers due to their present significant hardware and software development [1]. Multilevel structure of communication subsystem in NUMA multiprocessor computers makes former conclusions from distributed computing useful for that purpose.

Exploiting the full potential of parallel architectures requires a cooperative effort between the user and the computer system. Typically two approaches are possible: use of message passing and data-parallel programming [2].

In the reported study the *Pipe* algorithm for molecular dynamics (MD) was adopted, which was originally developed for vector computers [3, 4]. The algorithm concerns simulation of dynamics of particles confined in a long cylinder, thus is suitable for studies of fluid and mixtures in microcapillaries or growth of optical fibres. Interactions between argon particles are defined by the short-range 6/12 Lennard–Jones (LJ) potential. Only neighbouring particles are involved in computations of forces acting on each particle, so the order of the algorithm (experimentally confirmed) is $\sim o(N)$, where N is the number of particles. Simulation of the microcanonical ensemble is adopted.

On a base of the potential cutoff radius, R_C, the integer cutoff number, n_C is introduced, where n_C is a number of neighbours interacting potentially with a given particle. The Newtonian equations of motion are solved with the leap-frog scheme.

In the paper timing results are reported for the *Pipe Method* obtained on network of computers and on multiprocessor computers with message-passing environments as well as with data-parallel programming. Considerations about the effectiveness of the strategy of parallelisation and suitability of those two approaches are drawn.

2 Message-passing paradigm for Pipe Method

Current programming environments for parallel or distributed computing follow the explicit tasking languages, providing low-level constructs like message-passing primitives as their principal language constructs. Since such environments directly reflect the underlying hardware structure, such an approach allows users to effectively exploit the full potential of the parallel machine. The most popular is PVM, although others, like ParaSoft Express, Network Linda and p4 are available also. MPI [5] is expected as a message-passing standard of late 90's.

For the *Pipe Method* the domain decomposition is done with both: the fixed and with the scalable problem sizes [6, 7]. In the first case the number of particles, N, for the whole simulation is fixed and independend of the number of processing nodes. In the simulation of the scalable problem size, the number of particles depends linearly on the number of processing nodes. In the last case the initial number of particles for each node, N_K, is fixed – increase of the number of nodes, K, results in increase of the overall number of particles in simulation and greater statistical ensemble, since $N = K \times N_K$.

The whole cylindrical computational box is composed of domains allocated on different nodes loosely coupled in z-direction. The communication between nodes is needed only for particles moving from one domain to another (if determined by the solution of the Newtonian equations), for interparticle forces calculations at the boundaries of the domains and for computing of global physical results.

For forces calculations at the boundaries, the first n_C particles positions from the "upper" domain are sent to the "lower" one, increasing its number of particles by n_C, in order to direct forces calculations, which are in turn sent back to the "upper" domain.

For distributed computing we used PVM, ParaSoft Express, Network Linda and p4 interfaces. Also proprietary versions of message-passing environments were applied: ConvexPVM and NX-2. Since for the problem at hand performance with PVM (v2.4), Express (v3.2.5) and p4 (v1.2) is similar, while Linda (v2.4.6) performs at lower level [8], we report results for PVM only.

Using a cluster of heterogeneous computers a load-balancing (LB) procedure is profitable, especially for locally heavily used computers by other users. It could be useful also for multiprocessor computers in multiuser environment, in which the nodes are shared between several computational tasks. Although optimal definition of the LB procedure is not trival, some suboptimal estimates could be proposed. In the present work we used previously introduced citerion for LB, which is to get the same execution wall-clock time on each processing node [9]. The criterion takes into account the individual performance of each computer and its current load. To fulfil the criterion, particles are dynamically shifted between the nodes, according to the difference in execution wall-clock time for nodes in every timestep.

3 Data-parallel programming for Pipe Method

Worth approach to programming parallel architectures is direct compilation of conventional languages for parallel execution. To increase efficiency participation of the user in process of parallelisation is required at present. Commercialy developed languages are already available, among which TMC CM Fortran and C* are probably of most use. Other vendors provide for parallel computers possibility to use conventional languages directly compiling to parallel execution. An example of such environment is CONVEX Exemplar SPP-UX operating system offering virtual global shared memory programming. Early releases of HPF [10] are available for some multiprocessor computers also.

For the *Pipe Method* data-parallel programming with fixed problem size was applied. The program was written in CM Fortran and in Convex SPP Fortran 77. In both cases user directives were introduced for efficiency increase.

System (name)	Environment	Comment
Clusters		
SUN SPARC SLC (SLC)	C, PVM 2.4	ET, HM, SP, FX or SC
SUN SPARC 2 (SS2)	C, PVM 2.4	ET, HM, SP, FX or SC
IBM RS6000/320 (RS6)	C, PVM 2.4	ET or TR, HM, SP FX or SC
6×SUN SPARC 10, RS6000 950,55,530 (SSR)	C, PVM 3.2	ET, HT, LB, SC
Computers		
T800	C, Express 3.0 or Helios 1.1	FX
iPSC/860 (i860)	C, NX-2 (SDS 4.1.6)	SC
Intel Paragon XP/S (PG)	C, NX-2 (OSF/1 1.0.4)	SC
TMC CM5 (CM5)	CM Fortran 2.1	FX, DPP
SPP1000/XA (SPPP)	C, ConvexPVM 3.3.6	SC, SP
SPP1000/XA (SPPG)	Fortran, GSM 3.03	FX, SP, DPP

Table 1. Comparison of computer systems and environments

4 Computers and environments

Reported results were obtained with different kind of computers and environments. They are summarized in Table 1. For clusters Ethernet (ET) or Token-Ring (TR) were used for communication; the clusters are homogeneous (HM), separated (SP) from the external load (users) or heterogeneous (HT). Load-balancing (LB) is introduced for heterogeneous clusters or for multiprocessors with nodes used by other users. Results for the fixed (FX) problem size and for the scalable (SC) one are reported. Data-parallel programming (DPP) was applied in some cases. For Convex SPP1000/XA subcomplexes can be also separated or used simultaneously by other users.

5 Results

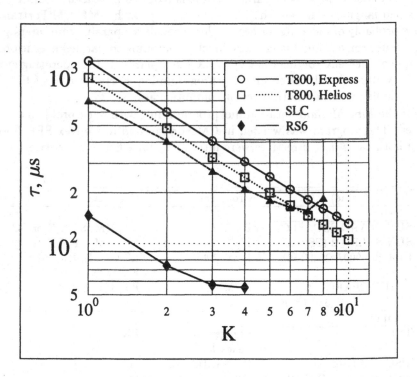

Fig. 1. Execution time, τ, per timestep and particle for fixed problem size on transputers and homogeneous clusters of SLC with ET and RS6 with TR for different number of processing nodes, K. $N = 16002$ (for transputers), $N = 32004$ (for workstations).

In Fig. 1 results for fixed problem size are presented. They confim well balanced performance/communication ratio in the network of T800 transputers. Although such a network could only hardly be used at present for numerically

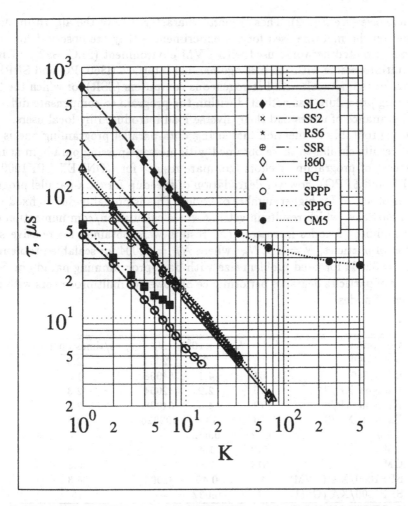

Fig. 2. Execution time, τ, per timestep and particle for fixed (CM5 and SPPG, $N = 32004$) and scalable (others, $N_K = 32004$) problem sizes for different number of nodes, K.

intensive computing due to low numerical performance, it would serve in some cases as a cost-efficient solution for development of parallel algorithms [11]. For much more powerful workstations connected via network that balance is not conserved, which results in degradation of performance with increase of number of processing nodes. The more powerful nodes, the worse characteristics of speedup is obtained (cf. results for SLC network with those for RS6 one, depicted in Fig. 1). No important difference in efficiency was created with two different network protocols (i.e. Ethernet and TokenRing) [11]. The results were obtained for $N = 16002$ for transputers and for $N = 32004$ for the clusters of workstations.

Higher efficiency and linear scaling properties were obtained with scalable problem size (see Fig. 2). Those general characteristics of the algorithm do not depend on the machine used for the experiment – they are observed for homogeneous, isolated networks used with PVM environment (SLC, SS2 and RS6), for multiprocessors with the message-passing interface (i860, PG and SPPP) as well as for the network with heterogeneous computers (SSR) for which the load-balancing procedure was activated within the program to compensate difference in performance of nodes and their diverse load introduced by local users.

It is profitable to mention that using data-parallel programing one is able to get results similar to those obtained with message-passing paradigm at much lower cost of programming effort (compare results for CONVEX SPP1000/XA – SPPP and SPPG), however degradation of speedup for data-parallel programming is observed for higher number of nodes. This is probably due to fixed number of particles, which results in getting worse calculation/communication ratio while number of nodes increases. The results were obtained for relative small number of particles, $N = K \times N_K$, where $N_K = 32004$ for scalable problem size and $N = 32004$ for fixed problem size with data-programming paradigm. Small number of particles degrades performance of powerful multiprocessors with large number of nodes.

Computer	Nodes, K	$N(\times 10^6)$	N_K	τ, μs/timestep/particle
iPSC/860	64	2	32004	2.3
iPSC/860	64	4.1	64008	1.3
Paragon XP/S	70	2.2	32004	2.3
Paragon XP/S	70	4.5	64008	2.3
Paragon XP/S	70	13.4	192024	2.3
CM5	512	0.032	—	26.
CM5	512	1.4	—	0.8
CM5	512	2.8	—	0.5
SPP1000/XA (PVM)	14	0.45	32004	4.3
SPP1000/XA (GSM)	7	0.032	—	13.

Table 2. Results for different number of particles

In Table 2 timing results for different number of particles are presented. Usually τ decreases with increase of number of particles. This feature reflects better performance/communication ratio in that case.

Some more investigations are needed for Paragon XP/S to explain constant value of τ, as well as data-parallel programming performance for Intel Paragon with HPF and for CONVEX SPP1000/XA for large N.

6 Conclusions

Scalable problem size from the user point of view turned out to be more profitable: it offers better statistical results due to greater ensemble and higher parallel efficiency. Powerful multiprocessor computers should be used for big problems; they start to offer efficient data-parallel programming tools which could make parallelisation easier with less programming effort.

7 Acknowledgments

Author is grateful to many people involved in the reported investigations. Thanks are due to Prof. Jacek Mościński, who stimulated those studies. Dr. M. Bubak, Mr. Mr. K. Boryczko and M. Pogoda are acknowledged for their contributions.

The work was supported by Polish Scientific Committee (KBN) Contract PB 2 P302 073 05 and by American-Polish Maria Sklodowska-Curie Joint Fund II (MEN/NSF-94-193).

References

1. Hertzberger B., Serazzi, G.: *Proc. of High-Performance Computing and Networking*, May 3-5, 1995, Milan, Springer-Verlag, 1995.
2. Decker, K.M., Dvorak, J.J., Rehmann, R.M., Ruhl, R.: Satisfying application user requirements: A next-generation tool environment for parallel systems, in: Hertzberger B., Serazzi, G. (eds.) *Proc. of High-Performance Computing and Networking*, May 3-5, 1995, Milan, Springer-Verlag, 1995, pp.206-228.
3. Mościński, J., Kitowski, J., Rycerz, Z.A., Jacobs, P.W.M.,: A vectorized algorithm on the ETA 10-P for molecular dynamics simulation of large number of particles confined in a long cylinder, *Comput. Phys. Commun.*, 54 (1989) 47-54.
4. Kitowski, J., Mościński, J.: Microcomputers against Supercomputers ? – On the geometric partition of the computational box for vectorized MD algorithms, *Mol. Simul.*, 8 (1992) 305-319.
5. Message Passing Interface Forum. MPI: A message passing interface standard. *Int. J. of Supercomputer Appl.*, 8, 3/4 (1994).
6. Gustafson, J.L.: Reevaluating Amdahl's Law, *Comm. of the ACM*, 31, 5 (1988) 532-533.
7. Karp, A.H., Flatt, H.P.: Measuring parallel processor performance, *Comm. of the ACM*, 33, 5 (1990) 539-543.
8. Boryczko, K., Bubak, M., Kitowski, J., Mościński, J., Słota, R.: Lattice gas automata and molecular dynamics on a network of computers, in: W. Gentzsch and U. Harms (eds.), *Proc. of High Performance Computing and Networking*, April 18-20, 1994, Munchen, Lecture Notes in Computer Science, vol. 796, Springer-Verlag, 1994, pp.177-180.
9. Boryczko, K., Kitowski, J., Mościński, J.: Load-balancing procedure for distributed short-range molecular dynamics, in: J. Dongarra and J. Waśniewski (eds.) *Proc. First Int. Workshop, PARA'94*, June 20-23, Lyngby, 1994, Lecture Notes in Computer Science, vol. 879, Springer-Verlag, 1994, pp.100-109.
10. High Performance Fortran Forum. High Performance Fortran Language Specification: Version 1, *Scientific Programming*, 2, 1-2 (1993).
11. Boryczko, K., Bubak, M., Kitowski, J., Mościński, J., Pogoda, M.: Molecular dynamics and lattice gas parallel algorithms for transputers and networked workstations, EUROMECH Coll.287, Sept. 21-25, 1992, Cagliari, Italy; also *Transport Theory and Statistical Physics*, 23, 1-3, (1993).

Lattice Field Theory
in a Parallel Environment

A. Krasnitz

The Niels Bohr Institute
Blegdamsvej 17, DK-2100 Copenhagen, Denmark

Abstract. Computations in lattice field theories account for a significant fraction of time allocated for parallel computing. Goals and methods of lattice simulations be briefly reviewed. The local nature of lattice field theory algorithms makes them good candidates for parallelization. I describe in some detail a MIMD implementation easily adaptable to various physical contents, portable to many parallel environments, and efficiently hiding internode communications. As an illustration, I show results from recent numerical simulations of matter nonconservation in the early Universe, performed on the UNI*C SP2 parallel machine.

High energy physics underwent a profound change in the early 70s with the emergence of the theory of strong interactions. For the first time physicists were presented with a theory which they believed described fundamental properties of Nature, yet the theory was too complex for many of these properties to be derived by analytical means. It turned out that the highly nonlinear nature of strong interactions severely limited the value of perturbative treatment which had been so spectacularly successful in the theory of electromagnetic interactions. From that time on, numerical methods have played an increasingly important role in theoretical high energy physics. Today the theory of strong interactions, more often called quantum chromodynamics (QCD) accounts for the bulk of computing time allocated for theoretical high-energy physics applications. At the same time, numerical methods have found applications in other theories in the field where non-perturbative treatment is required, most notably those of weak interactions and of gravity. In this talk I shall try to give a very brief general overview of the field followed by an application example run on the local SP2 [1].

The general theoretical framework of high-energy physics is quantum field theory. Fields are independent variables assigned to every point in space and to every instance in time. In fact, the space-time must be discretized and replaced by a four-dimensional (typically hypercubic) lattice. There are two reasons for doing so. First, the theory would give infinite predictions for observable quantities unless it is regularized by introducing a minimal distance in space-time, in this case the lattice spacing. The related second reason is that we wish to have a well-posed mathematical problem amenable to numerical treatment. For the same reason in a full quantum theory the real time must be replaced by an imaginary one. As a result, all distinction between space and time coordinates disappears. This trick makes possible computation of many important static properties of the theory, such as masses of elementary particles. On the other hand, it becomes very difficult to extract dynamical information, such as particle decay rates. As I shall discuss later on, this difficulty can sometimes be circumvented by the classical approximation. In these and some other cases the dimensionality of the problem reduces from 4 to 3.

I shall now try to formulate a generic problem in lattice field theory which embraces the majority of interesting applications. The problem is defined on a hypercubic lattice. Nearly every realistic theory of elementary particles requires presence of the so called gauge fields which mediate interactions among other degrees of freedom. It is appropriate for the mediating role of the gauge fields that they are usually defined on the links of the lattice. Most often the gauge fields also interact among themselves in a nonlinear fashion. In the lattice formulation, the gauge fields take values in a Lie group, the latter being SU(3) for QCD and SU(2) for the theory of weak interactions. Most often the fundamental representation of the group is used. It then takes $n \times d \times V$ floating-point numbers to represent a single copy of the gauge field configuration, where V is the lattice volume, d is the dimensionality of the problem, and n is the number of floating-point variables needed for a numerically convenient representation of a single gauge field. For the SU(2) group this number is 4, while for the SU(3) it grows to 18. In the following I denote by U_l the gauge link matrix residing on link l. The internal matrix indices of U_l are called the color indices (hence the name "chromodynamics" for the theory of strong interactions). More fields may have to be added, depending on the field content of the underlying problem. In case of weak interactions the theory includes two complex fields defined on the sites of a lattice, and in some approximations an additional triplet of real fields. I will collectively denote these fields by ϕ_j, where j is a site index. It is often necessary to take into account fermions, $i.e.$, fields which obey Pauli exclusion principle. Fermions cannot be naturally represented in terms of regular numbers. Insistence on such representation, needed for a computer simulation, leads to a serious complication, as we shall see in a moment. Typically, the total number of variables is very large. In a state-of-the-art QCD simulation on a 32^4 lattice an order of 10^8 variables are involved. Mathematically, the problem is that of computing, for any function $\mathcal{O}$ of the fields, the integral over all the fields

$$\langle \mathcal{O} \rangle \equiv \int \prod_{l,j} dU_l d\phi_j d\chi_j e^{-S(U,\phi)-|(\mathcal{D}(U))^{-1}\chi|^2} \mathcal{O}(U,\phi,\chi). \tag{1}$$

The specific functional form of S, called the action, is of little interest to us here. The only properties of S truly important numerically are that it is (a) bounded from below and (b) it is a local polynomial of U and ϕ whose every term only involves fields separated by no more than a fixed finite number (most often one) of lattice spacings. The second term in the exponent is necessary if fermions are to be taken into account. The complex field χ, carrying, like ϕ, both site and color indices, is an "unnatural", numerical representation for the fermions. The $\mathcal{D}$ matrix, depending on the gauge fields, has only nearest-neighbor couplings, but its inverse is, of course, highly nonlocal. The most important lattice gauge theories studied by large-scale computational methods are special cases of (1): in QCD the action does not involve the scalar ϕ field, whereas in the theory of weak interactions the non-fermion sector is most often considered. Under certain circumstances it is possible to neglect fermion contributions in QCD.

It is clear that, because of the large number of variables, (1) cannot be computed by standard numerical integration methods. Instead, the integrand of (1) is looked upon as a probability density in a multidimensional space, and averages of observables are found by a suitable importance sampling method, whereby the value of $\mathcal{O}$ is sampled with probability density $\exp[-S(U, \phi) - |(\mathcal{D}(U))^{-1}\chi|^2]$. There is a variety of importance sampling methods. I have no possibility here to discuss any of them. Let me only point out the essential features these methods possess. First, there is no need to recompute $\exp[-S(U, \phi) - |(\mathcal{D}(U))^{-1}\chi|^2]$ for every configuration in the sample. Only the increment of the probability density compared to the previous configuration is required. For a local quantity like S such an increment (called an update) is a sequence of local steps, each involving fields in a close vicinity of a single lattice site. With the inclusion of fermions, an update involves inversion of $\mathcal{D}$, which is usually done iteratively and again requires only local operations. This is welcome news: it means that the problem is a very suitable candidate for parallelization.

For the most interesting applications the numerical task is formidable. In order to approach the description of the physical world (1) must be calculated in a regime where the fields are correlated over large distances on the lattice. This in turn means that *(a)* large lattice volumes are required and *(b)* it takes many iterations of an importance sampling algorithm to produce every new statistically independent field configuration. With inclusion of fermions, the situation becomes even worse: approaching the real-world physics means making the fermion $\mathcal{D}$ matrix more and more ill-conditioned. With presently available resources it has been possible to extract from lattice simulations valuable information on both strong and weak interactions at high temperature. Some parameters of QCD have been determined accurately enough to allow direct comparison with experiment. But the bulk of work lies ahead. It has been estimated that, in order to perform a definitive fermionless QCD calculation, an effort on a teraflop-year scale is required. A realistic inclusion of fermions may be up to two orders of magnitude harder (assuming there will be no revolutionary improvement of the algorithms) [2].

The combination of high parallelizability and enormous resource demands explains why parallel computation was embraced very early on by the lattice community. Lattice applications have been run on nearly every commercially available parallel computer to date. Since lattice users develop their own software, they are usually among the first to run on a new platform and assess its performance, a quality welcomed by the vendors. Moreover, a number of lattice groups have constructed highly successful and cost-effective dedicated parallel computers. At the moment, the leading lattice dedicated machines are the ones at Columbia University (peak 16 Gf), QCDPAX at Tsukuba laboratory in Japan (peak 14 Gf), GF11 at IBM Yorktown Heights (peak 11 Gf), and APE at the

University of Rome (peak 6 Gf). As a spinoff from the lattice project, the APE machine is now commercially available. Several teraflop-scale lattice computers are currently under construction [3, 4, 5].

A number of codes have been developed over the years for lattice physics applications on general-purpose parallel computers. Some of these are public-domain software available from the FREEHEP database maintained by Stanford University [8]. I shall use an example of the one I am best familiar with, namely, the code developed, primarily for QCD applications, by physicists from MILC collaboration, perfoming state-of-the-art simulations of QCD on general-purpose machines. The code was initially developed for the IPSC machine produced by Intel and for the nCUBE, and was subsequently ported, with a modest effort, to the next generations of Intel parallel computers (the Touchstone Delta and the Paragon), to CM5, to T3D, and to SP2 [5, 7]. What makes the code easily portable is a cleverly written package of communication routines which utilize either proprietary message-passing libraries on each machine, or alternatively, the PVM. In fact, this package is the only machine-dependent part of the code. Let me give an example of how it all works. A user can choose, by specifying a precompiler directive, a lattice layout most suitable for a given application, *i.e.* the one that would minimize the internode communication traffic and achieve best possible load balance. All the variables (fields) assigned to a given location (site) of the lattice are members of a single structure. For instance, my most recent application defines

```
struct site {
    /* The first part is standard to all programs */
    /* coordinates of this site */
    short x,y,z;
    /* is it even or odd? */
    char parity;
    /* my index in the array */
    int index;
    /* Now come the physical fields, program dependent */
    su2_matrix link[2][3];
    efield_type efield[3];
    efield_type etemp;
    su2_matrix tempmat;
    su2_matrix staple[3];
};
typedef struct site site;
```

In this way, one only needs to establish a relation between a geometric location (site) on a lattice and a processor node where this site resides, without any reference to physical contents of the problem.

Let me describe a specific computational example. Suppose that, as is often the case, at a certain point the algorithm requires computing a product of all matrices on links making up an elementary square (plaquette), taking a trace of that product and summing up the result over the lattice. Since link matrices are assigned to a site from which they emanate in positive direction, this operation involves gathers from two neighboring sites, which may or may not require message passing. The nice feature of the MILC communication package is that the user need not know whichever is the case. The plaquette computation, using asynchronous sends and receives, is done as follows. First, buffers are allocated for data to be gathered (gen_pt[0] and gen_pt[1] below). Then gathers are performed: local data are simply pointed to, whereas the off-node data are exchanged by message passing. Since asynchronous sends and receives are used, the off-node data are waited for. Finally, when the computation is complete, the buffers are freed and the global sum is performed:

```
/* get umat[dir2] from direction dir1 */
tag0 = start_gather( F_OFFSET(umat[dir2]), sizeof(su2matrix),
    dir1, EVENANDODD, gen_pt[0] );
/* get umat[dir1] from direction dir2 */
tag1 = start_gather( F_OFFSET(umat[dir1]), sizeof(su2matrix),
    dir2, EVENANDODD, gen_pt[1] );
wait_gather(tag0);
wait_gather(tag1);
FORALLSITES(i,s){
    mult_su2_nn(&(s->umat[dir1]),(su2matrix *)gen_pt[0][i],
        &tmat1);
    mult_su2_na(&tmat1,(su2matrix *)gen_pt[1][i],&tmat2);
    realtrace_su2(&(s->umat[dir2]),&tmat2,ftemp);
    ss_sum += ftemp;
}
cleanup_gather(tag0);
cleanup_gather(tag1);
g_floatsum( &ss_sum );
```

In this way, internode communications are, to a large extent, hidden from the user.

Let me now proceed to describing the latest physics results obtained on the SP2 here in Copenhagen. The study we have been conducting is related to a very fundamental question of the origin of matter in the Universe. The matter around us and elsewhere in the Universe is made up mostly of protons and neutrons, or, on a deeper level, constituents thereof, called quarks. Corresponding to every such particle there is an antiparticle. Particles and antiparticles do not mix well and tend to annihilate each other, releasing large amounts of energy. The collective name for protons, neutrons, some other particles of this kind, and their constituent quarks, is baryons. The excess of baryons over anti-baryons is a quantity called baryon number.

A modern theory of elementary particles possesses a fundamental symmetry property which prohibits discrimination of antiparticles in favor of particles, or vice versa. However, our daily experience, as well as extensive astronomical observations, tell a different story: everywhere in the Universe matter is observed, but no sign of antimatter is seen. This is a puzzling fact awaiting explanation.

Thirty years ago the explanation would be: the baryon asymmetry was set, once and for all, at the beginning of times as an initial condition for the Universe. In this way, all the subsequent "whys" and "hows" would have to be answered by theology rather than by science. Since then, however, an entirely different picture has emerged. First of all, it turned out that such an initial asymmetry could not be accommodated by the most viable cosmological scenarios. Secondly, an important discovery was made by 't Hooft in 1976 [9]. He observed that the current Standard Model of elementary particles, which today stands on a solid experimental foundation, allows processes violating the baryon number. This is not something to worry about in everyday life: the probability of such processes depends on the temperature, and, even for temperatures 10^8 times higher than that of the Sun's interior, is exceedingly small. Under normal conditions (with a very liberal definition of normality) not a single proton decay would have been observed in the entire history of the Universe. But in the very early Universe far higher temperatures were encountered. It was conjectured that in a certain temperature range baryon-number violating processes are rapid enough to wipe out any primordial baryon excess. If so, high-energy theorists will face a new challenge: either find a trick to avoid primordial baryon erasure, or invent a mechanism for low-temperature baryogenesis [10]. These considerations indicate that the baryon number violation rate at high temperature is an interesting quantity.

Let me explain in some detail how this rate can be determined. At the heart of 't Hooft's celebrated work is an equation whose essence may be expressed as

$$\partial_t (B - N) = 0.$$

Here B is the baryon number and N is a known local functional of the gauge fields, i.e., the fields whose discrete version is given by the link matrices defined at the beginning. Thus in order to study baryon-number violation, we need to follow a real-time evolution of N. This, however, is easier said than done. First of all, the dynamics of N in the high-temperature regime is completely outside the scope of perturbation theory and has evaded any other analytical treatment so far. This leaves a lattice simulation as the only way to proceed. But then, as I have mentioned in the beginning, the existing lattice methods cannot handle *quantum* field theory in *real* time. In this situation, it was proposed some time ago to replace, for this particular purpose, quantum field theory by its classical approximation. At first, this sounds as a risky proposition: after all, it were precisely the shortcomings of the classical field theory at finite temperature which led to the discovery of quantum mechanics. Fortunately, it turns out that for a very restricted class of quantities, including N, the classical approximation is, in all likelihood, reliable.

Qualitative arguments suggest that at high temperatures N behaves as a random walk, i.e., for sufficiently long time t

$$\langle (N(t) - N(0))^2 \rangle = \Gamma V t, \tag{2}$$

where $\langle\rangle$ denotes thermodynamic average and V is the space volume. The "diffusion constant" Γ is temperature-dependent. It was expected, again on a qualitative level, that

$$\Gamma = \kappa c \left(\frac{kT}{30ch}\right)^4, \tag{3}$$

where T is the temperature, k is the Boltzmann constant, h is the Planck constant, c is the speed of light, and κ is a dimensionless numerical coefficient. A pioneering numerical work by NBI-CERN-Glasgow collaboration several years ago qualitatively confirmed (2) and (3) and estimated that κ's order of magnitude is one [11]. However, at that time adequate computer resources were not available, and, moreover, algorithms suitable for real-time simulation were yet to be developed.

The algorithmic difficulty with real-time classical approach was that classical gauge theories are locally constrained. In the usual electrostatics the local constraint obeyed by the electric field E is better known as Gauss' law: $\partial_i E_i = 0$. Analogous constraints, but more numerous and nonlinear, exist in the Standard Model. The importance sampling must be done carefully to avoid constraint violation. The algorithm for doing so was recently constructed [12].

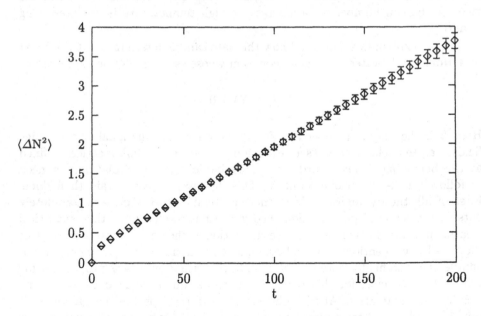

Fig. 1. Diffusion of N: mean square of $N(t) - N(0)$ as a function of the lag t for lattice inverse temperature $\beta = 12$ and for the lattice linear size $L_L = 32$.

The numerical work, aimed at confirming (2) and (3) and at establishing the value of κ with a 5% accuracy, required about 12000 node-hours on the UNI*C SP2 computer, with some additional work done on the Cray C92. Our parallel code ran at about 35 Mflops/node, which is close to performance reported by

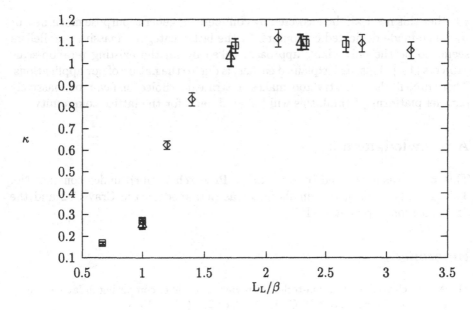

Fig. 2. The rate prefactor κ dependence on the dimensionless parameter L_L/β. The data points are for $\beta = 10$ (diamonds), 12 (squares), and 14 (triangles).

other lattice groups on a variety of platforms [5, 6]. We have been able to confirm that N exhibits a random-walk behavior (2), as illustrated by Figure 1. On a more quantitative level, we have been able to determine the coefficient κ of (3). The results, shown in Figure 2, are a strong evidence in favor of validity of the classical approximation for our problem. This is so because, as Figure 2 shows, κ exhibits very strong finite-size effects. Consequently, κ is a quantity sensitive to long-distance properties of the theory, for which the classical approximation is valid. Most importantly, our results show that, whatever the initial baryon number of the Universe was, it must have been wiped out by thermal fluctuations [13].

Our total numerical effort so far can be estimated as 400 gigaflop-hours, modest compared to major lattice applications amounting to many gigaflop-months. If, however, we wish to shed more light on the origin of matter, much more work is required. As I have already mentioned, thermal erasure of initial baryon number leads one to consider baryogenesis scenarios at low temperature. Estimates show that "low" in this context means some four times lower than the temperatures we have been working at. As figure 2 shows, this means that we will have to increase the lattice volume by a factor of 64 in order to keep finite-size effects in check. This by itself brings a needed effort into a gigaflop-year range. In addition, at low temperatures the power-law dependence of κ is likely to be replaced by an exponential decay: $\kappa \propto T^4 \exp(-M/T)$ with a constant M. This would entail further growth in computational demands.

While our project is small-scale compared to the major lattice applications, it reflects the dire need of more computing resources in the field. Time will tell whether this need will be answered by commercial, general-purpose machines, or by home-made dedicated computers. In the latter category, massive parallelism seems to be the only viable approach, taken by all the existing teraflop-scale projects [4, 5]. This, as I explained earlier, is due to the nature of our applications. Therefore, if the industry too makes a strategic choice in favor of massively parallel platforms, I think this will be good news for the lattice community.

Acknowledgments

This work was supported by the Danish Research Council under contract No. 9500713. The bulk of the simulations was perfomed on the Cray-C92 and the SP2 supercomputers at UNI*C.

References

1. A very clearly written non-technical review of parallel computing in lattice theory is by F.R. Brown and N.H. Christ, Science 239 (1988) 1393.

   An up-to-date information on the status of of the field is contained in Proceedings of yearly held symposia on lattice filed theory, published as Proceeding Supplement to Nuclear Physics B.

   A systematic textbook on lattice field theory is M. Creutz, *Quarks, gluons and lattices*, Cambridge University Press (1986).

   Recent publications on the subject are posted on hep-lat bulletin board, accessed through http://xxx.lanl.gov/archive/hep-lat.
2. C. DeTar, Nucl. Phys. A527 (1991) 547c.
3. D. Weingarten, Nucl. Phys. B (Proc. Suppl.) 26 (1992) 126.
4. Y. Iwasaki, Nucl. Phys. B (Proc. Suppl.) 34 (1994) 78.
5. R. Gupta, *Prospects of solving Grand Challenge problems*, Los Alamos technical report LA UR-94-3701.
6. C. Bernard *et al* (MILC collaboration), *Lattice QCD on the IBM scalable POWERParallel systems SP2*, University of Utah technical report UUHEP-95/1.
7. R.L. Sugar, Comp. Phys. Comm. 65 (1991) 268.
8. Accessed through http://www-spires.slac.stanford.edu/find/fhmain.html.
9. G. 't Hooft, Phys. Rev. Lett. 37 (1976) 8.
10. As the subject of baryogenesis has undergone a spectacular development in recent years, it is yet to be covered by a truly non-technical review. A determined reader can consult a review by A.D. Dolgov, Phys. Rep. 222 (1992) 309. Recent developments in the field are reflected in *Electroweak physics and the early Universe*, edited by J.C. Romao and F. Freire, Plenum Press, New York (1994).
11. J. Ambjørn, T. Askgaard, H. Porter and M. Shaposhnikov, Nucl. Phys. B353 (1991) 346; Phys. Lett. B244 (1990) 479.
12. A. Krasnitz, *Thermalization algorithms for classical gauge theories*, NBI report NBI-HE-95-25, hep-lat/9507025; Nucl.Phys. B (Proc. Suppl.) 42 (1995) 885.
13. J. Ambjørn and A. Krasnitz, *The classical sphaleron transition rate exists and is equal to* $1.1(\alpha_w T)^4$, NBI report NBI-HE-95-23, hep-ph/9508202.

Parallel Time Independent Quantum Calculations of Atom Diatom Reactivity

Antonio Laganà,[1] Stefano Crocchianti,[1] Guillermo Ochoa de Aspuru,[1] Ricardo Gargano,[1] and G.A. Parker[2]

[1] Dipartimento di Chimica, Università di Perugia, 06123-Perugia, Italy
[2] Department of Physics and Astronomy, University of Oklahoma, Norman, USA

Abstract. Some models for the parallel organization of quantum reactive computer codes are discussed. The need for articulating the parallelism at different levels is examined and possible solutions are worked out by analyzing the structure of related theoretical and computational approaches. For the reduced dimensionality program, for which the solution of the bound state problem needs little cpu time, a multilevel task farm parallelization over the angle at the upper level and over the propagation at the lower level was found to be appropriate. On the contrary, for the full dimensional method the solution of the bound state problem is time consuming and has to be parallelized. In this case, a different model has been adopted: the calculations of the surface functions relative to a given subset of sectors have been grouped together and the subsets have been distributed for parallel calculation using a single program multiple data model.

1 Introduction

One of the basic goals of computational chemistry is the construction of a molecular virtual reality (MVR) since this can significantly contribute to the development of chemical intuition. A rigorous construction of the MVR should be firmly based on first principles. Only in this way the exercise of human imagination would safely scale down into the molecular world where lenghts are of the order of 10^{-10} m, energies of the order of 10^{-14} erg and masses of the order of 10^{-27} kg.

To build a rigorous MVR we need efficient software tools able to model in real time the kinetics of complex chemical systems by describing their evolution using the rules of the electronic and nuclear dynamics. Though we are still far from both basing the construction of the MVR solely on first principles and making its different pieces match, the progress of computer technology is significantly accelerating these processes. In the case of gas phase systems, for example, the evolution of the computational procedures has now reached a point where one can combine together the treatment of the nuclear dynamics with the calculation of the electronic energy[1] or the treatment of the kinetic evolution of complex systems with the calculation of single collision properties of elementary processes.[2]

To a large extent, this progress is due to the development of parallel computing. Therefore, the design of chemical computer programs exploiting the innovative features of parallel architectures is a key step of this progress. Unfortunately, even for apparently embarrassingly parallel problems, as is the case of quasiclassical trajectories, there are no simple efficient parallel models.[3] The problem of finding optimum parallelization models is even more difficult when dealing with quantum approaches. Related algorithms, in fact, involve a strong coupling of the data and a large request of memory.

In this paper, we shall discuss the parallelization of two coupled channel (CC) time independent quantum reactive scattering codes. Of these computer codes we analyze possible parallelization schemes and illustrate the performances. To enhance the portability of the restructured codes use was made of PVM.[4]

The paper is organized as follows: In the second section, a sketch of the basic equations necessary to understand related numerical algorithms is given. In the third section, the models adopted to parallelize the reduced dimensionality program is illustrated and related performances discussed. In the fourth section, the work in progress to parallelize the accurate three dimensional quantum code is presented.

2 The computational techniques

The first computer code is based on the reduced dimensionality infinite order sudden (RIOS)[5] method. The RIOS approach reduces the dimensionality of an atom-diatom reactive scattering problem by decoupling both diatomic and atom diatom rotations. As a result, one has to solve for both reactant ($\lambda = \alpha$) and product ($\lambda = \beta$) channel the following fixed collision angle Θ_λ two mathematical dimension differential equation:

$$\left[\frac{\partial^2}{\partial R_\lambda^2} + \frac{\partial^2}{\partial r_\lambda^2} - \frac{A_l}{R_\lambda^2} - \frac{B_j}{r_\lambda^2} - \frac{2\mu}{\hbar^2}\left(V(R_\lambda, r_\lambda; \Theta_\lambda) - E\right)\right] \Xi(R_\lambda, r_\lambda; \Theta_\lambda) = 0 \quad (1)$$

at all values of Θ_λ for which at a given total energy E the two channels are open. In Equation 1, R_λ and r_λ are the mass scaled Jacobi coordinates, μ is the reduced mass of the system, $A_l = \hbar^{-2} l(l+1)$ and $B_j = \hbar^{-2} j(j+1)$ are the coefficients of the decoupled orbital and rotational terms of the Hamiltonian with l and j being the related quantum numbers.

Equation 1 can be solved using a CC technique. To this end the (R_λ, r_λ) plane is divided into two half planes using a straight line originating at $R_\lambda = r_\lambda = 0$ and following the ridge separating entrance and exit region. For each arrangement λ, the integration is carried out from the separation line to a large value of R_λ. To this end, each arrangement channel is segmented into many small sectors. Within each sector i the global fixed Θ_λ wavefunction $\Xi(R_\lambda, r_\lambda; \Theta_\lambda)$ is expanded in terms of the $\phi^i(r_\lambda; R_\lambda, \Theta_\lambda)$ eigenfunctions calculated by solving the one dimensional bound state equation

$$\left[\frac{\partial^2}{\partial r_\lambda^2} + \frac{2\mu}{\hbar^2}\left(V(r_\lambda; R_\lambda^i, \Theta_\lambda) - \epsilon_\lambda^i\right)\right] \phi^i(r_\lambda; R_\lambda^i, \Theta_\lambda) = 0 \quad (2)$$

at R_λ^i (the midpoint value of R_λ for sector i). By truncating the expansion to the first N_v terms, substituting the expansion into Equation 1 and averaging over r_λ, one obtains a set of N_v coupled equations of the type

$$\left[\frac{d^2}{dR_\lambda^2} - \mathbf{D}_\lambda^l\right] \psi_v^l(R_\lambda; \Theta_\lambda) = 0 \qquad (3)$$

where $\mathbf{D}_\lambda^l$ is the coupling matrix and $\psi_v^l(R_\lambda; \Theta_\lambda)$ are the coefficients of the expansion of $\varXi(R_\lambda, r_\lambda; \Theta_\lambda)$. After integrating Equation 3 through the different sectors to the asymptotes and imposing the appropriate boundary conditions (because of this hereafter we shall drop the label λ from the equations) one can estimate the detailed state v to state v' fixed Θ S matrix elements ($S_{lv,v'}(\Theta, E)$) (also the j label has been dropped from the notation because we take $j = 0$).

From $S_{lv,v'}(\Theta, E)$ elements, the state (v) to state (v') reactive cross section $\sigma_{v,v'}(E)$ can be computed by integrating over Θ and summing over l

$$\sigma_{v,v'}(E) = \frac{\pi}{k_v^2} \sum_l (2l + 1) \int_{-1}^{1} |S_{lv,v'}(\Theta, E)|^2 \, d\cos\Theta. \qquad (4)$$

The second computer code (ABM) is the first of a set of programs (APH3D) carrying out a full three dimensional quantum treatment of atom diatom reactive scattering. The method is based on the hyperspherical coordinate formalism of ref.[6]. In this formalism, the three internal coordinates are ρ, θ and χ and the Hamiltonian reads:

$$H = T_\rho + T_h + T_r + T_c + V(\rho, \theta, \chi) \qquad (5)$$

(the subscripts stand for hyperradius, hyperangles, rotation, and Coriolis, respectively). The individual terms are given by

$$T_\rho = -\frac{\hbar^2}{2\mu\rho^5} \frac{\partial}{\partial\rho} \rho^5 \frac{\partial}{\partial\rho}, \qquad (6)$$

$$T_h = -\frac{\hbar^2}{2\mu\rho^2} \left(\frac{4}{\sin 2\theta} \frac{\partial}{\partial\theta} \sin 2\theta \frac{\partial}{\partial\theta} + \frac{1}{\sin^2\theta} \frac{\partial^2}{\partial\chi^2}\right), \qquad (7)$$

$$T_r = AJ_x^2 + BJ_y^2 + CJ_z^2, \qquad (8)$$

and

$$T_c = -\frac{i\hbar\cos\theta}{\mu\rho^2 \sin^2\theta} J_y \frac{\partial}{\partial\chi}, \qquad (9)$$

with $A^{-1} = \mu\rho^2(1 + \sin\theta)$, $B^{-1} = 2\mu\rho^2 \sin^2\theta$ and $C^{-1} = \mu\rho^2(1 - \sin\theta)$.

To solve the scattering problem using a CC technique, the wavefunction $\varPsi_{tA}^{Jpn}$ for a given value of the total angular momentum J is expanded in products of Wigner rotation functions D_{AM}^J of the three Euler angles (α, β and γ) and surface functions Φ of the two internal hyperangles (θ and χ). The unknown functions ψ of the hyperradius ρ are the coefficients of the expansion of the global wavefunction. To carry out the numerical integration the ρ interval is divided into several small sectors. For each sector i a set of surface functions $\Phi_{tA}^{Jp}(\theta, \chi; \rho_i)$ is calculated at the sector midpoint ρ_i. These surface functions, which serve as a local basis set, are independent of ρ within the sector but change for different sectors.

Therefore, by analogy with the RIOS program, the first computational step is devoted to the calculation of the surface functions Φ_{tA}^{Jp}. To this end one has to integrate the two dimensional bound state equation

$$\left[T_h + \frac{15\hbar^2}{8\mu\rho_i^2} + \hbar^2 G_J + F\hbar^2 A^2 + V(\rho_i, \theta, \chi) - \mathcal{E}_{tA}^{Jp}(\rho_i) \right] \Phi_{tA}^{Jp}(\theta, \chi; \rho_i) = 0 \quad (10)$$

where $G_J = \frac{1}{2}J(J+1)(A+B)$ and $F = C - (A+B)/2$. Then, the second computational step is devoted to the integration, sector by sector, from small ρ values to the asymptote, of the following set of coupled differential equations

$$\left[\frac{\partial^2}{\partial\rho^2} + \frac{2\mu E}{\hbar^2} \right] \psi_{tA}^{Jpn}(\rho) = \frac{2\mu}{\hbar^2} \sum_{t'} < \Phi_{tA}^{Jp}(\theta, \chi; \rho_i) | H_i | \Phi_{t'A}^{Jp}(\theta, \chi; \rho_i) > \psi_{t'A}^{Jpn}(\rho),$$

$$(11)$$

where the internal Hamiltonian H_i reads

$$H_i = T_h + T_r + T_c + \frac{15\hbar^2}{8\mu\rho^2} + V(\rho, \theta, \chi). \quad (12)$$

Once the integration is performed, the solution is mapped into the space of the Jacobi coordinates, asymptotic boundary conditions are imposed and fixed J S matrix elements and related probabilities are evaluated.

3   The parallelization of the RIOS method

We first examine the work performed to parallelize the RIOS code. The scheme of the related numerical procedure is sketched below

SECTION I
Input data and calculate quantities of common use
SECTION II
LOOP on collision angles

subsection *a*
IF (first run) Calculate and store eigenvalues, eigenvectors and overlaps
subsection *b*
Read quantities necessary to assemble the coupling matrix
LOOP on energies
Embed the energy dependence into the coupling matrix
LOOP on l quantum number
Integrate fixed angle, fixed l scattering equations
Store detailed S matrix elements on disk
IF (converged with l) exit l loop

END the l loop
   Calculate and print the fixed angle contribution to the cross section
END the energy loop
   Integrate over the collision angle
END the angle loop

SECTION III

Perform final calculations and print the reactive cross section

Figure 1 - Scheme of the RIOS code.

As apparent from the scheme, the parallelization of the RIOS code can be carried out at different levels.

The level that was considered in the first instance (model I)[7] is the distribution on different processors of the tasks of calculating the fixed angle, fixed energy, single l propagation of the scattering wavefunction. The distribution was organized according to a task farm model.[8] The task farm model assigns to a master node (or to a host machine) the role of managing the distribution of the different computational tasks and to the other worker nodes the role of carrying out the computations associated with the assigned task. The choice of the task farm model was mainly motivated by the difficulty of adopting the simpler data partitioning scheme (single program multiple data, SPMD[8]). In fact, the need for checking the convergency of the calculation with l does not allow an *a priori* determination of the number of computational tasks that need to be assigned to the individual nodes. On the contrary, the use of a task farm model allows a dynamical distribution of the load among the available processors. This keeps the processors busy all the time since a new single l propagation (for the same angle and the same energy) is assigned as soon as a processor has ended its work. Then, when the convergency is reached, the master stops the current work and new single l propagations for a different energy are assigned.

This model gives a good scaling of the parallel performance on machines with a limited number of processors.[7] However, it is not suited for implementation on highly parallel architectures since the number of single l calculations may be smaller than the number of processors and the work lost when reaching convergency with l may be large. When implementing the program on a parallel machine having a large number of nodes in addition to parallelizing subsection (b) (the propagation) of the second section of the code we included in the same run all the angle values.

A first attempt was made by adopting a two level parallel model (model II). A way of constructing a two level parallelization model was to distribute different fixed angle calculations. In fact, since fixed angle calculations are also independent computational tasks (the RIOS method builds the 3D S matrix by averaging the independent fixed angle ones) when distributing fixed angle calculations the nodes can be clustered in subsets. Therefore, the two level task farm model consists of a master assigning different fixed angle calculations to the different clusters and a cluster submaster (as in the first parallelization model described above) distributing single l propagations.

A single level parallel model looping more rapidly on the energy than on the l value, can also be adopted (model III). That means that, for example, after distributing the first single l propagation for the first energy, the first single l propagation for the second energy is distributed next instead of the second l propagation for the first energy. In this way, the time lost in improductive single l calculations reduces significantly. In fact, the energy loop has no convergency check and the number of energies to be calculated is given as an input data. Therefore, a limited number of nodes (if not none) will still be carrying out calculations for the same energy when convergency with l is reached. The loss of single l propagation calculations, however, increases when approaching the end of the calculation. In this case, in fact, when the large majority of fixed energy calculations is completed, several nodes will be contemporarily running a single l propagation for the same energy. This means that more single l propagations will be stopped when convergency for that energy is reached.

Speedups measured for model II on an nCUBE 2 are 8.04, 13.22, 33.43 and 60.29 when using 16, 32, 64 and 128 nodes respectively.[9] The same measurements for model III led to the following speedups: 8.17, 10.16, 13.28 and 12.28. However, despite the poor performance, model III has some features of great interest for parallelization once its higher memory demand and more complex data management are mastered. As a matter of fact, more recent measurements carried out on the Cray T3D (after performing subsection (a) on a workstation) gave the following speedups 7.9, 15.2, 26.7 and 34.5 when using 8, 16, 32 and 64 nodes.

4 The parallelization of the ABM code

As already mentioned, the APH3D computational procedure is articulated into several programs. Of these, only the one performing the calculation of fixed ρ eigenvalues and eigenfunctions of the two internal hyperangles (ABM) as well as the one performing the fixed total angular momentum J propagation along the reaction coordinate ρ for a set of energy values (LOGDER) deserve to be considered for parallelization. The parallel restructuring of these two codes is an ideal test bed for the models and parallelization recipes we have developed for RIOS. ABM and LOGDER correspond, in fact, to subsections (a) and (b) of the second section of RIOS. In this paper, however, we shall confine the discussion to the parallelization of ABM.

Since, as already mentioned, the fixed ρ calculation of eigenvalues and eigenfunctions of the two hyperangles is a two dimensional bound state problem, ABM differs from subsection (a) of the second section of RIOS in several structural features. As a result, it is more memory demanding and time consuming.

The scheme of the program is:

Input data

Input data
Calculate quantities of common use
LOOP on sectors
   Calculate the value of ρ
   LOOP on Λ
      Calculate surface functions
      IF(not first ρ) calculate overlaps with previous sector
   END the Λ loop
   Calculate and store the coupling matrix
END the sector loop

Figure 2 - Scheme of the ABM code.

As apparent from the scheme, the external loop runs over the different sectors and has to deal with the surface functions of two adjacent sectors for all the Λ projections (inner loop). If the generation of surface functions was distributed among different processors, the calculation of overlaps and coupling matrix elements would require a significant amount of data transfer. In our case, a set of 277 surface functions was used. By properly handling related vectors and matrices, the amount of memory needed was kept slightly smaller than the Cray T3D node memory (8 Mw). This made it it possible to parallelize the program at the outer (sector) level. A difficulty in parallelizing at this level is the order dependency associated with the calculation of overlap integrals between surface functions of adjacent sectors (each sector needs the surface functions of the previous one). Such a dependency can be avoided by duplicating the surface functions calculation when the previous sectors is dealt by a different node. This makes a task farm model highly inefficient for this application since it requires a duplication of all surface function calculations. The SPMD model seems to be more convenient for ABM. Using the SPMD scheme a subset of sector calculations is assigned to every node and only for the first sector of the subset the calculation of the surface functions of the previous sector has to be repeated.

Performances measured when using the SPMD model are illustrated in Figure 3 where the individual node cpu time consumption for a 32 (upper panel) and a 64 (lower panel) run are plotted as a function of the node number. The plot for the 32 processor run shows that there is a strong load unbalance (of the order of 50%) for the first six nodes. The load unbalance is not due to duplicated surface function calculations. It is instead associated with the fact that, when distributing the calculation for 230 sectors over 32 nodes the first 6 nodes get 8 sectors (for a total of 48) while the remaining 26 get only 7 sectors (for a total of 192). This (one out of eight) extra calculation, however, cannot account for such a large load unbalance. A rationale for that can be found with the help of the lower panel plot relative to a 64 node run. It shows essentially two types of load unbalance. A small one (extending over the first 38 nodes) and a large one (affecting only the first 12 nodes). The small one is due to the fact that when using 64 nodes the first 38 nodes get 4 sectors each (for a total of 152) while the remaining 26 nodes get only 3 sectors each (one extra calculation out of four).

In the group of the first 12 nodes the first one, as seen also for the case of

32 nodes, consumes slightly less time because it needs not to (re)calculate the surface functions for the previous sector to compute overlaps. The extra load of these nodes is due to the fact that for the first 48 sectors the calculation of the surface functions is more time consuming because related to small ρ values. This explains also why the distinction between the two different contributions to the load unbalance did not show up in the upper panel. In that case, in fact, the number of nodes carrying out the extra work and the most time demanding work was the same (48) leading, as a result, to only one type of load unbalance.

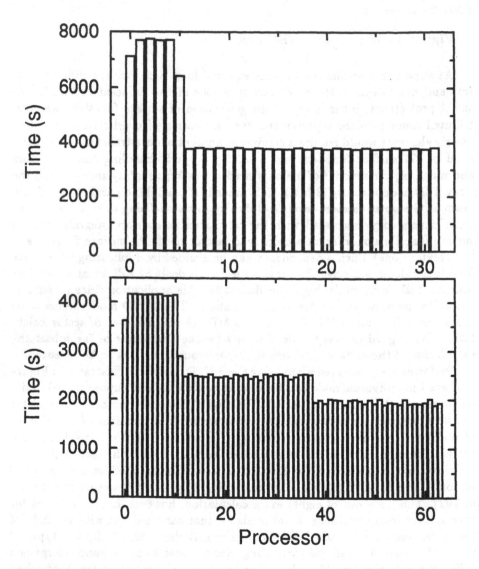

Fig. 3 - Plot of the individual computer time: run using 32 nodes (upper panel) and using 64 nodes (lower panel).

5   Conclusions

The investigation of the performances obtainable when using different parallel models to parallelize our RIOS and ABM three dimensional quantum reactive scattering codes has confirmed the conclusions drawn from the restructuring of the quasiclassical trajectory code: there are no embarrassingly parallel applications and significant effort has to be paid to design optimum parallelization schemes.

Our results, in fact, indicate that apparently good parallel models may not work as soon as either the problem or the machine are scaled up. Simplistic solutions may show not to work even for the simple cases for which they have been designed. On the contrary, when carrying out a careful analysis of the algorithm, limitations may often be overcomed. Unfortunately, there are no recipes valid for all situations. In our study, we found that for the reduced dimensionality approach for which the solution of the bound state problem needs only little cpu time, a multilevel parallelization of the tasks for the different loops (over the angle at the upper level, over the propagation at the lower level) with an attenuation of the order dependency at the convergency check (by inverting the hierarchy of energy and l loops) shows to be the model to adopt.

On the contrary, in the case of full dimensionality calculations the solution of the bound state problem needs a dedicated program and a dedicated restructuring work. In this case, a task farm model cannot be adopted. However, by duplicating the surface function calculation for the adjacent sector assigned to a different node, it has been possible to adopt an SPMD model. The model finds an obvious limitation both when increasing the number of nodes (the amount of duplicated work increases up to limit of 2; for a larger number of nodes additional nodes will remain idle) and when the dimension of the basis set becomes larger (the node memory becomes insufficient).

This confirms the conviction that our ability of mastering parallel computers for reactive scattering applications is still in its infancy and that the characteristics of available parallel architectures are not yet suitable for coping with currently used approaches to accurate quantum calculations of chemical reactivity. This means that we still need to learn how to estimate the limits of validity of the different parallel models and how to switch from a model to another when changing the parameters of the application.

6   Acknowledgments

Generous allocation of Cray T3D computer time by Cineca (Casalecchio di Reno, Italy) and EPCC (Edinburgh, United Kingdom) is acknowledged. Grants from Icarus (Cineca), Tracs (EPCC), Brazilian Science and Technology Council (CNPQ) and Basque Government are also acknowledged. Thanks are due to CNR and ASI for financial support and to Elda Rossi (Cineca) and Peter Maccallum (EPCC) for useful discussions and suggestions.

References

1. Bernardi, M., Olivucci, M., Robb, M.A.: Simulation of MC-SCF results on covalent organic multi-bound reactions: molecular mechanics with valence bond (MM-VB). J. Am. Chem. Soc. 114 (1992) 1606-1616.
2. Candler, G.V.: Interfacing nonequilibrium models with computational fluid dynamics methods. Nato ASI Molecular Physics and hypersonic flows, Maratea (1995) 72.
3. Laganà, A., Gervasi, O., Baraglia, R., Laforenza, D., Perego, R.: Where are embarrassingly parallel problems? the atom diatom quasiclassical reactivity Theor. Chim. Acta 84 (1992), 413-421
4. Beguelin, A., Dongarra, J., Geist, G.A., Manchek, R., Sunderam, V.S.: A user's guide in PVM Parallel Virtual Machine, Oak Ridge National Laboratory, TN, 1992.
5. Laganà, A., Garcia, E., Gervasi, O.: Improved infinite order sudden cross sections for the Li + FH reaction. J. Chem. Phys., 89 (1988) 7238-7241. Garcia, E., Gervasi, O., Laganà, A.: Approximate Quantum Techniques for Atom Diatom Reactions in Supercomputer Algorithms for Reactivity, Dynamics and Kinetics of Small Molecules, Laganà, A. ed., Kluwer, Dordrecht (1989) 271-294.
6. Parker, G.A., Pack, R.T and Laganà, A.: Accurate 3D quantum reactive probabilities of Li + FH. Chem. Phys. Letters 202 (1993) 75-81 ; Parker, G.A., Laganà, A., Crocchianti, S., Pack, R.T.: A detailed three dimensional quantum study of the Li + FH reaction. J. Chem. Phys., 102 (1995) 1238-1250.
7. Laganà, A., Gervasi, O., Baraglia, R., Laforenza, D.: Vector and parallel restructuring for approximate quantum reactive scattering computater codes in High Performance Computing, Delhaye, J.L., and Gelenbe, E., eds, North Holland, Amsterdam (1989) 287 - 298. Baraglia, R., Laforenza, D., Perego, R., Laganà, A., Gervasi, O., Fruscione, M., Stofella, P.: Porting of reduced dimensionality quantum reactive scattering code on a meiko computing surface, CNR Report 8/20 (1991).
8. Fox, G.C., Johnson, M., Lyzenga, G., Otto, S., Salmon, J., Walker, D., Solving problems on concurrent processors, Prentice Hall, Englewood Cliff (1988).
9. Baraglia, R., Laforenza, D., Laganà, A.: Parallelization strategies for a reduced dimensionality calculation of quantum reactive scattering cross sections on a hypercube machine. Lecture Notes in Computer Science, 919 (1995) 554-561.

Parallel Oil Reservoir Simulation

Jesper Larsen[1], Lars Frellesen[1],
John Jansson[2], Flemming If[2],
Cliff Addison[3], Andy Sunderland[3] and Tim Oliver[3]

[1] Math-Tech, Kildeskovsvej 67, 2820 Gentofte, Denmark[†]
[2] COWIconsult, Parallelvej 15, 2800 Lyngby, Denmark[†]
[3] Institute of Advanced Scientific Computation,
University of Liverpool, Liverpool, U.K[‡]

Abstract. In the EUREKA project PARSIM a large industrial simulation code is being parallelised in order to be able to perform the simulation interactively, giving the user immediate feedback through the simulators graphical user interface. Apart from the user interface, the simulator consists of two main components: A linear equation solver, and the simulation code itself. The design of the simulator is modular so that it has been possible to parallelise these two components independently. This paper describes how the simulator code and the linear equation solver has been parallelised, and gives some preliminary scalability results.

1 Introduction

Oil reservoir simulation is an established technique used by the petroleum industry to assess reservoirs and to plan production strategies. It is a very compute intensive application often requiring large amounts of cpu-time on state of the art supercomputers. The understanding of a reservoir and the optimisation of production planning depends on the number of different scenarios that can be simulated. To increase this number we introduce interactive reservoir simulation thus enhancing even further the need for compute power. A need that can only be met by parallel computing.

The paper is organised as follows: First, a short background on the simulator is given, and the reasons for wanting interactivity is discussed. Then choice of iterative algorithm for the linear equation solver and the design of the parallel simulator is discussed. Finally some scalability results for the solver is given.

2 COSI

The reservoir simulator COSI was originally developed by the Danish National Laboratory RISØ, The Technical University of Denmark and the consulting en-

[†] Math-Tech and COWIconsult gratefully acknowledge the support given by the Danish Agency for Development of Trade and Industry

[‡] The University of Liverpool gratefully acknowledges the support given by the UK Department of Trade and Industry.

gineering company COWIconsult. Since 1990 COWIconsult has been responsible for the development of the functionality of the simulator.

In the EUREKA project PARSIM, COSI is being parallelised in a collaboration between COWIconsult, Math-Tech and the Institute of Advanced Scientific Computing, Liverpool University (IASC).

COSI is a three-dimensional, three-phase, isothermal, numerical reservoir simulation model capable of running in black oil, wet gas or fully compositional mode. The simulator is especially good for modelling low permeable and dual porosity reservoirs.

The simulator is fully implicit, both in pressure and saturation. Space integration is performed by an integral finite volume technique, which enables an arbitrary combination of grid cell types, such as cartesian, cylindrical and irregular grid cells. The pressure of the oil phase and the specific component masses are chosen as primary variables. The non-linear equation system is solved by the Newton-Raphson technique, leading to a large sparse system of linear equations.

The oil reservoir is described as consisting of grid cells, boundaries connecting these grid cells and wells for injection into or production from the reservoir, as shown in Fig. 1.

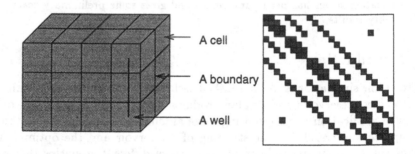

Fig. 1. The reservoir description and the structure of the corresponding matrix

The matrices created by the simulator consist of blocks of $n \times n$ elements (where n is the number of components plus one, typically 4) representing the interaction within one cell and between two neighbouring cells. The matrix is set up, so that all interactions, coming from the first cell, is placed in rows 1 to n, from the second cell in rows $n + 1$ to $2n$, etc. This is also shown in Fig. 1, where each block represents a n by n sub-matrix.

2.1 Interactive Reservoir Simulation

A reservoir simulation is normally performed as a batch job, meaning that the user submit an input file and several hours later receive an output file containing the simulation results. For sensitivity studies several simulations are run with different parameters. By comparing the results of the different simulations, the sensitivity of the different reservoir parameters can be assessed.

In order to aid the understanding of the reservoir behaviour and drive mechanisms interactive reservoir simulation is introduced. In this context interactive reservoir simulation means, that the user can monitor the simulation on-line and can interrupt the simulation at any time, modify parameters, and continue the simulation. Hereby the user can immediately see how the reservoir reacts to the imposed parameter change.

The above described form of the interactive reservoir simulation would normally increase the simulation time by at least one order of magnitude. To make this form of interactive simulation possible it is necessary to increase the computing speed accordingly by use of parallel computing.

2.2 Parallel Reservoir Simulation

In the original sequential simulation software a direct linear equation solver SESYS was used. This solver accounted for the majority of the overall time associated with the simulation. It followed, therefore, that significant improvements in the efficiency of the solver would have a similarly significant effect on the overall performance of the software.

During the PARSIM project a new sequential sparse iterative solver has been introduced, making the solver and the rest of the simulator use equal amounts of time during a simulation. Therefore the parallelisation effort must now focus more evenly on both the solver and the simulator in order for the application to scale. This implies that we need to find a parallelisation strategy which is well suited for both the equation solver and the simulation code.

3 The Parallel Linear Solver

At the core of the simulator lies the linear solver, where a sparse, unsymmetric system of equations $Ax = b$ needs to be solved for each time step. The parallel solver is being built on top of a set of parallel library routines, called the Sparse Distributed Data Library (DDL).

3.1 Parallel Library Routines

The DDL is a library system that permits a user to exploit the power of a distributed memory parallel computer from a single-threaded FORTRAN program. This data parallel programming model is achieved by letting the user call library routines which create, manage and operate on distributed objects. Currently, the DDL supports vector and sparse matrix distributed objects, but new types are easily added. A vector object is distributed by blocks, and a matrix object by blocks of rows (or columns).

It is often useful to be able to associate additional information or properties with a distributed object. The DDL implements a property list to allow the user to specify optional properties for an object. The DDL provides procedures to add, find and remove properties from the property list. Properties can include such things as permutations and communication patterns. The permutation property specifies a mapping between the physical storage index of an element of an object and the "logical" index to be used in computations. The communication pattern properties hold information about the communications required by DDL procedures, which improves the efficiency of these procedures. The property list also allows the user to extend the functionality of the DDL by defining new properties for distributed objects.

The DDL can also be extended by defining new high-level distributed object types. These new objects are defined hierarchically in terms of the existing distributed objects. Defining new high-level objects can be used to hide data complexity from the calling program. For example, if an application needs to manipulate a matrix in its LU factorised form then a new DDL sparse matrix distributed object can be defined which contains the sparse L and U sub-matrices.

The range of routines provided by the DDL is currently quite small and has been directed towards supporting sparse iterative solvers. There are level 1 BLAS operations, I/O operations, matrix-vector multiply and triangular solver routines along with some high-level preconditioning routines and the iterative solver. These routines are written mainly in C with some FORTRAN, and MPI is used for communication. User programs may be written in either C or FORTRAN.

3.2   The Solution Method

Iterative solvers based on the Conjugent Gradient method are well suited to solving large, sparse systems in a parallel environment, as the underlying operations in the algorithms are inherently parallel, and the fill-in associated with direct methods does not occur. After testing the suitability of various methods [1], it was found that preconditioned Transpose Free Quasi Minimal Residual method (TFQMR) [4] provided a sufficiently robust and efficient solver for problems generated from the COSI simulator. The iterates of the TFQMR solve are characterised by a quasi minimisation of the residual norm, leading to a smoother convergence curve than non-symmetric Conjugate Gradient based methods (CGS, BCG, BiCGStab).

3.3   Parallel Preconditioners

To facilitate satisfactory convergence rates preconditioning techniques have been incorporated into the solver. The current sparse DDL software contains routines to perform TFQMR iterations with a block Jacobi preconditioner based on a cell-based incomplete factorisation (ILU(0)) within each block. Setting up the preconditioner is undertaken completely independently by each processor as the preconditioner consists of simply the block diagonal D of the matrix A (number of blocks specified by the user). The system of equations to solve then becomes $D^{-1}Ax = D^{-1}b$, where the effectiveness of the preconditioner is measured by how closely D approximates A, generally the smaller $\|D^{-1}A - I\|$ the quicker the convergence.

Originally, D^{-1} took the form of an exact factorisation using the SESYS package. However, upon investigation it was found that ILU(0) methods provided a sufficiently close approximation to D^{-1} without the large overheads associated with each call to SESYS [2]. A cell-based ILU(0) factorisation is preferred to simple-point wise ILU(0), as tests have shown the problems to be extremely sensitive to the order of arithmetic operations when there is more than 1 equation per cell block.

The parallel preconditioned iterative solver has been implemented by using high level sequential Fortran code calling lower level DDL library routines which manage the parallelism [5]. As each processor is automatically allocated a sub-matrix the block diagonal preconditioner can be formulated and solved independently.

3.4 Alternative Parallel Preconditioners

Tests using the sequential code [1], [2] have shown global cell-based ILU(0) preconditioning to be generally more effective than block Jacobi in improving convergence rates. However in a parallel setting, cell-based ILU(0) preconditioning is unattractive due to the high levels of communication required during the associated triangular solves. In order to introduce greater parallel potential two methods have been investigated, both based on reorderings.

- Represent the inverses of triangular matrices as a product of several sparse factors whose inverses can be formed in place [3], [7].
- Use standard substitution methods for the triangular solves, but reorder the matrices to reduce inter-dependencies between nodes, thereby lowering inter-processor communication costs [6], [7].

Initial research has shown some promising results, and currently DDL support routines for the methods outlined above are under development.

4 Parallelising the Simulation code

The parallel version of the simulation code is built on top of the parallel equation solver, and has been designed to perform optimal with the parallel solver. The main demand with regards to integrating the parallel simulation code with the parallel solver, is to ensure that the data used by the solver is present on the processors where they is needed.

4.1 Distributing the Reservoir

In order to achieve the aim of having the data present where they are needed, the reservoir is parallelised using domain decomposition, where the grid cells and boundaries are split over the available processors. Cells and boundaries which have neighbours on other processors, will be duplicated onto those processors, making it possible to update the reservoir variables on each processor independently of the others.

When creating the vectors and matrices needed for the parallel solver, the DDL library underlying the solver determines the distribution of the datastructures. As there is a one-to-one mapping from reservoir data to the matrix and right hand side in the linear equation system, this determines to which processors the reservoir elements should go, if efficient support from the DDL library is to be achieved. The library makes a simple block oriented distribution, dividing the number of rows ($nrows$) with the number of processors ($nprocs$), placing the first $nrows/nprocs$ rows on processor 1, etc. Given the connection between the reservoir elements and the matrix rows, this implies that for the cells, the first $ncells/nprocs$ cells should be placed on the first processor, and so forth, where $ncells$ refers to the number of cells in the reservoir. This gives the distribution shown in Fig. 2, when distributing the previously shown reservoir over 6 processors. For processor 5, Fig. 2 also shows which cells have to be duplicated, either because they are neighbouring cells to the cells on processor 5, or because they are connected through the well (the cell in the upper, right corner).

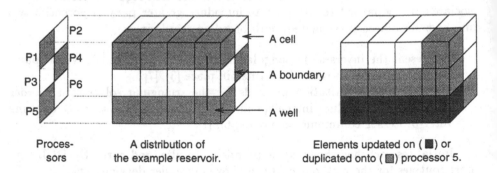

Processors | A distribution of the example reservoir. | Elements updated on (■) or duplicated onto (▨) processor 5.

Fig. 2. The reservoir distribution

As it can be seen, this distribution is also well-suited for the parallel simulation code, as it places blocks of adjacent cells on the same processor, thereby minimising the duplication of reservoir data and hence the communication overhead.

With the strategy described above there will be no computational overhead from the simulation itself, as each grid cell, boundary and well is updated on only one processor. There will be an overhead from the communication, dependent on, how many cells and boundaries are duplicated, and therefore have to be communicated in each iteration. But the chosen distribution should be near to optimal in most cases, indicating the simulation code will scale well.

4.2 A Modularised Approach

The development of enhanced functionality in the sequential simulation code and the development of a parallel version of the same code, has taken place concurrently in two different organisations. Therefore, it has been necessary to design the software for the parallel version as a separate add-on module, leaving most of the original sequential code untouched. Of course, some of the main subroutines, responsible for the flow of the program and for I/O have been changed, in order to handle the parallel setup, but none of the subroutines which update the reservoir variables has been touched.

This has been achieved by letting the simulation code on each processor believe that it only has to operate on the elements which has to be updated.

This is done, mainly by changing the meaning of the variables holding the number of elements on a processor, and by sorting the elements. This can be explained by using the cells as an example: When distributing the reservoir, the cells are sorted, so that the ones updated on a processor, gets to lie first in the arrays which hold information about cells. In the sequential version a variable holds the number of cells. In the parallel version, this variable is redefined to hold the number of elements *updated on a processor*. Thereby all loops referring to this variable will only loop over the elements that are to be updated, but are able to reference the duplicates when needed. This makes it possible to update only the elements that are to be updated, and still leaving most of the code untouched.

With the chosen approach, the main task for the parallel software is distributing the reservoir data, keeping duplicate reservoir elements up to date and gathering results.

4.3 Optimisation

The current version of the parallel simulation code uses a master-slave paradigm, where a master process is responsible for all I/O, and a number of slaves processes perform the computation. This has advantages with regards to reusing code from the sequential simulator, as all I/O routines can be used directly. However there is a performance penalty, in that one processor is left idle during calculations. Therefore the next version will remove this performance bottleneck, by re-implementing some of the I/O routines, making it possible to switch to a hostless programming model.

5 Current state

Currently the first version of the complete parallel simulator, with the parallel solver integrated, is undergoing testing. We therefore only present results from testing the parallel solver, stand-alone, on linear equations dumped from the simulator.

5.1 Scalability Results for the parallel solver

In preliminary tests on a CS-2 with 100 MHz HyperSPARC processors, configured with 2 processors and 128 MB of memory per node, we observed the speedup profile shown in Fig. 3 on a medium sized problem consisting of a $17 \times 16 \times 18$ discretisation with 4 equations per cell block (for a total of 19584 equations with 165717 non-zeros in the matrix).

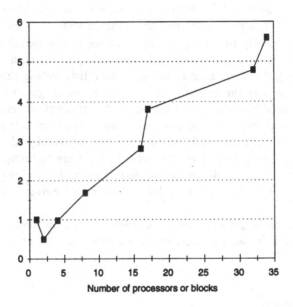

Fig. 3. Preliminary CS-2 Speed-up Curve (Preconditioner changes over processors)

The variations in speed-up are attributed partly to the fact that the number of blocks used in the underlying block-Jacobi / cell-based ILU(0) preconditioner equals the number of processors. Therefore, the 1 processor results use the generally superior ILU(0) preconditioner. In addition, performance varies with how well the number of blocks employed matches the underlying problem grid. Notice that the outer dimension of the problem is 17, which may explain why there is an additional speed-up when using 17 or 34 processors as opposed to 16 or 32.

Future experiments where the number of blocks in the preconditioner is fixed while the number of processors vary, will allow for more comprehensive assessment of the parallel performance. However the above results, obtained using unoptimised parallel code, do look encouraging. The results also illustrate the reasons for considering other parallel preconditioners.

6 Concluding Remarks

This paper has demonstrated the wide range of tools and disciplines used to parallelise the reservoir simulator: From basic parallel libraries, through parallel solvers, and up to domain decomposition of the reservoir. Furthermore it has showed the considerations necessary to perform such a multi-organisation development successfully.

The parallel interactive version of COSI is a successful example of how industry and academia from different countries can work together, producing state of the art software.

References

1. C. Addison. A preliminary investigation into sparse iterative solvers for oil reservoir simulation. *Proceedings of HPCN'94, Munich, Germany,* April 1994.
2. C. Addison. An update into sparse iterative solvers for oil reservoir simulation. *PARSIM Deliverable D3.3-1,* August 1994.
3. F.L. Alvarado and R. Schreiber. Optimal parallel solution of sparse triangular systems. *SIAM Journal of Scientific Computation,* 14 : 446-459, 1993.
4. R. W. Freund, A Transpose-Free Quasi-Minimal Residual Algorithm for non-Hermitian Linear Systems. *SIAM J. Sci. Comput,* 14: No. 2, 470-482, 1993.
5. T. Oliver. Sparse DDL version 2.1 User Guide. *Institute of Advanced Scientific Computation, University of Liverpool,* 1995.
6. R. Schreiber, W.P. Tang. Vectorizing the Conjugate Gradient method. *Unpublished Information, Dept. of Computer Science, Stanford University,* 1982.
7. A. G. Sunderland. Parallel Solution Strategies for Triangular Systems arising from Oil Reservoir Simulations. *Proceedings of HPCN'95, Milan, Italy,* May 1995.

Formal Specification of Multicomputers

José A. Gallud Lázaro

Universidad de Castilla-La Mancha, Escuela Politécnica,
Campus Universitario, 02071 Albacete, Spain
e-mail jgallud@info-ab.uclm.es
TEL 34+67+59 92 00

Abstract. Formal Methods of specification can play an important role in exploring the behaviour of complex systems, as distributed systems or parallel computing. Lotos, based on algebra of processes, has been chosen as an international standard for specifying many systems.

In this paper I show a model for a popular class of massively parallel machines: multicomputers. Such computers are considered the most promising architecture to achieve teraflops computational power. This paper shows a guide to employ the Lotos formal description technique in specifying problems on multicomputer architecture.

The attention is mainly focused in evaluating several features of these machines, in particular those related to the network used to interconnect nodes (processors). In the way of studying characteristics of multicomputers I use two Lotos tools: TOPO and LOLA, both were designed inside the Lotosphere Project. LOLA allows simulation, testing and transformation of Lotos specifications and TOPO is used to obtain a prototype. This work defines a new application of formal methods: computer architecture. The new formal description technique Lotos gives to hardware designers a rigorous method for the verification, validation and development of multicomputers.

1 Introduction

Future computer systems will need to add many innovative services demanded by new applications, increasing requirements, and advancing technologhy. Traditionally, such tasks have been specified in natural language, often with examples to clarify the differences between variants. Recently, computer world has seen a powerful trend toward formal methods, that is, using symbolism with precise mathematical meaning and rules and operation. Many formal models use the idea of regarding a system component as interacting with its environment by simultaneously performing actions.

Computer engineers know that an important problem in software design is the high cost related with developing large software systems. Several aspects affect in every development phase, from informal requirements to executable code. Formal Description Techniques let designers to give a precise relation between initial and final states of any application, without describing implementations details. When correctly applied, formal methods produce systems of the highest integrity and thus are especially recommended for security -and safety- systems.

Formal specification techniques are generally beneficial because a formal language makes specifications more concise and explicit. These techniques help us to acquire greater insights into the system design, they dispel ambiguities, they maintain abstraction levels, and they determine both our approach to the problem as well as its implementation.

Such techniques have been proved useful in developing, for example, a software architecture for a product line of osciloscopes, a model checking for the T9000 transputer[1], the design and verification of the Rollback chip [5], or distributed objects on multicomputers [9] and so on. The chief benefit of formal methods is that they allow certain questions about computer systems to be reduced to study. In this paper they are obtained several features of multicomputers allowing to designers apply formal methods with good results.

Multicomputers are a known class of massively parallel machines that have thousand of processors interconnected through a communication network. All of them use a small local memory, so that these machines belong to the class of memory distributed computers (MIMD). The main porpouse of this work is to establish a metodologhy for specifying different problems to solve on multicomputers. Therefore, we can stablish several steps:

a) choose an adequate formal technique
b) determine the relevant aspects of multicomputers
c) define a model of multicomputer
d) discuss some tests to prove different features

First at all we can see an overview on formal methods in section 2. There, it is chosen a specification language, Lotos, and it is explained why Lotos. The next section shows a case study: a proposal model for a generic multicomputer, with which we can study the main features of these computers. Finally, in section 4, it is showed the most known tools that every hardware designer have to use if he want to became a Lotos specifier. At last, I discuss the conclusions and the future work in section 5.

2 An Overview on Formal Methods

A formal specification is a specification expressed in a language whose vocabulary, syntax, and semantics are formally defined, and which has a mathematical basis.

The growing importance of formal specification methods is reflected in the each time more publications dealing with these methods. Furthermore, a number of authors have advocated the use of formal methods, not only in the software process, and also in the hardware design. Formal methods used in developing computer systems are mathematically based techniques for describing system properties. Such formal methods provide frameworks within which people can specify, develop, and verify systems in a systematic manner.

A method is formal if it has sound mathematical basis, typically given by a formal specification language. This basis provides the means of precisely defining notions like consistency and completeness and, more relevantly, specification, implementation, and correctness. It provides the means of proving that a specification is realizable, proving that a system has been implemented correctly, and proving properties of a system without necessarily running it to determine its behaviour. System designers use formal methods to specify a system's desired behavioral and structural properties.

One tangible product of applying a formal method is a formal specification. A specification serves as a contract, a valuable piece of documentation, and a means of communication among a client, a specifier, and a implementer. Because of their mathematical basis, formal specifications are more precise and usually more concise than informal ones.

A formal method is a methodologhy and not a computer program or language, it may or may not have tool support. This is an important aspect when you are going to decide what formal technique you might to use. However, it is each time more frequent to find associated to a formal language a set of tools. Thus you can find different formal methods: VDM, Z, Larch, Finite State Machines, Petri Nets, Temporal Logic, Transition Axioms, CSP, CCS, they are the most important ones.

Before explaining the choosen method, we can get a brief summary on formal methods: For a general introduction to formal methods see [10], and for more on the benefits of formal specification see [8]. On the distinction between a method and a language, and what specifying a computer means, you can find it in [7]. Two necessary papers about formal specification, will help you to get an overview on the classical myths of formal methods in [6,13]. And the software development point of view of this techniques can be found in [4]. In the last one it can be found an interesting sketch of formal specification.

Lotos is a formal specification language based on process algebras and equational specification of abstract data types. This language have been designed under the ISO working group on Formal Description Techniques. It can be said that Lotos have got the best ideas of CSP, CCS and ACT One. Lotos models are networks of processes that are activated concurrently and communicate through shared gates (channels). The activity of a process is modeled by a behaviour expression that specifies the temporal relationship between every synchronization on the gates. There are three synchronization cases: value matching, value passing and value negotiation. A readable tutorial is available [2].

This formal description technique have acquired a wide and good acceptation in last years. At the begining of Lotos, the main objective was to allow the writing of specifications of OSI standards that were unambiguous, precise, complete, and implementation independent. Nowadays, we can note the general need to apply formal methods not only in software development, or communication protocols as well as in the hardware design.

Four main criteria have determined the definition of Lotos: expressive power, formal definition, abstraction and structure. We can consider to multicomputers to be a complex system, at least we can think that producing correct, reliable software to this kind of computers is not trivial. The application of formal methods to hardware design is still an unexplored field. In the next section it is showed how modelling a basic multicomputer by means of Lotos technique.

3 Case study: Formal Description of the Router

Hypercubes, torus, and other structured nonshared memory MIMD machines, henceforth termed multicomputers, are winning wide acceptance, in part due to a natural match of their physical interconnection structures to the logical communication structures of many scientific applications. However, the most known multicomputers are hard to program for three reasons:

- Any parallel program is difficult to design, implement and debug.
- Multicomputers are harder to program than shared memory multiprocessors.
- Speedup depends on the efficient use of the distributed processors and memory.

More about these problems can be found in [9], and you can find a solution based on distributed objects.

When you want to model a system you must get information about it, so you have to study all the possible behaviours of the system. Therefore, it is necessary to obtain the most important features of multicomputers, as the topologhy of the communication network, the router algorithm,the number and the kind of nodes, the flow control algorithm and the switching techniques. Our Lotos model there will be able to describe the observable behaviour of the system. Then through a stepwise refinement manner, from highest level to down, we can define the complete system. This model will be focused on those features we can study.

Multicomputer rely on an interconnection network between nodes or processors to support the message-passing mechanism. The network plays a major role in determining the overall performance of a multicomputer. So that this work is focused in the communication network. The way the nodes are connected to one another varies among machines. In a directed network architecture, each node has a point-to-point, or direct, connection to some number of other nodes, called neighboring nodes. A survey of wormhole routing techniques in direct networks can be found in [11].

A general Lotos description of a multicomputer could be the next:

```
specification multicomputer :NOEXIT
behaviour
 processors[e1,..,en] || network[e1,..,en]
endspec
```

This first approach does not show a lot of information. However this specification has got high level of abstraction, without repairing in details of implementation. It can be saw as a formal definition of a multicomputer: several nodes or processes which use theirs synchronization gates to implement the message-passing mechanism. A brief description of a multicomputer with dynamic network can be found in [12].

In this description we could to obtain more details about the system by definning the two black boxes called processors and network. We are interested in the network point of view of the multicomputer because the time required to move data between nodes is critical to system performance, as it effectively determines what granurality levels of parallelism are possible in executing an application program.

Thus, each node in a multicomputer has a known architecture: router, processor, local memory and a functional unit. Each router supports some number of input and ouput channels. Normally, every input channel is paired with a corresponding output channel. Internal channels connect the local processor/memory to the router. Usually it can be found only two internal channels while it can be avoided a communication bottleneck. External channels are used for communication between routers, that is, between nodes.

In a first step it will must be modelled the system from the point of view of a node. After, we will compose in parallel a number of such nodes to form the multicomputer. This methodologhy reduces complexity of considering the global system. Initially, we only have to describe a quite small behaviour: a node can receive messages directed to himself or to other. Otherwise, a node can send a message.

Consider a torus with nine nodes and unidirectional channels. Each node can receive messages from two external channels (c_1 and c_2) and from the internal channel (c_i). Likewise, each node can send a message by two external channels (c_3 and c_4) and, potentially, one internal channel (c_i). This describes a behaviour that can be expressed using Lotos in this way:

```
specification router[ci,c1,c2,c3,c4] :EXIT

type Boolean is
 sorts bool
 opns
  true, false: ->bool
  not:bool->bool
 eqns
  forall x,y:bool
  ofsort bool
  not(not(x))=x;
  not(true)=false;
  not(false)=true;
endtype

type Nat14 is Boolean
sorts
 nat14
```

```
opns
 n1,n2,n3,n4: ->nat14
 succ,pred:nat14->nat14
 _plus_:nat14,nat14->nat14
 _eq_:nat14,nat14->bool
eqns
 ofsort nat14
 forall m,n:nat14
 n2=succ(n1);
 n3=succ(succ(n1));
 n4=succ(succ(succ(n1)));
 pred(m)=succ(succ(succ(m)));
 m plus succ(n)=succ(m plus n);
 m plus n1=succ(m);
 succ(succ(succ(succ(n1))))=n1;
 succ(succ(succ(succ(succ(m)))))=succ(m);
 ofsort bool
 forall m,n:nat14
 m eq n=false;
 m eq m=true;
endtype


type Mess is Nat14,Boolean
sorts
 mess
opns
 flit,flit1,flit2,flit3,flit4,end : ->mess
 cab       : mess->nat14
 _eq_ :mess,mess ->bool
 eqns
  ofsort bool
  forall x,y:mess
  x eq flit = true => x eq end = false;
  flit eq end =false;
  flit eq flit =true;
  flit1 eq flit1=true;
  end eq end=true;
 ofsort nat14
 forall z:nat14, x:mess
  x eq flit1=true =>cab(x)=n1;
  x eq flit2=true =>cab(x)=n2;
  x eq flit3=true =>cab(x)=n3;
  x eq flit4=true =>cab(x)=n4;
endtype
```

```
type Can is
sorts
 can
opns
 libre,ocup: -> can
endtype

behaviour
  enc[ci,c1,c2,c3,c4]
where

 process enc[ci,c1,c2,c3,c4] :EXIT :=
  ci?a:mess;(comp[ci,ci,c3,c4](cab(a)) ||| enc2[c1,c2,ci,c3,c4])
  []
  c1?b:mess;(comp[c1,ci,c3,c4](cab(b)) ||| enc2[ci,c2,ci,c3,c4])
  []
  c2?c:mess;(comp[c2,ci,c3,c4](cab(c)) ||| enc2[ci,c1,ci,c3,c4])
 endproc

 process enc2[ca1,ca2,ci,c3,c4] :EXIT :=
 ca1?a:mess;(comp[ca1,ci,c3,c4](cab(a)) ||| enc3[ca2,ci,c3,c4])
 []
 ca2?b:mess;(comp[ca2,ci,c3,c4](cab(b)) ||| enc3[ca1,ci,c3,c4])
 endproc

 process enc3[cb,ci,c3,c4] :EXIT :=
 cb?a:mess;comp[cb,ci,c3,c4](cab(a))
 endproc

 process comp[com1,ci,c3,c4](x:nat14):EXIT :=
 ([x eq n1]->ci!libre;ci!ocup;cross[com1,ci]
 [][not(x eq n1)]->(c3!libre;c3!ocup;cross[com1,c3]
                []
                c4!libre;c4!ocup;cross[com1,c4]
                ) )
 |[ci,c3,c4]|
  channel[ci,c3,c4]
 endproc

 process channel[ch1,ch2,ch3]:EXIT :=
  ch1!libre;ch1!ocup;chan[ch1]
  |||
  ch2!libre;ch2!ocup;chan[ch2]
  |||
```

```
  ch3!libre;ch3!ocup;chan[ch3]
endproc

process chan[canal]:EXIT :=
  canal!end;exit
  [] canal!flit;chan[canal]
endproc

process cross[x1,x2]:EXIT :=
x1?d:mess;([d eq flit]->x2!d;cross[x1,x2]
           [][d eq end]->x2!d;(hide ci,c3,c4 in enc3[x1,ci,c3,c4]))
  endproc

endspec
```

In this specification we have defined two basic processes: enc (enc, enc2 and enc3) and cross. The named process enc represents the router and the named process cross defines the crossing task. A router is moved between this two basic states: routing and cross a message. If a router is crossing from input channel A to output channel D, it is able to routing a message from input channel B to internal channel E. It can be noted the next notation:

```
External input channel: A and B (c1 y c2)
External output channel: C and D (c3 y c4)
Internal input and output channel: E (ci)
```

The two basic aspects of Lotos are clearly noted in the specification. The abstract data type give us the necessary types we are going to use. In this case we can see four sorts user defined: Bool, Nat14, Mess and Can. On the other hand we can observer the algebra of processes part. This part defines the behaviour of the system by means of processes and operators. The algebra of processes takes ideas of CSP and CCS [15,16]. Our specification shows the behaviour we wanted to the router. After the rigurous analysis of the router we can say that when a message arrives to the node, we need to know the destination. Then, process enc calls to process comp (computations) to obtain the output channel.

When we have to determine the output channel for input message, router calls to process cross (crossing). This process copy all flits from input channel to the output channel. It is possible that a new message arrives to the node. If this occurs, the process enc has to route the message in the same way. As this router has five (actually six) channel, we can have three communications as much.

The formal description gives me the possibility of validating my specification. Thus, we can employ a lot of Lotos tools like TOPO and LOLA. The explanation of this it can be found in the next section. The results obtained after applying such tools, will be used to define more precisely the specification.

4 Tools for a Suitable Framework

Lotos is a wide spectrum language which can provide formal support to a large part of the life cycle in the appropiate application domains. Its allows the formalization of software requirements, architectural designs, detailed designs and test cases. The formalization of a multicomputer router is an example of hardware application of formal methods.

The Lotos language and its associated mathematical theory can be used to formally support a stepwise refinement development process, as it has been showed in the last section.

The first design step has produced the most abstract description of the system. We have seen the specification of a multicomputer as a global system. Each new refinement is a more complete model of the multicomputer being designed, which is said at a new level of abstraction. The Lotos operator hide provides information hidding.

We have skipped some steps to define the behaviour of the router in a bottom-top way. This specification needs to be verified formally by employing the Lotos tools. There are a lot of tools: TOPO, LOLA, LITE, ARA and so on. In this paper it is showed only the most known TOPO and LOLA. LITE is an integrate environment developed inside the Lotosphere project and directed to industrial sights. A reduced version of the power tool can be found in different ftp sites. See the Lite User Manual for more information [14].

TOPO is a collection of tools to deal with LOTOS specifications. A full LOTOS implementation approach of a multicomputer can be based in the construction of an implementation independent abstract machine. This allows add to the specifications the design decisions, which can not be expressed in LOTOS, by means of annotating it.

We can begin the validation task by simulating the behaviour of the system. The Lotos laboratory Lola allow us to simulate step by step the dynamic behaviour. Lola goes through the transition tree and the user decides each time the action to be executed. It can be found an extended finite state machine associated to a Lotos specification. Lola has four expansion operations to calculate the EFSM from a specification. TOPO can obtain an executable prototype from an annotated specification.

An important step in validating a specification is to get a test of correctness. Lola can calculate the response of a specification to a test. By testing is possible to build systems which will accomplish all properties we want for it. The expansion and testing operations allow us to specify deadlock free, rigorous and formal systems.

5 Conclusions and Future Work

This paper describes how to apply formal specification in hardware design. In particular, the formal design of multicomputers has been suggested. Such architectures are the most promising ones to achieve teraflops computational power.

The communication network of the multicomputer plays an important role in determining the overall performance of the system. Formal description techniques give hardware designers the opportunity of solving easier and sooner the clasical problems of multicomputers. We have seen the power of formal methods in building rigorous systems.

A formal definition of the router has been presented in section 3. It is possible to note the abstraction and implementation independence of the system. In this example we have seen the two parts of any Lotos specification: Abstract data types and algebra of processes. The specification of the router shows the node point of view. We could have defined the overall point of view: thousand of nodes interconnected by means of a network.

Finally, two tools for specifying into Lotos have been presented. Lola is the Lotos laboratory and has three basic set of operations: simulating, testing and expansion. Topo allow us to get an executable prototype.

This work can be considered an introduction to the formal description of multicomputers. The possibility of obtaining deadlock free and rigorous subsystems in multicomputer achitecture is more realistic. Formal definitions of complex systems are a good basis to develop it more easily.

Future work will deal with the specification of the all multicomputer subsystems. It will be necessary to get correctness test of the systems. A first step will obtain a test for the specification showed in this paper. Formal verification can be done in different ways: Petri Nets, EFSM, Temporal Logic, and so on.

References

1. Barret, G.: Model Checking in Practice: the T9000 Virtual Channel Processor, IEEE Transactions on Software Engineering, Vol. 21, No. 2, pp. 69-78, February (1995)
2. Bolognesi, T., Brinsksma, E.: Introduction to the ISO Specification Language LOTOS, Computer Networks and ISDN Systems, Vol. 14, pp. 25-29,(1987)
3. Bowen, J.P., Hinchey, M.G.: Ten Commandments of Formal Methods, IEEE Computer,pp. 56-63, April, (1995)
4. Fraser, M.D., Kumar, K., Vaishnavi, V.K.: Strategies for Incorporating Formal Specifications in Software Development, Communications of the ACM, Vol. 37, No. 10, October (1994)
5. Gopalakrishnan, G., Fujimoto, R.: Design and Verification of the Rollback Chip Using HOP: A Case Study of Formal Methods Applied to Hardware Design,ACM Transactions on Computer Systems, Vol. 11, No. 2, pp. 109-145, May, (1993)
6. Hall, A.: Seven Myths of Formal Methods, IEEE Software, pp. 11-19, September,(1990)
7. Lamport, L.: A Simple Approach to Specifying Concurrent Systems, Communications of the ACM,Vol. 32, No.1, pp. 32-45, January, (1989)
8. Meyer, B.: On Formalism in Specification, IEEE Software, pp. 6-26, January,(1985)
9. Shwuan, K., Bo W.: "Topologies"-Distributed Objects on Multicomputers, ACM Transactions on Computer Systems, Vol. 8, No. 2, pp. 111-157,May (1990)
10. Wing, J.M.: A Specifier's Introduction to Formal Methods, IEEE Computer,pp. 8-24, September, (1990)

11. Ni, L.M., McKinley, P.K.: A Survey of Wormhole Routing Techniques in Direct Networks, IEEE Computer, pp. 62-76, February (1993)
12. Gallud, J.A., Garcia, J.M.: The Specification of a Generic Multicomputer Using Lotos, ACM Sigplan Notices, Vol.30, No. 2, pp. 21-24, February, (1995)
13. Bowen, J.P., Hinchey, M.G.: Seven More Myths of Formal Methods: Dispelling Industrial Prejudice, Proceedings FME 94, Second Formal Methods Europe Symposium, Lecture Notes in Computer Science, Vol. 873, pp. 105-117, (1994)
14. Caneve, M., Salvatori, E.: Lite User Manual, Lotosphere consortium, (1992)
15. Hoare, C.A.R.: Communicating Sequential Processes, Prentice-Hall, (1985)
16. Milner, R.: A Calculus of Communicating Systems, Lecture Notes in Computer Science, Vol.92, Springer-Verlag, (1980)

Multi-Million Particle Molecular Dynamics on MPPs

Peter S. Lomdahl[1] and David M. Beazley[2]

[1] Theoretical Division
Los Alamos National Laboratory
Los Alamos, NM 87545
[2] Department of Computer Science
University of Utah
Salt Lake City, UT 84112

Abstract. We discuss the computational difficulties associated with performing large-scale molecular dynamics simulations involving more than 100 million atoms on modern massively parallel supercomputers. We discuss various performance and memory optimization strategies along with the method we have used to write a highly portable parallel application. Finally, we discuss some recent work addressing the problems associated with analyzing and visualizing the data generated from multi-million particle MD simulations.

1 Introduction

Recently considerable interest has been devoted to application of Massively Parallel Processor (MPP) computer systems to significant unsolved problems in science and engineering. MPPs generally provides large amounts of cost effective memory that is rarely available on any other type of supercomputer systems. Large memory systems with tens of GBytes allow for simulations of a size that was not practical in the past. For example in materials science, the method of Molecular Dynamics (MD) [1, 2] running on MPPs is currently being used to study fracture and crack propagation on length-scales that was not possible just a few years ago. In order to make reasonable comparisons with experimental data it is often necessary to be able to simulate features on a micron scale. Realistic 3D MD simulations of this size requires at least hundreds of millions of atoms, preferably more.

In this paper, we describe recent advances in the development of our MD code for MPPs, SPaSM. Our basic algorithm has been described in details elsewhere [3, 4] as well as performance and scaling results [5]. Here we present our experience over the past year with performance optimization and code portability on a variety of MPP systems. While specific to our MD code, we believe that the lessons learned can be applied to any major scientific code on MPPs. The outline of the papers is as follows: in section 2 we briefly outline our cell based MD algorithm, in section 3 we present our recent optimization and performance strategy, in section 4 we outline our efficient memory management system, and

Figure. 1. Fracture experiment with 104 million atoms

in section 5 we deal with code portability. Finally in section 6 we comment on future directions for our MD work, including the daunting task of analyzing the data from 100 million atom MD simulations.

2 The Cell Algorithm

The general MD algorithm used in SPaSM has been presented in [3], we outline a brief overview and the important features here. The basic steps in an MD timestep involves the following three elements:

- Calculations of all forces on each atom.
- Numerical integration of the equations of $F = ma$.
- Updating the data structures.

In addition, we will periodically perform I/O to dump particle data and calculate important quantities like energy, etc. By far, the most time consuming step is number 2 above, it accounts for more than 90% of the total CPU time spent. We deal exclusively with this step below.

The data-layout used in SPaSM is based on an approach where space is decomposed into domains that are assigned to each processing node of the MPP. Each node then further subdivides its domain into a large collection of small cells.

The size of the cells is chosen equal to the maximum interaction distance for the interatomic potential, R_{max}. Atoms are assigned to a particular processing node and cell according the atom's coordinates. This particular approach allows forces to be readily calculated since it places atoms that are physically close into the same cell or the neighboring cell. To calculate the total force on a given atom we simply look at all of the atoms in the same cell and those in the neighboring cell. When neighboring cells are on different processing nodes, we use message passing to exchange the data needed to complete the force calculation.

3 Performance

Recently, there has been considerable interest in performing molecular dynamics simulations involving more than 100 million atoms. In a fracture simulation involving 104 million atoms, nearly 300 billion floating point operations were required for a single timestep. The entire simulation ran for 9000 timesteps and required 180 hours of compute time on a 512 processor CM-5. The need for efficient programming and high performance becomes clear. However, performance tuning is often a tricky and obscure game with many pitfalls. Having spent considerable efforts optimizing our MD code, we have found that gaining a basic understanding of computer architecture has been the most powerful performance tool available. In this section we discuss several aspects of optimizing tuning our MD code.

3.1 The CM-5 vector units (and what's wrong with them)

SPaSM was originally developed as a message-passing ANSI C code on the Connection Machine-5. Each CM-5 processing node consists of a 33 MHz SPARC processor with 5.1 Mflops peak performance and 4 vector units with a combined performance of 128 Mflops. Given that our C-code ran exclusively on the SPARC, it was clear that getting high performance would depend on utilizing the vector units. To use the vector units, we rewrote our force calculation in CDPEAC, a set of C macros for programming in the VU assembler language DPEAC [5]. In addition, we developed a special memory caching scheme, also written in CDPEAC, to reduce the number of memory accesses [1]. As a result of these modifications, we were able to achieve calculation rates between 25 and 50 Mflops/node[5].

Unfortunately, we have experienced a wide range of problems associated with using the vector units. Although we had Lennard-Jones and a tabulated pair-potential code written in CDPEAC, it has proven to be difficult to implement other types of potentials such as embedded-atom or Stillenger-Weber potentials in CDPEAC. When one begins to consider multiple atom types and complicated interactions, a CDPEAC implementation quickly becomes a hideous mess. This also raises an important issue—is time better spent running simulations (although at a somewhat slower rate) or spending several months writing and debugging CDPEAC?

This is not the only problem. Effective use of the VUs required a substantial amount of memory overhead (typically around 100%) which was simply intolerable when running large production runs. In addition, the SIMD nature of the VUs made it difficult to really implement the force calculation in the most efficient manner. In fact, in our fastest implementation nearly 50% of the floating point operations were unnecessary. Finally, while modular, it has been difficult to integrate the CDPEAC code into newer versions of the software in a reliable manner. From a software engineering standpoint, this creates many headaches.

3.2 Back to the SPARC

Is programming in CDPEAC the only way to get high performance from ANSI C on the CM-5? Surprisingly, the answer is no. We decided to take the force calculation from our original C code and analyze it to see how effectively we really used the SPARC processor in the first place [6]. Running a small test problem involving 256000 atoms interacting via a Lennard-Jones potential on a 32 processor CM-5, we used a combination of simulators, hand-analysis, and timings to measure how the processor spent its time [6]. This is shown in Table 1.

Table 1. SPARC Processor Utilitization

	Number of cycles	Time (sec)	%
Integer Instructions	28539072	0.865	9.4%
Memory Operations	156193792	4.733	51.2%
Floating Point	118053440	3.577	38.7%
Cache Miss	684642	0.021	0.1%
Total	303483682	9.196	99.4%
Observed CPU Time	305000000	9.25	100.0%

We see that the force calculation is spending more than 50% of its time accessing memory. It should be noted that the SPARC floating point unit can overlap instructions with integer and memory operations. The floating point time listed above takes this into account and represents the time that the SPARC is only processing floating point (no overlap). While 39% of the clock cycles are spent handling floating point, it should be noted that during 75% of these cycles (30% of the total time), the processor is stalled due to a data-dependency or is waiting for the FPU to finish a previous calculation. This table indicates several serious performance problems with our C-code–in particular the large number of memory operations.

3.3 Optimization Strategies

While it is not our goal to go into a detailed discussion of computer architecture, we would like to discuss the modifications made to our original MD code. Consider the code fragments Code 1 (Fig. 2) and Code 2 (Fig. 3) which calculate

```
pt1 = p1->ptr;
for (i = 0; i < p1->n; i++, pt1++) {
    pt2 = p2->ptr;
    for (j = 0; j < p2->n; j++, pt2++) {
        force(&pt1->r, &pt2->r, &f);
        pt1->f.x += f.x;
        pt1->f.y += f.y;
        pt1->f.z += f.z;
        pt2->f.x -= f.x;
        pt2->f.y -= f.y;
        pt2->f.z -= f.z;
    }
}
void force(Vector *r1, Vector *r2, Vector *f) {
    double dx,dy,dz,r2,fx,fy,fz;
    dx = r2->x - r1->x;
    dy = r2->y - r1->y;
    dz = r2->z - r1->z;
    r2 = dx*dx + dy*dy + dz*dz;
    if (r2 < Rmax2) {
        ...
        Calculate force
        ...
        f->x = fx;
        f->y = fy;
        f->z = fz;
    } else
        f->x = f->y = f->z = 0.0;
    return;
}
```

Figure. 2. Code 1. Original force calculation

forces between atoms in two different cells. Code 1 represents our original MD code while Code 2 is an optimized version. In Code 1, a user defined **force()** function is used whenever two particles interact. Typically, this function is located in a different module and is modified as necessary for different problems.

The nature of forces While modular, Code 1 presents several key performance problems. The first of these problems relates to the separation of the force calculation and the inner loop. Clearly, inlining will help performance, but there is a more subtle problem than this. In the inner loop, we simply call **force()** and add its return value to the total force on each particle. However, consider Figure 4. Since all forces are computed with a cutoff of r_{max} a large number of forces are simply zero. In 3D, each atom interacts with particles in a volume of $27r_{max}^3$, but only the atoms in a sphere of volume $4\pi r_{max}^3/3$ contribute to the

```
register double rx,ry,rz,fx,fy,fz;
register double dx,dy,dz,r2;
register int n1, n2;
n1 = p1->n;
n2 = p2->n;
pt1 = p1->ptr;
for (i = 0; i < n1; i++, pt1++) {
    rx = pt1->r.x;
    ry = pt1->r.y;
    rz = pt1->r.z;
    fx = fy = fz = 0;
    pt2 = p2->ptr;
    for (j = 0; j < n2; j++, pt2++) {
        dx = pt2->r.x - rx;
        dy = pt2->r.y - ry;
        dz = pt2->r.z - rz;
        r2 = dx*dx + dy*dy + dz*dz;
        if (r2 < Rmax2) {
        ...
        Calculate force
        ...
        fx += f.x;
        pt2->f.x -= f.x;
        fy += f.y;
        pt2->f.y -= f.y;
        fz += f.z;
        pt2->f.z -= f.z;
        }
    }
    pt1->f.x += fx;
    pt1->f.y += fy;
    pt1->f.z += fz;
}
```

Figure. 3. Code 2. Modified force calculation

total force. At uniform densities, we find that nearly 85% of the total forces calculated are zero and that for most of the time, Code 1 is adding or subtracting zeros from the total force. A simple fix for this is to add a return code to the force() function indicating a nonzero result. If we only accumulate forces when nonzero, we achieve a 58% speedup. While this is an obvious change in hindsight, it should noted that we didn't notice a problem until we performed a detailed code analysis and found that we had an unusually high number of floating point operations that were adding and subtracting zeros. It is often difficult to notice such pitfalls when related pieces of code are located in different modules.

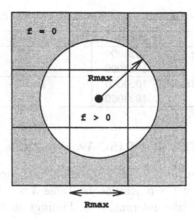

Figure. 4. Zero and nonzero force calculation.

Compilers For all of the advancements in compiler technology, the truth is that most compilers do a terrible job optimizing Code 1. On the CM-5, the force() function was not inlined into the inner loop. By inlining the function ourselves as in Code 2, we were able to achieve an additional 25% speedup. Another more serious problem is the extensive use of pointers. To prevent pointer aliasing problems, C compilers will always perform a memory access whenever anything is referenced by a pointer. Thus in Code 1, references such as p1->n in a **for** statement will always result in a memory access even though this quantity could easily be stored in a register for the duration of the calculation. In Code 2, we show how we have eliminated most pointer aliasing problems by introducing additional local register variables and using them in a carefully constructed manner. Taking this approach in combination with inlining results in a 6-fold reduction in memory operations.

3.4 Speedups on CM-5

Code 2 represents only a few minor changes to our original C-code, yet we achieve significant performance improvements as shown in Table 2. On the CM-5, the code runs 120% faster than before and is only 2.2 times slower than the fully optimized vector unit version. This is remarkable considering the peak performance of the VUs is 25 times greater than the SPARC. On the CM-5E, we see the immediate advantage of using a superscalar processor as the unmodified code is 3.75 times faster than before even though the clock rate is only 20% faster. Running the optimized code, we achieve an 88% speedup and code that runs faster on the SuperSPARC than the code that uses the vector units (which are 160 Mflops peak on the CM-5E). Clearly, this raises some serious questions about using CDPEAC on the CM-5. By making carefully selected modification to our C code, we have been able to get performance comparable to our VU code without any of the headaches associated with programming in assembler. For some simulations (especially with short cutoffs), we have found that the optimized C code runs better than the VUs even on the older CM-5.

Table 2. Performance on CM-5 and CM-5E (32 processors)

System	Particles	Unmodified	Optimized	Speedup
CM-5 (33 MHz SPARC)	1024000	42.63	19.54	119%
CM-5 (Vector Units)	1024000	-	8.87	- %
CM-5E (40 MHz SuperSPARC)	1024000	11.37	6.05	88%
CM-5E (Vector Units)	1024000	-	6.11	-

3.5 Speedups on T3D and RISC Workstations

An advantage of optimizing C code is that optimizations applied to the CM-5 can also be applied to other machines. In Table 3 we show the effects of using Code 2 on a variety of different machines. Timings on the CM-5 and T3D were done on 32 processor systems. Timings for the VPP500 were measured on a 16 processor system [1]. The "unrolling" column in the table represents the best timing we have achieved by manually unrolling the inner loop and reordering the force calculation. On all machines, we see substantial speedups. On newer superscalar machines, loop unrolling can be a big win as seen by the significant speedups on virtually all machines except on machines with 32-bit processors such as the CM-5 and Onyx. On the VPP500 which is a parallel-vector machine, we see that the same modifications result in enormous speedups of more than 300% in the fully optimized case.

Table 3. Time for simulating 32000 atoms/processor (speedup over Code 1 in paranthesis)

Machine	Compiler flags	Code 1	Code 2	Code 2 +Unrolling
CM-5 (33 MHz SPARC)	-O2	42.63	19.54 (119%)	23.53 (81%)
Cray T3D (150 MHz Alpha)	-O2	8.57	3.70 (132%)	3.18 (169%)
Fujitsu VPP500 [1]	-O -K2	21.4	6.15 (247%)	4.90 (337%)
HP-735 (99 MHz PA-7100a)	-O -Aa	4.68	1.89 (148%)	1.62 (189%)
IBM 590 (66 MHz Rios 2)	-O	4.98	2.58 (92%)	1.95 (155%)
SGI Onyx (150 MHz R4400)	-O2 -mips2	6.40	3.21 (99%)	3.78 (69%)
SGI Onyx (75 MHz R8000)	-64 -O2	7.73	3.47 (123%)	2.01 (285%)

3.6 A Pitfall

Trying to fully optimize code can lead to many mysterious results. As an example, the modification to the force calculation in Code 3 (Fig. 5) has been found in the MD literature as a method for reducing the number of floating point

[1] Timings reported for the VPP500 are for a 1 million atom simulation on 16 processors (64000 particles/processor). Code was compiled with the fvccpx compiler.

399

calculations required for cell-based MD algorithms [8]. When run on the T3D, this code runs 43% slower than the unrolled version of Code 2 even though it performs 13% fewer floating point operations and 32% fewer memory operations. In this case the performance degradation is due to additional control (branch) instructions created by the extra `if` statements. In fact, it we plot the number of control instructions versus execution time for various versions of optimized code, we get the graph in Figure 6 where the execution time is proportional to the number of control instructions. Clearly there are many performance problems to consider when optimizing for modern RISC microprocessors.

```
dx = x2 - x1;
r2 = dx * dx;
if (r2 < cutoff2) {
    dy = y2 - y1;
    r2 += dy*dy;
    if (r2 < cutoff2) {
        dz = z2 - z1;
        r2 += dz*dz;
        if (r2 < cutoff2) {
            /* Calculate forces */
        }
    }
}
```

Figure. 5. Code 3. Reduced FLOP count.

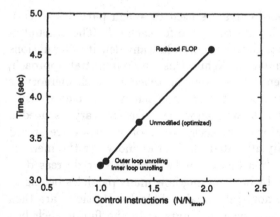

Figure. 6. Performance vs. Number of Control Instructions on T3D

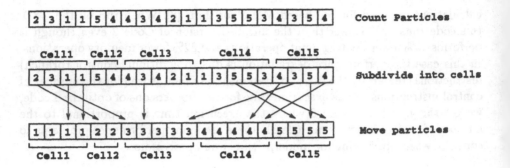

Figure. 7. Particle memory management on each processor.

4 Efficient Memory Management

Making efficient use of memory is critical for performing large-scale MD simulations. Each particle required 80 bytes (position, velocity, forces, and type information) of memory in double precision. Performing a 100 million particle MD simulation then requires a minimum of 8 Gbytes of RAM for particle data. Additional data structures to maintain the cell structure require additional memory. Finally, there must be space for I/O buffers, message-passing buffers and the executable itself. On a 512 processor CM-5 with only 16 Gbytes of memory, running a 100 million atom simulation easily pushes limits of available memory.

4.1 Particle management

As mentioned earlier, particle coordinates are used to assign particles to each cell. A simple approach is to preallocate the space for each cell. The amount of space allocated should be large enough to accommodate any density fluctuations that might occur during the simulation. While this was our initial approach, we found it to be grossly inefficient–often requiring twice as much memory as necessary to store all of the particles. A better approach is to make extensive use of dynamic memory management and allow cell sizes to vary as needed to accommodate density fluctuations. On each processor particles are stored sequentially in a large dynamically allocated block of memory. As the memory requirements on each node vary, this block can be increased or decreased in size as necessary. Cells are formed by determining how many particles belong to each cell and subdividing the memory into small subblocks. Particles are then moved into the proper subblock as shown in Figure 7. In the figure, each box represents a particle and the number in the box represents the the cell where that particle belongs. While the method may appear to move many particles around, in normal operation, not that many particles move between cells after each timestep. To improve performance we do not move particles to a new location if they are already in the correct cell.

Particle Redistribution Since particles move between processors, we perform a redistribution step after each timestep. Each processor checks all of the particle coordinates and determines which particles must be sent to other processors. Then, each processor performs a 6-way exchange of data similar to that described by Plimpton in [7]. This scheme only works when moving particles between neighboring processors, but with a relatively simple modification, we have changed the algorithm to perform particle routing. Under this scheme, we still exchange particle data with 6 neighbors, but we allow particles to move to *any* node in the entire system. Under this scheme, particles can be thought of as messages moving through a communications network with each node acting as a router. This requires very little overhead in normal use since particles only move to nearest neighbors. However, we have found this to be particularly useful for performing other operations. For example, it is now possible to change the processor geometry to get better load-balance during the middle of a simulation. We've also been able to perform fast parallel I/O independently of partition size by simply reading particles and allowing them to travel to the right processor before calculating forces. The steps performed in the global redistribution process are summarized below.

1. Check all particle coordinates. Put particles moving to another processor in a special buffer.
2. Move particles between processors and perform routing. Each processor checks received particles to see if they belong to its region. If so, the particle is placed into any available memory location. If not, the particle is put back in the message buffer and routed to the destination processor. This step is performed whenever the message buffers fill up in Step 1.
3. Figure out how many particles belong in each cell.
4. Form cells by subdividing memory.
5. Place particles as shown in Figure 7.

4.2 Message Passing and I/O

When calculating forces, cells along the boundaries of each processor must be sent to neighboring processors. Unlike some schemes which send entire faces of the computational subdomain to neighboring processors, we only send one cell at a time as necessary. Since each cell typically contains no more than a few dozen atoms, we usually only require a single buffer of no more 1024-4096 bytes per processor for all communication operations. Finally, we create a scratch buffer on each node for performing various operations such as writing output files, creating graphic images, and temporary calculations. This buffer is usually only a few hundred kilobytes in size. As a result, we are able to perform large-scale MD simulations with only a minimal amount of memory overhead. In test simulations, we have been able to run as many as 300 million atoms in double precision and 600 million atoms in single precision on a 1024 processor CM-5 with only 32 Gbytes of memory.

5 Code Portability

While developed on the CM-5, SPaSM has been ported to a variety of other machines by writing a custom message passing and I/O wrapper library. Rather than porting the code to any particular message passing system, this allows us to use a wide range of machines and systems.

5.1 Message Passing

Our message passing library is based on the functions we used in the CMMD message passing library on the CM-5. The entire library consists of the 11 function listed in Table 4. Having only 11 message passing functions makes porting the code very simple since most of these functions can be easily implemented in almost any message passing library. On the T3D, we have implemented the library in both PVM and the Cray shared memory library. On the Fujitsu VPP500, the library is implemented in P4. A message-passing emulator is also available for running and debugging the code on high performance single processor workstations. We also have a variation of this that allows us to run on multiple processors of a Sun SPARCcenter 2000 using Solaris multithreading.

Table 4. SPaSM message passing library

SPaSM_send_block()	SPaSM_receive_block()	SPaSM_send_and_receive()
SPaSM_reduce_int()	SPaSM_reduce_double()	SPaSM_sync_with_nodes()
SPaSM_set_global_or()	SPaSM_get_global_or()	SPaSM_bytes_received()
SPaSM_self_address()	SPaSM_partition_size()	

5.2 Parallel I/O library

Unfortunately, the existence of parallel I/O libraries is not as standardized as message passing. For this reason, we have implemented our own version of CMMD parallel I/O for running on other machines. The parallel I/O library tries to emulate CMMD as much as possible and contains the function in Table 5. Support for the three CMMD I/O modes, CMMD_independent, CMMD_sync_bc, and CMMD_sync_seq is provided. By writing wrappers, we are able to provide consistent file formats and I/O operations between machines without changing the main source code. Since the I/O library is built on top of our message passing wrapper library, we have often been able to provide I/O support on new machines with very little effort.

Table 5. SPaSM Parallel I/O functions.

SPaSM_open()	SPaSM_fopen()	SPaSM_set_io_mode()
SPaSM_fset_io_mode()	SPaSM_set_open_mode()	SPaSM_read()
SPaSM_write()	SPaSM_fprintf()	SPaSM_fscanf()
SPaSM_fgets()		

5.3 Some thoughts on portability

There has been considerable discussion in the parallel processing community about the problem of code portability on parallel machines. However, we have found that the message passing programming model to be widely available and flexible enough to use a variety of machines. Using our message passing and I/O wrapper libraries, we have a single version of source code that compiles on the following machines *without* any conditional compilation (except in the wrapper libraries).

- Connection Machine 5
- Cray T3D (PVM and Cray shared memory)
- Fujitsu VPP500 (using P4)
- Sun SPARCserver 2000 (Solaris Multithreading)
- Sun,IBM,HP,and SGI workstations.

This flexibility has been invaluable to the success of recent projects. Code can be developed on a workstation and later ported to a parallel machine without effort. With the unreliable nature of most parallel machines, we can use our parallel I/O library to read data files created on one machine and continue the run on another. By having one version of source code, we don't have to worry about managing several different code versions. As a few examples of recent successes we present the following :

- SPaSM ported from CM-5 to T3D in 3 days.
- Silicon modeling code developed on workstation and tested on a CM-5. Later is ported to T3D in 5 minutes by simply recompiling.
- Code ported to VPP500 in less than a week.
- Code ported to a parallel hardware simulator in only 2 days.
- Workstation port took one person 3 hours.

While much has been said about code portability, we feel that our approach has been highly successful.

6 Future Directions

One of the recent goals of large-scale molecular dynamics research has been to simulate more than a billion atoms. Recently, researchers at Oak Ridge National

Laboratory broke this barrier, but many questions still remain [9]. How does one go about analyzing data generated from a billion atom MD simulation? For that matter, how does one look at data from a 10 million atom MD simulation? From our experience, we have found that standard visualization and analysis packages are simply unable to deal with the volume of data generated by even a 10 million atom simulation. We also feel that unless useful data from multi-million particle MD simulations can be extracted, these simulations are not worth doing at all.

Presently, we are working on methods for data analysis and visualization of large-scale MD simulations. To illustrate the problem, the 104 million atom fracture experiment mentioned earlier performed more than 900 Gbytes of file I/O (checkpointing, restarting, generating intermediate output files). Each output file was approximately 1.6 Gbytes and contained only particle positions and kinetic energy in single precision. Restart files containing full information about the simulation were 5.8 Gbytes in size.

6.1 The wrong way to look at data.

Traditionally, we have looked at data as a post-processing step in which an SGI Onyx workstation would be used to generate images. Unfortunately, this process has proven to be almost intolerable for everyday work. Consider the following :

- Post-processing often occurs at a remotely located machine. Transferring a 1.6 Gbyte data file to our local network (via FDDI) takes between 30 and 40 minutes. Transferring that data from Los Alamos to the University of Utah would take more than 24 hours.
- Even if you successfully transfer the data, there's not enough local disk space to hold it.
- Visualizing 1 million atoms on an SGI Onyx takes about 5 minutes. Visualizing 10 million atoms, takes about 45 minutes and visualizing 38 million takes 3.5 hours (or 15 minutes if a single pixel is plotted for every atom).
- A picture alone is not that useful. We might want to analyze data in other ways.

We have also used custom sphere rendering code written by Mike Krogh in the Advanced Computing Laboratory. Running on 256 nodes, we have been able to generate images of 100 million atom data sets in less than 5 minutes. However, unlike the SGI program, this system is not interactive and it still suffers from the last point above. By dumping limited data to output files, it can be very difficult perform data-analysis as a post-processing step. If a visualization package has to start performing mathematical operations on 100 million atom data-sets, then it will encounter the same computational difficulties as those addressed in MD codes.

6.2 High performance analysis of 100 million atoms

We have finally given up on the graphics/parallel processing community to develop packages for analyzing large-scale MD simulations. Current packages seem to share a total disregard for the fact that if you're working with 100 million atoms, you can't afford to be adding visual programming systems, graphical user interfaces, using 24-bit ray-traced graphics, or making an arbitrary number of copies of the data. It's also clear that these packages can't address the bigger problem of looking at more than just making pretty pictures.

Presently, we are working on our own system for dealing with very large 3D molecular dynamics data by writing imaging and data analysis modules for SPaSM. First, we have written a text-based command interface to the code. This interface, which operates similarly to systems such as IDL, allows us to not only set up and run simulations, but also allows us to create variables, and access a wide range of C-functions. We have also completed a 2D imaging module for looking at surfaces or planar slices of large 3D datasets. Figure 8 shows two image generated by this system. In the images we are looking at an elliptical crack placed in a block of 152 million atoms at fixed temperature. All particle data is stored in double precision. In each image, we have displayed both the kinetic and the potential energy contribution of each particle for side-by-side comparison. The lower image shows a close up view of the crack tip. Each atom is represented by a pixel (which are enlarged in the close up view).

To minimize memory use, images are formed using 8-bit graphics. To improve network performance we convert image data into GIF format and send it through a Unix socket connection to a server running on a local workstation. This server takes the image and spawns a local viewer for image display (we have used **xv**). For the images shown in Figure 8, the total wallclock time to create the image on the CM-5, transfer the data over an ethernet connection, and display it on a Sun SPARCstation LX workstation takes less than 40 seconds. By comparison, the time required to perform a single timestep of the calculation is approximately 170 seconds (without VUs and in double precision).

It is important to note that dealing with 100 million particle data sets can not be accomplished using traditional approaches. When running with this many atoms, we simply do not have the memory to start including X11 libraries or using 24 bit graphic displays. So far, our image system is extremely lightweight. The simulation shown in Figure 8 required 13.7 Gbytes of memory. Of that, approximately 12 Gbytes was taken by particle data while only 0.3 Gbytes were required for generating the images shown.

We are also working on an efficient method for looking at data in 3D, but this work is in progress.

7 Acknowledgments

We would like to acknowledge the generous support of the Advanced Computing Laboratory for their continued support. We would also like to acknowledge Mike Krogh for his ongoing assistance and visualization assistance. We also thank Hidetoshi Konno, University of Tsukuba, Japan for providing access to the Fujitsu VPP-500. Finally, we would like to acknowledge our collaborators in MD research, Niels Grønbech-Jensen, Pablo Tamayo, Timothy Germann, Brad Ho-

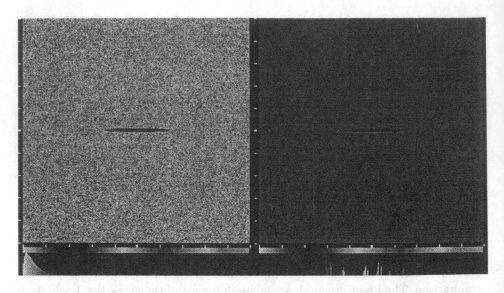

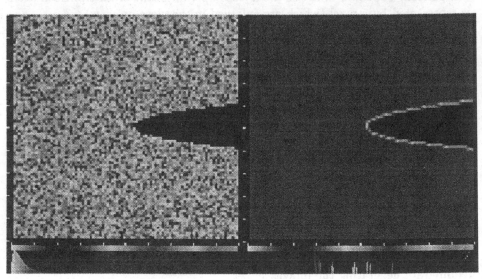

Figure. 8. 2D Images of 151,366,958 million atoms.

lian, and Shujia Zhou. This work was performed under th auspices of the U.S. Department of Energy.

References

[1] D.M. BEAZLEY, P.S. LOMDAHL, N. GRØNBECH-JENSEN, R. GILES, AND P. TAMAYO, *Parallel Algorithms for Short-range Molecular Dynamics*, Review article to appear in World Scientific's Annual Reviews in Computational Physics, vol. 3 (1995).

[2] P. TAMAYO, J.P. MESIROV AND B.M. BOGHOSIAN, *Proc. of Supercomputing 91*, IEEE Computer Society, 1991, pp: 462.

[3] D.M. BEAZLEY AND P.S. LOMDAHL, *Parall. Comp. 20 (1994) 173-195.*

[4] P.S. LOMDAHL, D.M. BEAZLEY, P. TAMAYO, AND N. GRØNBECH-JENSEN, *Multimillion Particle Molecular Dynamics on the CM-5*, International Journal of Modern Physics C, World Scientific (1993), p. 1075-1084.

[5] P.S. LOMDAHL, P. TAMAYO, N. GRØNBECH-JENSEN AND D.M. BEAZLEY, *Proc. of Supercomputing 93*, IEEE Computer Society (1993), pp: 520–527.

[6] D.M. BEAZLEY AND P.S. LOMDAHL, *Optimizing Large-Scale Molecular Dynamics Simulations for RISC-based MPPs and Workstations*, Submitted to IPPS'96, LA-UR-95-xxxx.

[7] S. PLIMPTON, *Fast Parallel Algorithms for Short-Range Molecular Dynamics*, Sandia National Laboratory Report, SAND91-1144, UC-705 (1993).

[8] S. CHYNOWETH, Y. MICHOPOULOS, AND U.C. KLOMP, *A Fast Algorithm for the Computation of Interparticle Forces in Molecular Dynamics Simulations.* Parallel Computing and Transputer Applications, IOS Press/CIMNE, (1992), p. 128-137.

[9] E.F. D'AZEVEDO, C.H. ROMINE, AND D.W. WALKER, *Shared-Memory Emulation Is Key to Billion-Atom Molecular Dynamics Simulation*, SIAM News, May/June, (1995).

Wave Propagation in Urban Microcells: a Massively Parallel Approach Using the TLM Method

Pascal O. Luthi[1], Bastien Chopard[1] and Jean-Frédéric Wagen[2]

[1] CUI, University of Geneva CH-1211 Geneva 4, Switzerland,
[luthi/chopard]@cui.unige.ch
[2] Swiss Telecom PTT, FE 412, 3000 Berne 29, Switzerland,
wagen_j@vptt.ch

Abstract. We consider a new approach to modeling wave propagation in urban environments, based on the Transmission Line Matrix (TLM) method. Two-dimensional simulations are performed using a map of a city A renormalization technique is proposed to convert the results to the three-dimensional space. Our approach provides good predictions for the intensity of a wave when compared with in-situ measurements and is appropriate to very fast massively parallel computations. In order to provide a performance analysis, the algorithm has been used as a benchmark on different parallel architecture (CM200, CM5, IBM SP2 and Cray T3D).

1 Introduction

The fast development and the growing importance of mobile, personal communication systems such as cellular phones require a detailed knowledge of wave propagation in heterogeneous media. Urban environments constitute a difficult problem which is studied by various authors[1, 2, 3]. Radio waves are absorbed, reflected, diffracted and scattered in a complicated way on the buildings and the amplitude pattern of a wave emitted by any antenna surrounded by such added obstacles may be quite complicated and beyond an analytical calculation.

Yet, the coverage region of an antenna is a crucial question in the development of the future mobile communication systems. As the number of users increases, several base stations should be used in a city in order to accommodate for a growing traffic. Each base station is in charge of a given region (called an urban microcell) whose size is defined so that the potential number of calls within this region matches the limited number of channels available at each station. For continuity reasons during the transmission, adjacent cells must have overlapping boundaries.

An optimized set of base stations should ensure a complete coverage with a minimum number of cells of appropriate area. Therefore the starting point for the planning or maintenance of any mobile communication network is based on a prediction of the radio wave attenuation around the transmitter (antenna) in the buildings layout. In this paper we propose a new approach to compute wave propagation in a urban cell, based on the Transmission Line Matrix (TLM) technique initially designed for studying the behavior of wave guides[4]. We show that this method is well suited to the present problem and offers promising prospects of development in this field.

The TLM method provides a natural implementation of a wave propagation dynamics within the framework of the Lattice Boltzmann Models (LBM)[5], as a discretized Huygens principle in which an incident flux creates scattered outgoing flux. LBM describe physical systems in terms of the motion of fictitious microscopic particles on a lattice. This method is inspired by the cellular automata modeling techniques[6, 7] and allows efficient simulations of many complex physical situations like fluid dynamics and reaction diffusion processes[8, 9, 10].

Thus, the TLM method is characterized by a simultaneous dynamics (all flux are updated at the same time) and a very simple numerical scheme. The interpretation of the dynamics in terms of flux (or flows) makes the boundary conditions easy to implement. As a result, the TLM method is a natural technique to simulate propagation phenomena in random media[11, 12] and is very appropriate to massively parallel computations as now possible on the new generation of scalable parallel computers. This allows fast interactive simulations which seems to be a basic requirement for an efficient tool in this domain.

This paper is organized as follows. In Sect. 2, we discuss the TLM method and its limitations. In Sect. 3, we present a way to renormalize the results of two-dimensional simulations (from a city map) in order to predict the intensity of a wave in the corresponding three-dimensional space. In Sect. 4, we give the results of numerical simulations in a complex urban environment and compare it with real measurements obtained by Swiss Telecom PTT in the corresponding situation. Sect. 5 presents a short performance analysis of the algorithm as implemented on different parallel architectures.

2 The TLM Approach

2.1 Discretizing of the Huygens Principle

According to the Huygens principle, a wave front consists of a number of spherical wavelets emitted by secondary radiators. The Transmission Line Matrix (TLM) method is a discrete formulation of this principle. For this purpose, space and time are represented in terms of finite, elementary units Δ_r and Δ_t which are related by the velocity of light such that $\Delta_t = \Delta_r/c_0\sqrt{2}$. Accordingly, two-dimensional space is modeled by a Cartesian grid of nodes with a mesh size of length Δ_r. Δ_t is the time required for an electromagnetic signal to cover a distance corresponding to the half of the network diagonal. Instead of looking at the field values ψ on the nodes as when the wave equation $\partial_t^2\psi = c_0^2\nabla^2\psi$ is discretized, we define flow values f_d on the arcs connecting the space-time network.

We Assume that an impulse flow is incident upon one of the nodes from the negative x-direction ($f_1 = 1$ at this node, see Fig. 1). Following Huygens ideas, this energy is scattered isotropically in all four directions, each radiated pulse carrying one fourth of the energy of the incident energy. The corresponding field quantities must then be 1/2 in magnitude. Furthermore, the reflection coefficient *seen* by the incident pulse must be negative in order to satisfy the requirement of field continuity at the node.

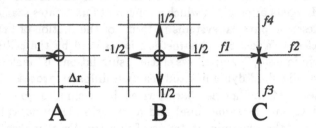

Fig. 1. The discretized Huygens method applied to a flow formalism (TLM). **A**: example of a flux of magnitude 1 entering a site along the first direction f_1. **B**: scattering effect of the TLM matrix on this flux, in direction and magnitude. **C**: labeling of the arc directions relative to a site.

The same arguments applied to the other direction of the lattice leads to the construction of the TLM matrix **W** which determines the flows coming out from a node as a linear combination of the incoming flows. Finally we write the evolution of the flow as following:

$$
\begin{pmatrix} f_1(x+\Delta_r) \\ f_2(x-\Delta_r) \\ f_3(y+\Delta_r) \\ f_4(y-\Delta_r) \end{pmatrix}_{t+\Delta_t} = \mathbf{W}\mathbf{f}_t = \frac{1}{2} \begin{bmatrix} 1 & -1 & 1 & 1 \\ -1 & 1 & 1 & 1 \\ 1 & 1 & 1 & -1 \\ 1 & 1 & -1 & 1 \end{bmatrix} \begin{pmatrix} f_1 \\ f_2 \\ f_3 \\ f_4 \end{pmatrix}_t \tag{1}
$$

We recover the field solution by noting that if $\psi = \sum_d f_d$ the above equation is equivalent to the discrete (finite differences) wave equation.

2.2 The Dispersion Relation

Assuming periodic boundary conditions, the solutions f_d of (1) can always be written as a discrete Fourier series

$$
f_d(\mathbf{r}, t) = \sum_{\mathbf{k} \in V_k} A_d(\mathbf{k}, t) \exp(i\mathbf{k}\mathbf{r})
$$

where

$$
A_d(\mathbf{k}, t) = \frac{1}{N_1 N_2} \sum_{\mathbf{r} \in V_r} f_d(\mathbf{r}, t) \exp(-i\mathbf{k}\mathbf{r})
$$

V_r denotes the set of all lattice points and the wave numbers $\mathbf{k} \in V_k$ are given by

$$
\mathbf{k} = (k_1, k_2) = \frac{2\pi}{\Delta_r} \left(\frac{n_1}{N_1}, \frac{n_2}{N_2} \right)
$$

with

$$n_j \in \left\{ -\frac{N_j}{2}, \cdots, 0, 1, \cdots, \frac{N_j}{2} - 1 \right\}$$

where N_1, N_2 are the number of sites (supposed to be even) along the dimensions of the square lattice. Introducing this in (1) we obtain

$$\mathbf{A}(\mathbf{k}, t + \Delta_t) = \mathbf{M}\mathbf{A}(\mathbf{k}, t)$$

where

$$\mathbf{M} = \begin{bmatrix} \frac{1}{2a} & -\frac{1}{2a} & \frac{1}{2a} & \frac{1}{2a} \\ -\frac{a}{2} & \frac{a}{2} & \frac{a}{2} & \frac{a}{2} \\ \frac{1}{2b} & \frac{1}{2b} & \frac{1}{2b} & -\frac{1}{2b} \\ \frac{b}{2} & \frac{b}{2} & -\frac{b}{2} & \frac{b}{2} \end{bmatrix}$$

where $a = \exp(ik_1 \Delta_r)$ and $b = \exp(ik_2 \Delta_r)$. If we write $\mathbf{A}(\mathbf{k}, t) = \mathbf{A_0} \exp(i\omega t)$ one is left with the condition that $\exp(i\omega t)$ is an eigenvalue of M and $\mathbf{A_0}$ the eigenvector. The four eigenvalues give the possible dispersion relation of the evolution:

$$e^{i\omega \Delta_t} = \pm 1$$
$$e^{i\omega \Delta_t} = \frac{1}{2}\left(\kappa \pm i\sqrt{4 - \kappa^2} \right) \tag{2}$$

where $\kappa = \cos(k_1 \Delta_r) + \cos(k_2 \Delta_r)$.

Clearly, the propagation corresponds to (2). After some algebra we see that a vector proportional to $[1, 1, 1, 1]$ belongs to the eigensubspace corresponding to the propagation eigenvalues. It is a matter of simple algebra to check that (2), in the limit $\Delta_r \to 0$, $\Delta_t \to 0(\Delta_r/\Delta_t = c_0\sqrt{2})$, is the expected dispersion relation for wave propagation: $\omega^2 = c_0^2(k_1^2 + k_2^2)$.

It is possible to vary, node by node, the refraction index of the medium we are propagating in, by adding a fifth arc connecting a node with itself and carrying a flow f_0. Now we write $(f_1, f_2, f_3, f_4, f_0)$ as the flows on one node and the TLM matrix then becomes

$$W = \frac{2}{Y_0 + 4} \begin{bmatrix} 1 & -\frac{Y_0+2}{2} & 1 & 1 & Y_0 \\ -\frac{Y_0+2}{2} & 1 & 1 & 1 & Y_0 \\ 1 & 1 & 1 & -\frac{Y_0+2}{2} & Y_0 \\ 1 & 1 & -\frac{Y_0+2}{2} & 1 & Y_0 \\ 1 & 1 & 1 & 1 & \frac{Y_0-4}{2} \end{bmatrix}$$

Looking for the dispersion relation using the same method as before we obtain the expected dispersion relation for a medium with a refraction index $n = \sqrt{Y_0 + 4}/2$. In this case the relation between the flow and the field solution takes the following and simply verifiable form: $\psi = f_1 + f_2 + f_3 + f_4 + Y_0 f_0$.

2.3 Source and Boundary Condition

Sources. We define a *sinusoidal wave source node* of the lattice as a site where all the flows f_d take the same value: $f_d = \gamma \sin(2\pi n/\tilde{T})$, where n indicates the successive times iterations; $\tilde{T} = T/\Delta_t$ is the dimensionless period of the source and γ is a normalization factor which depends of $\tilde{T}$. Such an isolated node on a two-dimensional lattice is appropriate to simulate a line source (cylindrical wave), whereas an arrangement of source nodes may generate a plane wave. In general, it is expected that a spatial distribution of such source points can be devised in order to reproduce a given antenna diagram. Good accuracy is obtained for large $\tilde{T}$ and it was observed that choosing $\tilde{T} < 6$ leads to poor propagation predictions[13].

Boundaries. In our simulation we consider a discrete map of a city. Obstacles (such as buildings) and the border of the system are represented as particular lattice sites and are characterized by different TLM propagators **W**, according to their physical required properties. Therefore, a site does not know whether its neighbor is an obstacle, a source, or whatever else; it is treated identically.

Besides the free propagators **W**, we consider several behaviors: partial or complete reflection and attenuated transmission. *Reflecting nodes* return all incoming flows with opposite sign and direction; An attenuation factor α is heuristically fixed in order to account for the reflexion coefficient relating incident to reflected energy at a particular boundary. As a result the easiest way to represent the buildings layout is to set on each lattice site belonging to a building:

$$\mathbf{W}_\alpha = \alpha \begin{bmatrix} 0 & -1 & 0 & 0 \\ -1 & 0 & 0 & 0 \\ 0 & 0 & 0 & -1 \\ 0 & 0 & -1 & 0 \end{bmatrix}$$

Attenuation nodes are regular TLM matrix with a multiplicative factor β: $\mathbf{W}_\beta = \beta\mathbf{W}$ where $\beta \in]0,1[$. Attenuation nodes do not prevent a wave from being transmitted through the obstacle but they lower its energy. Wave propagation at an attenuation node is always accompanied with reflection. This is due to the breaking of spatial homogeneity. For instance, on the border of the system, several layers of attenuated transmitters $\mathbf{W}_\beta$ allow a gradual adsorption of the energy with minimum reflection, which is crucial to mimic free propagation beyond the limits of our simulation.

3 Renormalization Method

When the transmitting antenna is below the rooftop, the essential features of a building layout is assumed to be captured by a two-dimensional simulation (which is equivalent to a line source and infinitely high buildings). But, real antenna are rather to be considered as point source in a three dimensional space. Moreover, for numerical reasons the wavelength used in the simulation usually does not match the experimental wave length. As a result, quantitative comparisons (or prediction) between simulations and real measurements become only possible after an *ad hoc* renormalization which takes into account the two unphysical effects of the model: the effective 2D propagation and the wrong wavelength.

In [14], we propose such a renormalization; accordingly, we suppose that the measured (real) amplitude A_{3D} of the wave can be approximated by the outcome a_{2D} of our simulation using the following relation:

$$A_{3D}(\mathbf{r}) \approx a_{2D} \Lambda (\lambda \delta(\mathbf{r}))^{-1/2}$$

where Λ is the actual wavelength used in the measurement and λ the simulated one. δ represents a typical distance between the transmitter and the receivers according to the buildings layout. One could think of the shortest distance which bypass the obstacles; indeed, for the renormalization of our simulations (presented in Sect. 4) we have used the so-called Manhattan distance which is the shortest distance following the lattice edges representing the discretized layout. A better definition for δ could be the length of the ray which mostly contributes to the propagation. Clearly, this path may avoid diffraction at the corners but includes reflections. Thus, it does not correspond to the shortest distance in the building layout. An estimate of this length using the relation $\delta(\mathbf{r}) = c_0 \tau_{\mathbf{r}}$, where $\tau_{\mathbf{r}}$ is the mean delay time for the impulse response at the receiver location, is currently under investigation. An overview of the different above-mentioned distance is presented in Fig. 2.

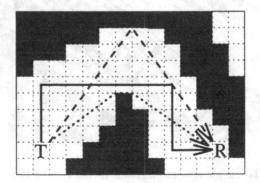

Fig. 2. Superimposed to the building layout (gray) and the lattice, we have represented in this figure different ways to compute the distance between a transmitter T and a receiver R. The solid line is one possible path representing the Manhattan distance, the dotted line is the shortest distance and the dashed line could be the path of strongest contribution.

4 Numerical Study and Results

The first step of our simulation procedure is the discretizing of the building layout. This can be simply achieved by scanning a map of a given urban area. Thus, it is not necessary to first calculate the coordinates of each corner. The result (see Fig. 3) is a two-dimensional array, typically of size 512×512 or 1024×1024, with four different kinds of elements: (i) The free space sites or Huygens radiators defined by the W matrix, (ii) the building sites defined either by W_α (reflecting nodes) or by W_β (attenuation nodes), (iii) the source node defined by γ and $\tilde{T}$ and (iv) the boundary of the array, defined by the absorbing layers with appropriate W_β's.

The simulation then consists of a synchronized updating of each site as a function of its nearest neighbors (except for the source) according to (1), until a steady states is reached. As will be shown in the next section, this procedure is well suited for an implementation on a massively parallel computer. Simulations have been intensively carried out on a CM-200. for which a typical experiment, including renormalization, takes less than 10 seconds.

Fig. 3. TLM Simulation of wave propagation in the city of Bern on a square lattice of dimension 512×512. The white blocks represent the buildings, the dot marks the position of the source and the white line corresponds to the measurement axis (see Fig. 4). The gray-scale indicates the simulated intensity of the wave; decreasing from white to black.

In Fig. 3 the building layout corresponds to a whole urban microcell in Bern (Switzerland) for which measurements were available. In Fig. 4 the amplitude of the wave along the measurement axis is ploted for two types of TLM building matrices: purely reflecting nodes (right) and attenuation nodes (left). For both situations the qualitative feature of our renormalized prediction (overall value, position and shape of the peaks) is in agreement with the measurement. However the introduction of attenuation nodes at the boundaries of the buildings gives better quantitative results. This might come from the fact that, unlike with reflecting nodes, the second situation allows part of the waves to penetrate and subsequently leave the buildings. We believe that this is a valuable technique (remind that our simulations are 2 dimensional) to take into account the real propagation over the buildings rather than through it.

The remaining discrepancies between predictions and measurements in Fig. 4 should be further investigated. They may partly arise from our choice concerning the distance term in the renormalization. A $\delta(\mathbf{r})$ based on the mean delay time may contribute to a better discrimination between points *nearby* and *far-off* the source. Moreover, the parameters of our simulations concerning the buildings (α and β) were chosen uniformly over the layout and correspond to heuristic averaged values because *coarse grained* measurements of the reflection coefficient are simply not yet available. However, the introduction of such local parameters would not require extra CPU time.

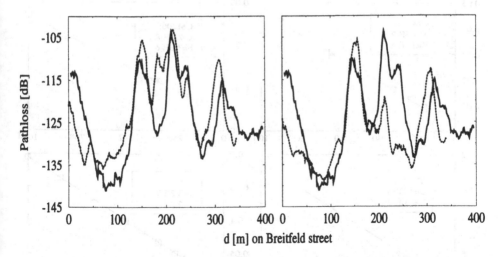

Fig. 4. TLM prediction (dashed line) and *in situ* measurements (thick lines) on the Breitfeld street, considering the full layout of the surrounding urban microcell. The parameters of the simulation are $\tilde{T} = 8$, $\Delta_r = 1.5m$, therefore $\lambda = 8.4m$, $f = 36$MHz (note that measurements have been done at $\lambda_0 = 0.16m$, $f = 1900$MHz). Two types of boundary condition were applied for the sites limiting the buildings in the discretized layout: reflecting walls defined by $\mathbf{W}_\alpha$, $\alpha = 0.39$ (right part of figure) and *permeable* walls defined by $\mathbf{W}_\beta$, $\beta = 0.39$ (left).

Note that other aspects of the TLM simulation presented here may introduce some effects: it is observed that changing the orientation of the building layout in the lattice or changing the wavelength of our transmitter affect somewhat the result (up to 10 dB locally for the problem described in Fig. 3). This is mostly due to the finite lattice size, as described in[13] and solutions to bypass these intrinsic difficulties of the TLM method are under investigation.

5 Performance Analysis and Comparison

In order to provide a performance analysis of our algorithm which is independent of the problem size n and the number of processors p we plot the execution time according to the following scaling law:

$$T_p = a\frac{n^2}{p} + bn$$

where T_p is the time to execute the code on p processors. The factor a represents the unit time of a site calculation; thus the effective Mflops per processor is proportional to $1/a$. The term bn accounts for inter-processor communications ($1/b$ is proportional to the bandwidth). Fig. 5 indicates a perfect agreement to our performance model for the CM5, CM200, CRAY T3D and IBM SP2.

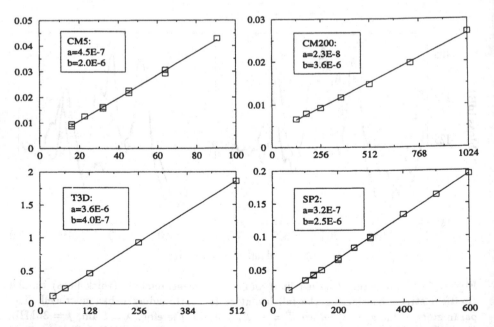

Fig. 5. Fit of the execution time according to our performance model $T_p = an^2/p + bn$. X-axis is n/p and Y-axis is $10^3 T_p/n$. Coefficient a and b are independent of problem and computer size and thus offer the possibility to compare different machines. Here $p = 32$ or $p = 64$ for the CM5, p is ranging from 2 to 32 for the CRAY T3D, from 2 to 12 for the SP2, and the CM-200 (8k processors) was considered as a whole: $p = 1$.

Focusing on a and considering our full CM200 (8k-processors with 256 FPU) as a reference machine we observe a decreasing effective per processor efficiency for this kernel, in the following order: CM200 with 700 Mflops, SP2: 50 Mflops (14 × smaller), CM5: 35 Mflops (20 × smaller), and T3D: 4.4 Mflops (156 × smaller). On the other hand, the coefficient b follows a reverse order: communication is 9 times slower on the CM200 than on the T3D; 6 times slower on the SP2 and 5 times slower on th CM5.

We have used a Data Parallel programming model (CM Fortran) on both CM machines and plain Message Passing (MPL) on the SP2. On the T3D we have used the Shared Memory model (CRAFT) which is known to be less effective than PVM; may this explain the low figure obtained on this machine?

6 Conclusion

The implementation of TLM method on a parallel computers offers a new perspective in the simulation of radio wave propagation in micro cellular urban environment. The method is based on a direct discretizing of the building layout onto a two-dimensional lattice. Providing a renormalization of the predicted results according to the distance between transmitter and receiver we have found a good agreement between predictions and measurements. Finally, the high speed of the algorithm allows, for the first time, a complete interactive simulation of the propagation, which in turn constitutes a powerful investigation tool.

Acknowledgment

The authors would like to acknowledge the Swiss National Science Foundation and Swiss Telecom PTT for financial support.

References

1. H. L.Bertoni, W. Honcharenko, L. R. Maciel, and H. Xia. UHF propagation prediction for wireless personal communications. In *IEEE Proceedings, Vol 82, No. 9*, pages 1333–1359, 1994.
2. T. Kürner, D. J. Cichon, and W. Wiesbeck. Concepts and results for 3d digital terrain-based wave propagation models: an overview. *IEEE Jour. on Selected Areas in Communications*, JSAC-11(7):1002–1012, Sept. 1993.
3. K. Rizk, J.-F. Wagen, and F. Gardiol. Ray tracing based path loss prediction in two microcellular environments. In *Proceedings IEEE PIMRC'94*, pages 210–214, The Hague, Netherlands, sep 1994.
4. Wolfgang J. R. Hoeffer. The transmission-line matrix method. theory and applications. *IEEE Transaction on microwave theory and techniques*, MTT-33(10):882–893, oct 1985.
5. R. Benzi, S. Succi, and M. Vergassola. The lattice boltzmann equation: theory and application. *Physics Reports*, 222(3):145–197, 1992.
6. H. J. Hrgovčić. Discrete representation of the n-dimensional wave equation. *J. Phys. A*, 25:1329–1350, 1991.

7. Bastien Chopard. A cellular automata model of large scale moving objects. *J. Phys.*, A(23):1671–1687, 1990.

8. G. Doolen, editor. *Lattice gas method for partial differential equations*. Addison-Wesley, 1990.

9. B. Chopard, P. Luthi, and M. Droz. Reaction-diffusion cellular automata model for the formation of liesegang patterns. *Phys. Rev. Lett.*, 72(9):1384–1387, 1994.

10. B. Chopard. Cellular automata modeling of hydrodynamics and reaction-diffusion proceses: Basic theory. In Plenum, editor, *Scale invariance, interface and non-equilibrium dynamics*, page , 1994. NATO Workshop, Cambridge, June 1994.

11. C. Vanneste, P. Sebbah, and D. Sornette. A wave automation for time-dependent wave propagation in random media. *Europhys. Lett.*, 17:715, 1992.

12. D. Sornette, O. Legrand, F. Mortessagne, P. Sebbah, and C. Vanneste. The wave automaton for the time-dependent schroedinger, calissical wave and klein-gordon equations. *Phys. Let. A*, 178:292–300, May 1993.

13. Pascal O. Luthi and Bastien Chopard. Wave propagation with transmission line matrix. Technical report, University of Geneva and Swiss Telecom PTT, 1994.

14. B. Chopard, P. Luthi, and J.-F. Wagen. Submitted to IEEE Transactions on Antennas and Propagation, june 1995.

The NAG Numerical PVM Library

Ken McDonald

The Numerical Algorithms Group Ltd., Wilkinson House, Jordan Hill Road, Oxford, OX2 8DR, UK

Abstract. The Numerical Algorithms Group Ltd has been actively involved in research and development into parallel software for many years and has contributed theoretical and practical expertise to such high-profile projects such as LAPACK. As a result of these efforts the NAG Fortran 77 Library incorporates state-of-the-art software that is efficient and portable across a wide range of scalar, vector and shared-memory computers.

A numerical library that is also efficient and portable across a range of distributed-memory machines has proven to be an elusive goal until recently. Encouraged by the promise shown by the ScaLAPACK project (a follow-on to the LAPACK project), and using the same public-domain message-passing systems (PVM and BLACS), NAG began developing a general-purpose numerical library for distributed-memory machines.

The NAG Numerical PVM Library described in this paper is of a modest size but demonstrates the concept of portable parallel software and has the potential for considerable expansion. Plans for the second release are already being implemented which will broaden the coverage and remove some limitations.

1 Library Design

One of the principal design aims for the NAG Numerical PVM Library was to make it simple for novice users to understand while retaining the flexibility to be useful to more experienced users. The interfaces to the library routines have been made as similar to the equivalent NAG Fortran 77 Library routines as possible to encourage established users of sequential NAG libraries to experiment with parallelism. This design philosophy has the additional benefit of allowing a development path from the NAG Numerical PVM Library to other NAG libraries in the future, for example a possible NAG HPF Library. The NAG Numerical PVM Library is divided into chapters, each deveoted to a branch of numerical analysis or statistics. Each chapter has a three-character name and a title, e.g. D01 - Quadrature). Furthermore, a set of utility routines to assist the user in parallel programming have been included in the library, mainly to manage the logical processors and information enquiry routines. With the facilities provided in the library it is possible to create parallel programs to solve quite complex problems very quickly.

1.1 Model of Parallelism

The NAG Numerical PVM Library was designed according to the Single Program Multiple Data (SPMD) model of parallelism; this model was chosen for several reasons:

- simplicity, but with the flexibility to mimic other, more complex models of parallelism,
- compatibility with the BLACS (Basic Linear Algebra Communications Subprograms) and ScaLAPACK material (see [1],[2]),
- easier conversion to HPF (see [4]) eventually.

Each instance of the user program that is executed forms one of the logical processors in a two-dimensional logical processor grid (as used by the BLACS and ScaLAPACK). By using the library utilities, the user can create a parallel program that looks remarkably similar to a conventional sequential program. The program is even executed in the same way as a sequential program — the necessary number of additional copies of the program to create the requested logical processor grid are automatically executed. No special set up is required for this step.

1.2 The Message Passing Mechanism

PVM (Parallel Virtual Machine, see [5]) and the BLACS were chosen for the message-passing environment due to their ease of use, ready availability and the large, established user-base. The majority of routines in the library make use of the BLACS for communication but, in certain exceptional circumstances, PVM is used where BLACS lack the required functionality: it should be remembered that the BLACS were designed for the regular communications to be found in most linear algebra problems — other areas such as adaptive quadrature have less predictable communication requirements and PVM's message-passing is more suitable.

A great deal of effort has been put into hiding the details of the use of PVM and the BLACS. Most of the details of managing the parallel processes can be left to the *utility* routines which are provided in the "Z01 Chapter" of the

NAG Numerical PVM Library (see [7]). However, enough flexibility has been built into the utility routines to allow users the possibility of spawning processes themselves using raw PVM calls and to declare further logical processor grids for other purposes (a technique known as multigridding).

1.3 Data Distribution

Arguments to NAG Numerical PVM Library routines are categorised in one of two classes for data distribution purposes, local or global. A global input argument must have the same value on all logical processors participating in a calculation on entry to a library routine. Global output arguments will have the same value on all logical processors participating in a calculation on exit from a library routine. An example of a global input argument is the size of a matrix; an example of a global output argument is the result of a numerical integration. Local arguments may assume different values on different processors, for example the elements of a distributed matrix on input and the factors of the same distributed matrix on output.

The most common data distribution used in the library is the "block scattered" distribution for matrices used by the ScaLAPACK routines. In this distribution a matrix is partitioned into rectangular blocks — the blocks along one matrix dimension are assigned to successive processors in the same processor grid dimension until the last processor is reached, then the block assignment returns to the first processor and continues in the same fashion until no blocks are left. An identical distribution strategy is carried out in the other matrix/grid dimension giving a "wrap-around" effect in both matrix dimensions. Data distribution is the responsibility of the user but support is given for input from and output to external files and also distribution using a user supplied procedure. These distribution support routines are described later.

Two routines for eigenvalue and singular value problems use a different data distribution for matrices — a "block column" distribution: each processor contains a block of consecutive columns of the matrix. Separate support routines are provided for this data distribution.

1.4 Error Trapping

When an error is detected in a conventional NAG sequential library, a user selected action takes place — typically an error message is output, possibly followed by program termination or continuation with an error flag to indicate the routine has failed. A similar mechanism for the NAG Numerical PVM Library was desirable but presented some fundamental difficulties to overcome. For example, imagine that during execution one logical processor discovers a fatal error; somehow this event must be passed to the other logical processors to prevent them from going any further with the current computation. To maintain efficiency, parallel numerical algorithms must minimize the amount of failure checking between processors since it introduces communication (and possibly synchronisation) between processors that does not contribute directly to the computation. Where it is necessary for processors to compare their error condition, it is carried out at a point where there is some other (unavoidable) reason for a synchronisation point in the code.

As far as possible the arguments supplied to any NAG routine are checked for consistency and correctness and this feature is maintained in the NAG Numerical PVM Library with some additional checking. A global argument must have the same value on entry to the library routine on every logical processor. Each library routine compares the values of its global arguments for consistency across the processor grid and treats any differences it finds as a failure condition. Local arguments may assume different values on entry to the library routine on each processor. Both global and local arguments are also tested in the conventional manner to see if they have meaningful values. The comparison of global arguments on logical processors introduces problems on a heterogeneous network of machines that are discussed later.

1.5 Arithmetic Issues

The first release of the NAG Numerical PVM Library is restricted to homogeneous networks of computers and true parallel machines due to an unresolved issue concerning differences in arithmetic between non-identical machines. The problem is best illustrated with an example: imagine an iterative process that terminates subject to some (floating-point) arithmetic condition — it is conceivable that in some circumstances on two different machines, one machine would terminate while the other would continue due to differences in their arithmetic. This could lead to communications deadlocking or some other fatal problem. Even specifying that the machines should conform to IEEE arithmetic is not sufficient to overcome these problems; variations in the way that different compilers order arithmetic expressions could cause discrepancies in the roundoff error of the results with the same consequences.

Checking the consistency of floating-point global arguments presents a similar problem. The numbers must be compared on each processor and once again machine arithmetic and the way that floating-point numbers are represented on different machines may cause two numbers or expressions that are apparently different machines may cause two numbers or expressions that are apparently identical to actually be slightly different.

NAG is currently studying how to overcome these difficulties, but for now the NAG Numerical PVM Library is restricted to networks of identical machines.

2 Library Contents

The majority of numerical routines in the NAG Numerical PVM Library are either contributed from external sources or developments of previous work carried out at NAG.

2.1 Quadrature

A suite of 5 quadrature routines are provided, covering one-dimensional, two-dimensional and multi-dimensional quadrature. Three of these quadrature routines solve one-dimensional problems and are adaptive, and suitable for badly behaved (i.e. singular or oscillating) integrands and integrands of the form $\sin(\omega x)g(x)$ or $\cos(\omega x)g(x)$ (the sine or cosine transform of $g(x)$). The two routines for singular and oscillating integrands both use Gauss-Kronrod rules, adapted from QUADPACK [10] routines, to compute the approximation of the integrals. The routine to calculate the sine/cosine transform, also a modified QUADPACK routine, uses either a Gauss-Kronrod rule or a modified Clenshaw-Curtis procedure to compute approximations to the integral. The two-dimensional quadrature routine approximates the integral over a finite region by Patterson's method for the optimum addition of points to a Gauss rule applied to both the inner and outer integrals. Increasingly accurate rules are applied until the estimated error meets a user-specified error tolerance. The final routine in the suite, for multi-dimensional quadrature, uses a Korobov-Conroy number-theoretic method to compute the integral with up to 20 dimensions. All of the quadrature routines require a user-supplied function to evaluate the integrand at a general point.

2.2 Optimization

In the first release there is only a single optimization routine for solving unconstrained nonlinear-least-squares problems. This routine uses a Gauss-Newton method to iterate towards the minimiser of the user-supplied objective function. On each iteration, the Jacobian of the function is estimated numerically with function evaluations being carried out in parallel (no functions to compute derivatives are required from the user). For objective functions that are expensive to compute, this is where the bulk of execution time is spent and gives high parallel efficiency. The new search direction for the minimiser is calculated by solving the system of equations generated using the QR factorization and solver in the ScaLAPACK. This is followed by a sequential one-dimensional minimisation along the search direction. This process is repeated until the user-requested accuracy is met or a limit on the number of iterations is reached.

2.3 Linear Algebra

The bulk of the linear algebra routines originally came from two public-domain packages, ScaLAPACK and PIM (Parallel Iterative Methods, see [8]), although the PIM software has been extensively modified. ScaLAPACK solves problems in dense linear algebra while PIM is suitable for sparse matrices.

The calling sequences of the ScaLAPACK routines have been maintained and it is possible for the NAG versions, the public-domain versions or manufacturer-supplied versions to be used interchangeably (entry points allow the use of either NAG or ScaLAPACK routine names to be used). One of the benefits of using NAG versions of the ScaLAPACK software is the additional security of extra argument checking and error mechanism which can help to pick up some errors common in parallel programming. Easy-to-use "black box" routines for solving systems of equations have been built around some of the ScaLAPACK routines, simplifying the calling sequences for novice users. In the first release, "black box" for solving systems of equations using LU factorization and Cholesky factorization have been provided.

The PIM material has been modified extensively; the original routines required user-supplied subroutines for calculating matrix-vector products, dot-products and preconditioning. The same result is achieved in the NAG Numerical PVM Library now by reverse communication. In the first release, routines are supplied for solving general unsymmetric systems of equations using a restarted GMRES method and for symmetric systems of equations using either a conjugate-gradient or SYMMLQ method.

The library also contains two routines for eigenvalue and singular value problems. One solves the symmetric eigenvalue problem for real matrices, and the other computes the singular value decomposition (SVD) of a real rectangular matrix. The parallel algorithms used by these routines are based on an extension to a one-sided Jacobi method due to Hestenes [3].

2.4   Random Number Generation

A routine for generating pseudo-random numbers is provided, which allows users to choose from a set of 273 Wichmann-Hill generators (see [6]). Each of these generators has been tested for statistical independence from any of the other generators. This makes them suitable for producing streams of random numbers on multiple processors with each stream showing no correlation to other streams. The generators have been designed to use 32-bit arithmetic, common to most current machines, while still producing pseudo-random sequences with the large period expected of much higher-bit arithmetic. This makes the Wichmann-Hill generators far more portable between machines than any of their predecessors. In the first release, only uniform distributions of numbers in the range (0.0,1.0) are generated.

2.5   Data Distribution Facilities

The correct distribution of data can be an error-prone area for novice users of distributed memory machines. In particular, the block-scattered distribution common to ScaLAPACK routines can be difficult to program correctly, with no easy way of checking and debugging. The NAG Numerical PVM Library contains a number of routines to assist users in the data-distribution task. Several routines exist for reading matrices from external files — each processor reads in the whole matrix *but only stores its local portion*; this means that no more storage space must be allocated than is necessary to hold the local part of the matrix and that each processor reads to the same point in the file (ie. the end of the matrix data). These input routines have their counterparts which output distributed data to an external file. To prevent write conflicts between processors, a single processor is designated to output to the external file and all other processors communicate their data to the designated output processor in turn.

In addition, where it is possible to code the value of matrix elements as a Fortran expression, library routines are available that accept a user-supplied subroutine that returns the value of specified matrix elements. These library routines calculate which matrix elements are to be stored locally on each processor for a given data distribution and use the user-supplied routine to fill the appropriate local arrays with correctly distributed data.

2.6   Parallel Programming Utilities

The NAG Numerical PVM Library provides a small set of parallel programming utility routines for managing logical processor grids and information enquiry routines.

The most important routines are for declaring and initialising a logical processor grid of a given size and destroying the grid when it is of no further use. In the first release, only one processor grid may exist for library use at any one time, although it can be destroyed and redeclared (possibly with a different size and shape). The user simply executes a single copy of his/her program and the library grid declaration routine executes enough additional copies to form the requested processor grid. The first call to a grid declaration routine fixes the maximum number of logical processors available to participate in a logical processor grid. Subsets of these logical processors may be used but attempting to declare a grid with a total number of processors greater than the first grid is an error.

The BLACS "context" for the library processor grid is normally passed as an argument to NAG Numerical PVM Library routines (although there are some exceptions). Passing the context of any other grid, or an invalid context is an error.

A number of enquiry routines are also provided. Useful information is returned identifying the processor designated for output, processors in the library context, information about the underlying PVM tasks and BLACS communication topologies.

3 Future Development

There will be two approaches taken in the future development of the NAG Numerical PVM Library: to remove some of the current restrictions associated with its use, and to extend the coverage of the numerical areas of the library further.

The principal extensions to the numerical areas of the library will be in dense and sparse linear algebra. Further ScaLAPACK routines will be introduced, including reductions to standard forms such as Hessenberg and bidiagonal matrices. In addition, more routines for the solution of eigenvalue problems will be added. More iterative methods will be added for solving sparse problems to give a wider choice and plans have been made to add common preconditioning routines to the library.

Some further quadrature and optimization material will be included. Current plans are for an adaptive multi-dimensional quadrature routine and a further routine based on a Monte-Carlo method. One more optimization routine to solve general unconstrained nonlinear problems will also be added.

At least one more chapter will be added to the library: "C06 - Summation of Series". This will initially contain two-dimensional and three-dimensional Fast Fourier transforms.

The first release of the NAG Numerical PVM Library contains linear algebra routines for real data only; routines for complex data will be added in future releases.

Further research will be carried out on the effect of slight differences in arithmetic on numerical software running on a heterogeneous network of machines with the intention of allowing the NAG Numerical PVM Library to be run on a heterogenous network of machines. The minimum extension achievable should be to allow the library to be run on a network of machines that conform to IEEE arithmetic. Techniques for "defensive" programming against differences in arithmetic on mixed networks, particularly for iterative methods are being developed at NAG.

The NAG Numerical PVM Library utilities will be slightly extended to allow the user to run library routines on multiple logical processor grids. The first release limits the user to running library routines on a single grid of processors at any time, although other grids may already be declared to run non-library material.

Although PVM is in widespread use, there is a newer message-passing environment that is gaining in popularity, the Message Passing Interface or MPI (see [9]). Implementations of the BLACS using MPI will enable NAG to rapidly produce a version of the material in the NAG Numerical PVM Library based on MPI.

References

1. Dongarra, J.J., Whaley, R.C.: A User's Guide to the BLACS v1.0. Technical Report CS-95-281 (1995), Department of Computer Science, University of Tennessee.
2. Choi, J., Dongarra, J.J., Ostrouchov, S., Petitet, A.P., Walker, D.W., Whaley, R.C.: The Design and Implementation of the ScaLAPACK LU, QR and Cholesky Factorization Routines. Technical Report CS-94-246 (1994), Department of Computer Science, University of Tennessee.
3. Hestenes, M.: Inversion of matrices by biorthogonalization and related results. J. SIAM 6 (1958) 51–90.
4. Koelbel, C.H., Loveman, D.B., Schreiber, R.S., Steele, G.L., Zosel, M.E.: The High Performance Fortran Handbook, The MIT Press (1994).
5. Geist, A., Beguelin, A., Dongarra, J.J., Jiang, W., Manchek, R., Sunderam, V.: PVM 3. A User's Guide and Tutorial for Networked Parallel Computing, The MIT Press (1994).
6. Maclaren, N.M.: The generation of multiple independent sequences of pseudorandom numbers. Appl. Statist. 38 (1989) 351–359.

7. NAG Numerical PVM Library Manual, Numerical Algorithms Group Ltd, Oxford (1995).
8. Dias da Cunha, R., Hopkins, T.: PIM 1.1 — the Parallel Iterative Methods package for Systems of Linear Equations. User's Guide (Fortran 77 version). Technical Report, Computing Laboratory, University of Kent (1994).
9. Gropp, W., Lusk, E., Skjellum, A.: Using MPI. Portable Parallel Programming with the Message Passing Interface, The MIT Press (1994).
10. Piessens, R., De Doncker-Kapenga, E., Überhuber, C., Kahaner D.: QUADPACK, A Subroutine Package for Automatic Integration, Springer-Verlag (1983).

Cellular Automata Modeling
of Snow Transport by Wind

Alexandre Masselot and Bastien Chopard

Parallel Computing Group
CUI, University of Geneva
24, rue du Général Dufour
1211 Genève 4 Switzerland
{masselot/chopard}@cui.unige.ch

Abstract. We present a lattice gas model to simulate snow transport
by wind and its deposition on a given ground profile. Our approach is
very well suited to a fine grained massively parallel computing.

1 Introduction

Massively parallel computing offers new approaches to many scientific problems
by allowing the processing of a large volume of data in an acceptable amount of
time. Here, we present a numerical simulation of snow transport by wind, which
is a phenomena involving many complex processes.

Predicting the location of snow deposit after a wind period is a crucial question to protect roads against snow drift or control the formation of wind-slab
(responsible for about 80% of accidently caused avalanches). Placing snow fences
at appropriate locations can substantially modify the shape of the deposit, by
screening out the wind. Snow fences are also used to store snow on some ski trails.
Thus, a reliable simulation technique of snow transport has a large impact in
exposed areas such as a mountain environment.

The numerical research on this subject is still not very developed and we
proposed a very intuitive 2-D approach to transport, deposition and erosion
of snow particles by wind. Our model considers a description at the particles
level: wind and snow "molecules" are the basic constituents and the dynamics
takes into account only the essential microscopic phenomena, namely those still
relevant at a macroscopic scale of observation.

Note that the transport of sand and the question of dune formation is a
closely related problem which can also be addressed in the present framework.

Our approach is based on the cellular automata technique to simulate a fluid
motion: the so-called FHP lattice gas model[1]. This method is ideally suited
to massively parallel computations and constitutes a promising alternative to
the more traditional numerical solutions of partial differential equations when
complex boundary conditions are involved.

In a first step, we briefly review the cellular automata method of fluid modeling. Then, we explain how snow particles in suspension can be added and
how they interact with wind and pile up on the ground. Finally we discuss the
implementation of this model on a Connection Machine CM-200.

2 The Model

2.1 The Cellular Automata Approach

In a cellular automata approach, the real physical world is represented as a fully discrete universe. Space is discretized as a regular lattice and particles move *synchronously* according to discrete time steps with a finite set of possible velocities. Mutual interactions are local.

In order to guarantee space isotropy, it has been shown that modeling a fluid flow[1] requires a hexagonal lattice (with the six directions $c_i, i \in \{0, \ldots 5\}$), as shown in Fig. 1.

Fig. 1. *The six lattice directions*

The description of the dynamics considers a site occupation variable n_i. At each site r and time t, there is $n_i(r, t) \in \{0, 1\}$ particle traveling along the ith direction (with velocity unity).

Each time step is made of two phases:

- Collision: when several particles meet at the same lattice site r, they interact so as to modify their direction of motion

$$n_i'(r, t) \leftarrow f_i(n_0(r, t), \ldots n_5(r, t)) \tag{1}$$

- Propagation: then, the particles move to a nearest neighbor site, according to their new direction of motion

$$n_i(r + c_i, t + 1) \leftarrow n_i'(r, t) \tag{2}$$

On a massively parallel computer, the spatial domain is partitioned and each part assigned to a different processor. In the data-parallel language, the strategy is to allocate a virtual processor for every lattice sites. All the processors can therefore simultaneously compute step (1). For step (2), each processor communicates with its six neighbors (north, west, south-west, south, east, north-east).

To build an hexagonal lattice from the processor interconnection topology, we transform the square lattice as shown in Fig. 2.

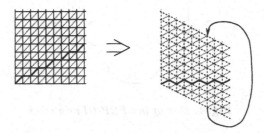

Fig. 2. *transformations to get an hexagonal lattice*

2.2 Wind Modeling

We shall represent the wind as a cellular automata fluid composed of many elementary particles. From the fundamental laws of physics, collisions between wind particle must satisfy mass and momentum conservation. This is the key ingredient in a microscopic model. Thus, we impose

- the mass ($\sum_{i=0}^{5} n_i$) is locally conserved
- so is the momentum ($\sum_{i=0}^{5} n_i c_i$)
- collisions rules are the same modulo $\frac{\pi}{3}$ rotations.

One of the simplest cellular automata model of fluid is the FHP model proposed in 1986[1]. If only one particle is entering a site, it goes straight to its neighbor in its direction of motion. If many particles are entering a site, they collide according to the rules given in Fig. 3

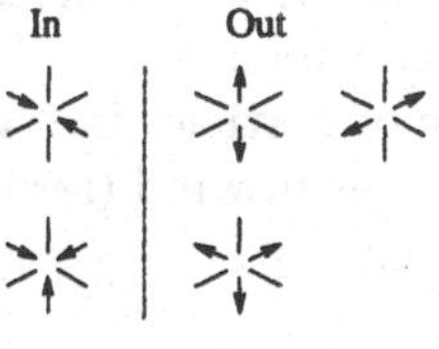

In   Out

Fig. 3. *FHP collisions rules; in the first case, we must choose randomly between the two outcomes. All the other configurations remain unchanged.*

A FHP fluid has a built-in viscosity, determined by the above collision rules. For many practical purposes, the viscosity of the FHP model is to large to simulate flows at high enough Reynolds numbers. In the so-called FHP-III model, rest particles and more sophisticated collision rules (for example, see Fig. 4) are considered to lower the viscosity.

Ultimately, we would like that our cellular automata fluid obeys the physical laws of fluid motion. The Navier-Stokes equation is the standard equation describing the flow of a real fluid. In the macroscopic limit and within some approximations, it can be shown[1] that the FHP dynamics indeed evolves according to the Navier-Stokes equation.

Fig. 4. *One of the FHP-III new rules*

Boundary Conditions For our simulation, we consider a two-dimensional hexagonal lattice of size $L_x \times L_z$, corresponding to a vertical slice of the real three-dimensional space. Wind is blowing from left to right. Since we have only a finite system, we must find appropriate boundary conditions to simulate an infinite system in the x-axis and semi-infinite in the z-axis.

Ground boundary: Ground is modeled by "solid" lattice sites. We either apply the slip (bounce forward) or the no-slip (bounce back) condition when a wind particle collides with the ground (or with a snow deposit).

Sky boundary: On the upper boundary, we simply inverse the vertical speed of a particle, without modifying its horizontal speed.

Left and right boundaries: Wind particles are injected, at a given speed from the left side. At the right side, they should behave as if the system were infinitely long. This problem is solved by considering a zero-gradient method: we add particles from the left to the right, and from the right to the left, according to three criteria:

- the density (number of particles) is conserved over the tunnel.
- The input speed has a given value.
- The vertical density profile of particles inserted backward is proportional to the profile observed a few sites upwind (zero-gradient method).

2.3 Snow Modeling

Snow particles are also represented as particles moving on the same lattice as wind particles. Without a wind flow, snow particles fall due to gravity. When they reach the ground, two kinds of phenomena can be considered

- There is a cohesion force and each particle sticks to the deposit, whatever the local configuration is.
- There is no cohesion and, at the landing, a particle rolls down until it finds a stable configuration (see Fig. 5).

2.4 Wind-Snow Interactions

The real wind-snow interactions are quite complex and still not fully understood. According to the spirit of our model, we consider only simplified interactions. Two main phenomena are taken into account in the present version of the model:

Fig. 5. *situation (1) is stable, but the (2) is not; the flake must go down.*

Transport: Flying snow is subject to gravity, but also to wind forces. On each lattice site, we compute the new direction of a snow particle by computing the mean wind speed in an hexagonal neighborhood. This driven force is combined with gravity. This "force" acts statistically, due to the restrictions imposed by a cellular automata approach.

Erosion: It is known that up to 80% of snow transport takes place due to the phenomena of saltation [3]: a micro-eddy pulls out a snow flake and, sometimes, a larger one makes it fly away. We model this phenomenon by assigning to each landed snow particle a counter of wind impacts; after a given number of impacts the snow particle can take off and, perhaps, fly if the wind is locally strong enough.

3 Computer Implementation

3.1 From Boolean...

On each site r, the wind configuration is given by $b = 6$ bits (7 with rest particles), which can be stored as an integer $n(r) = \sum_{i=0}^{b-1} 2^{n_i}$. Our two-phase algorithm (collision and propagation) discussed in Sect. 2.1 becomes:

- $n(r) \leftarrow f(n(r))$ where $f : \{0, 2^b - 1\} \mapsto \{0, 2^b - 1\}$; one notices that the function f can be computed once, and the results for all the values $f(i), i \in \{0, 2^b - 1\}$ stored in an array (lookup table) on each processor,
- we stream the ith bit of $n(r)$ in the direction c_i

The wind and snow configurations are both fully determined by 6 or 7 bits. The solid configuration also requires 7 bits per site (6 of them to know whether there is a solid site in direction c_i and the last one to know if the site itself is solid).

Therefore, the configuration at a given site is fully determined by 21 bits and we could make a general evolution function $F : \{0, 2^{21} - 1\} \mapsto \{0, 2^{21} - 1\}$. Problems occur when we try to store 2 mega-words on each processor, which cannot be done on our CM-200 (although each lookup table copy is shared among 32 processors). We must split the evolution in four interactions (wind-wind, wind-snow, wind-solid, snow-solid).

3.2 ... to Real Valued Quantities

Instead of describing the dynamics in terms of the absence $n_i = 0$ or presence $n_i = 1$ of a wind particle in every direction, we can also look at the probability $< n_i > \in [0, 1]$. That is the *Boltzmann approximation* which neglects N-body correlations and replaces the boolean operators *.or.*, *.and.* and *.not.* by the floating point operations $+, *, 1 - ..$ A direct computer simulation of the evolution equation for the $< n_i >$ is a technique known as the Lattice Boltzmann Method (LBM)[2]. It reduces statistical noise, and allows much more flexibility when adjusting the system parameters (such as density, entry speed, viscosity...).

4 A Few Results

The simulations we have performed so far with our model give quite promising results. Good qualitative agreement[4] is obtained with *in-situ* observations and wind tunnel experiments. The next two figures illustrate the snow deposit behind a fence and the wind behavior around it. Depending on the type of numerical experiment, the execution time typically ranges from several minutes to a few hours, for a system of size 128×1024 on a 8k processors CM-200.

In collaboration with the Swiss Institute for Snow and Avalanche Research, we are currently improving the model toward a more effective LBM scheme and a better modeling of snow-wind interactions.

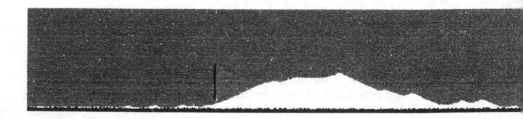

Fig. 6. *Cellular Automata model. Wind blows from left to right trough a tunnel. The fence has a ground clearance. Snow deposit is qualitatively close to reality.*

Fig. 7. *Reattachment wind isolines, when the wind is modeled by LBM. Same reattachment distances are observed in real wind tunnel.*

Acknowledgments

This research is supported by the Swiss National Science Foundation.

References

1. U. Frish B. Hasslacher et Y. Pommeau. *Phys. Rev. Lett. 56 (1986) 1505*; "Lattice gas method for partial differential equations," G. Doolen Edt., Addison-Wesley, (1990).
2. R. Benzi S. Succi M. Vergalossa. The lattice Boltzman equation : theory and application. *Physics Reports 222 No 3 (1992) p. 145-197*
3. Thierry Castelle. Transport de la neige par le vent en montagne: approche expérimentale du site du col du Lac Blanc. *PhD thesis, EPFL Lausanne, (1995)*
4. J.K. Raine D.C. Stevenson. Wind protection by model fences in a simulated atmospheric boundary layer. *Journal of Industrial Aerodynamics, 2 (1977) p.159-180*

Parallel Algorithm for Mapping of Parallel Programs into Pyramidal Multiprocessor*

O.G.Monakhov

Computing Centre, Sibirian Division of Russian Academy of Science,
Pr. Lavrentiev 6, Novosibirsk, 630090, Russia,
e-mail: monakhov@comcen.nsk.su

Abstract. A problem of mapping of an information graph of a complex algorithm into the pyramidal interprocessor network of a parallel computer system is considered. The parallel recursive algorithm for optimal or suboptimal solution of the mapping problem, the objective functions for mapping and experimental results for the pyramidal multiprocessor system MEMSY are presented.

1 Introduction

There is considered a problem of mapping of the information graph of a complex algorithm into the pyramidal interprocessor network of the parallel computer system (CS), which consists of processor nodes (PN) with distributed shared memory. The mapping problem [1-4] is known as NP-complete problem, and it is reasonable to develop a parallel algorithm for solving the problem. In this paper the parallel recursive algorithm is presented for optimal or suboptimal solution of the mapping problem, the objective functions for mapping are developed. It is shown by experiments on multiprocessor system MEMSY that the proposed algorithm is faster as compared to the sequential centralized algorithm.

2 Optimal mapping problem

Let a model of a parallel program be the graph $G_p = (M, E_p)$, where M is a set of modules (processes), E_p be a set of edges, representing information connections between modules. Let t_i be defined as weight of the module $i \in M$, representing the execution time (or the number of computational steps) of the module i. Let v_{ij} be defined as weight of the edge $(i, j) \in E_p$, representing the number of information units passed from the module i to the module j.

A model of the multiprocessor system with distributed memory is an undirected graph $G_s = (P, E_s)$ representing the network topology (structure) of the system, where P is a set of PN, and edges E_s represent interconnection links between PN.

* This work is supported by RFBR project N94-01-00682

The distance between the nodes i and j of the graph $G_s(G_p)$ is denoted as d_{ij}. The neighbourhood of the node i with the radius $\rho \geq 1$ is the set $L_\rho(i) = \{j \in M \mid d_{ij} \leq \rho\}$. Let $L(i) = L_1(i)$.

Let $\varphi : M \rightarrow P$ be the mapping of an information graph of the parallel program G_p into the structure G_s of CS. Let the mapping φ be represented by the vector $X = \{x_{ik}; i \in M, k \in P\}$, where $x_{ik} = 1$ if $\varphi(i) = k$ and $x_{ik} = 0$ if $\varphi(i) \neq k$.

Let the quality of the mapping parallel program graph into the structure of CS for the given vector X is described by the functional: $F(X) = F_E(X) + F_C(X)$, where $F_E(X)$ represents computational cost (the overall module execution time of a parallel program on the system or the load balancing of PN for given X) and $F_C(X)$ represent the interprocessor communication cost (the overall interaction time between modules, which are distributed in different PN, or the distance between adjacent modules of the program graph on CS structure for given X). The optimal mapping problem of a parallel program graph into CS structure consists in optimization of the functional $F(X)$ by means of the parallel recursive algorithm.

3 Cost functionals of mapping

Now let us describe two examples of the objective cost functions which represent the computational load balance of PN and the communication cost for the given mapping X.

$$F_1(X) = \sum_{k=1}^{n} [(\sum_{i=1}^{m} x_{ik} - Mx_k)^2 + \sum_{p=1}^{n} \sum_{i=1}^{m} \sum_{j \in L(i)} d_{kp} x_{ik} x_{jp}],$$

$$\text{where } Mx_k = \sum_{p \in L_\rho(i)} \sum_{i=1}^{m} x_{ip} / \mid L_\rho(k) \mid .$$

The first term in this expression describes deviation of PN_k load from average load in the neighbourhood of PN_k with the radius $\rho \geq 1$, the second term describes the distance between PN_k and processors, which contain the adjacent modules to the modules embedded into PN_k, $1 \leq k \leq n$.

$$F_2(X) = \sum_{k=1}^{n} [\sum_{i=1}^{m} t_i x_{ik} + \sum_{p=1}^{n} \sum_{j \in L(i)} c_{ij}(v_{ij}) d_{kp} x_{ik} x_{jp})],$$

where $c_{ij}(v_{ij})$ is the time needed to transfer v_{ij} data units between the modules i and j, when they are located in the neighbouring PN. The first term in this expression describes the overall module execution time, the second term describes the overall interprocessor communication time, $m = \mid M \mid, n = \mid P \mid$ and $\sum_{k=1}^{n} x_{ik} = 1$.

Thus, the optimization of the mapping φ consists in minimizing the nonlinear function $F(X)$ with linear restrictions and integer variables. Let Z denotes this task. The optimal solution of the task Z will be found by the following parallel recursive mapping algorithm.

4 Parallel recursive mapping algorithm

In this section, there is presented a hierarchical recursive parallel mapping algorithm for partitioning the modules of a parallel program, described by the graph G_p, into $n \geq 4$ sets and allocating each set on its own processor of the pyramidal multiprocessor system. The goal of the algorithm is to produce an allocation with the minimal communication cost and the computational load balance of processors.

At the first step algorithm divides a given parallel program graph into four parts and finds the optimal partition according to the given cost function. Then, each part of the graph is allocated to each node on the upper plane of the system. Each node on the upper plane divides its part of the program graph into five parts and allocates these parts to its own node and to four down nodes on the lower plane. Then, this algorithm is recursively repeated on the lower plane and on each other plane of the system until the bottom plane. In the end each node of the system has its own part of the program graph (Fig. 1).

The algorithm contain a basic procedure - bisection, which divides the modules of the program graph into two sets with the minimal communication cost between them and roughly equal computation load. The bisection algorithm consists of following steps.

1. All modules are divided into two sets: $\mid P_1 \mid = n - 1$ and $\mid P_2 \mid = 1$.

2. A module $i \in P_1$ is removed to P_2, where the module i has the maximal gain $D_i = \max_i \{E_{12}(i) - I_{12}(i); i \in P_1\}, E_{12}(i) = \{\sum_j d_{ij}; j \in P_2\}, I_{12}(i) = \{\sum_j d_{ij}; j \in P_1\}$.

3. If $\{\sum_j t_j; j \in P_2\} < \{\sum_j t_j; j \in P_1\}$, then go to step 2, else end.

The hierarchical recursive mapping algorithm divides the given parallel program graph and allocates the parts of the program graph on processors of a system with pyramidal topology. The upper level (plane) of the system contains 4 nodes and other levels contain 4^k nodes, where k is the number of levels. The recursive algorithm consists of the following steps.

1. The modules of the program graph are divided by bisection algorithm into 4 parts (at first - into two parts, then each part - again into two parts) and are allocated to 4 processors of the upper plane.

2. Each processor of the upper plain divides its part of the program graph into five parts and allocates these parts to its own node and to four down nodes on the lower plane connected with this node. Each processor $p_j \in \{p_1, p_2, p_3, p_4\}$ of the upper plane executes (in parallel with others) the following algorithm.

2.1. All modules allocated to p_j are divided into two sets: S_1 and $S_2 = \emptyset$. Let $M_2 = M \backslash S_1$.

2.2. A module $i \in S_1$ is removed to S_2, where the module i has the maximal gain $D_i = \max_i \{E_{12}(i) - I_{12}(i); i \in S_1\}, E_{12}(i) = \{\sum_j d_{ij}; j \in M_2\}, I_{12}(i) = \{\sum_j d_{ij}; j \in S_1\}$.

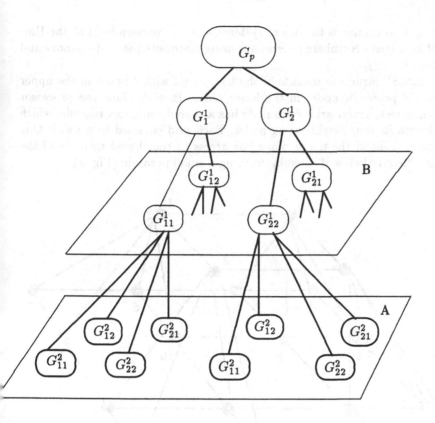

Fig. 1. Graph of the recursive mapping algorithm

2.3. While $\{\sum_j t_j; j \in S_2\} < \{\sum_j t_j/N; j \in M\}$ go to step 2.2.

2.4. Modules of the set S_2 are allocated to the processor p_j and modules of the set S_1 are divided by the bisection algorithm into 4 parts and are allocated to 4 processors of the lower plane (as at Step 1).

2.5. In parallel, each processor of lower plain applies the algorithm recursively from Step 2.1.

The algorithm is recursively repeated until the bottom plane of the system and can map of a program graph into the pyramidal system with a given number of levels.

5 MEMSY - pyramidal multiprocessor system

MEMSY (Modular Expandable Multiprocessor System) [5] is an experimental multiprocessor system with a scalable architecture based on locally shared memory between a set of adjacent nodes and other communication media. The

MEMSY system continues the line of systems which have been built at the University of Erlangen - Nurnberg (Germany) using distributed shared-memory and pyramidal topology.

The MEMSY structure consists of the two planes with 4 nodes in the upper plane and 16 processor nodes in the lower plane. In each plane the processor nodes form a rectangular grid. Each node has a shared - memory module, which is shared with its four neighbouring nodes. Each grid is closed to a torus. One processing element of the upper plane has access to the shared memory of the four nodes directly below it, thereby forming a small pyramid (Fig. 2).

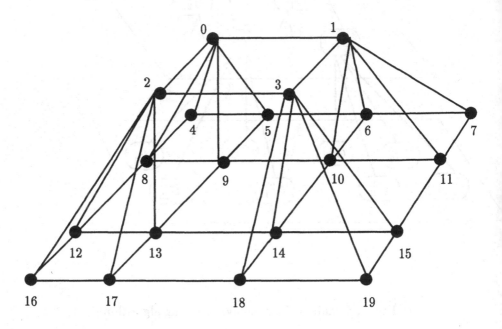

Fig. 2. Structure of MEMSY system

The MEMSY consists of following functional units: 20 processor nodes, one shared-memory module (communication memory - 4 Mbytes) at each node, the interconnection network between processor nodes and communication memories, a special optical bus (FDDI net) connecting all nodes, a global disk memory (1,57 Gbytes). Each node of the MEMSY consists of four processors $MC'88000$ with 25 Mflops performance, 32 Mbytes local memory, 500 Mbytes local disk memory.

The programming model of the MEMSY was designed to give a direct access to the real structure and the power of the system. The application programmer can use a variety of different mechanisms for communication and coordination defined as a set of system library calls which can be called from C and $C++$ languages. There are the following mechanisms for communication and coordination: shared communication memory between neighbouring nodes, message passing mechanisms, semaphores and spinlocks, FDDI net for fast transfer of high volume data. The operating system of MEMSY (MEMSOS) is based on Unix adapted to the parallel hardware. The multitasking/multiuser feature of Unix and traditional I/O library calls for local and global data storage are supported. The MEMSOS allows different applications (single user parallel program) to run simultaneously and shields from one another.

6 Experimental results on MEMSY system

The proposed parallel mapping algorithm has hierarchical recursive structure and suits for the pyramidal topology of MEMSY. This algorithm was implemented on MEMSY for one, two and four nodes with communication via shared memory. In experiments there were allocated simple program graphs - square grids with a given size, but arbitrary weighted graphs of parallel programs can be also allocated. There were obtained mapping of the given graphs on all MEMSY nodes, the computing time, speed up and efficiency of the mapping algorithm and values of the objective cost function. The results are presented in Tables 1 and 2.

Table 1. Computing time (TN), speed up $(SN = T1/TN)$ and efficiency $(EN = SN/N)$ of the parallel mapping algorithm executed on N nodes for mapping of square grid $(a \times a)$ into MEMSY architecture

Size of grid	T1 sec.	T2 sec.	T4 sec.	S2	S4	E2	E4
10 x 10	8	5	4	1.6	2.0	0.8	0.5
15 x 15	97	57	40	1.7	2.43	0.85	0.61
20 x 20	529	310	201	1.7	2.63	0.85	0.66
25 x 25	2058	1190	874	1.72	2.35	0.86	0.59
30 x 30	6145	3558	2510	1.72	2.45	0.86	0.61
35 x 35	15839	9421	5931	1.68	2.67	0.84	0.66
40 x 40	35253	20359	13162	1.73	2.68	0.865	0.67

Table 2. The objective cost function values (FN) obtained by the parallel mapping algorithm on N nodes for mapping of square grid ($a \times a$) into MEMSY architecture

Size of grid	F1	F2	F4	F1/F2	F1/F4
10 x 10	840	840	844	1.0	0.995
15 x 15	1778	1770	1676	1.004	1.06
20 x 20	2488	2488	2546	1.0	0.997
25 x 25	4326	4230	4184	1.02	1.033
30 x 30	5786	5786	5726	1.0	1.01
35 x 35	7884	7716	7628	1.02	1.033
40 x 40	9600	9600	9546	1.0	1.005

Thus, the results show that the proposed parallel recursive algorithm is efficient and produces an optimal or good suboptimal solution of the mapping problem. The algorithm has hierarchical recursive structure and suits well for application on the multiprocessor system MEMSY.

Acknowledgments

I would like to thank Professor H.Wedekind for supporting this work. I also thank MEMSY system group and especially T.Thiel and S.Turowski for helpful consultations and providing access to MEMSY system.

References

1. N.N.Mirenkov, *Parallel programming for multimodule computer systems*, Radio i svyas, Moscow, 1989, 320p.
2. F.Berman, L.Snyder, On mapping parallel algorithms in parallel architectures.-*J. Parallel Distrib. Comput*, 4 (1987) pp.439-458.
3. D.Fernandez-Baca, Allocating modules to processors in a distributed system.-*IEEE Trans. Software Eng*, 15 (1989) pp.1427-1436.
4. O.G.Monakhov, Parallel mapping of parallel program graphs into parallel computers.- Proc. Internat. Conf. "Parallel Computing 91".-Elsevier Science Publishers, Amsterdam, 1992. - pp.413-418.
5. F.Hofman, M.Dal Cin, A.Grygier, H.Hessenauer, U.Hildebrand, C.-U.Linster, T.Thiel, S.Turowski, MEMSY: a modular expandable multiprocessor system. - Technical report. Univerity of Erlangen-Nurnberg, 1992, 18p.

Data-Parallel Molecular Dynamics with Neighbor-Lists

Ole H. Nielsen[1,2,3]

[1] UNI•C, Technical University of Denmark, Bldg. 304, DK-2800 Lyngby, Denmark,
[2] Center for Atomic-scale Materials Physics (CAMP), Physics Dept., Technical
University of Denmark, Bldg. 307, DK-2800 Lyngby, Denmark,
[3] Joint Research Center for Atom Technology (JRCAT), NAIR, 1-1-4 Higashi,
Tsukuba, Ibaraki 305, Japan.

Abstract. We report on a data-parallel classical molecular dynamics algorithm and its implementation on Connection Machines CM-5 and CM-200 using CM-Fortran, a Fortran-90 dialect. A grid-based spatial decomposition of the atomic system is used for parallelization. Our previous algorithm on the CM's calculated all forces in the nearby region. A different algorithm using classical Verlet neighbor lists is more efficient, when implemented on the CM-5 with its indirect addressing hardware. The code has been used for extensive simulations on a 128-node CM-5E, and performance measurements are reported.

1 Introduction to parallel molecular–dynamics

Molecular Dynamics (MD) simulations of atomic or molecular microscopic systems are very important in several fields of science. The present project at Center for Atomic–scale Materials Physics of the Technical University of Denmark, in collaboration with Joint Research Center for Atom Technology in Tsukuba deals with the materials–physics properties of "large" metallic solid–state systems, consisting of typically 10.000 to 1 million atoms.

The basis of all MD studies is a theoretical model of the interactions between the constituent atoms. In the present work a theoretical framework called the *Effective Medium Theory*[1] (EMT) for atomic interactions has been used. In this theory, charge density "tails" from neighboring atoms are accumulated by every atom, and a non-linear functional is used to derive the interatomic energies and forces. Electrostatic energies are added to the density term, and one-electron energies may be included as well if required. Owing to the physical properties of metals, these interactions have a short range of 4–5 atomic radii. Some forms of the EMT theory resemble semi-empirical methods such as the *Embedded Atom Method*[2].

Short ranged interactions dictate the types of algorithms chosen for MD simulations: Only well localized interactions need to be taken into account, in contrast to other types of systems requiring global interactions. Hence MD algorithms must be considered which allow for efficient access to atomic data in localized regions of space. A review of current research in parallel short-range MD algorithms is presented in ref. [3].

The parallel algorithm chosen in the present work is based upon a decomposition of the physical system into identical cells laid out on a regular 3–dimensional grid. Thus the algorithm only needs to consider data that reside in nearby cells on the grid. This is naturally reflected in the implementation on distributed–memory parallel computers such as the Connection Machines. This was briefly described in our previous work[4], which also reported performance measurements for CM-200 and CM-5E supercomputers.

2 The parallel Verlet neighbor-list algorithm

Even if the implementation cited above is using the Connection Machine hardware with extremely high efficiency, there is an algorithmic problem: When forces are computed between atom pairs in neighboring grid-cells, many of the forces are in fact computed for pairs whose interatomic distance is greater than the cutoff-radius of the interactions.

A kind of "algorithmic efficiency" can be estimated as the number of "relevant" non-zero interactions compared to the *total* number of calculated interactions as follows: If the cutoff-radius (or rather the radius at which the smooth cutoff-function employed in the EMT is considered to be effectively zero) is equal to the grid-cell side length D, the volume of the sphere of non-zero interactions around a given atom is $\frac{4}{3}\pi D^3$, whereas the interactions are computed inside a volume $(3D)^3$ for local $3 \times 3 \times 3$ interaction subgrids. The ratio of these volumes, which is an estimate of the algorithmic efficiency of the force computation, is approximately 0.155. If the code uses local $5 \times 5 \times 5$ subgrids and a cutoff-radius of $2 \times D$, the algorithmic efficiency is a much better 0.268. We implemented[4] a skipping of some cells in the $5 \times 5 \times 5$ subgrid, leading to about 20% improvement of this algorithmic efficiency. Larger local subgrids could be used, at the cost of more communication and a decreased number of atoms per cell. This would lead to decreased load-balance, however, so there is an upper limit to proceeding along this line.

Thus we conclude that a fairly low upper bound to the algorithmic efficiency of using local $n \times n \times n$ subgrids exists, when all interactions are computed. One possible resolution of this low efficiency is to use the classical Verlet neighbor-list method[5], but in the context of a parallel algorithm. One implementation of such an algorithm by Tamayo is discussed in ref. [3] (section 5).

We have implemented a standard Verlet neighbor-list algorithm in the context of our previously described parallel MD algorithm[4], with the aim of achieving the fastest possible execution time at the expense of using large amounts of RAM memory for storing neighbor-list related data in each parallel node.

Thus we are explicitly *not* aiming for simulating the largest possible systems, but rather on the fastest solution of more modest-sized systems (of the order of 1 million atoms). This tradeoff was decided in view of the physics that we want to investigate. Another aspect that influences algorithmic choices is the fact that our EMT interactions, which are more appropriate for simulating metallic systems, require more than 10 times the number of floating-point operations per pair-interaction, than for the case of Lennard-Jones interactions which are usually employed in MD studies. Hence communications overheads in our algorithm will be intrinsically much less significant than for the simpler Lennard-Jones interactions.

The implementation of a Verlet neighbor-list algorithm in a data-parallel CM-Fortran MD code is relatively straightforward. The task that distinguishes the data-parallel implementation from the usual serial one is the access to data in non-local processors. Our choice is to simply *copy* the data (atomic (x, y, z) coordinates, plus some interaction data) from the nearby $3 \times 3 \times 3$ local subgrid into the memory of the local processor. The CSHIFT Fortran-90 intrinsic is used for copying the data. This task is sketched in Fig. 1, and it is carried out whenever the atoms are moved.

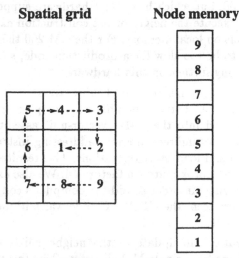

Spatial grid   **Node memory**

Figure 1. Copying of of 2-dimensional spatial grid data (left) onto the node's local memory (right). Data from cells 2-9 are replicated in the local memory of the node representing cell 1. In 3-D there will be a total of 27 cells in a $3 \times 3 \times 3$ subgrid.

Of course, a smaller part of the data will not represent neighbors of any atom in the central cell of Fig. 1, and hence this data could in principle be omitted. Also, it can be argued that it is a waste of memory to replicate atomic data from 26 neighboring cells. This is a deliberate choice that we have made: We want the most regular (and hence the fastest) communication pattern of data between the "vector-unit" memory of the CM-5E nodes. We also want all atomic data in a linear list in memory so that a Verlet neighbor-list algorithm can be implemented simply and efficiently. Our aim is to get the fastest possible execution at the cost of using extra memory. Since the 128-node CM-5E machine at our disposal has 128 MB of memory per node, and we want to simulate systems of only up to 1 million atoms, we can easily afford to use memory so freely.

With all relevant data available in the local memory of each node, there remains to construct the neighbor lists for each atom. Owing to all the atomic data from the central and neighboring cells being arranged in a simple linear list in local memory, it is a simple matter to traverse the list in a data-parallel fashion for each atom in the central cell and add any neighbors to a list of pointers (the *neighbor-list*). In CM-Fortran the atom-atom distance is calculated by a FORALL statement, and the decision whether to include the atom in the neighbor-list is implemented by a MERGE Fortran-90 intrinsic function. The pointers to atoms are added to the list by a "scatter" operation, which was first coded as a FORALL statement, but since the present CM-Fortran compiler (version CM5 VecUnit 2.2.11) hasn't implemented indirect memory access efficiently for 3-D arrays, we decided to use the faster CM-Fortran library CMF_ASET_1D routine in stead.

The performance-critical part of the above construction is the "scatter" operation, which requires local indirect-memory addressing that is emulated in software on the CM-200, but which has direct hardware support on the CM-5, so that the CM-5 can execute the neighbor-list construction exclusively with efficient vector operations on local memory. For the CM-200 the indirect-memory addressing was found to be too slow for a production code, so the present algorithm wasn't pursued any further on this hardware.

Since the CM-Fortran compiler generated sub-optimal code for local indirect-memory addressing, the CM-Fortran neighbor-list loop (a total of 7 Fortran statements) was translated into the CM-5 vector-unit assembler "CDPEAC", taking optimal advantage of indirect-memory addressing instructions and overlapping of computation and load/store operations. This resulted in a speedup of the neighbor-list loop by approximately a factor of 4. We also experimented with CM-Fortran "array aliasing" in order to aid the compiler's code generation, and while improvements were seen, the CDPEAC code was still more than twice as fast as the compiler's code.

After this optimization, an update of the neighbor-lists consumes a time roughly equal to the time for a single MD timestep. This means that the classical trick of increasing the cutoff-radius by adding a *skin* radius must be used in order to re-use the neighbor-list during several MD timesteps. We monitor exactly the atoms' deviations from their positions at the time of the neighbor-list construction, and when the maximum deviation exceeds half of the skin radius, an update of the neighbor-list is carried out. We believe that it is important not to postpone updates beyond this point, since the MD simulations would then not conserve energy. Atoms will migrate between cells during the simulation, and such migration-updates are performed prior to the neighbor-list updates.

When atomic forces are calculated and summed in a loop over the atoms in the neighbor-list, a simple parallel "gather" operation of the neighbor-atoms' data is performed using FORALL on the data structure sketched in Fig. 1. When the atomic data are in CM-5 vector-registers, the forces can be calculated by the EMT method as in ref. [4]. Again, the efficiency of the force computation depends partially on the compiler-generated code for indirect-memory access, but less so than for the neighbor-list generation because many floating-point operations are executed for each "gather" operation. We have considered CDPEAC coding of the "gather" operation, but so far this hasn't been done.

3 Performance measurements on CM-5E

The code described above, being a relatively limited extension of the code in ref. [4], was used for large-scale production on the 128-node CM-5E at the JR-CAT computer system in Tsukuba. Scientific results of using the code for studies of kinematic generation of dislocations in multi-grained copper crystals under strain have been reported in ref. [6].

Here we present performance results for this code on a typical dataset from ref. [7], which can be compared to the results of our previous paper[4]. A copper cluster with 178.219 atoms was simulated using a spatial grid of $16 \times 16 \times 16$ cells. A near-melting-point temperature of about 1400 K was used as a rather difficult test-case, giving rise to significant diffusion of atoms and hence frequent updates of the neighbor-list and migration of atoms between cells.

The performance of the new algorithm compared to the previous one[4] is shown in Table 1:

Part of the code	Present work	Old work (ref. [4])
Atomic forces kernel	0.91	6.52
Kernel: communication	0.15	1.46
Atomic rearrangement	0.35*	0.50
Neighbor-list update	1.39*	
Parallel disk I/O	0.10	0.11
Total	$\approx$1.8*	8.62

Table 1. Elapsed time (seconds) for a single MD time-step, as measured on a 128-node CM-5E.
(* updates only performed every $\approx$3 timesteps, see the text below)

Since the algorithms with/without neighbor-lists differ significantly, one cannot compare all items in the table directly. However, the **atomic forces kernel** is now faster by a factor of more than 7 owing to the introduction of neighbor-lists. The kernel's computational speed is 5.1 GFLOPS, compared to the 11.7 GFLOPS for the "old" code[4], the difference being mainly due to ineffective code for indirect memory addressing through FORALL statements. Thus, the old code was more efficient, but due to the algorithm used, it still performed much less useful work per unit of time than the code implementing the present algorithm. The 5.1 GFLOPS corresponds to 25% of the CM-5E vector-units' theoretical peak performance, and this a very good absolute performance number.

The parallel load-balance due to fluctuations in atomic density is about the same for the new and old algorithms, and presents no problem for the present metallic solids. The system simulated in this work contained a significant fraction of cells containing vacuum (no atoms), whereas the present grid-based algorithms are more efficient for "full" systems such as bulk materials. There are imbalances in the present system's neighbor-lists: A neighbor-list contains up to about 83 atoms with an average of 68 atoms, when a skin-radius of 10% of the cutoff-radius is used. In a data-parallel code the longest neighbor-list determines the loop-length, so we have only about 80% load-balance here.

The **kernel: communication** item in Table 1 accounts for the data copying sketched in Fig. 1, constituting only 14% of the total force computation time. The **parallel disk I/O** using the CM-5E parallel SDA disks is a negligible part of the MD timestep.

The combined **atomic rearrangement** and **neighbor-list** updates require a time of about 1.6 times that of the the force computation. Fortunately, the neighbor-list is usually performed every 5-20 MD timesteps. In the present system at near-melting, however, updates are done about every 3 MD timesteps. This leads to an average MD timestep of about 1.8 seconds, or almost 5 times faster than the "old" algorithm for the same problem. For lower temperatures the neighbor-list updates would be less frequent, and the algorithmic speedup would be larger.

4 Conclusions

The present work reports on a data-parallel molecular dynamics algorithm using Verlet neighbor-lists for the local force computations, which achieves 5.1 GFLOPS on a 128-node CM-5E. The force computation is as efficient as any serial implementation would be, except for a 20% reduction due to variations in neighbor-list length. The parallel code does local indirect memory addressing less than optimally due to compiler shortcomings, so truly optimum efficiency would require an improved compiler or a recoding in CDPEAC assembler for the CM-5. For the case of neighbor-list construction we used the latter approach. The construction of neighbor-lists takes about 1.6 times the force computation, but it only needs to be carried out after several MD timesteps.

The final load-balancing issue is related to distributing the system in a uniform grid, but for the case of fairly homogeneous solid systems such as the metals we're studying this issue is not significant.

The parallel Verlet neighbor-list code is about 5 times faster than our previous parallel code, which did not emply neighbor-lists.

In conclusion, we only need the CM-Fortran compiler to be able to generate efficient code for local indirect-memory addressing on the CM-5E vector-units in order for our data-parallel CM-Fortran code to be optimally efficient. If this were achieved, the code's performance should be as high as that of any serial neighbor-list algorithm, except for the minor load-imbalances discussed above.

Our CM-Fortran code is not restricted to Connection Machines, but should be portable with a limited effort to High-Performance Fortran (HPF) compilers that support the `FORALL` statement.

5 Acknowledgments

Dr. Jakob Schiøtz provided important feedback in the optimization process of the present CM-Fortran code. We are grateful to Palle Pedersen of Thinking Machines Corp. for CDPEAC assembler optimization of the CM-Fortran neighbor-list subroutine. CAMP is sponsored by the Danish National Research Foundation. The author would like to acknowledge computing facilities at JRCAT Supercomputer System in Tsukuba, Japan, as financially supported by New Energy and Industrial Technology Development Organization (NEDO).

References

1. K. W. Jacobsen, J. K. Nørskov, and M. J. Puska, Phys. Rev. B **35**, 7423 (1987); K. W. Jacobsen, Comments Cond. Mat. Phys. **14**, 129 (1988).
2. S. M. Foiles, M. I. Baskes, and M. S. Daw, Phys. Rev. B **33**, 7983 (1986)
3. David M. Beazley, Peter S. Lomdahl, Niels Grønbech-Jensen, Roscoe Giles, and Pablo Tamayo, *Parallel Algorithms for Short-Range Molecular Dynamics*, to appear in Annual Review of Computational Physics **3**, Ed. D. Stauffer (World Scientific, 1995). Available on Internet at URL:ftp://think.com/users/tamayo/md
4. Ole H. Nielsen, *Molecular dynamics: experiences with Connection Machines CM-200 and CM-5*, in proceedings of *Workshop on Parallel Scientific Computing (PARA94)*, ed. J. Wasniewski, (Springer-Verlag, Berlin, 1994), Lecture Notes in Computer Science, vol. 879.
5. L. Verlet, Phys. Rev. **159**, 98 (1967).
6. J. Schiøtz, K. W. Jacobsen and O. H. Nielsen, *Kinematic generation of dislocations*, Phil. Mag. Lett., in press.
7. O. H. Nielsen, J. P. Sethna, P. Stoltze, K. W. Jacobsen and J. K. Nørskov, *Melting a copper cluster: Critical droplet theory*, Europhysics Letters, **26**, 51 (1994).

Visualizing Astrophysical 3D MHD Turbulence

Åke Nordlund[1,2], Klaus Galsgaard[1,3] and R.F. Stein[4]

[1] Astronomical Observatory / NBIfAFG, Øster Voldgade 3, 1350 Copenhagen K, Denmark
[2] Theoretical Astrophysics Center, Blegdamsvej 17, 2100 Copenhagen Ø, Denmark
[3] University of St. Andrews, The Dept. of Math., St Andrews, FIFE, KY16 9SS, Scotland
[4] Dept. of Physics and Astronomy, Michigan State University, East Lansing, MI 48824, USA

Abstract. Some results from a recently developed magnetohydrodynamics code for massively parallel processing are presented. Although primarily intended for massively parallel computers such as the Connection Machine and the IBM SP-2, this Fortran-90 code also parallelizes and vectorizes well on traditional supercomputers such as the Cray C-90. Techniques for visualizing and animating vector fields are described, and factors that contribute to the overall scientific productivity of a computing project are discussed.

1 Introduction

Magnetic fields are ubiquitous in astrophysics, where the large scales imply that even plasmas with relatively high electrical resistivities must be considered as "almost perfectly conducting". Astrophysical plasmas on every conceivable scale between the magnetospheres of planets and large scale galactic superclusters are known to have, or are suspected to have, dynamically active magnetic fields.

Plasmas with magnetic fields are often extremely intermittent, either in the magnetic field itself or, in the case of magnetically dominated plasmas, in the electrical current associated with the reconnection and dissipation of the magnetic field. Such plasmas have therefore been difficult to model and understand with analytical methods. Conditions even remotely similar to the ones relevant in astrophysics are difficult to achieve in laboratory experiments, and numerical experiments in the form of massive computer simulations are thus an attractive alternative.

We are currently engaged in a number of such experiments. In one set of experiments we are studying dynamo action; i.e., the spontaneous creation of magnetic energy in turbulent plasmas [14, 15, 1]. In these experiments, the turbulence is sub-sonic, and the magnetic energy is initially small compared to the kinetic energy of the plasma. In a similar set of experiments, we study MHD/turbulence under the more extreme conditions characteristic of the cold molecular clouds found in star formation regions [18, 19]. A third set of experiments is aimed at obtaining a qualitative understanding of dissipation in magnetically dominated plasmas driven by boundary motions [9, 8]. Such a situation arises when low-density plasmas are magnetically connected to (anchored in) near-by turbulent plasmas with much higher densities. The best studied example is the thin solar corona, where violent events ("flares") and general heating are being generated by shearing of the coronal magnetic field by motions in the much denser solar photosphere.

The assumptions and equations employed are summarized in the next Section. Some newly developed visualization modules are described in Section 3 and applied to results from some of the numerical experiments in Section 4. Concluding remarks concerning the overall scientific productivity of such a computing and graphics endeavor are offered in Section 5.

2 Astrophysical MHD

Many astrophysical plasma systems may be described with reasonable accuracy in the so called magneto-hydro-dynamics (MHD) approximation, which is obtained by neglecting the displacement current in the Maxwell equations, and then combining the remaining equations with continuum equations for mass density ρ, velocity $\mathbf{u}$, and (internal) thermal energy e. The resulting partial differential equations are

$$\frac{\partial \rho}{\partial t} = -\nabla \cdot \rho \mathbf{u}, \tag{1}$$

$$\frac{\partial \mathbf{B}}{\partial t} = -\nabla \times \mathbf{E}, \tag{2}$$

$$\mathbf{E} = -(\mathbf{u} \times \mathbf{B}) + \eta \mathbf{J}, \tag{3}$$

$$\mathbf{J} = \nabla \times \mathbf{B}, \tag{4}$$

$$\frac{\partial \rho \mathbf{u}}{\partial t} = -\nabla \cdot (\rho \mathbf{u}\mathbf{u} + \underline{\tau}) - \nabla p + \mathbf{J} \times \mathbf{B} - g\rho, \tag{5}$$

$$\frac{\partial e}{\partial t} = -\nabla \cdot (e\mathbf{u}) - p\nabla \cdot \mathbf{u} - \rho(T - T_{\text{ref}})/t_{\text{cool}} + Q_{\text{visc}} + Q_{\text{Joule}}, \tag{6}$$

$$\tag{7}$$

where ρ, $\mathbf{u}$, $\mathbf{B}$, $\mathbf{E}$, η, $\mathbf{J}$, $\underline{\tau}$, p, g, $p = (\gamma - 1)e = \frac{2}{3}e$, $T = p/\rho$, T_{ref}, t_{cool} Q_{visc} and Q_{Joule} are the density, velocity field, magnetic field, electric field, electric resistivity, electric current, viscous stress tensor, gas pressure, constant of gravity, internal energy, temperature, reference temperature, cooling time, viscous and Joule dissipation, respectively [16].

This set of partial differential equations allow for such varying and numerically challenging phenomena as hydrodynamic and magnetohydrodynamic shocks, vortex tubes and magnetic flux tubes, and electrical current sheets. In addition there are a multitude of linear waves such as pressure waves, internal gravity waves, torsional Alfvén waves, and fast and slow mode magnetic waves.

Discontinuities in the form of shocks and electrical current sheets are not only allowed by these equations, they occur ubiquitously in many astrophysical systems. Granular convection in stellar surface layers easily develops supersonic patches which cause shocks [13], the turbulence in cold molecular clouds is highly supersonic (rms Mach numbers of the order of 10) [18], and the boundary driving of coronal magnetic fields causes a hierarchy of fragmented current sheets to form [6, 9, 8].

2.1 The Fortran-90 Code

We have developed and thoroughly tested a Fortran-90 code for solving the MHD equations [16, 20]. Since most of the phenomena we are interested in occur on time scales comparable to the fastest wave modes of the problem, we adopted a straightforward explicit time stepping technique [10], and concentrated most of our efforts on minimizing the numerical diffusion in the bulk of the model (using high order spatial differencing on a staggered mesh), while at the same time providing sufficient dissipation in regions where the solution spontaneously develops discontinuities, by using discontinuity capturing techniques.

The main advantage of implementing this code in Fortran-90 comes from the use of array-valued functions. All the 24 differentiation and interpolation operators necessary when writing the MHD equations on a staggered mesh may be cloned out of an implementation of two of them. The details of the implementation may vary, depending for example on the use of specially tuned stencil libraries [2], without affecting the code at the "physics" level where, for example, the continuity equation is written as

$$\text{drhodt} = - \text{ddxup(px)} - \text{ddyup(py)} - \text{ddzup(pz)}. \tag{8}$$

px, py and pz are the components of the mass flux ρu, centered on cell faces, and ddxup, ddyup and ddzup are 6th order derivative operators that return their results at cell center, where the density ρ is centered.

The full version of the equations of motion, with (hyper-) viscosity, Lorentz force, Coriolis force, and gravity is considerably more complicated, but it is still relatively straight-forward to implement, given the set of differentiation and interpolation operators.

2.2 Performance

The performance of the code as a whole is, on most machines, determined by the performance of the staggering operators. Thus, the task of optimizing for a particular machine boils down to optimizing two (or possibly six—if the three directions must be treated differently) of the staggering operators; the rest of the staggering routines are clones and may be generated with a stream editor.

The performance of the staggering operators on various supercomputers is shown in Table 2.2. The speed on a single C-90 CPU is roughly the same as that of a CM-200/8k, or a CM-5/32, using the standard CSHIFT operator. The specially tuned stencil library is almost a factor of two faster, and delivers over 4 GFl on a CM-5/128. For small experiments, the SGI R8000 runs at 150 Mfl per CPU, but for larger problem sizes the performance drops due to cache bottlenecks. The figure for the IBM SP/2 is an estimate, based on early testing.

Table 1. The performance of 6th order staggering operators, as optimized for various supercomputers. The entries marked CSHIFT use standard Fortran-90 circular shifts. PSHIFT is the CMSSL library implementation of circular shifts. The C-92 "f77 unrolled" entry refers to hand unrolled Fortran-77 loops. The entries marked "stencil library" use a stencil library by Bourbonnais [2].

Machine	method	CPUs	size	GFl
C-92	f90 CSHIFT	1	128^3	0.35
C-92	f77 unrolled	1	128^3	0.49
C-92	f77 unrolled	2	128^3	0.98
CM-200/8k	CSHIFT	256	128^3	0.39
CM-200/8k	PSHIFT/CMSSL	256	128^3	0.44
CM-5	CSHIFT	32	128^3	0.51
CM-5	PSHIFT/CMSSL	32	128^3	0.55
CM-5	stencil lib.	32	128^3	0.88
CM-5	CSHIFT	128	256^3	2.14
CM-5	PSHIFT/CMSSL	128	256^3	2.26
CM-5	stencil lib.	128	256^3	4.08
SGI, R8000	f77 unrolled	1	64^3	0.15
SGI, R8000	f77 unrolled	1	128^3	0.075
SGI, R8000	f77 unrolled	2	64^3	0.30
SGI, R8000	f77 unrolled	2	128^3	0.14
IBM SP-2	f77 unrolled	16	128^3	0.5

3 Visualizing Vector Fields

The complexity of solutions to the MHD equations has its roots in the non-linear interaction of several vector quantities, driven by terms such as $u \times B$ and $j \times B$. It is clearly important to be able to visualize these vector fields. One possibility is to show 3D renderings of vectors or glyphs, whose directions and shape encode properties of the vector field. Such techniques have serious limitations, however. Current numerical simulations employ millions of mesh points, which renders modules that display a vector at every mesh point virtually useless. Severely limiting the number of vectors displayed, by sub-sampling and/or by only displaying vectors above or below certain thresholds, can be a useful technique [12].

With such methods, however, little or no information is conveyed about the topology and distant connectivity of a vector field. Much of the qualitative behavior of solutions to the MHD equations is indeed intimately related to the topology of the magnetic field, and that of the vorticity. In the limit of weak magnetic fields, for example, isotropic turbulence is characterized by the formation of vortex tubes and magnetic flux tubes, at locations determined by the local flow topology, and the local availability of vorticity and magnetic field lines. And in the limit of strong magnetic fields, the formation of current sheets is the inevitable consequence of particular combinations of magnetic field line topology and motion patterns [9].

Information about field topology and connectivity is best displayed by rendering field lines, either just as thin lines or as material lines of finite width ("wires"), to enhance the 3D impression. IDL [5] procedures for wire-rendering have been developed by Galsgaard [4].

3.1   Explorer Modules

We have developed `Fieldlines` and `AnimFieldlines` modules [17] for NAG / Explorer [3] that can render a vector field as a set of field lines. The default behavior is to trace the vector field from start points chosen randomly, with a probability density proportional to the field strength. Along each field line, the brightness of the field line is proportional to a power of the field strength (the square of the magnetic field, for example, corresponds to the magnetic energy density). A low cut-off (for example at 5–10 % of the maximum brightness) is useful to prevent too much crowding, and a high cut-off may be used to selectively probe weak-field regions (sometimes more dynamically significant than strong-field regions). Field lines starting points may be chosen randomly throughout the volume, or in a sphere of variable radius around an interactively chosen center point. One may also choose to add individual field lines through selected points, a technique which is often very useful for studying localized regions with critical topology; e.g., reconnection in electrical current sheets (cf. Fig. 6). It is a good idea to complement the field line rendering with isosurfaces, since the isosurface patches tend to enhance the 3D effect.

The `AnimFieldlines` module is similar to the `Fieldlines` module, but also has provisions for generating animation sequences. In addition to the controls and input ports of the `Fieldlines` module, `AnimFieldlines` has an input port that accepts a 3D velocity field, and a parameter that controls a time step. The module generates a sequence of field line renderings. For the first rendering, starting points for field line rendering are generated with the methods described above. Between subsequent renderings, the starting points for the field lines are moved in accordance with the velocity field. For an animation of magnetic field lines, one applies a velocity field $u_B = E \times B / B^2$, which is the perpendicular velocity of magnetic field lines.

4   Examples

Some results from recent numerical experiments serve to illustrate the usefulness of the visualization modules discussed in the previous section.

4.1   Dynamo Action in Isotropic Turbulence

Isotropic turbulence may be generated by driving a triple-periodic system in s spherical shell of wavenumbers. We have used such simulations to study the spontaneous (dynamo) amplification of a weak seed magnetic field, and to study the diffusion of a large scale magnetic field by intermediate and small scale turbulent motions [15]. Figure 1 shows a 64^3 sample from a 128^3 isotropic turbulence experiment where the driving is confined to a shell with $k \leq 2$ (box size 2π). The magnetic field is illustrated with isosurfaces of constant magnetic energy, and with magnetic field lines generated by the AnimFieldlines module. Most of the magnetic energy is confined to tube-like structures, whose thickness are of the order of a few mesh points. Figure 2 shows a close-up of a few such *magnetic flux tubes*. The vorticity is concentrated in similarly thin *vortex tubes*.

Fig. 1. A 64^3 sample from a 128^3 isotropic turbulence experiment. Isosurfaces of constant vorticity (blue) and magnetic field strength (red—for color versions of all figures, see [11]). The field lines show the magnetic field, with a brightness proportional to the magnetic energy density.

Fig. 2. A smaller sample from the same dataset as in Fig. 1.

Vortex tubes and magnetic flux tubes sometimes overlap in space (Fig. 2 upper right). Tubes of concentrated magnetic field or vorticity may also occur in isolation (Fig. 2 upper left), and occasionally one finds a magnetic flux tube that is being wrapped around a vortex tubes (Fig. 2 lower middle). Each of these situations may give rise to an increase of the total magnetic energy, and it is of great interest to be able to follow the time evolution of the magnetic field in animations of magnetic isosurfaces and magnetic field lines [11].

As illustrated by such animations, isolated magnetic flux tubes may grow by accreting field lines from surrounding weak field plasma, a process that increases the total magnetic energy in the short term, but requires the weak magnetic field to be replenished. A simple mechanism for such a replenishment is caused by the relative (bulk) motion of magnetic flux tubes and the general turbulence; this causes the somewhat weaker magnetic field connecting the stronger flux tubes to spread into the background.

The wrapping of a magnetic flux tube around a vortex tube that is taking place in the lower parts of Fig. 2 also corresponds to an increase of the magnetic energy; the length of that magnetic flux tube is increasing as it is being wound up, and hence the total magnetic energy in the wavenumber range from the inverse width of the flux tube to the inverse diameter of the flux tube spiral increases due to this process.

Another cause of increase of the total magnetic energy is illustrated by the time evolution of the structure in the upper right of Fig. 2 where, because of the overlap of a magnetic flux tube with a vortex tube, the relatively weak magnetic field that extends out from the "ends" of the flux tube isosurface is being stretched into a spiral shape. This corresponds to an increase of the magnetic energy at intermediate scales, and thus is one of the mechanisms that maintains the "inertial" range of wavenumbers (those intermediate between the smallest and the largest scales). The power spectrum of the magnetic energy is consistent with a $k^{1/3}$ power law in the inertial range, although the range of wavenumbers is too small to empirically determine the inertial range exponent.

4.2 Magnetic Dissipation in Fragmented Current Sheets

Figure 3 shows a rendering of surfaces of constant electric current, from a 136^3 numerical experiment [9] where a magnetically dominated plasma is being driven by shearing motions on two boundaries (to the left and right in Fig. 3). This is an idealized model of situations such as the one in the optically transparent solar corona, which is dominated by a magnetic field that has its footpoints in the much denser solar photosphere (the optical surface of the Sun). The turbulent motions in the solar interior cause the footpoints of the coronal magnetic field to perform a random walk on the solar surface, and this causes a complicated braiding of the coronal magnetic field lines.

In the numerical models, a similar braiding is caused by the randomly chang-ing shearing motions on the boundary. We have shown [9] that the braiding in-evitably causes the formation of electrical current sheets, and that these current sheets are responsible for dissipating the energy flux (Poynting flux) associated with the work done on the magnetic field at the boundaries.

Figure 4 shows a blow-up of a small part of the volume displayed in Fig. 3. The current isosurfaces have been made semi-transparent, and magnetic field lines have been added on both sides of the current sheets, to illustrate the discontinuity in magnetic field line orientation across a current sheet.

Figure 5 shows part of the same region at closer range, to illustrate the intricate, hierarchical nature of the current sheets. A bridge-like current sheet protrudes out from the main current sheet, twisting slightly to the left, and connecting to the current sheet at the far left. As illustrated by the overlap-ping systems of field lines, each current sheet separates regions of discontinuous magnetic connectivity.

In the initial phases of this experiment, a large, smoothly twisted current sheet extends between the two boundaries. The formation of this current sheet is

Fig. 3. Isosurfaces of constant electric current density, illustrating the currents sheets formed when boundary work is performed on a magnetically dominated plasma [9].

caused by perpendicular shear patterns on the two boundaries. Magnetic reconnection in the current sheet causes intense dynamic activity, with jets of plasma shooting out along the edges of the current sheet (Fig. 6). The reconnection is driven by the tension force along magnetic field lines, as two independent magnetic flux systems are pressed against each other. A video sequence illustrating the reconnection was shown at the meeting [11].

5   Concluding Remarks

The scientific conclusions that may be drawn from these experiments are discussed elsewhere [9, 15, 8, 19]. In stead, a few remarks that are more directly related to the theme of this conference deserve to be made here:

– The relative ease of programming massively parallel computers in the SIMD mode of operation, using Fortran-90 or High Performance Fortran, should be emphasized. Array-valued function representation of numerical staggering operators leads to particularly compact and easy-to-read numerical codes,

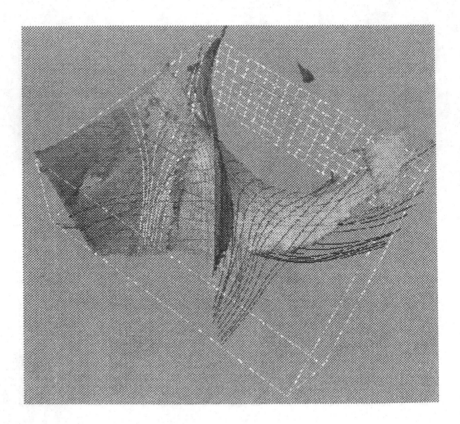

Fig. 4. A blow-up of a subvolume near the bottom of Fig. 3. The isosurfaces have been made semitransparent, and magnetic field lines are rendered on both sides of the current sheets.

that are easier to debug, maintain and extend than codes where the parallelism has to be explicitly programmed, using for example PVM constructs.

Note that nothing in principle prevents a multi-CPU shared memory machine with local caches, or a (loose or tight) cluster of CPUs with distributed memory, from being considered (by the compiler) to be a SIMD system with the local caches (in the case of a shared memory machine) taking the place of the memory local to each FPU in a Connection Machine. Future High Performance Fortran compilers may be expected to do an increasingly good job at this, thus relieving the scientist from the "human compiler mode of programming that to some extent has been forced upon us by the migration of supercomputing to low-cost clusters of workstations.

– The importance of visualization in general, and vector field visualization in particular, in the analysis of multi-dimensional simulations such as for example the MHD-turbulence simulations discussed here, also deserves to be emphasized.

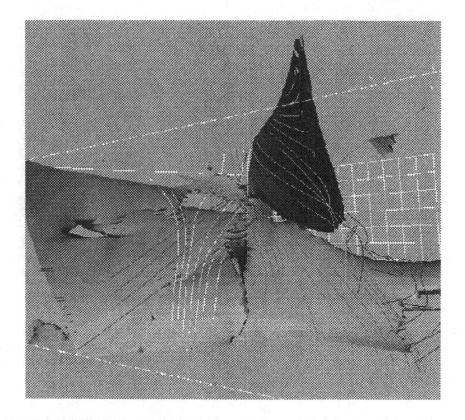

Fig. 5. A close-up of the complex field lines topology near the current sheets, from a view point near the lower left corner of Fig. 4

For vector fields which are divergence free (e.g., the magnetic field and the vorticity), or nearly divergence free (e.g., the mass flux in nearly anelastic flows), visualization of field lines are preferable over vectors or glyphs, since field lines bring out the topology (connectivity) of the fields. A visual appreciation of the topology of vector fields such as velocity and magnetic field is often crucial for a qualitative understanding of the dynamical processes at work.

– Time animated visualizations serve to illustrate and clarify the dynamical aspects of these systems.

The ability to follow the time evolution of the topology of vector fields is particularly important in MHD-simulations, since the reconnection of magnetic field lines is crucial both to dynamo action and magnetic dissipation. The video sequences that were shown at the meeting, and other animation sequences from some of the experiments discussed above, are available as MPEG files [11].

In summary, a number of factors contribute to the overall scientific productivity of a certain computing environment. In addition to the raw performance

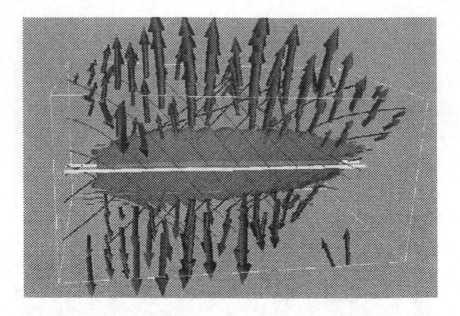

Fig. 6. A rendering of a large scale current sheet that forms early in the experiment. The magnetic field lines illustrate the tangential discontinuity of the magnetic field across the current sheet, and the arrows show the jet sheets created along the edges of the current sheet.

(and price) figures, such factors as ease of programming, debugging and maintenance, and access to powerful and easily extendable visualization tools, should certainly not be forgotten.

Acknowledgments

This work was supported in part by the Danish Research Foundation, through its establishment of the Theoretical Astrophysics Center (Å.N), by NASA, through grant NAGW-1695 (RFS), and by the Danish Center for Applied Parallel Processing (CAP) through a PhD grant (KG). CM-200 and Cray C-90 computing time was provided by the Danish Natural Science Research Council. Additional computing was done at the Centre National de Calcul Parallele en Sciences de la Terre in France. The `AnimFieldlines` module was developed in part by Thomas Lund.

References

1. Brandenburg, A., Jennings, R. L., Nordlund, A., Rieutord, M., Stein, R. F., Tuominen, I. 1996, J. of Fluid Mech., 306, 325
2. Bourbonnais, R. 1995, roch@think.com, in these proceedings
3. NAG Explorer, http://www.nag.co.uk:70

4. Galsgaard 1993, IDL procedures for "wire" rendering, available from kg@astro.ku.dk
5. Interactive Data Language, http://sslab.colorado.edu:2222/projects/IDL/idl_ssl_home.html or http://www.rsinc.com
6. Galsgaard, K. 1995, J. Geophys. Research, (in preparation)
7. Galsgaard, K., Nordlund, Å. 1995a, J. Geophys. Research, (submitted)
8. Galsgaard, K., Nordlund, Å. 1995b, J. Geophys. Research, (in preparation)
9. Galsgaard, K., Nordlund, Å. 1995a, J. Geophys. Res., (in press)
10. Hyman, J. 1979, in R. Vichnevetsky, R. S. Stepleman (eds.), Adv. in Comp. Meth. for PDE's—III, 313
11. Nordlund, Å. 1995, Color figures and MPEG movie accompanying this paper, available through http://www.astro.ku.dk/~aake/papers/para95
12. Nordlund, Å. 1995, An Explorer module for vector rendering with cut-offs, http://www.astro.ku.dk/Explorer/VectorsCutOff.html
13. Nordlund, Å., Stein, R. F. 1991, in D. Gough, J. Toomre (eds.), Challenges to Theories of the Structure of Moderate Mass Stars, Vol. 388, Springer, Heidelberg, p. 141
14. Nordlund, A., Brandenburg, A., Jennings, R. L., Rieutord, M., Roukolainen, J., Stein, R. F., Tuominen, I. 1992, ApJ, 392, 647
15. Nordlund, Å., Galsgaard, K., Stein, R. F. 1994, in R. J. Rutten, C. J. Schrijver (eds.), Solar Surface Magnetic Fields, Vol. 433, NATO ASI Series
16. Nordlund, Å., Galsgaard, K. 1995, Journal of Computational Physics, (in preparation)
17. Nordlund, Å. and Lund, T. 1995, Explorer modules for field line rendering and animation, http://www.astro.ku.dk/Explorer/Fieldlines.html
18. Padoan, P. 1995, Monthly Notices Roy. Astron. Soc., (in press)
19. Padoan, P., Nordlund, Å. 1996, Phys. Fluids, (in preparation)
20. Stein, R. F., Galsgaard, K., Nordlund, Å. 1994, in J. D. B. et al. (ed.), Proc. of the Cornelius Lanczos International Centenary Conference, Society for Industrial and Applied Mathematics, Philadelphia, p. 440

A Parallel Sparse QR-Factorization Algorithm

Tz. Ostromsky[1], P. C. Hansen[1] and Z. Zlatev[2]

[1] UNI•C, Danish Computer Centre for Research and Education,
Technical University of Denmark, Bldg. 304, DK-2800 Lyngby, Denmark
e-mail: Tzvetan.Ostromsky@uni-c.dk
e-mail: Per.Christian.Hansen@uni-c.dk
[2] National Environmental Research Institute, Frederiksborgvej 399,
DK-4000 Roskilde, Denmark
e-mail: luzz@sun2.dmu.dk

Abstract. A sparse QR-factorization algorithm for coarse-grain parallel computations is described. Initially the coefficient matrix, which is
assumed to be general sparse, is reordered properly in an attempt to
bring as many zero elements in the lower left corner as possible. Then
the matrix is partitioned into large blocks of rows and Givens rotations
are applied in each block. These are independent tasks and can be done
in parallel. Row and column permutations are carried out within the
blocks to exploit the sparsity of the matrix.

The algorithm can be used for solving least squares problems either directly or combined with an appropriate iterative method (for example,
the preconditioned conjugate gradients). In the latter case, dropping of
numerically small elements is performed during the factorization stage,
which often leads to a better preservation of sparsity and a faster factorization, but this also leads to a loss of accuracy. The iterative method is
used to regain the accuracy lost during the factorization.

An SGI Power Challenge computer with 16 processors has been used
in the experiments. Results from experiments with matrices from the
Harwell-Boeing collection as well as with automatically generated large
sparse matrices are presented in this work.

1 Introduction

1.1 Sparse Linear Systems and the Least Squares Problem

The least squares problem can be defined as follows: Solve

$$\min \|Ax - b\|_2 \tag{1}$$

or equivalently, solve the system of linear equations

$$Ax = b + r \ \wedge \ A^T r = 0 \tag{2}$$

providing we are given the matrix $A \in \mathcal{R}^{m \times n}$, $m \geq n$ and the right-hand side
vector $b \in \mathcal{R}^m$. We assume that the matrix A has the following properties:

- A is **large**,
- A is **sparse** (most of its elements are equal to zero),
- A has **full column rank** (i.e. $rank(A) = n$).

We also assume that A has neither any other special property nor any special structure. In other words, A is *general sparse*.

Sparse linear least squares problems appear in various technical and scientific areas like tomography, photogrametry, geodetic survey, cluster analysis, molecular structure, structural analysis, and so on. More details on these and other applications can be found in [17].

The long history and the practical importance of the least squares problem explain why several well developed methods for its solution have gradually been developed, for example:

- The method of normal equations
- The QR-factorization method
- The Peters–Wilkinson method
- The augmentation method
- Iterative methods

(see [3, 4, 19]). One of them, the QR-factorization method, is discussed in more detail below.

1.2   The Method of Orthogonal Factorization

One of the most popular methods for solving numerically the sparse linear least squares problem is by orthogonal factorization,

$$A = QR, \tag{3}$$

where $Q \in \mathcal{R}^{m \times n}$ has orthogonal columns ($Q^T Q = I$) and $R \in \mathcal{R}^{n \times n}$ is an upper triangular matrix. Usually the matrix Q is obtained as a product of elementary orthogonal matrices and is not calculated explicitly (Q is likely to be very dense compared to A). The QR-factorization is considered as an expensive process. This drawback can be compensated by efficient preservation of sparsity, which will be discussed in §1.3.

The major advantage of the orthogonal factorization is its stability. Some stable orthogonalization methods are discussed below. We can exploit extensively block parallelism in these algorithms without stability problems. In general, no pivoting for stability is required [3], so we can use it to preserve better the sparsity.

The method of *Householder* [14] is based on elementary Householder reflections. Each reflection creates zeros in an entire column of the active submatrix. The method is stable and very efficient for dense matrix factorization. Applied to sparse matrices it tends to create more fill-ins than the Givens method (see below). The method of Householder has some other useful properties and it has

---

[3] Pivoting is necessary if one wants to compute a rank-revealing QR-factorization; this is a topic for future work.

been successfully used in some multifrontal techniques for sparse matrix factorization, exploiting small dense submatrices [1, 15, 16].

The *Givens method* uses elementary plane rotations. By each rotation a single element of the active submatrix is annihilated. It is as stable as Householder's method and best in preserving the sparsity [19]. Two rows only take part in each rotation, which means that there is a lot of potential parallelism in this method. These reasons were the most essential in the decision to select this method for our parallel algorithm. There are two versions – *Classical Givens* [13] and *Givens-Gentleman* [9]. The first one has been used by George and Heath in the package SPARSPAK-B [10, 12]. Gentleman's version (also known as *fast Givens*) is more economic with regard to floating point operations, but has some overhead and requires some kind of pivoting in an attempt to prevent from a possible overflow. It has been used in the code LLSS by Zlatev [19].

The method of *Gram-Schmidt* is the only one that calculates explicitly the orthogonal basis Q (which is necessary in some applications). It is unstable in its classical variant. An improved version, the *Modified Gram-Schmidt method* has extensively been studied and proved to be satisfactory stable (see e.g. [18]).

1.3 Preserving the Sparsity

We can very often save large amounts of work that would have little effect on the solution by considering as zeros all matrix elements and fill-ins that are relatively small (in absolute value), according to a certain criterion. As a result we obtain an incomplete QR-factorization and an inaccurate initial solution. The incomplete factor R can be used as a preconditioner in the system of *semi-normal* equations [2], as it is shown in [19] for the Givens-Gentleman method. For the classical Givens method we have:

$$
\begin{aligned}
C &= (R^T)^{-1} A^T A R^{-1} \\
z &= Rx \\
d &= (R^T)^{-1} A^T b
\end{aligned}
\tag{4}
$$

From (2) and (4) we obtain the linear system

$$
Cz = d
\tag{5}
$$

with symmetric and positive definite matrix C. It should be mentioned that matrix C is never calculated explicitly. The conjugate gradients method (denoted CG throughout this paper) is used to solve the preconditioned semi-normal system (5).

2 SPARQR – a Sparse Parallel QR-Factorization

The algorithm consists of five main stages as follows:

Stage 1 Reordering

Stage 2 Block partitioning

Stage 3 Factorization using Givens rotations and (optionally) dropping relatively small elements

Stage 4 Back substitution (finding the initial solution)

Stage 5 Improving the solution by an iterative method (Preconditioned CG)

Both the dropping of small elements, performed during the factorization stage, and the iterative method used in Stage 5 are optional and dependent on each other (the iterative method is activated only when dropping has been performed). This is in fact the computer-oriented manner of exploiting the sparsity [19]. For unstructured matrices, this device is one of the most efficient ways to exploit the sparsity as much as possible. By varying the user-driven parameter TOL, called *drop-tolerance*, one can obtain a more or less accurate incomplete QR-factorization, that respectively costs more or less time. If the factorization is completed (which means we have obtained a non-singular upper triangular factor R), then R will be used during Stage 5 as a preconditioner in the system of semi-normal equations, see (4). An appropriate initial approximation of the solution vector is found in Stage 4. In most cases, the preconditioned CG algorithm used in Stage 5 converges reasonably fast and finally gives a solution with the required accuracy. Otherwise we must reduce the drop-tolerance and restart the process from Stage 1, trying to obtain more accurate preconditioner.

For many matrices this hybrid procedure solves the problem faster, and sometimes much faster (up to a factor of 20) than the pure direct method (without any dropping). On the other hand, finding the optimal drop-tolerance can be a difficult task.

We have parallelized Stage 3, which is the most time-consuming (see the results in Table 1), and partially Stage 2, while Stage 1 is inherently sequential. Stages 4 and 5 are in principal parallelizable, but they are not parallelized yet. This is one of our tasks for the future.

More details about the first three stages of **SPARQR** are given below.

2.1 The Reordering Algorithm

This is the first stage of our algorithm (and of many other algorithms for sparse matrices). The purpose is to obtain an appropriate structure of the matrix by performing row and column permutations. The reordered matrix is exploited later for efficient parallel factorization. The solution of the least squares problem is invariant upon any row permutation, but this is not so for the column permutations. Thus, while we can just "do and forget" the row permutations, we must store information about the column permutations and use it to restore the solution of our original problem.

The *Locally Optimized Reordering Algorithm* (**LORA**) [7, 8] is used in **SPARQR**. It attempts to move as many as possible zeros to the lower left corner of the matrix. This has two positive effects. First, it tends to reduce the number of non-zeros below the main diagonal and, thus, the number of potential fill-ins. Second, the structure obtained by LORA (see Fig. 1) allows well balanced block partitioning and, consequently, efficient block-parallel factorization.

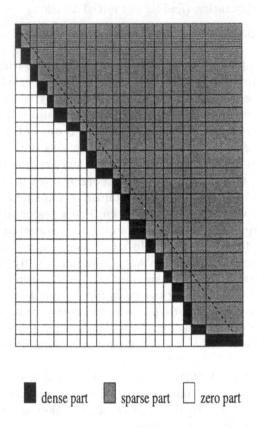

■ dense part   ▨ sparse part   ☐ zero part

Fig. 1. The structure of a matrix, reordered by LORA

The square matrix version of LORA is described in detail in [8], where it was used to reorder a square matrix before its LU-factorization by Gaussian elimination. The generalization for rectangular matrices is quite straightforward; the basic properties and the complexity estimate, $O(NZ \log n)$, where NZ is the number of non-zeros, remain unchanged. The numerical experiments with both SPARQR and the LU-factorization code confirm that LORA is a cheap and efficient reordering scheme. It is applicable to a wide class of problems in which general sparse matrices are involved.

2.2 Partitioning into Blocks

During the second stage both the set of rows and the set of columns of the matrix (as ordered by LORA) are divided into k parts, called respectively *block-*

rows and *block-columns*. The goal is to represent the reordered matrix in a $k \times k$ block upper triangular form with rectangular blocks (where each block is the intersection of a block-row and a block-column). The block-rows are processed in parallel during the most time-consuming Stage 3. Therefore it is advantageous to try to obtain block-rows of approximately the same size.

The pattern created by LORA (see Fig. 1) has already block upper triangular form with dense diagonal blocks. The corresponding block-rows and block-columns are usually too small to be used directly as a partitioning. However, by gathering several such blocks we can easily obtain a partitioning with the desired properties.

2.3 Factorization

The factorization stage consists of k steps, where k is the same as in §1.2. During each step the diagonal blocks of the active submatrix are triangularized by Givens rotations. The transformations are local within the corresponding block-row, therefore the block-rows can be processed in parallel. Because the diagonal blocks are rectangular, at the end of the i-th step each block-row can be divided into two parts: a *head* (the rows that form the upper triangular part of the diagonal block) and a *tail* (the non-trivial rows that have only zeros in the diagonal block). Before the next step a rearrangement of the block-rows is performed so that the tail of each block-row is gathered with the head of the next block-row, thus forming a new block-row. The last block has obviously no tail, so during the $(i+1)$-th step the number of block-rows decreases by one. The head of the first block-row is already a part of R, so it is excluded from the active part of the matrix. Thus we have k block-rows during the first step, $k-1$ during the second, and so on. If the algorithm is run on a parallel machine with p CPUs ($k \geq p$), then during the last $p-1$ steps the average utilization of the processors will gradually decrease. To increase the average utilization (for the whole factorization stage), the number of block-rows k should be considerably larger than p.

3 Numerical Results

The algorithm SPARQR has been implemented and tested on an SGI Power Challenge computer with 16 processors. This is a shared memory machine, but it should be mentioned that each processor has its own cache. The results of some experiments performed on this machine are given in this section.

In our experiments we have used rectangular test matrices from the Harwell-Boeing collection [6], type *rra*. These matrices are very well known and they have been used extensively for developing and testing new algorithms. These matrices come from practical problems and applications related to various technical and scientific areas, which makes them very popular. Unfortunately most of them are not sufficiently large when modern high-speed computers are used. That is why we included in our experiments one larger matrix, obtained by using one of the sparse matrix generators described in [19]. This matrix is called matrf2.

Stage	Time on 1 CPU		
	well1850	amoco1	bellmedt
$TOL= 2^{-5}$			
1 Reordering	0.025 (1.6%)	0.062 (0.4%)	0.135 (0.2%)
2 Partitioning	0.004 (0.3%)	0.007 (0.1%)	0.009 (0.0%)
3 Factorization	1.398 (89.6%)	10.987 (78.1%)	53.833 (98.8%)
4 Back subst.	0.001 (0.1%)	0.002 (0.0%)	0.004 (0.0%)
5 Precond. CG	0.131 (8.4%)	2.992 (21.4%)	0.468 (0.9%)
Total	**1.560** (100%)	**14.047** (100%)	**54.450** (100%)
$TOL= 10^{-12}$			
1 Reordering	0.026 (1.0%)	0.062 (0.4%)	0.136 (0.0%)
2 Partitioning	0.004 (0.1%)	0.006 (0.0%)	0.009 (0.0%)
3 Factorization	2.685 (98.7%)	17.301 (99.5%)	998.018 (99.9%)
4 Back subst.	0.001 (0.0%)	0.005 (0.0%)	0.079 (0.0%)
5 Precond. CG	0.003 (0.1%)	0.012 (0.1%)	0.253 (0.0%)
Total	**2.720** (100%)	**17.386** (100%)	**998.495** (100%)

Table 1. Times (in seconds and percentages) spent in the main stages

The first experiment was carried out in order to compare the relative cost of the five main stages of SPARQR and to see if the incomplete factorization strategy by dropping numerically small elements pays off. We solved sequentially several linear least squares problems with matrices of different size. Partitioning into $k = 128$ blocks was used in all these runs. For the particular matrices used this is a good choice for parallel execution as well. Table 1 contains the results of this experiment for three rectangular Harwell-Boeing matrices: well1850, amoco1 and bellmedt. The times for the five stages (both in seconds and in percentages of the total time) are given in the table. The upper half of the table presents the results for a relatively large value of the drop-tolerance, while in the lower half the drop-tolerance is very small (practically zero).

The comparison of the computing times for the five stages shows that in all the experiments the factorization stage is strongly time-dominant. In case of a small drop-tolerance this stage consumes almost 100% of the time; the remaining stages are negligible with respect to their times. In case of a large drop-tolerance, as one can expect, the iterative Stage 5 often also has a significant deal in the total time.

Usually a large drop-tolerance leads to a better preservation of sparsity in the factorization stage (which reduces the factorization time), but results in a more inaccurate factorization (which requires more iterations at Stage 5). In our experiments it was very advantageous to use a large drop-tolerance, because the reduction of the factorization time was larger than the increase of the iteration time. For some matrices that create a lot of fill-ins, the use of a large drop-tolerance is extremely helpful. Such an example is the matrix bellmedt, which requires a lot of time if factorized directly. The use of a large drop-tolerance $TOL = 2^{-5}$, however, leads to a reduction of the computing time of more than 18 times.

Results of parallel execution of our code are presented in Tables 2 and 3. Partitionings with various number of blocks were applied to each problem in order to examine the relations between the size of the matrix, the number of blocks and the best execution time (and, eventually, the speed-up). Each of the speed-ups, given in the tables, is calculated as the ratio between the time on 1 CPU and the parallel time for the corresponding number of CPUs, given in the same row.

MATRIX	no. of	Time (*Speed-up*)		
(size)	blocks	1 CPU	8 CPUs	16 CPUs
	8	60.74	25.54 (*2.4*)	
bellmedt	16	55.36	15.08 (*3.7*)	13.34 (*4.1*)
	32	53.95	10.90 (*5.0*)	8.43 (*6.4*)
$m = 5831$	64	52.32	9.14 (*5.7*)	6.24 (*8.4*)
$n = 1033$	128	54.16	9.12 (*5.9*)	5.82 (*9.3*)
$NZ = 52012$	256	62.37	10.69 (*5.8*)	6.87 (*9.1*)

Table 2. The Harwell-Boeing matrix bellmedt with drop-tolerance 2^{-5}.

Table 2 presents the results of testing one of the biggest Harwell-Boeing matrices, bellmedt, wth drop-tolerance 2^{-5}. For this matrix (which has 1033 columns) we obtained satisfactory results by using partitionings with up to 256 block-rows (and 256 block-columns respectively, so in the last example there are on average 4 columns per block-column). Increasing the number of block-rows above this number leads to a quick degradation in the results, as the blocks become too small. The best results for this matrix were obtained by partitioning into 128 blocks.

Table 3 contains the results for a considerably bigger matrix, created by the sparse matrix generator CLASSF described in [19]. The speed-ups are significantly greater than those for bellmedt (especially on 16 nodes). As the matrix is larger, partitioning into more blocks is possible. The best number of blocks in this case is 256.

MATRIX	no. of	Time *(Speed-up)*		
(size)	blocks	1 CPU	8 CPUs	16 CPUs
	8	313.66	61.25 *(5.1)*	
matrf2	16	302.57	53.03 *(5.7)*	35.24 *(8.6)*
	32	292.42	47.22 *(6.2)*	28.32 *(10.3)*
$m = 32000$	64	276.36	42.49 *(6.5)*	25.77 *(10.7)*
$n = 5000$	128	260.89	37.11 *(7.0)*	21.99 *(11.9)*
$NZ = 160110$	256	245.41	35.23 *(7.0)*	19.32 *(12.7)*
	512	284.60	41.14 *(6.9)*	23.63 *(12.0)*

Table 3. An automatically generated sparse matrix matrf2 with drop-tolerance 10^{-12}.

4 Conclusions

In general, **the bigger the matrix, the better the speed-up.** Starting parallel tasks on a shared memory machine requires a more or less constant overhead. If the tasks are too small, the result of the parallelization can be even negative. That is why even for a fixed number of blocks (that potentially allows the same degree of parallelism), the speed-up is better for the bigger matrix. To have blocks of the same size in both cases means that the larger matrix should be divided into larger number of blocks, which implies better overall load-balance on a fixed number of CPUs. As a result, for larger matrices we usually have better speed-up both for a fixed number of blocks and for the optimal blocking.

A larger number of blocks usually implies better load-balance and better speed-up. After smooth improvement up to a certain point called *the best number of blocks*, the time begins to deteriorate very quickly. If we denote the best number of blocks for given matrix $A \in \mathcal{R}^{m \times n}$ on p processors by $K_p(A)$ and if $n^{2/3}$ is considerably larger than p, then $K_p(A) \approx n^{2/3}$ is true in most cases.

The above statement was confirmed by many experiments we carried out on the SGI. It is not clear yet whether it remains valid on other machines and for an arbitrary sparse matrix. Nevertheless it gives us a useful practical rule for finding a good partitioning strategy and it is of great importance for catching the best performance of our algorithm.

No attempt to improve the utilization of the cache has been made yet. On a machine like SGI this can have significant effect on the performance. This is one of our tasks for the future.

In many cases **dropping small elements, combined with an appropriate preconditioned iterative method, is a powerful technique for accelerating the factorization.** Some matrices are very sensitive to the choice of the drop-tolerance. The problem of prediction the best drop-tolerance needs further investigation. However, to find a good (but not the optimal) value of the drop-tolerance is usually not very difficult.

Acknowledgements

This research was partially supported by the BRA III Esprit project APPARC (# 6634) and by research grant 9500764 from the Danish Natural Science Research Council.

References

1. P. R. Amestoy, I. S. Duff and C. Puglisi, *Multifrontal QR-factorization in a multiprocessor environment*, TR/PA/94/09, CERFACS, **1994**.
2. A. Björck, *Stability analysis of the method of semi-normal equations for least squares problems*, Linear Algebra Appl. **1988/89**, pp. 31–48.
3. A. Björck, *Least squares methods*, in P. G. Ciarlet and J. L. Lions /editors/, *Handbook of Numerical Analysis*, vol.1: *Finite Difference Methods – Solution of Equations in $\mathcal{R}^n$*, Elsevier/North-Holland, Amsterdam, **1990**.
4. A. Björck, R. J. Plemmons and H. Schneider, *Large-Scale Matrix Problems*, North-Holland, New York, **1981**.
5. I. S. Duff, A. M. Erisman and J. K. Reid, *Direct Methods for Sparse Matrices*, Oxford University Press, Oxford-London, **1986**.
6. I. S. Duff, R. G. Grimes and J. G. Lewis, *Sparse matrix test problems*, ACM Trans. Math. Software, 15 (**1989**), pp. 1–14.
7. A. C. N. van Duin, P. C. Hansen, Tz. Ostromsky, H. Wijshoff and Z. Zlatev, *Improving the numerical stability and the performance of a parallel sparse solver*, Computers and Mathematics with Applications (to appear).
8. K. A. Gallivan, P. C. Hansen, Tz. Ostromsky and Z. Zlatev, *A locally optimal reordering algorithm and its application to a parallel sparse linear system solver*, Computing, vol.54 No.1, bf 1995, pp. 39–67.
9. W. M. Gentleman, *Least squares computations by Givens transformations without square roots*, J. Inst. Math. Appl. 12, **1973**, pp. 329–336.
10. J. A. George and M. T. Heath, *Solutiom of sparse linear least squares problems using Givens rotations*, Linear Algebra Applications 34, **1980**, pp. 69–73.
11. J. A. George, M. T. Heath and E. G. Y. Ng, *A comparison of some methods for solving sparse linear least squares problems*, SIAM J. Sci. Stat. Comput. 4, **1983**, pp. 177–187.
12. J. A. George and E. G. Y. Ng, *SPARSPAK: Waterloo sparse matrix package user's guide for SPARSPAK-B*, Research Report CS-84-47, Department of Computer Science, University of Waterloo, Ontario, **1984**.
13. J. W. Givens, *Computation of plane unitary rotations transforming a general matrix to a triangular form*, J. Soc. Ind. Appl. Math. 6, **1958**, pp. 26–50.

14. A. S. Householder, *Unitary triangularization of a nonsymmetric matrix*, J. Assoc. Comput. Mach. 5, **1958**, pp. 339–342.
15. P. Matstoms, *The Multifrontal Solution of Sparse Linear Least Suares Problems*, Thesis No.293, LIU-TEK-LIC-1991:33, Linköping, Sweden, **1991**.
16. C. Puglisi, *QR-factorization of large sparse overdetermined and square matrices using the multifrontal method in a multiprocessor environment*, TR/PA/93/33, CERFACS, **1993**.
17. J. R. Rice, *PARVEC workshop on very large least squares problems and supercomputers*, Report CSD–TR 464, Purdue University, West Lafayette, IN, **1983**.
18. Xi. Wang, *Incomplete Factorization Preconditioning for Linear Least Squares Problems*, Ph.D. thesis, UIUC, Urbana, Illinois, **1993**.
19. Z. Zlatev, *Computational Methods for General Sparse Matrices*, Kluwer Academic Publishers, Dordrecht-Toronto-London, **1991**.

Decomposing Linear Programs for Parallel Solution*

Ali Pınar, Ümit V. Çatalyürek, Cevdet Aykanat   Mustafa Pınar**
Computer Engineering Department   Industrial Engineering Department
Bilkent University, Ankara, Turkey   Bilkent University, Ankara, Turkey

Abstract. Coarse grain parallelism inherent in the solution of Linear Programming (LP) problems with block angular constraint matrices has been exploited in recent research works. However, these approaches suffer from unscalability and load imbalance since they exploit only the existing block angular structure of the LP constraint matrix. In this paper, we consider decomposing LP constraint matrices to obtain block angular structures with specified number of blocks for scalable parallelization. We propose hypergraph models to represent LP constraint matrices for decomposition. In these models, the decomposition problem reduces to the well-known hypergraph partitioning problem. A Kernighan-Lin based multiway hypergraph partitioning heuristic is implemented for experimenting with the performance of the proposed hypergraph models on the decomposition of the LP problems selected from NETLIB suite. Initial results are promising and justify further research on other hypergraph partitioning heuristics for decomposing large LP problems.

1 Introduction

Linear Programming (LP) is currently one of the most popular tools in modeling economic and physical phenomena where performance measures are to be optimized subject to certain requirements. Algorithmic developments along with successful industrial applications and the advent of powerful computers have increased the users' ability to formulate and solve large LP problems. But, the question still remains on how far we can push the limit on the size of large linear programs solvable by today's *parallel processing technology* .

The parallel solution of block angular LP's has been a very active area of research in both operations research and computer science societies. One of the most popular approaches to solve block-angular LP's is the Dantzig–Wolfe decomposition [1]. In this scheme, the block structure of the constraint matrix is exploited for parallel solution in the subproblem phase where each processor solves a smaller LP corresponding to a distinct block. A sequential coordination phase (the master) follows. This cycle is repeated until suitable termination criteria are satisfied. Coarse grain parallelism inherent in these approaches has been exploited in recent research works [5, 8]. However, the success of these approaches depends only on the existing *block angular* structure of the given constraint matrix. The number of processors utilized for parallelization in these

---

* This work is partially supported by the Commission of the European Communities, Directorate General for Industry under contract ITDC 204–82166
** Supported in part through grant no. 9500764 by the Danish Natural Science Council.

studies is clearly limited by the number of inherent blocks of the constraint matrix. Hence, these approaches suffer from *unscalability* and *load imbalance*.

This paper focuses on the problem of decomposing irregularly sparse constraint matrices of large LP problems to obtain block angular structure with specified number of blocks for scalable parallelization. The literature that addresses this problem is extremely rare and very recent. Ferris and Horn [2] model the constraint matrix as a bipartite graph. In this graph, the bipartition consists of one set of vertices representing rows, and another set of vertices representing columns. There exists an edge between a row vertex and a column vertex if and only if the respective entry in the constraint matrix is nonzero. The objective in the decomposition is to minimize the size of the master problem while maintaining computational load balance among subproblem solutions. Minimizing the size of the master problem corresponds to minimizing the sequential component of the overall parallel scheme. Maintaining computational load balance corresponds to minimizing processors' idle time during each subproblem phase.

In the present paper, we exploit hypergraphs for modeling constraint matrices for decomposition. A hypergraph is defined as a set of vertices and a set of *nets* (hyperedges) between those vertices. Each net is a subset of the vertices of the hypergraph. In this work, we propose two hypergraph models for decomposition. In the first model—referred to here as the *row–net* model—each row is represented by a net, whereas each column is represented by a vertex. The set of vertices connected to a net corresponds to the set of columns which have a nonzero entry in the row represented by this net. In this case, the decomposition problem reduces to the well-known *hypergraph partitioning* problem which is known to be *NP-Hard*.

The second model—referred to here as the *column–net* model—is very similar to the row–net model, only the roles of columns and rows are exchanged. The column–net model is exploited in two distinct approaches. In the first approach, hypergraph partitioning in the column–net model produces the dual LP problem in primal block angular form. In the second approach, dual block angular matrix achieved by hypergraph partitioning is transformed into a primal block angular form by using a technique similar to the one used in stochastic programming to treat non-anticipativity [9].

2 Preliminaries

A hypergraph $\mathcal{H}(\mathcal{V}, \mathcal{N})$ is defined as a set of vertices $\mathcal{V}$ and a set of nets (hyperedges) $\mathcal{N}$ between those vertices. Every net $n \in \mathcal{N}$ is a subset of vertices. The vertices in a net are called *pins* of the net. A graph is a special instance of a hypergraph such that each edge has exactly two pins. $\Pi = (P_1, \ldots, P_k)$ is a *k-way partition* of $\mathcal{H}$ if the following conditions hold: each part $P_\ell, 1 \le \ell \le k$ is a nonempty subset of $\mathcal{V}$, parts are pairwise disjoint, and union of k parts is $\mathcal{V}$.

In a partition Π of $\mathcal{H}$, a net that has at least one pin (vertex) in a part is said to *connect* that part. A net is said to be *cut* if it connects more than one part, and *uncut* otherwise. The set of uncut (*internal*) nets and cut (*external*) nets for a partition Π are denoted as $\mathcal{N}_I$ and $\mathcal{N}_E$, respectively. The cost of a partition Π (*cutsize*) is defined by the cardinality of the set of external nets, i.e., $cutsize(\Pi) = |\mathcal{N}_E| = |\mathcal{N}| - |\mathcal{N}_I|$. A partition Π of a hypergraph $\mathcal{H}$ is said to be feasible if it satisfies a given balance criterion $V_{avg}(1-\varepsilon) \leq |P_i| \leq V_{avg}(1+\varepsilon)$ *for* $i = 1, \ldots k$. Here, ε represents the predetermined maximum *imbalance ratio* allowed on part sizes, and $V_{avg} = |\mathcal{V}|/k$ represents the part size under perfect balance condition. Hence, we can define the hypergraph partitioning problem as the task of dividing a hypergraph into two or more parts such that the number of cut nets (cutsize) is minimized, while maintaining a given balance criterion among the part sizes.

Hypergraph partitioning is an NP-hard combinatorial optimization problem, hence we should resort to heuristics to obtain a good solution. However, especially in this application, heuristics to be adopted should run in low-order polynomial time. Because, these heuristics will be executed most probably in sequential mode as a preprocessing phase of the overall parallel LP program. Hence, we investigate the fast Kernighan–Lin (KL) based heuristics for hypergraph partitioning in the context of decomposing linear programs. These KL-based heuristics are widely used in VLSI layout design.

The basis of the KL-based heuristics is the seminal paper by Kernighan and Lin [6]. KL algorithm is an iterative improvement heuristic originally proposed for 2–way graph partitioning (bipartitioning). KL algorithm performs a number of passes over the vertices of the circuit until it finds a locally minimum partition. Each pass consists of repeated pairwise vertex swaps. Schweikert and Kernighan [11] adapted KL algorithm to hypergraph partitioning. Fiduccia and Mattheyses [3] introduced vertex move concept instead of vertex swap. The vertex move concept together with proper data structures, e.g., bucket lists, reduced the time complexity of a single pass of KL algorithm to linear in the size of the hypergraph. Here, size refers to the number of pins in a hypergraph. The original KL algorithm is not practical to use for large graphs and hypergraphs because of its high time complexity, and so the partitioning algorithms proposed after Fiduccia-Mattheyses' algorithm (FM algorithm) have utilized all the features of FM algorithm. Krishnamurthy [7] added to FM algorithm a *look-ahead* ability, which helps to break ties better in selecting a vertex to move. Sanchis [10] generalized Krishnamurthy's algorithm to a multiway hypergraph partitioning algorithm so that it could directly handle the partitioning of a hypergraph into more than two parts. All the previous approaches before Sanchis' algorithm (SN algorithm) are originally bipartitioning algorithms.

3 Hypergraph Models for Decomposition

This section describes the hypergraph models proposed for decomposing LP's. In the row–net model, the LP constraint matrix A is represented as the hypergraph $\mathcal{H}_R(\mathcal{V}_C, \mathcal{N}_R)$. The vertex and net sets $\mathcal{V}_C$ and $\mathcal{N}_R$ correspond to the columns and rows of the A matrix, respectively. There exist one vertex v_i and one net n_j for each column and row, respectively. Net n_j contains the vertices corresponding to the columns which have a nonzero entry on row j. Formally, $v_i \in n_j$ if and only if $a_{ji} \neq 0$. A k-way partition of $\mathcal{H}_R$ can be considered as inducing a row

$$A_B^p = \begin{pmatrix} B_1 & & & \\ & B_2 & & \\ & & \ddots & \\ & & & B_k \\ R_1 & R_2 & \ldots & R_k \end{pmatrix} \qquad A_B^d = \begin{pmatrix} B_1 & & & C_1 \\ & B_2 & & C_2 \\ & & \ddots & \vdots \\ & & & B_k & C_k \end{pmatrix}$$

Fig. 1. Primal (A_B^p) and dual (A_B^d) block angular matrices

and column permutation on A converting it into a primal block angular form A_B^p with k blocks as shown in Fig. 1. Part P_i of $\mathcal{H}_R$ corresponds to block B_i of A_B^p such that vertices and internal nets of part P_i constitute the columns and rows of block B_i, respectively. The set of external nets $\mathcal{N}_E$ corresponds to the rows of the master problem. That is, each cut net corresponds to a row of the submatrix $(R_1, R_2, \ldots, R_k)$. Hence, minimizing the cutsize corresponds to minimizing the number of constraints in the master problem.

The proposed column–net model can be considered as the dual of the row–net model. In the column–net model $\mathcal{H}_C(\mathcal{V}_R, \mathcal{N}_C)$ of A, there exist one vertex v_i and one net n_j for each row and column of A, respectively. Net n_j contains the vertices corresponding to the rows which have a nonzero entry on column j. That is, $v_i \in n_j$ if and only if $a_{ij} \neq 0$. A k-way partition of $\mathcal{H}_C$ can be considered as converting A into a dual block angular form A_B^d with k blocks as shown in Fig. 1. Part P_i of $\mathcal{H}_C$ corresponds to block B_i of A_B^d such that vertices and internal nets of part P_i constitute the rows and columns of block B_i, respectively. Each cut net corresponds to a column of the submatrix $(C_1^t, C_2^t, \ldots, C_k^t)^t$.

Dual block angular form of A_B^d leads to two distinct parallel solution schemes. In the first scheme, we exploit the fact that dual block angular constraint matrix of the original LP problem is a primal block angular constraint matrix of the dual LP problem. Hence, minimizing the cutsize corresponds to minimizing the number of constraints in the master problem of the dual LP.

In the second scheme, A_B^d is transformed into a primal block angular matrix for the original LP problem as described in [2, 9]. For each column j of the submatrix $(C_1^t, C_2^t, \ldots, C_k^t)^t$, we introduce multiple column copies for the corresponding variable, one copy for each C_i that has at least one nonzero in column j. These multiple copies are used to decouple the corresponding C_i's on the respective variable such that the decoupled column copy of C_i is permuted to be a column of B_i. We then add column-linking row constraints that force these variables all to be equal. The column-linking constraints created during the overall process constitute the master problem of the original LP.

In this work, we select the number of blocks (i.e., k) to be equal to the number of processors. Hence, at each cycle of the parallel solution, each processor will be held responsible for solving a subproblem corresponding to a distinct block. However, a demand-driven scheme can also be adopted by choosing k to be greater than the number of processors. This scheme can be expected to yield better load balance since it is hard to estimate the relative run times of the subproblems according to the respective block sizes prior to execution.

4 Hypergraph Partitioning Heuristic

Sanchis's algorithm (SN) is used for multiway partitioning of hypergraph representations of the constraint matrices. Level 1 SN algorithm is briefly described here for the sake of simplicity of presentation. Details of SN algorithm which adopts multi-level gain concept can be found in [10]. In SN algorithm, each vertex of the hypergraph is associated with $(k-1)$ possible moves. Each move is associated with a *gain*. The *move gain* of a vertex v_i in part s with respect to part t $(t \neq s)$, i.e., the gain of the move of v_i from the home (source) part s to the destination part t, denotes the amount of decrease in the number of cut nets (cutsize) to be obtained by making that move. Positive gain refers to a decrease, whereas negative gain refers to an increase in the cutsize.

Figure 2 illustrates the pseudo-code of the SN based k-way hypergraph partitioning heuristic. In this figure, $nets(v)$ denotes the set of nets incident to vertex v. The algorithm starts from a randomly chosen feasible partition (Step 1), and iterates a number of passes over the vertices of the hypergraph until a locally optimum partition is found (*repeat–loop* at Step 2). At the beginning of each pass, all vertices are *unlocked* (Step 2.1), and initial $k-1$ move gains for each vertex are computed (Step 2.2). At each iteration (*while–loop* at Step 2.4) in a pass, a feasible move with the maximum *gain* is selected, tentatively performed, and the vertex associated with the move is *locked* (Steps 2.4.1–2.4.6). The locking mechanism enforces each vertex to be moved at most *once* per pass. That is, a locked vertex is not selected any more for a move until the end of the pass. After the move, the move gains affected by the selected move should be updated so that they indicate the effect of the move correctly. Move gains of only those unlocked vertices which share nets with the vertex moved should be updated.

```
1     construct a random, initial, feasible partition;
2     repeat
2.1      unlock all vertices;
2.2      compute k − 1 move gains of each vertex v ∈ V
            by invoking computegain(H, v);
2.3      mcnt = 0;
2.4      while there exists a feasible move of an unlocked vertex do
2.4.1       select a feasible move with max gain gmax of an unlocked vertex v
               from part s to part t;
2.4.2       mcnt = mcnt + 1;
2.4.3       G[mcnt] = gmax;
2.4.4       Moves[mcnt] = {v, s, t};
2.4.5       tentatively realize the move of vertex v;
2.4.6       lock vertex v;
2.4.7       recompute the move gains of unlocked vertices u ∈ nets(v)
               by invoking computegain(H, u);
2.5      perform prefix sum on the array G[1 . . . mcnt];
2.6      select i* such that Gmax = max1≤i*≤mcnt G[i*];
2.7      if Gmax > 0 then
2.7.1       permanently realize the moves in Moves[1 . . . i*];
         until Gmax ≤ 0;
```

Fig. 2. Level 1 SN hypergraph partitioning heuristic

```
computegain(H, u)
1           s ← part(u);
2           for each part t ≠ s do
2.1             g_u(t) ← 0;
3           for each net n ∈ nets(u) do
3.1             for each part t = 1, ..., k do
3.1.1               σ_n(t) ← 0;
3.2             for each vertex v ∈ n do
3.2.1               p ← part(v);
3.2.2               σ_n(p) ← σ_n(p) + 1;
3.3             for each part t ≠ s do
3.3.1               if σ_n(t) = |n| - 1 then
3.3.1.1                 g_u(t) ← g_u(t) + 1;
```

Fig. 3. Gain computation for a vertex u

Gain re-computation scheme is given here instead of gain update mechanism for the sake of simplicity in the presentation (Step 2.4.7).

At the end of each pass, we have a sequence of tentative vertex moves and their respective gains. We then construct from this sequence the *maximum prefix subsequence* of moves with the *maximum prefix sum* (Steps 2.5 and 2.6). That is, the gains of the moves in the maximum prefix subsequence give the maximum decrease in the cutsize among all prefix subsequences of the moves tentatively performed. Then, we permanently realize the moves in the maximum prefix subsequence and start the next pass if the maximum prefix sum is positive. The partitioning process terminates if the maximum prefix sum is not positive, i.e., no further decrease in the cutsize is possible, and we then have found a locally optimum partitioning. Note that moves with negative gains, i.e., moves which increase the cutsize, might be selected during the iterations in a pass. These moves are tentatively realized in the hope that they will lead to moves with positive gains in the following iterations. This feature together with the maximum prefix subsequence selection brings the *hill–climbing* capability to the KL–based algorithms.

Figure 3 illustrates the pseudo-code of the move gain computation algorithm for a vertex u in the hypergraph. In this algorithm, $part(v)$ for a vertex $v \in \mathcal{V}$ denotes the part which the vertex belongs to, and $\sigma_n(t)$ counts the number of pins of net n in part t. Move of vertex u from part s to part t will decrease the cutsize if and only if one or more nets become internal net(s) of part t by moving vertex u to part t. Therefore, all other pins ($|n| - 1$ pins) of net n should be in part t. This check is done in Step 3.3.1.

5   Experimental Results

Level 2 SN hypergraph partitioning heuristic is implemented in C language on Sun 1000E (60MHz SuperSparc processor) for experimenting the performance of the proposed hypergraph models on the decomposition of LP problems selected from NETLIB suite [4]. Table 1 illustrates the properties of the LP problems used for experimentation. Tables 2–4 illustrate the performance results for the row-net model (RN), column-net model with dual LP approach (CN-D), and

Table 1. Properties of the constraint matrices of the selected NETLIB LP problems

name	M	N	Z	z^r_{max}	z^r_{avg}	z^c_{max}	z^c_{avg}
perold	625	1376	6018	37	9.63	16	4.37
sctap2	1090	1880	6714	24	6.16	6	3.57
ganges	1309	1681	6912	84	5.28	13	4.11
ship12s	1151	2763	8178	49	7.10	6	2.96
sctap3	1480	2480	8874	31	6.00	6	3.58
bnl2	2324	3489	13999	82	6.02	8	4.01
ship12l	1151	5427	16170	75	14.05	6	2.98

Table 2. Average decomposition results for the row-net model (RN)

name	k	Master Problem $M\%$ (σ)	Master Problem $Z\%$ (σ)	Sub-Problems min $M\%$	Sub-Problems max $M\%$	Sub-Problems min $N\%$	Sub-Problems max $N\%$	Sub-Problems min $Z\%$	Sub-Problems max $Z\%$	exec. time (secs)
perold	2	19.2 (2.79)	37.7 (7.26)	35.1	45.7	45.4	54.6	25.1	37.1	1.40
	4	47.3 (4.83)	73.6 (3.82)	7.8	18.8	22.4	27.5	4.1	9.5	1.80
	6	59.0 (4.06)	81.8 (2.66)	3.8	10.7	14.9	18.4	1.6	5.0	3.23
	8	68.8 (2.19)	87.8 (1.35)	1.0	7.8	11.1	13.9	0.4	3.2	3.27
sctap2	2	9.7 (2.13)	31.1 (6.58)	41.2	49.1	46.1	53.9	30.6	38.3	1.88
	4	15.6 (0.57)	46.3 (0.72)	19.3	22.9	22.7	27.4	12.1	14.7	3.25
	6	17.0 (0.84)	47.8 (0.75)	12.4	15.2	14.9	18.4	7.7	9.7	5.83
	8	19.0 (1.24)	49.6 (1.00)	9.0	11.3	11.2	13.8	5.5	7.1	8.40
ganges	2	10.0 (1.32)	23.1 (1.72)	41.1	48.8	45.6	54.4	34.6	42.3	1.30
	4	15.2 (1.70)	27.8 (2.00)	18.4	23.7	22.5	27.4	14.6	21.4	3.90
	6	18.1 (1.73)	30.3 (2.30)	11.4	16.0	14.9	18.4	8.4	14.8	6.42
	8	20.7 (2.55)	33.8 (3.86)	7.5	12.1	11.1	13.8	4.9	11.3	9.20
ship12s	2	15.8 (0.52)	71.3 (1.54)	40.4	43.9	45.1	54.9	12.1	16.6	1.38
	4	22.9 (2.05)	80.4 (2.48)	16.3	25.2	22.5	27.5	3.9	6.2	3.35
	6	29.1 (1.70)	87.4 (1.88)	9.4	18.6	14.9	18.4	1.7	2.6	3.52
	8	31.7 (0.56)	90.2 (0.60)	6.4	15.9	11.2	13.7	0.9	1.5	2.75
sctap3	2	8.3 (1.35)	29.7 (4.23)	41.9	49.8	45.9	54.1	31.4	38.9	3.58
	4	15.1 (0.77)	43.1 (0.84)	19.2	23.2	22.6	27.4	12.6	15.9	4.50
	6	17.5 (1.06)	45.7 (1.18)	12.3	15.4	15.0	18.3	7.9	10.4	8.62
	8	19.4 (1.51)	47.5 (1.38)	8.8	11.4	11.2	13.8	5.6	7.7	11.85
bnl2	2	14.0 (1.71)	41.6 (5.04)	38.8	47.2	45.2	54.8	24.6	33.8	5.75
	4	21.9 (0.80)	60.5 (1.31)	15.3	24.5	22.5	27.4	7.8	11.8	9.35
	6	24.6 (1.95)	64.8 (2.22)	7.4	17.1	14.9	18.4	3.7	7.6	15.18
	8	28.5 (2.31)	69.6 (2.53)	4.1	13.2	11.2	13.8	1.8	5.7	20.98
ship12l	2	16.7 (0.14)	70.3 (0.19)	40.1	43.2	45.0	55.0	13.2	16.6	3.67
	4	25.2 (3.79)	74.7 (2.00)	13.6	24.8	22.5	27.3	4.7	7.5	9.62
	6	59.9 (2.76)	92.4 (1.40)	3.0	14.7	15.0	18.4	0.3	2.2	11.90
	8	66.9 (2.54)	95.8 (1.27)	1.9	12.5	11.2	13.8	0.1	1.1	14.32

column-net model with block transformation (CN-T), respectively. In Table 1, M, N and Z denote the number of rows, columns, and nonzeros in the constraint matrices, respectively. Here, z^r (z^c) represents the number of nonzeros in the rows (columns) of a constraint matrix.

The proposed hypergraph representations of the selected constraint matrices are partitioned into $k = 2, 4, 6, 8$ parts by running the level 2 SN algorithm. The

Table 3. Decomposition results for column-net model with dual LP approach (CN-D)

name	k	Master Problem		Sub-Problems						exec. time
				min	max	min	max	min	max	
		$M\%$ (σ)	$Z\%$ (σ)	$M\%$	$M\%$	$N\%$	$N\%$	$Z\%$	$Z\%$	(secs)
perold	2	19.5 (2.23)	26.5 (3.46)	33.7	46.8	45.3	54.7	30.2	43.3	0.97
	4	29.5 (2.21)	39.4 (3.16)	13.5	21.4	22.4	27.6	10.8	19.3	1.93
	6	33.4 (2.07)	44.6 (3.07)	7.3	14.9	14.7	18.5	5.2	13.2	3.35
	8	36.2 (1.83)	48.7 (2.43)	5.1	11.1	11.1	13.9	2.9	9.5	4.58
sctap2	2	16.0 (2.34)	21.9 (3.19)	36.8	47.2	45.4	54.6	32.8	45.3	1.02
	4	32.5 (2.52)	44.2 (3.31)	14.0	19.6	22.4	27.5	10.6	17.0	2.25
	6	37.8 (2.62)	51.3 (3.39)	7.7	12.5	14.8	18.4	5.1	10.5	3.42
	8	40.8 (2.14)	55.2 (2.68)	5.3	9.1	11.1	13.8	3.2	7.3	5.03
ganges	2	9.4 (3.11)	13.0 (7.16)	40.4	50.2	45.7	54.3	36.7	50.3	1.50
	4	30.6 (1.63)	58.5 (4.32)	14.8	21.2	22.5	27.4	7.5	16.8	2.42
	6	33.8 (1.62)	63.8 (4.07)	9.2	13.0	14.9	18.3	4.6	10.2	3.88
	8	35.8 (1.27)	66.5 (2.96)	6.5	9.7	11.1	13.8	3.2	7.1	5.85
ship12s	2	9.5 (2.19)	10.1 (2.23)	33.9	56.6	38.1	52.4	33.7	56.1	1.27
	4	16.1 (5.30)	17.0 (5.36)	13.3	28.5	17.2	27.2	13.2	28.1	3.33
	6	17.9 (6.38)	19.0 (6.48)	5.9	20.3	9.8	18.3	5.8	20.0	5.20
	8	19.8 (6.19)	21.0 (6.28)	3.1	15.7	6.8	13.8	3.1	15.5	7.70
sctap3	2	16.7 (2.71)	22.9 (3.70)	36.7	46.6	45.8	54.2	33.4	43.7	1.82
	4	31.9 (1.99)	43.4 (2.65)	13.9	20.0	22.5	27.6	10.7	17.5	3.33
	6	36.9 (2.18)	49.9 (2.84)	8.1	12.4	14.9	18.3	5.8	10.4	4.92
	8	39.6 (1.69)	53.5 (2.16)	5.6	9.3	11.1	13.8	3.5	7.6	7.25
bnl2	2	11.5 (2.85)	13.2 (3.53)	37.8	50.7	44.1	54.0	34.3	52.4	3.75
	4	19.7 (2.71)	23.4 (3.65)	14.8	25.9	21.8	27.3	12.0	26.8	9.07
	6	23.3 (3.26)	27.9 (4.33)	8.4	17.7	14.4	18.3	6.0	18.4	14.38
	8	26.4 (3.48)	32.0 (4.56)	5.3	13.6	10.6	13.8	3.4	14.4	22.70
ship12l	2	1.8 (1.68)	2.0 (1.69)	40.7	57.6	38.8	51.7	40.6	57.4	3.65
	4	8.1 (5.77)	8.5 (5.80)	15.1	29.7	17.6	26.9	15.1	29.6	7.17
	6	8.9 (5.14)	9.4 (5.18)	8.5	20.6	10.7	18.2	8.4	20.5	10.95
	8	12.5 (4.13)	13.0 (4.16)	4.2	16.0	6.7	13.8	4.2	16.0	15.50

maximum imbalance ratio is selected as $\varepsilon = 0.1$. In Tables 2–4, SN heuristic is executed 40 times for each hypergraph partitioning instance starting from different, random, initial partitions. Tables 2–4 display the averages of these runs. In Tables 2–4, $M\%, N\%$, and $Z\%$ denote the percent ratios of the number of rows, columns, and nonzeros of the master problem (subproblems) to the total number of rows, columns and nonzeros of the overall constraint matrix, respectively. Minimum and maximum values of these percent ratios are displayed for the subproblems. In Tables 2 and 3, σ values denote the standard deviations of the respective averages. In Table 4, $+M\%, +N\%$ and $+Z\%$ denote the percent increases in the number of rows, columns, and nonzeros, respectively, due to the column-linking rows added during the block transformation. Hence, $M\%, N\%$, and $Z\%$ values in Table 4 correspond to the percent ratios to the respective sizes of the enlarged constraint matrix.

As seen in Table 2, RN model yields promising results for sctap2, ganges and sctap3 problems. In the decomposition of these problems, $M\%$ values for the master problems remain below 21% for all k. As seen in Table 3, CN-D model

Table 4. Decomposition results for column-net model with transformation (CN-T)

name	k	Increase in the Problem Size			Master Problem		Sub-Problems						exec. time
		+M%	+N%	+Z%	M%	Z%	min M%	max M%	min N%	max N%	min Z%	max Z%	(secs)
perold	2	43.3	19.7	9.0	30.1	8.2	31.9	38.0	44.5	55.5	40.2	51.6	1.00
	4	84.3	38.3	17.5	45.6	14.9	12.2	15.0	20.9	29.1	17.0	25.9	1.75
	6	109.0	49.5	22.6	52.1	18.5	7.1	8.9	13.6	19.7	9.7	17.4	3.23
	8	127.6	58.0	26.5	56.0	20.9	4.9	6.1	10.0	15.5	6.4	13.7	4.60
sctap2	2	28.4	16.5	9.2	22.1	8.4	35.5	42.4	45.5	54.5	40.6	51.0	1.15
	4	82.9	48.1	26.9	45.2	21.2	12.3	15.1	22.7	27.7	16.5	22.7	2.27
	6	109.8	63.6	35.6	52.2	26.2	7.1	8.8	14.7	18.6	9.7	14.4	3.50
	8	128.0	74.2	41.5	56.1	29.3	4.9	6.1	10.8	14.2	6.5	10.8	5.33
ganges	2	11.4	8.9	4.3	10.2	4.1	41.1	48.7	45.8	54.2	41.9	54.0	1.52
	4	84.6	65.9	32.0	45.8	24.3	12.2	14.8	22.4	27.9	14.1	27.4	2.55
	6	120.0	93.4	45.4	54.5	31.2	6.8	8.3	14.5	18.9	8.2	18.9	3.80
	8	146.9	114.4	55.7	59.5	35.7	4.5	5.6	10.5	14.8	5.6	14.6	6.10
ship12s	2	23.1	9.6	6.5	18.5	6.1	36.9	44.5	41.1	58.9	37.8	56.2	1.45
	4	42.3	17.6	11.9	28.8	10.5	16.0	19.6	16.7	33.4	14.4	30.0	3.33
	6	44.7	18.6	12.6	30.2	11.1	10.3	12.9	8.9	23.6	6.8	21.2	5.12
	8	53.8	22.4	15.1	34.3	13.0	7.3	9.1	5.7	19.4	4.3	16.8	7.97
sctap3	2	27.4	16.3	9.1	21.4	8.4	36.0	42.6	46.1	53.9	41.1	50.5	2.05
	4	77.2	46.1	25.8	43.5	20.5	12.7	15.6	22.3	27.7	16.6	23.0	3.17
	6	100.7	60.1	33.6	50.1	25.1	7.4	9.2	14.8	18.8	10.0	14.8	5.08
	8	116.5	69.5	38.8	53.8	28.0	5.2	6.4	10.8	14.3	6.8	10.9	8.05
bnl2	2	16.8	11.2	5.6	14.3	5.3	39.3	46.5	43.8	56.2	38.4	56.3	3.92
	4	39.7	26.4	13.2	28.3	11.6	16.2	19.7	20.2	29.9	15.6	28.5	9.10
	6	53.4	35.6	17.7	34.7	15.0	9.8	12.0	12.2	21.5	8.1	20.4	14.80
	8	62.1	41.4	20.6	38.1	17.0	6.9	8.5	9.0	16.9	5.5	16.0	22.88
ship12l	2	6.9	1.5	1.0	5.8	1.0	43.5	50.7	42.7	57.3	42.2	56.8	3.58
	4	35.0	7.4	5.0	23.6	4.6	17.2	21.0	17.2	32.2	16.3	30.7	7.40
	6	39.3	8.3	5.6	26.1	5.2	11.0	13.6	11.4	22.2	10.4	21.0	11.28
	8	56.4	12.0	8.0	34.3	7.3	7.2	9.1	6.7	17.1	6.0	15.9	16.00

gives promising results for ship12s and ship12l problems. In the decomposition of these problems, $M\%$ values for the master problems remain below 20% for all k. As expected, CN-T model produces master problems with large $M\%$ values but small $Z\%$ values in general. The results of CN-T model for ship12s, bnl2 and ship12l problems seem to be promising. These experimental results do not favor any model, since the performance of different models vary on different problem instances due to their inherent structures.

A close examination of Tables 1–4 reveals a correlation between the performance of the SN algorithm and the net degrees of the hypergraph models of the constraint matrices besides their inherent structures. Here, degree d_n of a net n is the number of pins (vertices) connected to net n. In Table 1, z_{avg}^r and z_{avg}^c correspond to the average net degrees of the constraint matrices in the RN and CN models, respectively. For example, in the RN model, the average net degrees of perold ($d_{avg} = 9.63$) and ship12l ($d_{avg} = 14.05$) problems are much larger than those of the other problems displayed in Table 1. As seen in Table 2, the performance of the SN algorithm deteriorates on these two problems. Similarly, in the CN-D model, the average net degrees of ship12s ($d_{avg} = 2.96$) and ship121 ($d_{avg} = 2.98$) problems are much smaller than those of the other problems displayed in Table 1. As seen in Table 3, SN algorithm shows much better performance on these problems than the other problems.

It is well known that the performance of the KL-based algorithms deteriorates on hypergraphs with large net degrees. In fact, multi-level gain concept in SN algorithm is proposed as a remedy to this problem. In SN algorithm, higher level gains should be used in tie-breaking with increasing net degrees. However, memory requirement of SN algorithm drastically increases with increasing level number. The aim of this paper was an initial experimentation of the proposed hypergraph models for decomposition. We are currently investigating the performance of other hypergraph partitioning heuristics for this application.

6 Conclusion

Decomposition of constraint matrices of LP problems was investigated to obtain block angular structures for scalable parallelization. Hypergraph models proposed to represent LP constraint matrices reduce the decomposition problem to the well-known hypergraph partitioning problem. A Kernighan-Lin based multiway hypergraph partitioning heuristic was implemented for experimenting with the proposed hypergraph models. Promising results were obtained in the decomposition of the LP problems selected from NETLIB.

References

1. G. B. Dantzig and P. Wolfe. Decomposition principle for linear programs. *Operations Research*, 8:101–111, 1960.
2. M. C. Ferris, and J. D. Horn. Partitioning mathematical programs for parallel solution. Technical report TR1232, Computer Sciences Department, University of Wisconsin Madison, May 1994.
3. C.M. Fiduccia and R.M. Mattheyses. A linear-time heuristic for improving network partitions. In *Proceedings of the 19th ACM/IEEE Design Automation Conference*, pages 175–181, 1982.
4. D. M. Gay, "Electronic mail distribution of linear programming test problems" *Mathematical Programming Society COAL Newsletter*, 1985.
5. S. K. Gnanendran and J. K. Ho. Load balancing in the parallel optimization of block-angular linear programs. *Mathematical Programming*, 62:41–67, 1993.
6. B.W. Kernighan and S. Lin. An efficient heuristic procedure for partitioning graphs. Technical Report 2, The Bell System Technical Journal, Feb. 1970.
7. B. Krishnamurthy. An improved min-cut algorithm for partitioning VLSI networks. *IEEE Transactions on Computers*, 33(5):438–446, 1984.
8. D. Medhi. Bundle-based decomposition for large-scale convex optimization: error estimate and application to block-angular linear programs. *Mathematical Programming*, 66:79–101, 1994.
9. S. S. Nielsen, and S. A. Zenios. A massively parallel algorithm for nonlinear stochastic network problems. *Operations Research*, 41(2):319–337, 1993.
10. L. A. Sanchis. Multiple-way network partitioning. *IEEE Transactions on Computers*, 38(1):62–81, 1989.
11. D. G. Schweikert and B. W. Kernighan. A proper model for the partitioning of electrical circuits. In *Proceedings of the 9th ACM/IEEE Design Automation Conference*, pages 57–62, 1972.

A Parallel Computation of the Navier–Stokes Equation for the Simulation of Free Surface Flows with the Volume of Fluid Method

Z.A. Sabeur

The Civil Engineering and Building Division
School of Construction and Earth Sciences
Oxford Brookes University, Oxford OX3 0BP, UK

1 Background

Experimental and theoretical studies of wave dynamics at coastal structures are important in civil engineering. The building of sustainable sea defences and walls needs a good knowledge of both static and dynamic pressure loads at structures. Most numerical techniques, which simulate free surface flows, fail to model the full dynamics of wave impact with structures [1], [2]. The complex interaction of the fluid interface with other boundaries leads to the numerical instability and diffusion of the discretised pressure and velocity fields. Hence, the study of impact pressure distributions in space and in time, at or in the structures, is incomplete.

The volume of fluid method (VOF) of Hirt and Nichols [3], however, is capable of simulating the full dynamics of wave impacts at coastal structures. The possibility for distortion of the fluid interface and its interaction with solid boundaries makes the VOF technique as one of the most powerful methods for solving engineering problems of fluid flows with moving interfaces. The pressure and velocity field distributions during the whole dynamic process of impact can be computed and, therefore give valuable information to the coastal engineers for building breakwaters and sea defences which sustain powerful waves during bad weather.

The elegant mathematical formulation of the VOF technique is based on the association of a normalised function $F(x, y, z, t)$ in space and in time with the fractional volume of fluid in a computational cell. For instance, a cell full of fluid corresponds to a value of one for the F function while an empty cell is associated with a zero value for the F function. Hence, an intermediate value of F between zero and one represents a free surface cell. The accurate representation of the F function in space and in time, for a given computational grid domain, leads to a sharp definition of the fluid interface in space and in time. Moreover, information on nearest neighbour values of the F function, uniquely determines the orientation of the fluid interface and the side of the flow domain on which the fluid lies.

As a consequence of the above, the VOF technique enables the simulation of overturning and breaking waves, as well as the interaction of free surfaces with external boundaries [4].

2 Fundamental Fluid State Equations

2.1 Governing Transport Equation for the F Function

The F function is governed by a transport equation of the fluid. In a 2D case, it is defined as follows,

$$\frac{\partial F}{\partial t} + U\frac{\partial F}{\partial x} + V\frac{\partial F}{\partial y} = 0 \tag{1}$$

which means that the F function moves with the fluid in time and in space [5].

2.2 Governing State Equation for the Fluid

The Navier–Stokes equations (NS) are the governing equations for a fluid under transient flow state. In this study, the fluid flow is assumed to be incompressible but not irrotational. In a 2D space one has the following set of equations,

$$\frac{\partial U}{\partial t} + U\frac{\partial U}{\partial x} + V\frac{\partial U}{\partial y} = -\frac{\partial P}{\partial x} + v\nabla^2 U + g_x$$

$$\frac{\partial V}{\partial t} + U\frac{\partial V}{\partial x} + V\frac{\partial V}{\partial y} = -\frac{\partial P}{\partial y} + v\nabla^2 V + g_y \tag{2}$$

supplemented with the continuity equation which assures fluid mass conservation.

$$\frac{\partial U}{\partial x} + \frac{\partial V}{\partial y} = 0 \tag{3}$$

Note that U, V are the velocity fields in the x and y directions respectively, whereas P is the reduced pressure field.

Equations (2) and (3) are discretised in space and in time by means of a chosen finite difference scheme, then combined together in order to derive a Poisson equation as shown below:

$$\frac{P_{i+1,j}^{n+1} - 2P_{i,j}^{n+1} + P_{i-1,j}^{n+1}}{\delta x^2} + \frac{P_{i,j+1}^{n+1} - 2P_{i,j}^{n+1} + P_{i,j-1}^{n+1}}{\delta y^2}$$

$$= \frac{1}{\delta t}\left(\frac{\Psi_{ij}^n - \Psi_{i-1j}^n}{\delta x} + \frac{\Phi_{ij}^n - \Phi_{ij-1}^n}{\delta y}\right) \tag{4}$$

For each computational fluid cell $\{ij\}$, $\{P_{ij}\}$ is the reduced pressure variable at time level $n + 1$. Ψ and Φ are finite difference operators which involve the velocity fields at time level n. The pressure field solution at time level $n + 1$ enables the computation of the velocity fields at the new time level $n + 1$, hence the fluid position can be advanced by means of the Donor-Acceptor (DA) principle [5]·and the F field at time level $n+1$ evaluated. The transport equation (1) is automatically solved. More details about the numerical procedure are described in references [3] and [4].

3 Parallelisation of the Poisson Equation Solver

3.1 Introduction

Preliminary work on the numerical solver of the Poisson equation involved an iterative Gauss–Seidel algorithm [6] which was efficient up to a maximum number of unknown pressure variables. The build-up of rounding errors and the convergence of the algorithm deteriorates as the mesh is refined and the number of computational cells increased. Consequently, the newly developed VOF code has been ported to the Connection Machine, based at the University of Edinburgh Parallel Computing Centre (EPCC), and the parallelisation of subroutines achieved by re-writing the code in CM-Fortran [7]. The most important feature of CM-Fortran, which treats an array as a fundamental variable for operations and storage allocation, is applied in matrix multiplication. An array object can be referenced by name in an expression or assignment or passed as an argument to any fortran intrinsic function. The operation is performed on every element of the array, which is contrary to the subscription of array elements and their treatment as individual scalars in fortran77.

3.2 Mathematical Approach

The Poisson equation (4) can be written in matrix form by arranging the pressure variables in a column vector and in a conventional lexicographical order.

$$
(P_m) \cdot
\begin{pmatrix}
P_1 \\
P_2 \\
\cdot \\
\cdot \\
P_{i-1} \\
P_i \\
P_{i+1} \\
\cdot \\
\cdot \\
P_{N^2}
\end{pmatrix}
=
\begin{pmatrix}
W_1 \\
W_2 \\
\cdot \\
\cdot \\
W_{i-1} \\
W_i \\
W_{i+1} \\
\cdot \\
\cdot \\
W_{N^2}
\end{pmatrix}
\tag{5}
$$

For instance, in a given computational cell (i, j), pressure value P_{ij} corresponds to the kth vector element of the associated pressure vector $\mathbf{P}$. Thus,

$$k = N \cdot (j - 1) + i \tag{6}$$

where $i, j = 1, 2, \ldots N$.

The same convention applies for the associated vector $\mathbf{W}$ on the right hand side of equation (4).

In an $(N \times N)$ rectangular grid, the above convention leads to an $(N^2 \times N^2) P_m$ matrix. However, it can be reduced substantially in size if only the cells full of fluid are considered. This can be achieved by checking the F function distribution during the construction of the associated Poisson matrix.

In a rectangular computational grid, the Poisson matrix is sparse because each row has a maximum number of five non zero elements. Nevertheless, some rows have less than five non zero elements and correspond to a boundary region. For instance, the row which refers to the fluid free surface region has two non zero elements if the Dirichlet boundary condition is applied. For instance, if for a given computational cell the free surface orientation is horizontal and the rest of the fluid lies below the interface, the Poisson equation is defined as follows;

$$P_{ij} - (1 - \kappa) \cdot P_{ij-1} = \kappa \cdot (\sigma \times P_s) \qquad (7)$$

where κ is the ratio of the distance between cell centres to the distance between the free surface edge and the centre of the interpolation cell $(i, j - 1)$. σ is the surface tension and P_s is the pressure at the surface. Hence, the associated row elements of the Poisson matrix are constructed in the following way;

$$(P_m)_{kk} = 1$$
$$(P_m)_{kl} = -(1 - \kappa) \qquad (8)$$

with $k = N.(j - 1) + i, l = N.(j - 2) + i$.

In the present work, the free surface tension is set equal to zero, therefore $W_{\kappa\kappa}$ is also equal to zero.

The Poisson matrix is real and non symmetric and its band structure changes continuously as the fluid flow configuration varies in time and in space. This is due to the fact that the type of fluid cells, involved in the solver, varies from free surface cells to full or empty cells. On the other hand, the Poisson matrix may vary in size, however the variation is kept small due to the conservation of the total volume of fluid and the choice of small time steps by the numerical stability criterion. Nevertheless, the inverse of the Poisson matrix is calculated for each time step, which makes the solver subroutine the most time consuming part of the VOF code.

3.3   Connection Machine Scientific Library Solver

Access to the Connection Machine Scientific Library (CMSSL release version 3.) at EPCC was possible and a preliminary test of the mathematical approach was achieved by calling the gen_gj_invert(_) subroutine [8] as follows;

$$\text{PIVOT=gen\_gj\_invert (MATRIX, Nsize, CMSSL\_total\_pivoting, ier)} \qquad (9)$$

which is a parallelised CM Fortran total pivoting inversion method of an unstructured non symmetric matrix.

The Poisson matrix and MATRIX share the same memory space, hence there is no need to declare the inverse matrix. The elements of the inverse matrix are overwritten on MATRIX. ier is an integer which takes values zero or one. It is equal to one when a non zero value of the PIVOT in double precision is chosen and the inversion of the Poisson matrix successfully performed. Otherwise the calculation stops and returns a zero value for ier. Nsize is the size of the Poisson matrix and its new value, if different from the old one, is declared for each time step.

The associated Pressure vector is then obtained by performing matrix multiplication of the inverted Poisson matrix with vector **W** and using the CM fortran array multiplication function MATMUL.

$$P = (P_m)^{-1} \cdot W \text{ leads to } P = \text{MATMUL}((P_m)^{-1}, W) \qquad (10)$$

Finally, the new pressure vector elements are transferred into their original matrix element form in order to compute the new velocity fields, then the fluid is advanced in time with the new F function distribution by the use of the DA principle. Nearest neighbour cells to a given fluid cell, their respective upstream and downstream cells are taken into consideration for passing the fluid both in the vertical and horizontal directions. Therefore, the DA principle makes the advancement of the fluid routine difficult to parallelise. However, the exploitation of the FORALL statement which is a parallel version of a DO-LOOP in CM fortran was used for gaining speed whenever possible.

The use of the inversion routines from CMSSL version 3., for the Poisson solver, is at its early stage, and the gen_gj_invert(_) subroutine is used for checking the correctness of the proposed mathematical scheme in order to parallelise the Poisson solver and compare the physical results with early calculations based on the serial Gauss–Seidel iterative method.

Also, the fact that the associated Poisson matrix does not need to be stored and can be inverted column by column, suggests that algorithms such as the bi-conjugate gradient [9] or Lanczos [10] for real and non-symmetric matrices are the prime candidates for gaining maximum parallelisation in the VOF code.

4 Numerical Computations

A cubic metre of a standing column of water in a (2.5m × 2.5m) rectangular tank, sloped by five degrees to the vertical, was set to a free fall in order to generate an impact at the right hand side of the wall. Hydrodynamic pressures along the bed and at the vertical wall are then computed. A 60x50 uniform and rectangular mesh is used in the computation. However, an average of 550 cells are used in the computations of the Poisson matrix. The time steps varied from 0.01 to 0.002 seconds depending on the overall maximum value of velocity magnitudes. Free slip boundary conditions were applied at the walls and at the bed, and the Diriclet Boundary condition, with no surface tension, at the free surface. On the CM machine, 2.4 minutes per time step were required for the computation of the pressures, velocities and the F function distribution.

This is a gain in speed of not more than 20% but one believes that the future implementation of the Lanczos or the Bi-conjugate gradient algorithms would improve the parallelisation rate substantially.

5 Results and Discussions

The generated wave hit the right hand side of the tank at around $t = 0.414$ seconds with a maximum velocity jet, reaching $4.8 m.s^{-1}$ approximately, and an impact dynamic pressure of $19.649 N.m^{-2}$. Fig. 1 shows the flow configuration before impact while Fig. 2 shows the distortion of the free surface, after impact, and its interaction with the vertical boundary. Fig. 3 shows the continuous vertical acceleration of the jet velocities at later time.

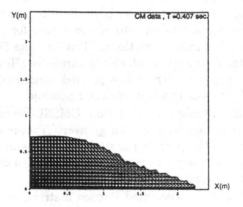

Fig. 1. Flow configuration and velocity field

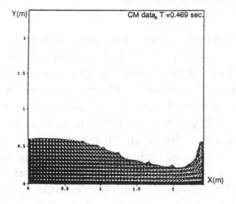

Fig. 2. Flow configuration and velocity field

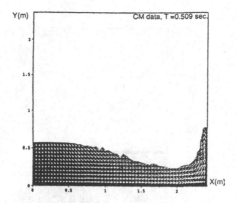

Fig. 3. Flow configuration and velocity field

At impact, violent pressures occur at the wall where the flow is accelerated upwards in a very short length of time. So far, one has not found an upper limit for the impact pressure peaks and believes that, instead, the computation of pressure impulse which is an integration of the pressure over a time period is useful to compute for the physical study of wave impact at coastal structures. [11], [12]

Nevertheless, one shows that by solving the full NS equations, the non linear terms are included in the computation and the prediction of the velocities and pressures during the whole process of impact is possible with the VOF technique.

As shown in Fig. 4, the pressure gradients around impact time are very large and an acceleration from zero to $25.0 m.s^{-2}$ in 0.1 second is calculated.

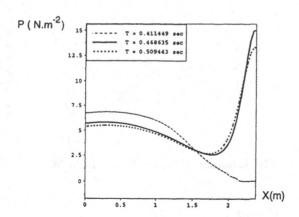

Fig. 4. Hydrodynamic pressures along the bed (CM data)

Equally, both the u and v velocity gradients are large. Figs. 5 and 6 show clearly the jump of the maximum isocontours of V-velocities from a zero value at the wall to a maximum reaching $5.2 m.s^{-1}$ in less than 0.1 sec.

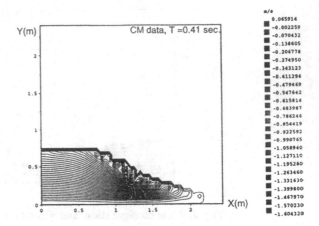

Fig. 5. Isocontour for the V-velocity field

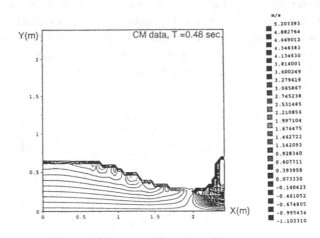

Fig. 6. Isocontour for the V-velocity field

Similarly, in Figs. 7 and 8, the U-velocity isocontours maximum value drops from a $4.84 m.s^{-1}$ to zero.

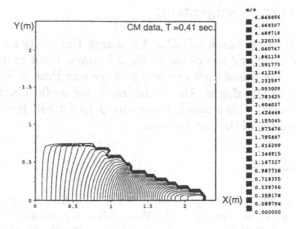

Fig. 7. Isocontour for the U-velocity field

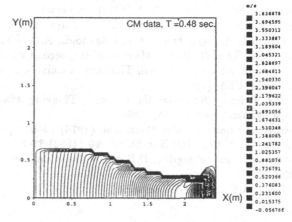

Fig. 8. Isocontour for the U-velocity field

6 Conclusion

The NS equations were solved by the use of a CMSSL library subroutine which directly inverts the Poisson matrix. The mathematical approach which sets a scheme for the parallelisation of the Poisson solver subroutine leads to results which are compatible with schemes based on iterative methods used in early calculations such as the Gauss–Seidel. Despite the change of sparsity and size of the Poisson matrix, as the fluid advances in space, one believes that the Bi-conjugate gradient and the Lanczos algorithms are the best candidates for maximum parallelisation of the solver subroutine.

Also, the extension of the problem to 3D is possible because the mathematical approach is still applicable with the VOF technique.

7 Acknowledgements

I would like to thank EPCC at Edinburgh University for the generous allocation of CM time and use of the CMSSL Library. I am grateful to HR Wallingford Ltd for the use of their graphics package and Prof. A.E.P. Veldman of the University of Groningen, The Netherlands, for useful discussions on wave impact phenomena. This research is supported by Oxford Brookes University and the Department of the Environment.

References

1. Vinje, T. and Brevig, P.: Adv. Wat. Res. **4** No 2 (1981) 77–82
2. Grilli, S and Svendsen, I.A.: Water Wave Kinematics. Ed by Tørum, A and Gudmetad, O.T. Kluwer Academic publisher (1989) 387–412
3. Hirt, C.W. and Nichols, B.D.: J. Comp. Phys. **39**/1 (1981) 201–225
4. Sabeur, Z.A., Roberts, W. and Cooper, A.J.: 4th ICFD Conference on Numerical Method in Computational Fluid Dynamics, Oxford University Press, Edited by Morton, K.W. and Baines, M.J. (1995) (in press)
5. Ranshaw, J.D. and Trapp, J.A.: J. Comp. Phys. **21** (1976) 438–453
6. Burden, R.L., Douglas Faires, J. and Reynolds, A.C.: Numerical Analysis (Prindle, Weber and Schmidt, Boston Massachussets) second edition (1981)
7. CM Fortran: Reference Manual, Thinking Machine Corporation, Cambridge, Massachussets, USA, (1991)
8. CMSSL Release: *Notes for the CM-200*, Thinking Machine Corporation, Cambridge, Massachussets, USA, (1993)
9. Fletcher, H.: Lecture Notes Math. **506** (1976) 73–98 Springer–Verlag
10. Lanczos, C.: J. Res. Nat. Bur. Stand. **49** (1952) 33–53
11. Cooker, M.J. and Peregrine, D.H.: 22nd Int. Conf. on Coastal Eng. Netherlands (1990) ASCE
12. Veldman, A.E.P.: private discussion

Improving the Performance of Parallel Triangularization of a Sparse Matrix Using a Reconfigurable Multicomputer*

José L. Sánchez[1], José M. García[2], Joaquin Fernández[1]

[1] Universidad de Castilla-La Mancha, Escuela Politécnica
Campus Universitario, 02071 Albacete, Spain
{jsanchez,joaquin}@info-ab.uclm.es
[2] Universidad de Murcia, Facultad de Informática
Campus de Espinardo, 30071 Murcia, Spain
jmgarcia@dif.um.es

Abstract. Many applications require the solution of a least-squares (LS) problem from a coefficient matrix and a measurement vector. In some cases, the solution must be obtained within a short period of time, requiring great computation power, as is the case of state estimation in electric power systems. In some cases, the coefficient matrix is a large and sparse one, requiring special techniques to reduce the computation-time and storage requirements. In these situations, fast Givens rotations are very well suited for parallel computers because they exhibit a great potential parallelism.

In this paper we improve the performance of the fast Givens rotations algorithm for sparse matrices by means of using a reconfigurable multicomputer. A reconfigurable multicomputer is a message-passing multiprocessor in which the network topology can change during the execution of the algorithm. In this way, the interconnection network can match the communication requirements of a given algorithm.

In this paper we show the improvement for applying this novel technique to improve the performance of the parallel fast Givens rotations, and we present general concepts related with it. This technique consists basically in placing the different processors in those positions in the network which, at each computational moment and according to the existing communication pattern among them, are more adequated for the development of such computation.

1 Introduction

Many numeric calculation problems need to be solved in the shortest possible time, therefore requiring the availability of computers with high calculation power. In this sense, massively parallel computers have become the best alternative to achieve this objective. Their high processing speed is based on parallel execution of different processes which, properly combined, will produce the solution required.

---

* This work was supported in part by CICYT under Grant TIC94-0510-C02-02

Parallel algorithm implementation is not immediate, being in many cases necessary a great programming effort and specially when these problems are specifically of sequential type. Another problem related with the parallel implementation is the election of the best parallel algorithm. Sometimes, the sequential and parallel behaviour is not the same. So, it is necessary to choose the algorithm with the best parallel execution.

In this paper, we solve in parallel the least-squares problem by means of numerical methods based on orthogonal transformations. This problem is very usual in many scientific applications, as lineal system resolution or eigenvalue problems. Among QR descomposition, Householder transformations are usually preferred when programming a serial computer, due to their higher speed. However, Givens rotations are very well suited for parallel computers, because they exhibit a great potencial parallelism. Moreover, there is an improved version of Givens rotations, known as fast Givens rotations. Finally, in some cases the coefficient matrix is a large and sparse one.

In this paper, we present a novel technique to improve the performance of parallel triangularization of a sparse matrix using fast Givens rotations. This parallel algorithm is executed in a multicomputer. Among all computer architectures developed to this moment, multicomputers present the highest performance for resolution of sparse matrices problems. In this class of machines, the communication between processors relies on an interconnection network, generally with a point-to-point topology. The main problem presented by multicomputers is precisely the saturation of the interconnection network. When a process requires one datum which is being calculated by other processes executed in different processor, communication is established between both. Another possible cause which can produce communication between two processes is when they must be synchronized.

In order to reduce the negative effect due the communication, a novel technique used to improve the performance in multicomputers is the dynamic reconfiguration of the interconnection network. This technique consists basically in placing the different processors in those positions in the network which are more adecuated for the development of the computation and communication.

In this paper we show the performance improvement in fast Givens rotations using a reconfigurable network multicomputer. We are encouraged by than to continue our researches in depth. The rest of the paper is organized as follows. In the next section we briefly introduce Givens rotations and the parallel algorithm. In section 3 we present the dynamic reconfiguration of the interconnection network, and in section 4 we show and analyse the evaluation results. Finally, in section 5 we give some conclusions and ways for future work.

2 Fast Givens Rotations

A Givens rotation can be defined by a transformation matrix $J(i, k, \theta)$, where θ is the rotation angle. The definition of $J(i, k, \theta)$ is well known and can be found in [8]. The application of an $m x n$ transformation matrix $J(i, k, \theta)$ to an $m x n$ matrix A annihilates the element A_{ki}, choosing the appropiate value of θ.

The transformation of A into an upper triangular matrix can be achieved by calculating and applying to A a sequence of Givens rotations, which annihilate the elements below the diagonal. In several problems, including LS, the rotations are also applied to an $m x 1$ coefficient vector b. The result of the transformation is a triangular equation system, which can be solved by backward substitution.

The classical algorithm for matrix triangularization on serial computers nullifies all the elements below the diagonal sequentially. A pair of rows is selected in each iteration, calculating a rotation and applying it to all the elements of the selected rows. This algorithm has a complexity $O(mn^2)$.

To obtain a parallel implementation of the algorithm, we must take into account that the rotations of rows of A are totally independent and can be applied in any order. The only requirement for applying rotation to a pair of rows is that the first nonzero elements of both rows occupy the same column position. So, any pair of rows of the same type can be processed in parallel with any other pair. Processes do not need to communicate during the rotation process. However, after each rotation one of the rows must be transferred to another process in most cases.

The algorithm to triangularizate a matrix A and its associated coefficient vector b, based on fast Givens rotations, the theoretical proof of the algorithm and more information about Givens rotations can be found in [8]. We have followed the studies developed by Duato [4] in order to apply fast Givens rotations in multicomputers.

2.1 Parallel Algorithm

The parallel implementation of the algorithm requires as many processes as columns the sparse matrix has. If we define the type of a row as the column position occupied by its leftmost non-zero element, then it is well known that only rows of the same type can be rotated together. Then we distribute the rows among processes in such a way that each process stores all the rows of the same type. After a pair of rows has been rotated, one of them increases its type, being sent to the corresponding process to be rotated again. Processes are mapped to processors depending on a predefined cost function, in order to minimize the communication cost. In this paper, we have ussualy taken into account a round robin or cyclic distribution.

Empty rows are discarded and the algorithm finishes when there is at most a single row in each process. As the rotation of a pair of rows cannot produce a row of a lower type, a token is passed through all the processes to determine when the triangularization program has finished.

Our machine model allows that the communication is carried out asynchronously and in parallel with the processing.

The basic algorithm executed by a given process, is the following:

```
repeat
     if received()
          then begin
                    receive rows from other processes
                    insert rows in local matrix
               end
     if rows_counter>1
          then begin
                    extract two rows from local matrix
                    calculate and apply a rotation
                    insert the first row in the local matrix
                    calculate destination process of second row
                    send second row to destination process
               end
     until finished
```

It is interesting to note here that the algorithm has not a regular communication pattern. The communication pattern is not fixed and is varying along the time. Moreover, there is no locality of communications. So, it is very difficult to choose a good topology that can adjust this communication pattern. A deep study of this algorithm can be found in [9].

3   A Reconfigurable Multicomputer

Multicomputers rely on an efficient interconnection network. The network is a critical component because performance is very sensitive to network latency and throughput. In our work, we consider a multicomputer with a control-flow mechanism called wormhole routing [3]. Interconnection networks with wormhole routing mechanism are insensitive to the communication distance, but they are fairly sensitive to conflicts on the same link, that is, the congestion problem. Several techniques have been proposed to reduce or avoid congestion, such as virtual channels, random routing or message combining. The technique here presented is related to the reconfiguration capacity of the interconnection network topology.

The dynamic reconfiguration of the interconnection network is a solution adopted in order to reduce the cost of the communication. Basically, consists in placing the different processors in those positions in the network which, at each computational moment and according to the existing communication pattern among them, are more adequated for the development of such computation.

The basic idea is the following: when messages arriving by a given channel to their destination nodes have supported an important delay, the reconfiguration algorithm will try to put the destination node close to the site that is producing those delays, by exchanging its position with its neighbour more closed to the conflict zone.

A reconfigurable network has the following advantages:

- Programming a parallel application becomes more independent of the target architecture because the architecture adapts to the application.

- This feature provides the flexibility required for an efficient execution of various applications. Moreover, in this way it is easy to exploit the locality in communications.
- Finally, this feature is very well suited for parallel applications where communications pattern varies over time.

A reconfigurable network is controlled by a reconfiguration algorithm. There are two types of reconfiguration: static or dynamic. In this paper, we focus on dynamic reconfiguration, that is, the topology can change almost arbitrarily at run-time. We present the results we have obtained with this reconfiguration algorithm [6] for a numerical problem: the fast Givens rotations.

The algorithm we have developed for the dynamic network reconfiguration has the following properties: *uses global reconfiguration* (several changes can be carried out in one step), *preserves the topology* (after reconfiguration, the network has the same topology), *is based on contention network* (a node can reconfigurate the network taking into account information about the contention in the network), *produces a small alteration* (a node can only make an exchange with one of its neighbour nodes) and *uses three thresholds for network reconfiguration*.

Figure 1 shows the effects of this technique for a very elemental situation. In (a) and (b) messages sent by processors must in some situations go through the same channels until they reach their destination, originating logical delays in communication. Once situation (c) is reached, this problem disappears, thus accelerating emission and reception of these messages. The processor which receives the messages manages to place itself in its best position in the network. This is achieved by changing the situation of this processor by means of small alterations in the network, that is, exchanging its position with a neighbour processor.

4 Performance Evaluation

In this section we are going to evaluate and analyse the results obtained after we have applied the algorithm for the dynamic reconfiguration of the interconnection network at the parallel triangularization of a sparse matrix using fast Givens rotations.

Fast Givens rotations algorithm produces communication among the different processors of the multicomputer through its interconnection network which is mainly due to the movement of rows in the matrix from one process to another. It seems therefore obvious to try to reduce the negative effect introduced by that communication in the total time of execution of the algorithm.

This algorithm has been chosen because we cannot know a priori the communication pattern between nodes, because it depends on the structure of the sparse matrix, and therefore a suitable topology cannot be selected. Moreover, the communication pattern will vary over time.

For dense matrices, the rotation of two rows of type t produces two rows of types t and $t + 1$ respectively. Then, this algorithm performs very well on a ring.

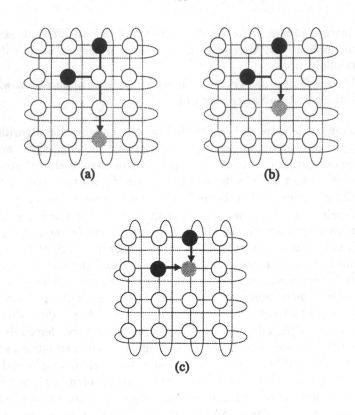

(a)   (b)

(c)

● Source nodes   ● Destination node

Fig. 1. An example of the reconfiguration algorithm effects

For sparse matrices, on the other hand, we cannot know a priori the communication pattern. As processing advances, matrix fill-in increases, approaching the behaviour of denses matrices.

Simulation has been used because it is difficult to analyse theoretically the arithmetic and communication complexities when sparse matrices are processed. This is specially true when the matrix structure changes dynamically during the processing, as is the case of Givens rotations. In this case, experimental results are needed to evaluate the performance of the algorithm.

The evaluation methodology we have used is summarized in the following sections.

4.1 Programming and Simulation Environment

The results we present have been obtained with Pepe, a Programming Environment for Parallel Execution [7]. Pepe takes a parallel program as input and generates intermediate code for an execution on a multicomputer. The most important parameters of this multicomputer can be varied by the user. Pepe generates performance estimates and quality measures for the interconnection network.

Pepe provides a user-friendly visual interface for all phases of parallel development. This graphical interface has been designed with the aim of keeping it really immediate and comfortable to the user, following the styles adopted nowadays by most of the human-oriented interfaces. In our environment, the user gets tools for easy experimentation both different parallelization possibilities and different network parameters. With this methodology the programmer can change very quickly parallelization strategies and evaluate this parallelization with analysis tools.

Pepe has two main phases and several modules within it. The first phase is language-oriented, and it allows us to code, simulate and optimize a parallel program. In this phase, interactive tools for specification, coding, compiling, debugging and testing were developed. This phase is architectural independent and it is the front-end of our environment. The second phase has several tools for mapping and evaluating the reconfigurable architecture. We can vary several parameters as different interconnection topologies or routing algorithms. The link between two phases is an intermediate code that is generated as possible result of the first phase. This allows that the user can handle our environment as whole or each phase singly. For example, we can execute only the first phase for testing the parallel behaviour of an algorithm on an ideal multicomputer. The other possibility is to obtain an intermediate code from a key parallel algorithm. Then, we can execute several times the second phase with different network parameters to evaluate them for this key algorithm.

4.2 Characteristic of the Testing Matrices

The testing matrices have been generated at random by making a homogeneous distribution of their non-zero elements. Large size matrices have been selected to produce a large message traffic to better appreciate in this way the advantages of the reconfiguration algorithm. For a specific number of columns in the matrix, tests with different rectangularity factors have been carried out for a minimum value of two, since smaller factors can hardly produce traffic in intermediate nodes of the network. An average number of two non-zero elements per row has been taken in order to ensure we are dealing with sparse matrices. Most of the results included in this work refer to matrices of 2400 rows and 1200 columns.

In order to increase the efficiency of the parallel algorithm, a column rearrangement in the matrix is usually made before applying Givens rotations. This is directed to attain that the first columns in the matrix are those with the lowest number of non-zero elements.

4.3 Performance Measures

The most important performance measures are delay and throughput. Delay is the additional latency required to transfer a message with respect to an idle network. It is measured in clock cycles. The message latency lasts since the message is introduced in the network until the last flit is received at the destination node. An idle network means a network without message traffic and, thus without channel multiplexing. Throughput is usually defined as the maximum amount of information delivered per time unit. It is measured in flits per clock cycle.

4.4 Results

In this section, we show some evaluation results. These results have been obtained making use of a hipercube topology. We choose this topology because is that we have obtained the best results for the static case.

Figure 2 shows the throughput as a function of network size. It can be seen that the algorithm achieves a higher throughput for the whole range of network sizes which have been analysed. In similar way we have obtained other graphs showing the average message delay as a function of network size. All of them clearly show that the dynamic reconfiguration of the network scales well with network size. Figure 3 shows the average message delay versus message length.

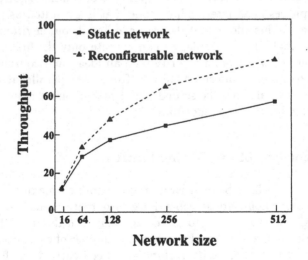

Fig. 2. Throughput as a function of network size

The curves correspond to different values for the number of virtual channels on a 256-node network: Static algorithm (Static 1vc), dynamic algorithm (Dynamic 1vc), static algorithm with two virtual channels (Static 2vc) and dynamic algorithm with two virtual channels (Dynamic 2vc).

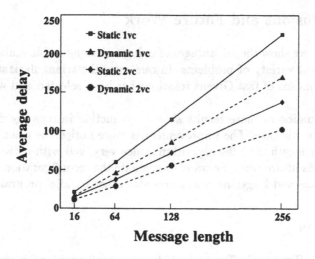

Fig. 3. Average message delay as a function of message length

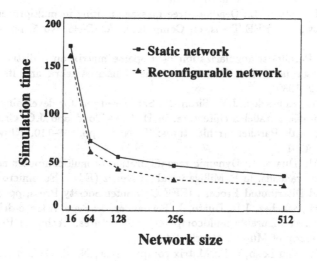

Fig. 4. Total simulation time as a function of network size

The results show a significant reduction in message delay, especially for long messages.

Finally, figure 4 shows the simulation time as a function of the network size. It is noticed that the reduction of the total time of the computation in the triangularization process is kept for the different sizes of the network.

5 Conclusions and Future Work

In this paper we show the advantages of using a reconfigurable multicomputer for solving a great variety of problems. In our case, the triangularization of sparse matrices by means of fast Givens rotations has been selected and we present the results here.

The evaluation of these results shows a reduction in message delay with respect to the static case. The improvement is more noticeable when the messages have a larger length, and the algorithm scales very well with network size.

As regards future work we want to evaluate the reconfiguration algorithm for larger matrices and larger networks and also other parallel programs.

References

1. Adamo, J., Bonello, C.: Tenor++: A dynamic configurer for Supernode machines. Lecture Notes in Computer Science. No. 457, pp. 640–651, Springer Verlag (1990)
2. Bauch, A., Braam, R., Maehle, E.: DAMP: A dynamic reconfigurable multiprocessor system with a distributed switching network. 2nd European Distributed Memory Computing Conference, Munich, April, 1991
3. Dally, W.J., Seitz, C.L.: Deadlock-free message routing in multiprocessor interconnection networks. IEEE Trans. on Computers, Vol. C–36, No. 5, pp. 547–551, May, 1987
4. Duato, J.: Parallel triangularization of a sparse matrix on a distributed–memory multiprocessor using fast Givens rotations. Linear Algebra and its Applications, 121:582–592, 1989
5. Fraboul, Ch., Rousselot, J.Y., Siron, P.: Software tools for developing programs on a reconfigurable parallel architecture. in D. Grassilloud and J.C. Grossetie (Eds.), Computing with Parallel Architectures: T. Node, pp. 101–110, Kluwer Academic Publishers, 1991
6. García, J.M., Duato, J.: Dynamic reconfiguration of multicomputer networks: Limitations and tradeoffs. in P. Milligan and A. Nunez (Eds.), Euromicro Workshop on Parallel and Distributed Proces., IEEE Computer Society Press, pp. 317–323, 1993
7. García, J.M., Sánchez, J.L., Duato, J., Fernández, J.: Pepe: A trace-driven simulator to evaluate reconfigurable multicomputer architectures. Technical Report DIS TR 4–95, University of Murcia, March, 1995
8. Golub, G.H., Van Loan, C.F.: Matrix computations, North Oxford Academic, 1983
9. Sánchez, J.L., García, J.M.: Estudio de la reconfiguración dinámica de la red. Evaluación de nuevas propuestas para mejorar su eficiencia. Technical Report DIS TR 13–94, University of Murcia, October, 1994

Comparison of Two Image-Space Subdivision Algorithms for Direct Volume Rendering on Distributed-Memory Multicomputers

Egemen Tanin, Tahsin M. Kurç, Cevdet Aykanat, Bülent Özgüç

Dept. of Computer Eng. and Information Sci.
Bilkent University, Bilkent 06533, Ankara – TURKEY

Abstract. *Direct Volume Rendering* (DVR) is a powerful technique for visualizing volumetric data sets. However, it involves intensive computations. In addition, most of the volumetric data sets consist of large number of 3D sampling points. Therefore, visualization of such data sets also requires large computer memory space. Hence, DVR is a good candidate for parallelization on distributed-memory multicomputers. In this work, image-space parallelization of Raycasting based DVR for unstructured grids on distributed-memory multicomputers is presented and discussed. In order to visualize unstructured volumetric datasets where grid points of the dataset are irregularly distributed over the 3D space, the underlying algorithms should resolve the *point location* and *view sort* problems of the 3D grid points. In this paper, these problems are solved using a *Scanline Z-buffer* based algorithm. Two image space subdivision heuristics, namely *horizontal* and *recursive rectangular* subdivision heuristics, are utilized to distribute the computations evenly among the processors in the rendering phase. The horizontal subdivision algorithm divides the image space into horizontal bands composed of consecutive scanlines. In the recursive subdivision algorithm, the image space is divided into rectangular subregions recursively. The experimental performance evaluation of the horizontal and recursive subdivision algorithms on an IBM SP2 system are presented and discussed.

1 Introduction

Direct Volume Rendering (DVR) is a technique to create an image from the three-dimensional volume data without generating an intermediate geometrical representation. Usually, volume data is represented by 3D voxels which constitute the atomic pieces of the overall data structure. One of the approaches used in direct volume rendering for visualizing these 3D voxel based data sets is *Raycasting*, which is the basis of this research. This approach is an image-space approach and mainly uses ray shooting from each pixel of the image plane and sampling along its way [6, 7, 9]. In order to visualize unstructured volumetric data sets, where grid points of the data set are distributed irregularly over 3D space, the underlying algorithms should resolve the *point location* and *view sort* problems. The point location problem refers to the determination of the location of the point of intersection of the ray with an individual voxel in the whole data set. The view sort problem refers to the determination of the correct intersection order of the ray with the voxels along its way. In this paper, the point location and the view sort problems are solved using a *Scanline Z-buffer* based algorithm, which is introduced by [1, 2] in volume rendering domain. The reason for this is to solve the problem of fast point location determination process and view sorting efficiently for unstructured grids.

Although DVR is a common tool for visualization, it operates on volume data representation that requires a large amount of memory. DVR is also very slow since it requires massive computations for each image generation. So interactive speed rates are very hard to achieve. An important approach to solve the speed and memory problems of volume rendering is to employ parallel processing. Furthermore, many engineering simulations are usually run on parallel architectures because of simulation time and memory constraints. Therefore, visualizing the results on the same parallel architecture avoids the cost of transferring large volume of data to sequential graphics workstations for visualization process.

Most of the work for parallelization of DVR has been carried out using structured grids [3, 4, 5]. Recently some parallel methods have been developed for unstructured grids [1, 2] which form a basis for this research. These recent approaches are done on shared memory architectures which does not solve many of the problems of volume rendering on distributed-memory architectures. These approaches use scattered decomposition for task distribution and hence can be effectively used in shared memory architectures. However, such decomposition schemes will incur large communication overhead in distributed-memory architectures.

Although parallel volume rendering of rectilinear grids has been accomplished to a great degree, domain mapping problem for unstructured grids is a crucial problem to be solved. This paper investigates the parallelization of Raycasting based DVR for unstructured grids. In this work, an image-space parallel algorithm for DVR on distributed-memory MIMD architectures is presented and two load balancing heuristics are proposed to distribute the load evenly in rendering phase. Experimental results on an 8 node IBM SP2 architecture are presented and discussed. This target architecture is a distributed-memory multicomputer. The nodes of the IBM SP2 are connected by a multi-stage interconnection network.

2 Raycasting Based DVR

The high quality of the images produced by the raycasting approach makes it a desirable choice for DVR. In this approach a ray is shot from each pixel and traversed throughout the whole volume to determine the list of voxel intersections. Each voxel intersection means an entry/exit point of the ray with the voxel. For each voxel intersection, a sampling is computed at the midpoint of the ray between its entry and exit points by interpolating the scalar values at the grid points of the intersected voxel. The voxel intersections should be traversed in a predetermined order (front-to-back in our case) for the composition of sampled color and opacity values. Ray shooting, sampling, and finally composition steps require the detection of the position of the sampling point in the whole data and finding the next (therefore previous) voxels to be intersected with the ray for composition. These operations bring tremendous amount of computation to the process of raycasting. Therefore, finding the consecutive intersections and the locations of the sampling points should be done efficiently, which we refer to as efficient *point locating* and *view sorting* operations.

In this work, we exploit the Scanline Z-buffer algorithm widely used in polygon rendering to resolve the point location and view sort problems. The same idea has been recently used in the parallelization of DVR on shared-memory multiprocessors [1, 2]. The standard Scanline Z-buffer algorithm is used for rasterizing the polygons but the only deviation occurs in the rendering step where in volume rendering each pixel keeps a linked list of polygons for compositing and finding a final pixel color value. Therefore, the overall algorithm needs only polygonal data to be rendered. Note that volume data composed of tetrahedrals is already in polygonal form where each tetrahedral is composed of four triangles. The set of distinct triangles can be extracted easily from this tetrahedral data set. Therefore, for such data sets traversing triangles along the ray is equivalent to traversing the voxel data itself. For other types of data sets, triangulation should also be done by just connecting original sample points in a way that we will have a set of polygons in the final dataset. This operation once completed can be saved and used forever. Algorithms exist for triangulating a given volumetric dataset [8]. Here, and hereafter, we will mainly assume triangles as our inputs for the sake of simplicity. The Scanline Z-buffer based algorithm is given in Fig. 1 as a flow-chart.

In the proposed algorithm, we move from scanline to scanline and from pixel to pixel one-by-one. Initially, each polygon is placed into the *y-bucket* list of the first scanline that the polygon is intersected with. The polygons intersected by the current scanline are put into the *active polygon list*. The x-intersection points of polygons with the current scanline generate spans of the polygons on that scanline. The generated spans are put into the *x-bucket list* and sorted with respect to minimum x-intersections using x-bucket sort. The calculations of intersections of polygons with the scanline, insertion and deletion operations on the active polygon list are done incrementally using *inter-scanline* coherency. For each pixel on the current scanline, the intersection of the ray shot from the pixel and spans that cover that pixel are determined and put into the *z-list*, which is a sorted linked list, in the order of increasing z-intersection values. The z-intersection calculations, sorting of z-intersection values, insertion to and deletion from z-list are done incrementally using *intra-scanline* coherency. As we know the location of each intersection of the active polygon with the ray and as the list is built in an incremental fashion, we can say that we just have an array of sorted intersections with a three dimensional line and a set of planes. For a single pixel, after the intersections are found, we can go through the list and take samples between each successive pair of triangles and composite it into the pixel color.

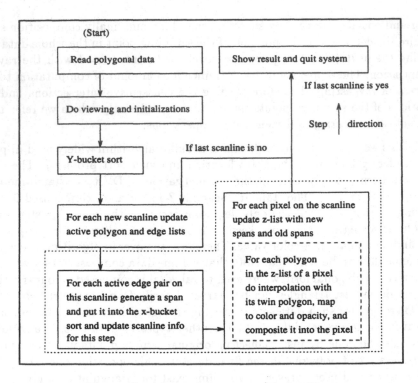

Fig. 1. Flow chart of the sequential algorithm.

3 Parallel Implementation

The parallel algorithm presented in this section is an image-space parallel algorithm. In image-space parallelism, the image plane is partitioned among the processors. Then, each processor runs a sequential volume rendering algorithm to generate the image for its local image plane sections. Each processor needs the volume data which is covered by the view volume of the local image plane regions. Therefore, the volume data is also partitioned and distributed to the processors according to the partitioning of the image plane. The parallel algorithm consists of four main steps.

At the first step, the global volume data is evenly partitioned and distributed among the processors so that each processor receives a distinct set of N_T/P triangles (polygons), where N_T is the number of triangles and P is the number of processors.

At step 2, the image space is partitioned into P regions and each region is assigned to a processor. The partitioning of the image space should be done to achieve a good computational load balance. Two strategies to achieve this goal is presented and discussed in the next sections. The triangle set received at the first step is utilized to perform an adaptive division of the image plane.

After the partitioning and assignment of the image plane, each processor needs triangles which fall into the view volume of the local image plane partition.

The local triangle data set may contain triangles that belong to image plane partitions assigned to other processors. Similarly, some of the triangles that are covered by the view volume of local image plane section may reside in the local memories of other processors. Therefore, at step 3, some of the triangles – hence, volume data – should be exchanged among the processors. Each processor finds the image plane region a triangle belongs to by performing projection and clipping operations and sends the triangle to the corresponding processor and receives triangles that fall into its local image plane region. Note that the triangles at the boundaries of image plane partitions will be shared by two or more processors. Hence, such triangles may be transmitted more than once.

At step 4, each processor runs the sequential volume rendering algorithm for its local image plane section using new local volume data without further inter-processor communication.

3.1 Load Balancing

In many scientific applications, the volume data to be visualized is not regularly sampled and distributed in 3D space. Hence, the computational work load on the image space will also be irregularly distributed. In addition, different viewing locations will result in different work load distribution on the image space. Hence, a straightforward division of image plane into equal size regions may result in very poor load balance among the processors due to the nature of the volume data. Therefore, an adaptive division of image plane will generate better work load distribution and better processor utilization.

There are three parameters that affect the computational work load in a image plane section. First one is the number of triangles, because the total work load due to clipping of a triangle to boundaries and insertion operations into y-bucket and active polygon lists are proportional to the number of triangles in a region. The second parameter is the number of scanlines each triangle spans. This parameter represents the computational work load associated with the construction of *x-spans*, and insertion of these spans into x-bucket list. The total number of pixels generated by rasterization of x-spans of a triangle is the third parameter affecting the computational load in a region. Each pixel generated adds computations required for sorting, insertions to and deletions from z-list, interpolation and composition operations. The y-span of each triangle gives the number of scanlines covered by the triangle. Hence, the number of triangles at each scanline can easily and correctly be calculated using y-span of each triangle. However, the length of the x-span at each individual scanline requires rasterizing edges of the triangle. This computational overhead can be decreased by using the bounding box of the triangle instead of triangle itself. The *x-span* length at each scanline is approximated by the *x-span* of the bounding box.

The two subdivision heuristics presented in the following sections use these parameters to estimate the work load (WL) in a region using Eq. (1).

$$WL = aN_T + bN_S + cN_P \qquad (1)$$

where N_T, N_S, and N_P represent the number of triangles, spans, and pixels, respectively, to be processed in a region. The values a, b, c represent the relative computational costs of operations associated with triangles, spans, and pixels, respectively.

3.2 Horizontal Subdivision

In this scheme, the image plane is subdivided into P horizontal bands of consecutive scanlines such that each band has almost equal work load. The division of the image plane is carried out using the distribution of work load in y-dimension of the image plane. Dividing the image plane into bands of consecutive scanlines preserves the inter-scanline coherency to some extent.

Each processor calculates local work load distribution using local triangle information. The work load distribution for each parameter (triangle, span and pixel) is stored in separate arrays. The global work load distribution is calculated by performing a global sum operation on these arrays. This global sum operation can be done in $log_2 P$ steps for each array using the communication structure of the IBM SP2. At the end of global sum operation, each processor has global work load distribution in y-dimension of the image plane. Using this work load distribution information, the image plane is divided horizontal bands. An example of horizontal subdivision scheme is illustrated in Fig. 2. After the division operation, each processor exchanges local triangle information with other processors according to the subdivision of the screen. The local triangles that project to the region of another processor are transmitted directly to that processor.

In this scheme, the atomic task is defined to be a scanline, i.e., scanlines are not divided. In this way, intra-scanline coherency is not disturbed. However, the image plane is partitioned in only one dimension, namely y-dimension. Due to this restriction, the scalability of horizontal division scheme is limited by the number of scanlines. In addition, the work load at each region is determined by the work load at each scanline. Hence, if there are large differences in the work loads of scanlines, the load imbalance between regions may still be large. These limitations of the horizontal subdivision can be eliminated to some extent by using subdivision in both dimensions of the screen. The recursive rectangular subdivision scheme that implements this idea is described in the following section.

3.3 Recursive Rectangular Subdivision

In recursive rectangular subdivision (recursive subdivision) scheme, the image plane is divided into P rectangular regions in $log_2 P$ steps. In this work, the number of processors (P) is assumed to be a power of two. This scheme is similar to the horizontal division scheme. The same data structures and load balancing parameters are used. However, unlike horizontal division scheme, load distribution in two dimensions are used to obtain a division. Therefore, data structures

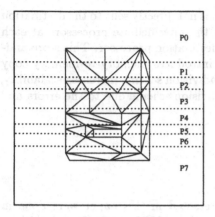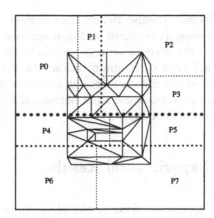

Fig. 2. An example of horizontal (left) and recursive (right) subdivision for eight processors. The regions are separated by dotted lines.

used for load distribution in y-dimension are duplicated for load distribution in x-dimension.

At the first step, each processor is assigned the whole screen as its local image region. Each processor, then, updates its local copy of the global work load arrays using the local object data. The work load distribution in two dimensions are obtained by performing global sum operations on these arrays. Then, each processor divides its local image region into two subregions either horizontally or vertically. The division that achieves better load balance is chosen. After the division, half of the processors are assigned to one of the regions, and the other half of the processors are assigned to the other region. After each processor is assigned a image region, it exchanges volume data with processors assigned to the other region. After this exchange operation, each processor has some portion of the volume data that projects onto its local image region. The division operation is repeated for new image region using new local volume data. After $log_2 P$ steps, each processor is assigned a unique rectangular region of the screen. An example of recursive subdivision is given in Fig. 2.

Although there are similarities between two subdivision schemes, the overheads introduced are not the same. In horizontal subdivision scheme, only inter-scanline coherence is disturbed while intra-scanline coherence is preserved. However, recursive subdivision scheme may disturb both the inter-scanline and intra-scanline coherence. The disturbance in the inter-scanline and intra-scanline coherency introduces run-time overheads which are hard to estimate a priori. In addition, the use of bounding box approximation may introduce more errors in recursive subdivision scheme than horizontal subdivision scheme. The number of triangles and number of spans in a region calculated by bounding box approximation may differ from the actual amounts for a rectangular region created by vertical subdivision. However, bounding box approximation does not introduce these errors in horizontal subdivision. Furthermore, unlike horizontal subdivision scheme, the volume data is exchanged during the division phase of the recursive subdivision scheme. Hence, the volume data is not directly sent to the destination processors. It is stored in the memories of the intermediate processors at each step of the division until it reaches to the destination processor. This store-and-forward type of communication structure introduces memory-to-memory copy overhead. Therefore, although recursive subdivision scheme increases scalability, it may perform worse than horizontal subdivision scheme due to the errors and overheads introduced.

4   Experimental Results

The implementations of the algorithms presented in this paper were done in C language using the message-passing constructs of IBM SP2. The algorithms were tested on a data set composed of 409600 distinct triangles for two screen resolutions, 256x256 and 512x512. The timing results presented in the graphs are the average of viewings at three different view points for each of the screen resolutions.

The speedup graphs for total execution time and rendering-phase are given in Fig. 3. As is seen from the figure, the horizontal subdivision scheme achieves higher speedup values than the recursive scheme. Although recursive scheme utilizes subdivision in both dimensions of the screen, vertical divisions of the screen regions disturb intra-scanline coherency. The intra-scanline coherency is a crucial factor in the execution time, because it involves incremental sorting, insertion, and deletion operations into z-lists. As is also seen from the figures, the speedup values increase with increasing resolution of the screen due to the increase in the accuracy of work load distributions. Speedup values of 4.78 for total execution time and 6.34 for rendering-phase are achieved using the horizontal scheme for screen resolution of 512x512 on 8 processors.

Figure 4 illustrates the load-balance graphs for rendering phase. The load balance in the graphs is calculated as $100 \times (1 - ((t_{max} - t_{min}) / t_{max}))$, where t_{max} and t_{min} denote the execution times of the maximally and minimally loaded processors, respectively. As is seen from the figure, the recursive subdivision scheme achieves better load balancing with respect to estimated run times, which are calculated using Eq. (1). On the other hand, the horizontal scheme achieves better load balancing performance in real execution times. This is due to the disturbance of both inter-scanline and intra-scanline coherence in recursive subdivision scheme. Note that the load balance achieved in real execution times is less than the one in estimated times. This is because of the fact that the disturbance of both inter-scanline and intra-scanline coherency results in run-time overhead that cannot be predetermined before execution. As is seen from the figure, load balance in estimated run times increases with increasing screen resolution due to better accuracy. As is also seen from the figure, the load balance for real run times also increases with increasing screen size for this data set. However, the run-time overheads introduced due to disturbance in inter-scanline and intra-scanline coherency may affect this behavior.

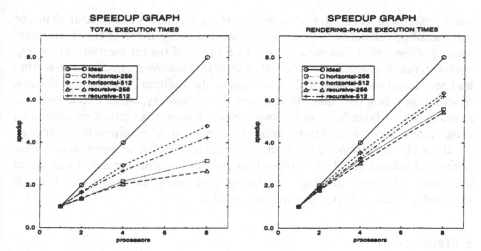

Fig. 3. Speedup graphs of the horizontal and recursive division schemes for total and rendering-phase execution times.

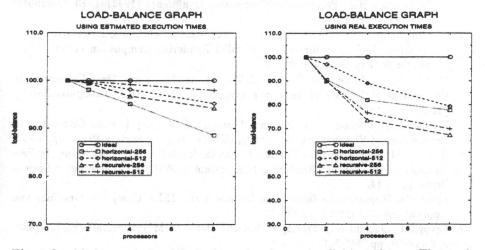

Fig. 4. Load-balance graphs of the horizontal and recursive division schemes. The graph on the left represents the load-balance using estimated rendering times. The graph on the right represents the actual load-balance obtained from the experimental results.

5 Conclusions

In this study, two image space subdivision methods for direct volume rendering on distributed-memory multicomputers are compared. It is experimentally observed that horizontal subdivision scheme achieves better performance results than the recursive subdivision scheme. The horizontal subdivision scheme only disturbs inter-scanline coherency whereas recursive subdivision scheme disturbs intra-scanline coherency as well. Hence, recursive subdivision method introduces run-time overhead in the rendering phase more than the horizontal subdivision. Unfortunately, the costs of these run-time overheads cannot be determined before execution. The intra-scanline coherency is much more crucial than the inter-scanline coherency. The intra-scanline coherency involves the incremental sorting of the linked lists with respect to z values of the intersections of the ray with volume elements. Hence, disturbance of this coherency results in re-sorting and re-calculation of these linked list values. In addition, a horizontal division of an image region only disturbs inter-scanline coherency, which affects only the scanlines at the boundaries of the two regions. However, a vertical division of an image plane disturbs the intra-scanline coherency at all scanlines in the region.

Upon the deductions drawn from experimental results, we can conclude that horizontal scheme should be prefered to recursive scheme for small number of processors. However, for large number of processors, the recursive scheme is expected to achieve better performance results.

References

1. Challinger, J. Parallel Volume Rendering for Curvilinear Volumes. In Proceedings of the Scalable High Performance Computing Conference (1992), IEEE Computer Society Press, pp. 14-21.
2. Challinger, J. Scalable Parallel Volume Raycasting for Nonrectilinear Computational Grids. In Proceedings of the Parallel Rendering Symposium (1993), IEEE Computer Society Press, pp. 81-88.
3. Corrie, B., and Mackerras, P. Parallel Volume Rendering and Data Coherence. In Proceedings of the Parallel Rendering Symposium (1993), IEEE Computer Society Press, pp. 23-26.
4. Elvins, T. T. Volume Rendering on a Distributed Memory Parallel Computer. In Proceedings of Visualization '92 (1992), IEEE Computer Society Press, pp. 93-98.
5. Hsu, W. M. Segmented Ray Casting For Data Parallel Volume Rendering. In Proceedings of the Parallel Rendering Symposium (1993), IEEE Computer Society Press, pp. 7-14.
6. Levoy, M. Display of Surfaces From Volume Data. IEEE Computer Graphics and Applications 8, 3 (1988), pp. 29-37.
7. Levoy, M. Efficient Ray Tracing of Volume Data. ACM Transactions on Graphics 9, 3 (1990), pp. 245-261.
8. Shirley, P., and Tuchman, A. A Polygonal Approximation to Direct Scalar Volume Rendering. Computer Graphics 24, 5 (1990), pp. 63-70. In Proceedings of San Diego Workshop on Volume Visualization.
9. Upson, C., and Keeler, M. VBUFFER: Visible Volume Rendering. Computer Graphics 22, 4 (1988), pp. 59-64. In Proceedings of SIGGRAPH '88.

Communication Harnesses for Transputer Systems with Tree Structure and Cube Structure

Ole Tingleff

Institute of Mathematical Modelling,
The Technical University of Denmark,
DK-2800 Lyngby, Denmark
e-mail: ot @ imm.dtu.dk

Abstract. This paper describes two simple and easy to use communication harnesses for transputer systems: one for a system with ternary tree structure and one for a system with cube structure. We tell about the purpose of a general communication harness and present 2 implementations. We demonstrate the performance of the harnesses on some applications. The transputers are installed in a PC using 2 B008 motherboards. The programming language is occam2 running under the occam2 toolset.

1 Introduction

Many computational problems can be split into subtasks which can be treated independently. Such problems are suited for parallel computations in a processor farm with slave processsors controlled by a master processor. A tree structured processor farm is well suited for such a situation.

In other problems the subtasks are not mutually independent, they have to exchange data during the parallel computations. In this situation we should choose a processor farm where the distance from any processor to any other processor is small. A cube structure is a good choice here.

Transputer systems where the processors are housed in so called motherboards have the advantage that the actual configuration of the connections between the processors can be altered at will. Thus it is possible to examine and use very many different connection structures.

It is a characterestic of the transputer that it was developed ab initio as a component in a parallel computing system. Thus, communication hardware is integral in the processor and *read* and *write* are in the set of fundamental operations for the processor. It has 4 communication links in the processor. The programming language is occam2, which is charaterised by communicating sequential processes constituting the parallel programme. The programmes were run under the occam2 toolset [1].

2 Why is a Reconfigurable System useful?

In most computations on parallel systems a major part of the total time is spent on communication and on processors waiting for communication. The users consider this as a loss, the communication loss.

In this connection it is very usefull to have a parallel system where the structure of the connections between the different processors can be changed. Thus we can examine, e.g. how much connections with tree structure are better than a pipe line structure for our computation task. It is important that the actual connections between the processors can be adjusted so as to form a structure, appropriate for our problem.

The transputer system on IMM consist of 19 transputers mounted in 2 B008 motherboards. Each of these has a programmable multiswitch enabling (nearly) any communication link on one transputer to be connected to a link on any other transputer.

3 Why is a Communication Harness useful?

The purpose of the communication harness is to handle the sending and receiving of data parcels between the different processes in the parallel programme. Using the harness, we can separate the parts of the programme that perform the computations from those that handle communication. This leads to several advantages. Firstly we can develop the harness and test it on trivial computations. More importantly we can develop our computation sections independently of the communication sections, and it is very easy to develop a parallel programme for a new problem using a well proven communication harness. We can also move the computational sections from a programme for one communication structure into another communication structure.

```
{{{  middlepr.occ
#INCLUDE "comprot.inc"
...  Comments
PROC MiddleProcess(CHAN OF COMMUNICATION up.from, up.to,          --parent
                                         left.from, left.to,      --left child
                                         middle.from, middle.to,--middle child
                                         right.from, right.to,   --right child
                                                             ) --who am I ?
                        VAL INT  proc.num
    #USE "supercomm.t8h"
    #USE "compute.t8h"
    CHAN OF COMMUNICATION compute.from, compute.to :
    PRI PAR
       Communicate (left.from, left.to, middle.from,
                    middle.to, right.from, right.to,
                    up.from, up.to,
                    compute.from, compute.to        )  -- my own  Compute
         Compute (compute.from, compute.to,              -- to & from the world
                  proc.num                 )             -- where am I ?
    :        -- Bottom of      M I D D L E P R O C E S S
}}}  middlepr.occ
```

Fig. 1. Process running on a typical transputer in the ternary tree.

As an example, Figure 1 gives a listing of the process runing on a typical of processors in an processor farm with ternary tree structure. In occam the processes which run in parallel exchange data along communication channels only, so the parameters of the processes are (nearly) always such channels. The process *Compute* which performs the actual computations receives all necessary data along channel *compute.to* and delivers all results along channel *compute.from*. The purpose of process *Communicate* is to receive all data ariving at this processor and send on all data meant for other processors, or send the data to *Compute*, if they are meant for this processor. Likewise all results, either from *Compute* or from other processors are transmitted towards the user.

It is important to notice that *Compute* and *Communicate* run in parallel with *Communicate* in highest priority. This gives *Communicate* the right to interupt *Compute*, whenever a data parcel arrives meant for another processor. If the 2 processes run in the same priority, the communication loss increases considerably.

4 The Tree Structure and its Harness

The tree structure is a good choice when our computation can be split into subtasks which can be performed in parallel without the subtasks exchanging data between them. Thus all communication consist of data sent from the user via a master processor to one of the slave processors and results sent from a slave processor to the user (via the master). We use a ternary tree because this utilises all 4 communication links on the transputers. In Figure 2 you see a ternary tree with 3 layers. Each transputer has a label. The master processor has label 1, and the slaves have labels 2 − 40. You can see that data, sent from the master to any of the slaves only passes at most 2 processors before arriving at its destination.

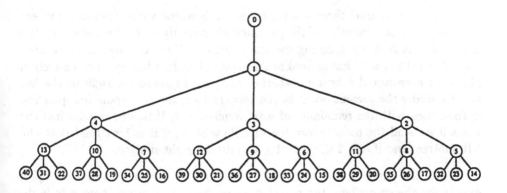

Fig. 2. The ternary tree structure with transputer labels shown.

```
File up2down.occ, State = Editing!, Language = Occam
{{{   Receives , treats command of 3 types: start, activate or stop
ALT
  up.from ? CASE
    ...  Start command
    {{{  Run command
    Run.Cmd;
      u.tp;                                    --target processor
      ...  Relevant data
    SEQ
      tp.rem := u.tp REM 3
      {{{  Transmits data
      IF
        u.tp = 1                               --message for this node
          down.to ! Run.Cmd;          --treat locally
          u.tp;                       --address here
          ...  Relevant data
        tp.rem = 2                             --message for right child
          right.to ! Run.Cmd;         --send to right
          (u.tp / 3) + 1;             --address to right
          ...  Relevant data
        tp.rem = 0                             --message for middle child
          middle.to ! Run.Cmd;        --send to middle
          u.tp / 3;                   --address to middle
          ...  Relevant data
        tp.rem = 1                             -- message for left child
          left.to ! Run.Cmd;          --send to left
          u.tp / 3;                   --address to left
          ...  Relevant data
        TRUE
          SKIP
      }}}
    }}}
    ...  Stop command
    stop.it ? stop.var
      running := NOT stop.var
}}}
```

Fig. 3. The routing section of PROC Communicate.

The labels of the master and the 3 first slaves form a basis for the labelling method. To get the labels of the next level in the tree we multiply the labels of the present level with 3, and put those on the children of the middle third of the present level. From these labels we subtract 1 and put them as labels of the right-hand third of the new level. Likewise we add 1 to the "middle labels" and put these on the left-hand third of the new level.

In each data parcel there is a parameter $u.tp$ whose value determines where the parcel is sent. The value of the parameter is initially set to the address on the parcel, and it is changed during the transmission. When the parameter reaches the value 1, the parcel has arrived at its destination. In a binary tree, a question of odd or even would determine whether the parcel goes to the right or the left subtree under the present node. In the ternary tree, the corresponding question is concerned with the remainder of $u.tp$ divided by 3. If this remainder has the value 0 we send the parcel down the middle subtree, if it is 1 we send it to the left subtree, and if it is 2 the parcel goes into the right subtree.

When the parcel goes into the next level of the tree, $u.tp$ is divided by 3, expect for the case where the parcel goes to the right subtree; here $u.tp$ is divided by 3 and 1 is added. Thus the controlling value is a "pseudo remainder" having the values 1, 0 and -1 for left-, middle and right subtree, respectively. The relevant part of the *Communicate* process is given in Figure 3.

5 The Cube Structure and its Harness

Not all parallel computations can be performed as mutually independent sub-tasks; often the subtasks have to exchange intermediate results during the computations. In many cases a subtask on one of the processors must broadcast data to all the other slave processors. Thus the distance from any of the processors to any other of the processors must be minimal. A processor farm with cube structure (or hypercube structure) is appropriate for this situation.

We chose a cube structure for the transputer farm, see Figure 4. Here we can see that the master, labelled 0 uses all its 4 communication links. It is possible to build a transputer farm with a 4 dimensional hypercube, but that is more complicated because we shall need an extra transputer sitting on one of the edges of the cube in order to establish a connection to the user. A cube structure is far more simple to build.

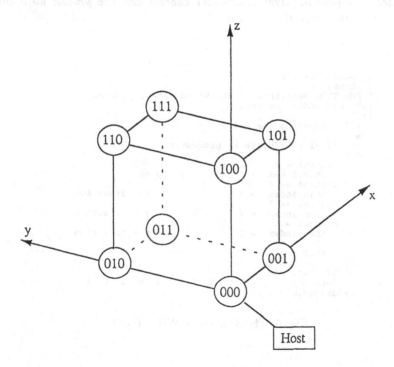

Fig. 4. The Cube Structure with the transputer labels (0,1,2,...,7) given in binary representation.

When the transputers are labelled as in Figure 4 we can use the well-known grey codes to controll the routing of data parcels through the network. A fundamental operation in these is the binary exclusive or, XOR or in occam: ><. This operator compares the bit patterns of its 2 integer arguments and produces an integer which has 0 where the bits are equal and 1 where the bits are different.

For example   7 >< 5 is 2
because   111 XOR 101 is 010.

We use this operator on the address on the parcel and the label on the transputer. If the result has 1 in the left-most position the parcel has to be sent in the z-direction (see Figure 4); if the result has 1 in the middle position the parcel goes in the y-direction, and if it has 1 in the right position the parcel is sent on in the x-direction. If the result is 0, the parcel address and the transputer label are equal, and the parcel has come to its destination. Thus a parcel with address 7 and sitting at transputer 5 must be sent on along the y-direction, see Figure 4.

Because of its symmetry with respect to its arguments, the XOR operation can be used in many instances in the harness, e.g. telling which operator is the x-neighbour to transputer 3, or which transputers is the space opposite of transputer 5, and all routing operations. Central in the latter is FUNCTION *Send.index* in Figure 5. Given the parcel *address* and the *procssr.no* it determines the relevant transmission channel.

```
{{{   sendindex.occ
...   Comments
INT FUNCTION Send.index (VAL INT address , procssr.no )
  INT chan.index, control :
  VALOF
    {{{   Computing the index
    SEQ
      control := address >< procssr.no
      IF
        control = 0
          chan.index := 0          -- to me
        control >= 4
          chan.index := 3          -- to z - direction
        control >= 2
          chan.index := 2          -- to y - direction
        control = 1
          chan.index := 1          -- to x - direction
        TRUE
          SKIP
    }}}
    RESULT chan.index
;   --  Bottom of        S E N D . I N D E X
}}}  sendindex.occ
```

Fig. 5. The Send.index FUNCTION.

6   Applications using the Tree Structure

The first real application of the tree structure was BLAS3 type computations, the matrix-matrix product:

$$\underline{C} = \underline{A} \times \underline{B}.$$

If we slice the matrix $\underline{A}$ so that each transputer get (about) the same number of complete rows and also the entire matrix $\underline{\underline{B}}$, then the transputer can determine the corresponing rows of the result completely independent of the computations on the other transputers.

Our tree harness runs with first level (master and 3 slaves) or with first and second level, master and 12 slaves. In the large tree there are some rather small transputers, so here we can at most multiply a $(48 \cdot 24)$ matrix with a $(24 \cdot 24)$ matrix. In Figure 6 you see the speedup achieved on these small matrices. The master does not do any matrix computations when $\underline{A}$ has a multiple of 12 (the number of slaves) as number of rows and this gives extra efficiency. In the worst case the master treats 11 rows of $\underline{A}$ and $\underline{C}$, whereas the slaves compute for 4 rows at most. This is the explanation for the variation of the speedup, seen in Figure 6. In the small tree the best speedup was 3.76.

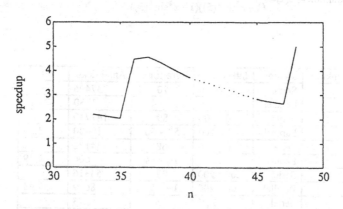

Fig. 6. Speedup for the matrix-matrix product computed on the master and 12 slaves. Matrix sizes are (n · 24) and (24 · 24).

M. Christiansen and S. Munk made a small project (see [2]), using the tree structure for Romberg Integration. A week and 2 days after starting (without prior knowledge of the occam language) they had the Romberg integrator running on 13 transputers using the communication harness. Consider

$$I = \int_a^b f(x)dx \ .$$

In Romberg integration this integral is approximated by the area under a linear interpolation of $f(x)$, the interval being divided in 2, 4, 8, 16, ... parts of equal length. The final result derives from extrapolation on these approximations. Figure 7 gives an idea about the speedup achieved. The reason for the speedup being more than 13 in some cases is that the sequential version of Romberg uses the same steplength in the whole of $[a,b]$, whereas in the parallel version, each transputer can determine its optimal steplength.

In his M.Sc. thesis, Chr. Würtzen (see [3]) described how he made a programme for a parallel version of Sturmian sequence computations determining all eigenvalues of a symmetric tridiagonal matrix. Each transputer determines the same number of eigenvalues. The computations on the transputers are independent, but at the beginning there is some duplication of the computations.

Funktion 1, $f(x) = e^{\sin(2x)\cos(3x)}$.

Funktion 2, $f(x) = \dfrac{1}{1+x^2}$.

Funktion 3, $f(x) = e^x - x^3$.

Funktion 4, $f(x) = \dfrac{x\mathrm{Arcsin}(x)}{(1 - x^2\sin^2(5))(1 - x^2\sin^2(3))}$.

Funktion	Metode	Interval	Iterationer	Antal ticks	SPU
1	Seriel	-10 ; 10	70	37496	
	Parallel	-10 ; 10	34 - 41	1380	27.17
1	Seriel	0 ; 100	89	142730	
	Parallel	0 ; 100	53 - 60	10680	13.36
2	Seriel	0 ; 100	68	34172	
	Parallel	0 ; 100	14 - 34	606	56.39
2	Seriel	-200 ; 200	93	54316	
	Parallel	-200 ; 200	19 - 58	8659	7.43
3	Seriel	-5 ; 5	35	273	
	Parallel	-5 ; 5	13 - 19	82	3.33
3	Seriel	-100 ; 100	74	35445	
	Parallel	-100 ; 100	5 - 42	764	46.39
4	Seriel	0 ; 0.75	35	308	
	Parallel	0 ; 0.75	14 - 19	143	2.66
4	Seriel	0 ; 0.95	64	10418	
	Parallel	0 ; 0.95	14 - 35	503	20.69

Fig. 7. Speedup for Romberg integration, run on 13 transputers.

His version of the method ran on a transputer farm with ring structure, and I moved his computation parts into the tree harness. In Figure 8 you see some of the speedups achieved.

The Sturmian sequence method is iterative, incorporating bisection and Newton's method. Thus some eigenvalues pose more trouble some than others, and this accounts for the irregular curves.

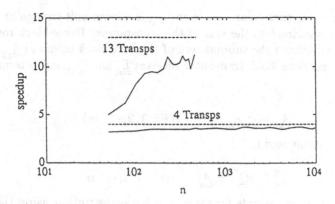

Fig. 8. Speedup for Sturmian sequences run in a tree structure with 4 and 13 transputers. The matrix order is n.

7 Applications using the Cube Structure

The first major programme running in the cube harness is a solver for equation systems with non-symmetric, full matrices. We have chosen the Gauss-Jordan method which consists of row operations producing 0 in the coefficients below the diagonal position (as in Gauss eliminations) and also in coefficients above the diagonal position. This decouples the system completely and the components of the solution can be found independently.

For the parallel implementation of this method we section the matrix in blocks in the same way as in the matrix · matrix programme, i.e. the system

$$\underline{\underline{A}}\,\underline{x} = \underline{y}$$

is sectioned horizontally and each of the transputer (numbers 0,1,..,m; m=7) receives and treats (about) the same number of equations. The corresponding block representation of the equation system is

$$\begin{bmatrix} \underline{\underline{A}}_{00} & & \cdots & & \underline{\underline{A}}_{0,m} \\ & \underline{\underline{A}}_{11} & & & \underline{\underline{A}}_{1,m} \\ & & \cdot & & \cdot \\ & & & \cdot & \cdot \\ & & & & \cdot \\ \underline{\underline{A}}_{m,0} & \underline{\underline{A}}_{m,1} & \cdots & & \underline{\underline{A}}_{m,m} \end{bmatrix} \begin{bmatrix} \underline{x}_0 \\ \underline{x}_1 \\ \cdot \\ \cdot \\ \cdot \\ \underline{x}_m \end{bmatrix} = \begin{bmatrix} \underline{y}_0 \\ \underline{y}_1 \\ \cdot \\ \cdot \\ \cdot \\ \underline{y}_m \end{bmatrix}$$

In the first step of the elimination transputer 0 makes an LU decomposition of the diagonal submatrix $\underline{\underline{A}}_{00}$:

$$\underline{\underline{A}}_{00} = \underline{\underline{L}}_0 \underline{\underline{U}}_0 \tag{1}$$

where $\underline{\underline{U}}_0$ is an upper triangular matrix and $\underline{\underline{L}}_0$ is a lower unit triangular matrix. These are then broadcast to the rest of the transputers. Using block row operations we now transform the submatrices of the first block column $(\underline{\underline{A}}_{10}, \underline{\underline{A}}_{20}, \cdots$ $\cdots, \underline{\underline{A}}_{m,0})$ into 0-matrices. Each transputer receives $\underline{\underline{L}}_0$ and $\underline{\underline{U}}_0$, and determines $\underline{\underline{C}}$, so that

$$\underline{\underline{A}}_{i0} - \underline{\underline{C}}\,\underline{\underline{A}}_{00} = 0 \qquad (i = 1, 2, \cdots, m)$$

Trivial manipulations lead to

$$(\underline{\underline{U}}_0^T \underline{\underline{L}}_0^T)\underline{\underline{C}}^T = \underline{\underline{A}}_{i0}^T \qquad (i = 1, 2, \cdots, m) \tag{2}$$

and this gives us $\underline{\underline{C}}$ via simple forward- and backsubstitution using the triangular factors $\underline{\underline{U}}_0^T$ and $\underline{\underline{L}}_0^T$. With the $\underline{\underline{C}}$ matrix the transputer then performs the nescessary block row operations

$$\underline{\underline{A}}_{ij} = \underline{\underline{A}}_{ij} - \underline{\underline{C}}\,\underline{\underline{A}}_{0j} \qquad (i = 1, 2, \cdots, m \text{ and } j = 1, 2, \cdots, m)$$

$$\underline{y}_i := \underline{y}_i - \underline{\underline{C}}\underline{y}_0 \qquad (i = 1, 2, \cdots, m) \tag{3}$$

In the next step of the elimination $\underline{\underline{A}}_{11}$ is decomposed by transputer 1 like in (1) and block row operations similar to (2) and (3) produce 0- matrices in the block column below $\underline{\underline{A}}_{11}$ and also above $\underline{\underline{A}}_{11}$. This process goes on till we reach the final result, a system of equations whose coefficient matrix has 0 everywhere except in the diagonal blocks. These contain $\underline{\underline{L}}_0$ and $\underline{\underline{U}}_0$, $\underline{\underline{L}}_1$ and $\underline{\underline{U}}_1, \cdots$, $\underline{\underline{L}}_m$ and $\underline{\underline{U}}_m$. All operations (2) and (3) are done in parallel.

Now, the system has been block decoupled and the subvectors of unknowns, i.e. $\underline{x}_0, \underline{x}_1, \cdots, \underline{x}_m$ can be determined independently using forward - and backsubstitution with the relevant $\underline{\underline{L}}_k$, $\underline{\underline{U}}_k$ and $\underline{y}_k$. This is done in parallel.

Test runs with the method show encouraging results. The parallel Gauss-Jordan programme is faster than a sequential programme with the most efficient equation solver for this situation, the Doolittle method. For larger systems it even beats a sequential Cholesky programme, built for the symmetric case.

8   Future work

At present, I am working on parallel implementation of a Quasi-Newton method, determining

$$\underline{x}^* = argmin_{\underline{x}} f(\underline{x})$$

using the first derivatives of f. An other programme implements Gauss-Newton's methods for non-linear least squares problems, also using the first derivatives:

$$x^* = argmin_{\underline{x}} \sum_{j=1}^{m} (f_j(\underline{x}))^2$$

Programmes which can check the user's parallel implementation of the first derivatives are running.

The 2 methods above are iterative methods and in each iteration the expensive parts are: Computation of the function and its derivatives and the solution of a linear equation system. Besides that, Gauss-Newton also has a matrix-matrix product. These are parts, I expect to be appropriate for parallel implementation, whereas the rest of each step in the iteration and the control of whole iteration are better performed sequentially by the master in the transputer farm.

Now, the function evaluation (including first derivatives) and the matrix-matrix products are best suited for the tree structure, the parallel computations being mutually independent. On the other hand, the cube structure is best suited for the equation solver, so that is the one chosen.

A bit further in the future are parallel implementations of Peters and Wilkinson's method for linear least squares problems and of Householder's transformations, to be used in determination of eigenvalues of symmetric full matrices.

References

1. INMOS Ltd: Occam 2 toolset, user manual. INMOS (1991)
2. Christiansen, M., Munk, S.: Romberg Integration på Transputer Netwærk (in Danish). Numerisk Institut, DTH (1993)
3. Würtzen, C.E.: Parallelberegning af Egenværdier (in Danish). Numerisk Institut, DTH (1993)

A Thorough Investigation of the Projector Quantum Monte Carlo Method Using MPP Technologies

Peter Valášek[1] and Ralf Hetzel[2]

[1] Max-Planck-Institut für Plasmaphysik, Euratom Association
85748 Garching bei München, Germany
[2] Institut für Theoretische Physik, Technische Universität Dresden, Germany

Abstract. The projector quantum Monte Carlo method is one of the most important numerical techniques in solid state physics to study the ground-state properties of many-electron systems. It can be applied to a variety of systems, but is most often used to study models for the recently discovered high-temperature superconducting compounds. We give a survey of the algorithm and discuss technical details and general problems that occur. Furthermore, we present a performance and error analysis for our implementation of the algorithm when applied to the 2-dimensional Hubbard model. It appears that massive parallelization can speed up and stabilize the final results, and more detailed studies will give rise to an improved parallel algorithm.

1 Introduction

Monte Carlo simulations have become an established and useful method for treating problems in classical statistical physics. The first attempts to apply these ideas to quantum physical problems trace back to the years of 1950-1960 when numerical calculations for three- and four-body nuclei and liquid helium were attempted [1]. Today Monte Carlo simulations provide a powerful tool when applied to quantum many-body systems which typically resist a solution by analytic and perturbational means.

Quantum Monte Carlo (QMC) techniques are of statistical nature, and their realm are strongly correlated many-body problems, where the number of basis states for a reasonably sized system is extremely large prohibiting any exact numerical solution. In order to study systems of even larger sizes and at lower temperatures, increasingly large amounts of computational power are required.

Several different QMC algorithms have been developed in the course of the years. Although they are very different with respect to their numerical implementation, the basic ideas are very similar to all of the algorithms [2]. Here, we employ the projector method which extracts the ground-state of a model to compute low energy and low temperature properties.

This paper desribes our efforts of porting an optimized vector code to a massively parallel computer, the Connection Machine CM-5. We demonstrate that the usage of a massively parallel computer can speed up and enhance convergence of computed quantities if carefully designed.

The article is outlined as follows: First, we introduce the most important model for strongly correlated electronic systems, i.e. the Hubbard model. Then we analyze the projector quantum Monte Carlo method and will apply it to the Hubbard model. Finally, an overview of our parallel implementation of the algorithm is given, followed by a discussion of its properties.

2 Model

The simplest model to describe electrons on a lattice is the tight-binding model. The Hamilton operator for this model, neglecting the Coulomb repulsion between electrons, reads in second quantization

$$H_{TB} = \sum_{i,j,\sigma} t_{i,j} \, c^\dagger_{i,\sigma} c_{j,\sigma} - \sum_{i,\sigma} \varepsilon_i \, n_{i,\sigma} \quad , \tag{1}$$

which is the standard notation in many-body physics. The operators $c^\dagger_{i,\sigma}$ create and $c_{i,\sigma}$ annihilate an electron of spin $\sigma = \uparrow, \downarrow$ in an orbital centered at site i. The quantum nature of the electrons is taken into account by special anticommutation rules for the c-operators. The hopping integral $t_{i,j}$ between two sites allows an electron to move from site j to site i. The second term introduces a local potential ε_i with the occupation number operator $n_{i,\sigma} = c^\dagger_{i,\sigma} c_{i,\sigma} \in \{0, 1\}$. This model can easily be solved exactly by Fourier transformation in momentum space.

If the Coulomb repulsion between electrons is taken into account, the model can no longer be solved analytically. In a crude approximation, one therefore takes only *local* repulsion between electrons into account; this may be justified due to screening effects. Two electrons of different spins may occupy the same orbital; they now feel a repulsion of strength U. This approximation was first introduced by Hubbard [3] and is given (without local potentials) by the following Hamiltonian

$$H = H_0 + H_1 = \sum_{i,j,\sigma} t_{i,j} \, c^\dagger_{i,\sigma} c_{j,\sigma} + U \sum_i n_{i,\uparrow} n_{i,\downarrow} \quad . \tag{2}$$

The Hubbard model is widely used to describe electronic features of transition metals, e.g. Fe, Co, Ni, magnetic ordering, and metal insulator transitions. A vast amount of work has recently dealt with this model in an attempt to describe high-temperature superconductors; this is one of the major challenges in todays physics.

Since the Hubbard model is not solvable exactly in general — only in a few limiting cases —, various perturbative analytical and numerical methods are employed to calculate its physical properties. Our choice will be the projector quantum Monte Carlo method which we will present in the next section.

3 Algorithm

We are interested in the ground-state properties of the Hubbard model. In physical terms, it means that we will consider very low temperatures ($T \to 0$) and will study low energy excitations of the order of $\mathcal{O}(\text{eV})$. The method we are going to employ is the projector quantum Monte Carlo (PQMC) algorithm which makes no approximations to the model as compared to pertubational approaches. The basic idea behind this algorithm is to use a projector operator to extract the ground-state wavefunction $|\Psi_0\rangle$ from a trial wavefunction $|\Psi_T\rangle$ which is non-orthogonal to it. A useful projector which was first proposed by Sugiyama and Koonin [4] reads

$$e^{-\beta H} |\Psi_T\rangle = e^{-\beta E_0} |\Psi_0\rangle \langle \Psi_0|\Psi_T\rangle \; + \; \sum_{n=1}^{\infty} e^{-\beta E_n} |\Psi_n\rangle \langle \Psi_n|\Psi_T\rangle$$

$$= e^{-\beta E_0} \left(|\Psi_0\rangle \langle \Psi_0|\Psi_T\rangle \; + \; \sum_{n=1}^{\infty} e^{-\beta(E_n - E_0)} |\Psi_n\rangle \langle \Psi_n|\Psi_T\rangle \right) .$$

In the limit of large β the second term vanishes and the result will be proportional to the ground-state. Expectation values which contain the relevant information for any measurement, are defined in the usual way as

$$\langle \mathbf{A} \rangle = \frac{\langle \Psi_T | e^{-\beta H} \mathbf{A} e^{-\beta H} |\Psi_T\rangle}{\langle \Psi_T | e^{-2\beta H} |\Psi_T\rangle} . \tag{3}$$

To compute the expectation value of any operator, there are three different steps to deal with: First we have to choose a trial wavefunction and a convenient representation of it. Then one calculates the effect of the projector on the trial wavefunction and, finally, the expectation values itselves. As a trial wavefunction one usually employs the exactly known solution of the tight-binding model for independent electrons. The multi-electron function can be written in terms of a Slater determinant of single-electron orbitals ψ for N sites and n_σ electrons

$$|\Psi_T^\sigma\rangle = \prod_{\lambda=1}^{n_\sigma} \left(\sum_{l=1}^{N} \psi_{\lambda,l}^\sigma c_{l,\sigma}^\dagger \right) |0\rangle = |\det\{\psi_{\lambda,l}^\sigma\}\rangle . \tag{4}$$

The wave function is characterized by coefficients $\psi_{\lambda,l}^\sigma$ and is orthonormal with respect to the λ-th single-particle orbital ($\sum_l \psi_{\lambda,l}^\sigma \psi_{\mu,l}^\sigma = \delta_{\lambda,\mu}$).

Due to the anti-commutation relations for fermion operators, it is impossible to calculate the effect of the projector $\exp(-\beta H)$ on the trial wavefunction $|\Psi_T\rangle$ directly. But the projector can be approximated by a factorization into independent propagations

$$e^{-\beta H} \simeq \left(e^{-\frac{\beta}{2m} H_0} e^{-\frac{\beta}{m} H_1} e^{-\frac{\beta}{2m} H_0} \right)^m . \tag{5}$$

The error of this Trotter-Suzuki decomposition [5] is controlled by the total number of Trotter-slices m which should take values $m \gg \beta U$. Accordingly, we have to apply a sequence of propagators to the trial wavefunction. The propagation with $\exp\left(-\frac{\beta}{2m}H_0\right)$ is always a multiplication with the same matrix [2] $|\Psi^{(i+1)}\rangle = \mathbf{H}_0 |\Psi^{(i)}\rangle$. The many body interaction H_1 is transformed [6] into a sum over Hubbard-Stratonovich fields $\sigma_l \in \{-1, 1\}$

$$\exp\left(-\frac{\beta}{m}H_1\right) \quad \propto \quad \prod_{l=1}^{N} \sum_{\sigma_l=\pm 1} \exp\left(2\lambda\left(n_{l,\uparrow} - n_{l,\downarrow}\right)\sigma_l\right) \quad , \tag{6}$$

which can be treated the same way as $\mathbf{H}_0$, namely as a multiplication with a matrix depending on the $\{\sigma\}$-configuration. Repetition of this procedure leads to a summation over all σ_l-fields for all trotter-slices. An expectation value as defined in Eq.(3) is now a sum over a large phase space volume with an effective propagator $\mathcal{F}$ depending on σ

$$\langle\mathbf{A}\rangle \;=\; \frac{\displaystyle\sum_{\{\sigma\},\{\sigma'\}} \overbrace{\langle\Psi_T|\mathcal{F}(\sigma)}^{\langle\Psi_{ex}(\sigma)|}\,\mathbf{A}\,\mathcal{F}(\sigma')\,|\Psi_T\rangle}{\displaystyle\sum_{\{\sigma\},\{\sigma'\}} \underbrace{\langle\Psi_{ex}(\sigma)\,|\Psi_{ex}(\sigma')\rangle}_{\omega(\sigma,\sigma')}} \;=\; \frac{\displaystyle\sum_{\{\sigma\},\{\sigma'\}} \mathcal{A}(\sigma,\sigma')\,\omega(\sigma,\sigma')}{\displaystyle\sum_{\{\sigma\},\{\sigma'\}} \omega(\sigma,\sigma')} \quad . \tag{7}$$

In the last step, we have extended the nominator by $\omega(\sigma,\sigma')/\omega(\sigma,\sigma')$ and have defined $\mathcal{A} = \langle\Psi_{ex}|\,A\,|\Psi_{ex}\rangle/\omega(\sigma,\sigma')$.

Even for a small system with only 16 sites and a moderate value of m, it is impossible to sum up over all configurations. Therefore we apply the Monte Carlo method to estimate the sum. Accordingly, the sign of ω is factored from ω to ensure positive definiteness for the probabilities $\omega(\sigma,\sigma')$. Finally, we employ importance sampling to get an estimate for

$$\langle\mathbf{A}\rangle \;=\; \frac{\langle\mathcal{A}\cdot\mathrm{sign}\,(\omega)\rangle_{|\omega|}}{\langle\mathrm{sign}\,(\omega)\rangle_{|\omega|}} \quad , \tag{8}$$

where the expectation value $\langle\cdots\rangle$ is defined in the usual way as

$$\langle g\rangle = \frac{1}{N}\sum_{i}^{N} g\left(\{\sigma\}_i\right) \tag{9}$$

One way to find configurations of σ-fields with the right statistical distribution is to use a Markov-Chain of configurations $\{\sigma\}$. The probability to accept a new configuration is then given either by the Metropolis ratio $R = \omega^T/\omega^i$ of a trial configuration ω^T to the current configuration ω^i or by the heat-bath ratio $R/(1+R)$. Fig. 1 gives a global view of the algorithm.

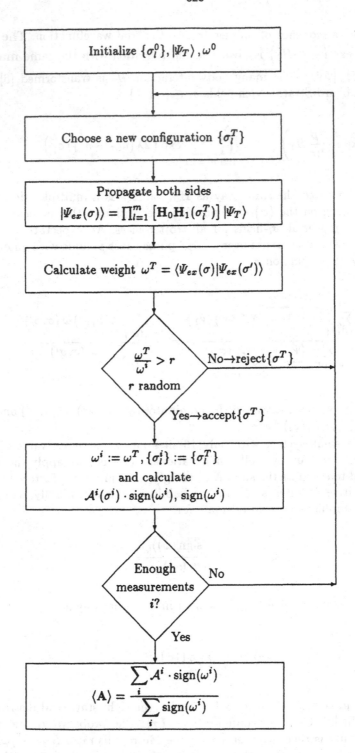

Fig. 1. Flowchart of the basic algorithm for a projector Monte Carlo calculation

4 Implementation

4.1 Overview of the CM-5 architecture

Before describing our massively parallel implementation of the PQMC method, it will be useful to give a brief survey of the architecture of the CM-5 computer. Further information may be found in the *Connection Machine Model CM-5 Technical Summary [7]*.

The CM-5 is a scalable, massively parallel supercomputer with up to 16k Sun Sparc processores. Each processor controls four independent vector-units with up to 32MB memory each. The control structure of the program runs on a front-end processor while the data are handled by a processor array consisting of individual nodes layed out in a tree-like structure.

There are two different ways of programming the CM-5, the nodal and the global view. In the former view, it is possible to run independent programs on each node, the user being responsible for the inter-nodal communication. Contrary, in the latter programming model the user runs a single data parallel program and is not concerned with the number of available nodes or the communication between them. The CM-5 Fortran (CMF) compiler covers most features of the FORTRAN 90 language as well as some extensions from the forthcoming high-performance FORTRAN (HPF) standard, allowing a nearly machine independent code for the global programming model.

4.2 Porting of the PQMC algorithm

To ensure portability in the future, we decided to use the F90 language with only a few non-standard data parallel statements special to HPF. The data layout is set by compiler directives in order to reduce inter-nodal communication. Besides, we made extensive use of optimized numerical library routines for the large amount of matrix manipulations inherent to the algorithm.

Previously, sequential code has been developed which is easily optimized for vector computers. Sequential programs face, in principle, two problems: Importance sampling is restricted to a limited number of configurations compared to the total volume of phase space of the Hubbard-Stratonovich fields $\{\sigma\}$. This is due to the fact that the calculation of the transition probability between one configuration and the next one is a computationally intensive task, leading to time-consuming runs to achieve a statisfactory sampling of phase space. The other problem encountered is that, due to a limited memory size, only small systems, typically of the order of $\mathcal{O}(10)$ lattice sites, can be handled.

Our goal was to extent both limitations by using a parallel computer. It turned out that the most effective change was simply to replicate the physical system, up to one system for each vector unit, thereby effectively increasing the number of Markov chains generated. This amounts to a parallelization of the loop over propagations using independent $\{\sigma\}$-fields for each replicated system. After completion of the propagation steps, expectation values are collected from all systems and, finally, the average value is computed. This change was implemented by introducing an additional, parallel dimension for every quantity, vector or matrix, used in the propagation loop. Inquiry functions and compiler directives map the subarrays to every vector-unit depending on the hardware available. While keeping the communication overhead low, the additional average over replicated systems dramatically reduces the statistical error for any quantity measured yielding much better statistics.

Care has to be taken when enlarging the system size. Since the system size determines the dimensions of many matrices, one may easily exceed the local memory of each node. In such a case, the number of replicated systems has to be reduced and a different data layout has to be chosen. The trial wavefunctions and the matrices should now be distributed throughout the whole lattice while library routines will manage the communication. Using the memory of all processors, even larger systems may now be handled.

Finally, we were able to reduce the length of the actual program from over 7500 lines for the sequential version down to 3500 lines for the parallel version.

5 Discussion

Let us first discuss the problems associated with a Monte Carlo (importance) sampling of phase space. The generated Markov chain should, in principle, span the whole phase space, but detailed studies in a single-site system show increasing non-ergodic behaviour with correlation strength U [8]. The Hubbard-Stratonovich fields σ_i become more and more correlated in the Trotter direction while barriers will most likely separate regions of phase space. It means that the probability of flipping all $\{\sigma_i\}$ values for one site decreases with increasing Trotter time, and the whole Monte Carlo random walk becomes more and more localized. One way to circumvent this problem is to start with a number of different configurations and generate a whole set of Markov chains.

In principle, there are two opposite limits between which one has to choose the right strategy. On one hand, one can do a lot of short, independent runs each of which generates its own Markov chain. In the extreme case of a chain of length one, only random σ-configurations are sampled which is known not to be very accurate. On the other side, one very long Markov chain will, for intermediate times, become trapped in just one region of phase space.

It was our goal to balance the number of Markov chains and the minimum length for them to compute average values most efficiently. One quantity which sensitively displays the behaviour of the system in phase space, is the autocorrelation function for the weight $\omega(\sigma, \sigma')$ (see Eq.(7)). It shows a similar behaviour independent of the choice of input parameters for the program, e.g. of the repulsion or correlation strength U. Starting from unity, it exponentially decays to and then fluctuates around zero. We interpret this behaviour in the following way: Initially, the Markov-chain accepts configurations with a high probability until it reaches a region of high statistical weight where it stays trying to step further by accepting configurations with a lower weight. The first steps needed to find a region of high weight in phase space, are called warmup steps and are discarded before any measurement is taken. The minimal number of steps needed for any run is the number of warmup steps which strongly depends on the correlation strength as shown in Fig. 2. Besides, the acceptance rate is depicted which is the ratio of accepted to totally attempted σ-configurations. Again, it depends on the correlation strength and on the way the trial σ^T-fields are chosen.

We have studied several ways to choose new σ-configurations resulting in different numbers of warmup steps $\tilde{N}_w$ and acceptance rates r_a. When we define a normalized number of warmup steps $N_w = \tilde{N}_w * r_a$, we observe a surprisingly simple relation between N_w and the correlation strength U

$$N_w = 2 * U \tag{10}$$

It should be mentioned that after the warmup steps are completed, measurements should be taken over a sufficiently large number of configurations to achieve the desired accuracy.

Let us turn to the question of how the error depends on the number and length of Markov chains. We present a few physical results for the two-dimensional Hubbard model of size 4x4 sites with $n_\uparrow = n_\downarrow = 8$ electrons and $U = 4$, using 64 Trotter time slices. We computed the energy $E_0 = \langle H \rangle$ and the spin-spin correlation function $\langle S(0,0)S(2,2) \rangle$ which measures the coherence of the spins at site $(0,0)$ and site $(2,2)$. In Fig. 3 we depict the relative error, defined as the standard deviation devided by the mean value, as a function of the total number of Monte Carlo steps. We observe an algebraic decay of the relative error of the total energy and the spin-spin correlation, respectively. A minimal number of Monte Carlo steps is needed to obey the $1/\sqrt{N}$ law. For the same number of Monte Carlo steps, the relative error of the energy is always smaller than the one of the correlation function. This clearly indicates that it takes much more steps to sample long range order to sufficient accuracy than a global quantity as the total energy. Being aware of that Markov chains are computed in parallel, it is obviously a better strategy to increase the number of Markov chains while shortening their length than doing one very long run.

6 Summary

We have demonstrated that the quantum Monte Carlo algorithm can efficiently be implemented on a massively parallel architecture. We discussed how improved performance is obtained by parallizing over the number of Monte Carlo steps. This strategy allows us to reduce interprocessor communication and to increase processor utilization, the two major considerations for efficient massively parallel computation. Applying these ideas to the two-dimensional Hubbard model, we presented a few physical results indicating that the size of the error bars can significantly reduced by parallelization. Keeping the total number of Monte Carlo steps fixed, it appears to be of advantage to cut one long Markov chain into pieces which are executed in parallel.

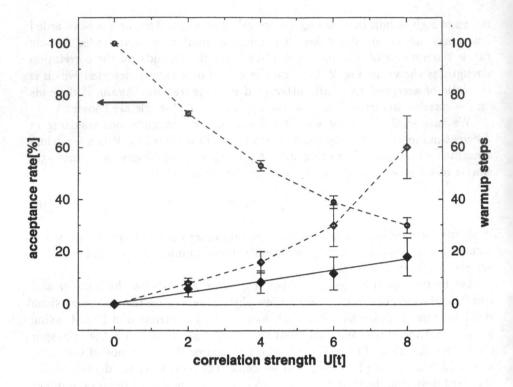

Fig. 2. The acceptance rate (circles), warmup steps (open diamonds) and normalized warmup steps (filled diamonds) versus the correlation strength U for the 4x4 Hubbard model at half filling. We computed 64 Markov chains in parallel and used 64 Trotter slices. Trial configurations were chosen by flipping 25% of the σ-fields for one Trotter slice. Lines are guide to the eye.

Acknowledgements

We would like to thank W. von der Linden and M. Frick for many useful and stimulating discussions. The calculations were performed at the CM-5 of the Gesellschaft für Mathematik und Datenverarbeitung mbH (GMD) at Sankt Augustin, Germany; we would like to acknowledge GMD for their support.

References

1. J. W. Negele and H. Orland, *Quantum Many-Particle Systems*, edited by Frontiers in Physics, (Addison-Wesly Publishing Company, 1987, p.400 ff.); M. H. Kalos, Phys. Rev. **128** (1962) 1791; M. H. Kalos, D. Levesque and L. Verlet, Phys. Rev. A **9** (1973) 2178.
2. W. v.d. Linden, Phys. Rep. **220** (1992) 53.

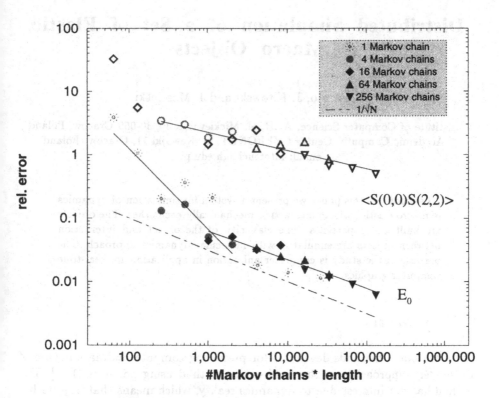

Fig. 3. The relative error of the total energy (filled symbols) and the spin-spin correlation (open symbols) for different numbers of total Monte Carlo steps. The same symbols belong to the same number of Markov chains varying only their length. We used the same Hubbard model as in Fig. 2. Lines are guide to the eye.

3. J. Hubbard, Proc. R. Soc. **A 276** (1963) 238.
4. G. Sugiyama and S. E. Koonin, Ann. Phys. **168** (1986) 1.
5. H. F. Trotter, Prog. Am. Math. Soc. **10** (1959) 545; M. Suzuki, Prog. Theor. Phys. **56** (1976) 1454; M. Suzuki, J. Math. Phys. **26** (1985) 601; M. Suzuki, Phys. Rev. Lett. **146** (1990) 319.
6. J. Hubbard, Phys. Rev. Lett. **3** (1959) 77; J. E. Hirsch, Phys. Rev. **B 28** (1983) 4059.
7. Connection Machine Model CM-5 Technical Summary, November 1993.
8. R. T. Scalletar, R. M. Noack and R. R. P. Singh, Phys. Rev. **B 44** (1991) 10502.

Distributed Simulation of a Set of Elastic Macro Objects

R. Wcisło, J. Kitowski and J. Mościński

[1] Institute of Computer Science, AGH, al. Mickiewicza 30, 30-059 Cracow, Poland
[2] Academic Computer Centre CYFRONET, ul. Nawojki 11, Cracow, Poland
email: kito@uci.agh.edu.pl

Abstract. In this paper we present a system for simulation of dynamics of macro elastic objects interacting mechanically each other. The objects are built up of particles. The elasticity of the object and interaction between objects are simulated with molecular dynamics approach. The purpose of the study is computer animation in application for real-time computer graphics.

1 Introduction

There are many methods developed for producing computer animation. One of the modern approaches in this field is the method using particles [1, 2]. The method has got interest due to dynamics reality, which means that objects behave according to physical laws (contrary to observer imagination only). Other features of the method include speed and fluency for producing smooth animation. Modern parallel architectures are powerful enough at present to reproduce the scene dynamics in real-time.

In the paper we present main ideas and sample results from the system which goal is real-time animation and on-line visualisation of elastics macro objects which can interact each other. The objects can be deformed or broken during the time of animation. The movement of every object is simulated independently using molecular dynamics method; the same technique is used for simulation of mutual interactions between the objects.

In order to speed up the simulation, calculations are organized in parallel. One of the architectures applied for those purposes can be a cluster of workstations. Due to difference of load needed for simulation of dynamics of each object load-balancing is necessary.

2 Base assumptions of the system

In the system the following elements are introduced:

- macroscopic objects,
- walls,
- external gravitational field,

which form the scene of animation.

The objects are characterized by:

- mass and uniform or nonuniform density distribution,
- initial velocity,
- elasticity by means of Young's modulus,
- interactions with other objects or with the walls,
- object dynamics influenced by the gravitational field,
- similar size of the objects.

Granularity of the object is not taken into account.

The walls are described by the following characteristics:

- infinite plane,
- infinite mass,
- motion with constant velocity perpendicular to the system coordinates,
- interactions with objects only.

3 System description

The operation of the system consists of several stages (see Fig.1).

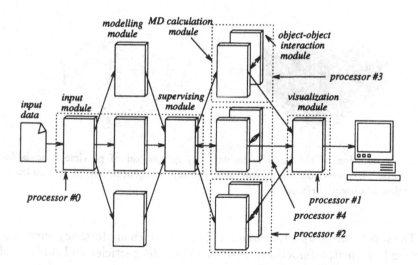

Fig. 1. Operation stages of the system executed in parallel.

In the modelling stage the introduced objects get particle representation (particle model). This representation is applied further in simulation stage. On-line visualisation is invoked during the simulation. For particle model generation, complex objects (see Fig.2 as an example) are divided into subobjects of cuboid shape (see Fig.3).

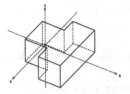

Fig. 2. Example of the complex object.

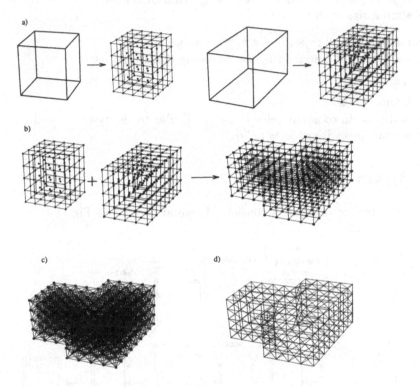

Fig. 3. Generation of the particle model. a) generation of particle models for the subobjects, b) consolidation of the subobjects, c) generation of interactions, d) generation of object surfaces.

The subobjects get particle models (Fig.3a), which are further joined together (see Fig.3b). Next, interactions between the model particles and surfaces of the object are generated (Fig.3c,d).

In the simulation stage four kinds of interactions are applied:

1. interactions within each object represented by the particle model,
2. interactions between objects,
3. object–wall interactions,
4. gravity.

The interactions are evaluated each timestep, which simulates dynamics of the scene.

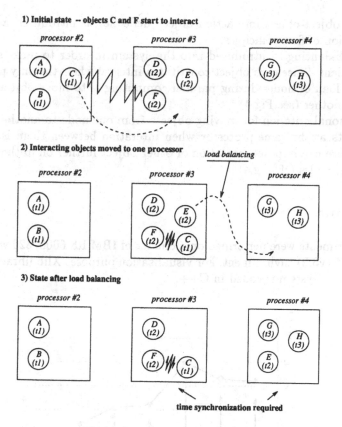

Fig. 4. Load-balancing scheme.

Evaluation of interactions between particles within each object model is based on the list of particle links generated at the modelling stage. Two particle model of interactions is adopted with a given potential

$$V_{i,j}(r_{i,j}) = E_{i,j}(r_{i,j} - \frac{r_{i,j}^2}{2r_{0;i,j}}) \tag{1}$$

where i, j – particle indices, $r_{i,j}$ – distance between i and j particles, $E_{i,j}$ – constant, representing elasticity, $r_{0;i,j}$ – initial distance between i and j particles. Thus the pair force is harmonic one.

Taking $E_{i,j} = E = const$ for every pair (i, j) then perpendicular force acting on the object surface, S, is $F = SE\frac{\Delta r}{r_0}$, thus E states for Young's modulus.

Newtonian equations of motion are solved taking into account not only forces acting on each particle but also other elements of animated scene. Gravitational field is introduced via modification of particle velocities in direction of the field. Interactions with walls are simulated using the ideal particle deflection from the wall. For object-object interaction object deflection is introduced according to conservation of first principles.

Load-balancing is introduced into the system in order to make simulations more efficient. Since each object constitutes integral whole the only possiblity of changing load of nodes during parallel computing is to move objects from one node to another (see Fig.4).

Additional criterion for moving objects from one node to another is to keep the objects at the same processor when interation between them is confirmed. This feature can impose translation of other object further on in the network of processing nodes (cf. Fig.4).

4 Results

The experiments were performed on a cluster of IBM RS 6000/320 workstations with PVM (v3.2) environment. For visualisation purposes Xlib library calls were applied. The system is coded in C++.

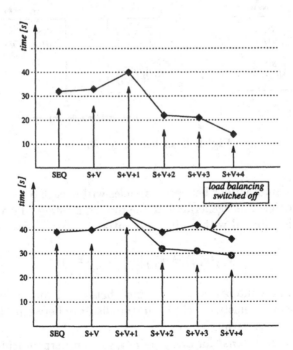

Fig. 5. Execution time for 4 objects with no mutual interactions (upper) and with interactions including load-balancing (down).

In Fig.5 comparison of execution wall-clock time for distributed computing is shown for scene consisting of 4 objects. Two cases are investigated: with objects which do not mutually interact and with objects interacting each other. In addition, performance of load-balancing is also depicted. *SEQ* represents sequential

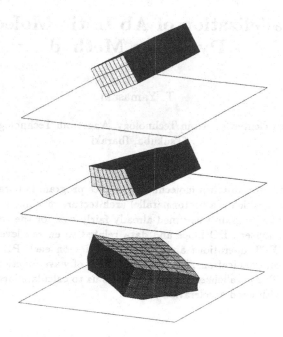

Fig. 6. Example of an elastic cuboid droping on the wall.

version of the code. $S+V+k$ means that the program uses $2+k$ processors, from which one processor (mentioned by S) is a supervising processor (processor #0 in Fig.1) which role is also to generate particle models at the modelling stage, second processor (mentioned by V) executes visualisation module (processor #1 in Fig.1) and k processors are used for object simulation (processors #2-4 in Fig.1). For $k = 0$ one processor is used for both: supervision and simulation. Gain of distributed computing for object interacting each other was obtained in this case for $k > 1$ using the load balancing algorithm.

Results of the real-time animation of simple cuboid are presented in Fig.6 in chosen timesteps. The particles are put in nodes of the shown grid. One can verify, that even for small number of particles the realistic effect was obtained. In future medium with viscosity will be introduced into the system which could improve reality of animation.

Acknowledgments
The work was partially supported by KBN Grant PB 2 P302 073 05.

References

1. Sinkwitz R., C.: Particle motion simulation - a parallel distributed workstation application for real-time, *Future Generation Computer Systems*, **8** (1992) 43-47.
2. Wcisło R., Dzwinel W., Kitowski J., Mościński J.: Real-time animation using molecular dynamics methods, *Machine Graphics and Vision*, **3** 1/2 (1994) 203-210.

Parallelization of Ab Initio Molecular Dynamics Method

T. Yamasaki

Joint Research Center for Atom Technology, Angstrom Technology Partnership,
Tsukuba, Ibaraki

Abstract. An ab-initio molecular dynamics program is parallelized on
VPP500/32 which is a vector parallel architecture machine with 32 processors. The program was tuned already fairly well for the conventional
vector computers. DO loops and data related to energy levels are partitioned. FFT operations are executed locally on each PE for saving
network communications, but multiple FFTs of wave functions are parallelized. The parallelized program enables us to calculate large material
systems with good acceleration ratios.

1 Introduction

A large computer system was introduced into Joint Research Center for Atom
Technology (JRCAT) in 1994. The system composed mainly of powerful four
computers: (1) a vector parallel computer VPP500, (2) a massively parallel computer CM-5E,(3) a general purpose server DEC10000/660, and (4) a graphic
work station IRIS ONYX which is connected to a video recording system. Features of these computers are displayed on the Table.1. Our main targets are to
investigate atomic-scale physics of materials like defect and growth of Si(001)
surface [1, 2], organic molecules [3], etc. by using a first principles molecular
dynamics method.

Table 1. Features of JRCAT computers

	VPP500	CM-5E	DEC10000/660	IRIS ONYX
The number of CPUs, nodes, or PEs	32 PEs	128 nodes	6CPUs	4CPUs
Peak performance	51.2 GFLOPS	20.48 GFLOPS	641 SPECint92 1179 SPECfp92	328 SPECint92 344 SPECfp92 2.1 M 3D vec./sec
Main memory	8.0GB	16.0GB	1GB	1GB
Extended memory	8.0GB			
Disk storage	75GB	44GB	161.7GB	18GB
Remarks	· Vector-parallel	· Massively parallel	Server machine	· AVS · Video system

After the introduction of the computer system, we struggled to parallelize on the VPP500 a program which was existing and tuned fairly well for conventional vector architecture computers. In this article we will describe details of the parallelization.

2 Ab Initio Molecular Dynamics Method

2.1 Algorithm

The starting point of the Car-Parrinello [4] method is a Lagrangian as follows.

$$L = \mu \sum_{i=1}^{N_e} < \dot{\Psi}_i | \dot{\Psi}_i > + \frac{1}{2} \sum_{n=1}^{N_a} M_n |\dot{\mathbf{R}}_n|^2$$

$$-E[\{\Psi_i\}, \{\mathbf{R}_n\}] + \sum_{j=1}^{N_e} \lambda_{j,i}(< \Psi_i | \Psi_j > - \delta_{i,j}) \tag{1}$$

Where,

μ fictitious mass of an electron

$|\dot{\Psi}_i >$ fictitious velocity of a wave function

N_e the number of electronic states

N_a the number of ions

M_n mass of an ion

$\mathbf{R}_n$ coordination of an ion

$\lambda_{j,i}$ Lagrange's multiplier

which is introduced for orthonormalization of wave functions,

and $E[\{\Psi_i\}, \{\mathbf{R}_n\}]$ represents total energy of a system.

From Eq.(1), we can derive Euler-Lagrange's equations of motion for electrons and ions as follows.

$$\mu |\ddot{\Psi}_i >= -H|\Psi_i > + \sum_{j=1}^{N_e} \lambda_{j,i}|\Psi_j > \tag{2}$$

$$M_n \ddot{\mathbf{R}}_\mathbf{n} = -\nabla_n E[\{\Psi_i\}, \{\mathbf{R}_n\}] \tag{3}$$

Here, H is a Kohn-Sham Hamiltonian [5] and is a functional of charge density $\rho(\mathbf{r})$ which is composed of the wave functions.

$$\rho(r) = \sum_{i=1}^{N_e} f(i)|\Psi_i(r)|^2, \tag{4}$$

where $f(i)$ is an occupation number for the state labeled with i.

When electronic states reach to their most stable points, their artificial accelerations $|\ddot{\Psi}_i>$ are zero and if the λ matrix is diagonalized Eq.(2) becomes to be equivalent to a conventional eigen value equation of

$$H|\Psi_i> = \epsilon_i|\Psi_i> . \tag{5}$$

As gradually reducing kinetic energies of electrons and ions, time evolution of wave functions and ionic coordinates according to Eq.(2) and Eq.(3) will bring the system to its global minimum point. Actually speaking, other types of methods, for example a conjugate gradient or a steepest descent, are used often in spite of Eq.(2) to update the wave functions because of efficiency. We adopted a method proposed by Williams and Soler [6],

$$\mu|\dot{\Psi}_i> = -(H - \lambda_{i,i})|\Psi_i> \tag{6}$$

which is a modification of the steepest descent method. We orthonormalize wave functions by using the Gram-Schmidt algorithm in compensation for the omission of non-diagonal elements of the λ matrix.

The Car-Parrinello method saves CPU time and memory greatly compared to the traditional first principles methods which diagonalize the Hamiltonian matrix in Eq.(5) to obtain electronic structures.

We use non-empirical pseudopotentials [7, 8] which have separated forms according to the prescription of Kleinman and Bilander [9]. We can use two kinds of pseudopotentials: a norm-conserving type [10] and the Vanderbilt type [11]. Wave functions are expanded with plane waves. The advantage of plane wave expansion is efficiency of vector operations and applicability of fast Fourier transformation (FFT). By using FFT operations we can save further the CPU time used for the calculation of Eq.(2) or Eq.(6) and Eq.(4).

2.2 Time-Consuming Operations

FFT operations are executed frequently and share dominant part of a whole CPU time. For example, in a case of diamond surface(C(001)2x2) FFT operations consume about 40% of the whole CPU time. In cases of Si system, where a norm conserving pseudopotential is used, the share of FFT operations is much more than the case of diamond surface. Sometimes it reaches to 80%. The Gram-Schmidt orthonormalization is another costly operation. Parallelization of this operation is a difficult problem, which will be described in later part of this article.

3 Parallelization

3.1 VPP500

The VPP500 (Fig.1) introduced into the JRCAT has 32 processing elements(PEs) and 2 control processors(CPs). Each PE consists of a data transfer unit, a memory unit of 256MBytes, a scalar unit, and a vector unit of 1.6GFLOPS processing speed. The PEs are connected together via a crossbar network with high bandwidth of 400 and 800 MB/second for unidirectional and bidirectional data exchange, respectively. The CPs control communications among the back-end PEs and between the front-end machine(= Global System Processor:GSP) and the back-end PEs. The GSP contains a semiconductor memory (= a system storage unit:SSU) of 8GB and an array disk of 75GB.

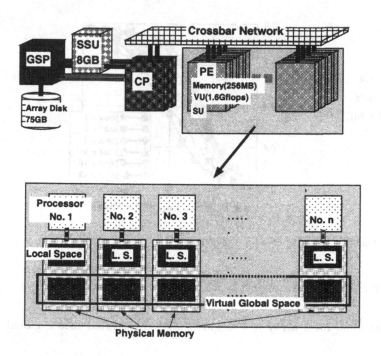

Fig. 1. Physical architecture of the VPP500 (above) and logical architecture of processor elements (below).

We use a parallelization language of VPP FORTRAN 77, which is an extension of the standard FORTRAN 77. By means of insertions of compiler directives (which begin with !xocl or *xocl) we can parallelize conventional FORTRAN programs, and make them concurrent ones for conventional architecture computers and the VPP500, because the compiler directives are recognized as comments by conventional systems. Although the VPP500 is a machine of distributed memory architecture, the VPP FORTRAN 77 allows us to use a global memory space also. The global memory is virtual and is physically distributed onto PEs (Fig.1).

Each PE is equivalent to a vector processing supercomputer. Fig.2 shows results of CPU timing for a kind of two dimensional vector operations. The vector processing speed of each PE almost reaches to the catalog peak performance in most efficient cases and reaches to 1 GFLOPS when the inner do-loop length exceeds 2048 even though outer do-loop length is 1. We measured CPU timing of other kinds of dot product operations and found that 2048 is a magic number of the vector lengths. In most cases the processing speed of a dot product operation reaches to its maximum when the inner DO loop length is around 2048. In other words, one can expect the dot product operations whose do-loop lengths exceed 2048 can be processed efficiently enough.

```
do 1 j = 1, jmax
  do 2 i = 1, imax
    a(i,j) = t(j) + s(j)*b(i,j)
2   continue
1 continue
```

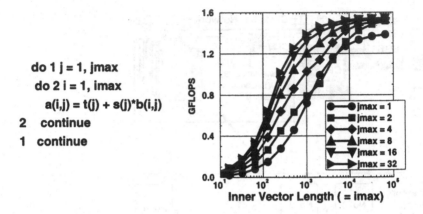

Fig. 2. Relationship between DO loop lengths and vector processing speeds

We use three kinds of data transportation via the crossbar network.

1. Transpose
   A two dimensional local array is transposed and substituted into a global array.
2. Broadcast
   A local array or a variable on a specified PE is broadcasted over all other PEs.
3. Global Summation
   Data among PEs are summed over and unified.

Bandwidth for network communications is about 400 MB/second which is broad enough, but the latencies of the communications cost about 30 to 100 micro seconds. This fact means data size of network communications is preferable to be large enough and the number of the communications should be minimized, otherwise efficiency of the communications would be poor. We measured time used for the communications and deduced the bandwidths and the latencies. Results for the global summations with 4 PEs are plotted in Fig.3. The graph of left hand side means the latency is about 40 micro seconds, and that of right hand side shows data transportation is not at all efficient when data size is below 1 KB. Similar tendencies are observed for the other network communications.

We need to reduce the number of times of network communications as small as possible. Following remedies are useful.

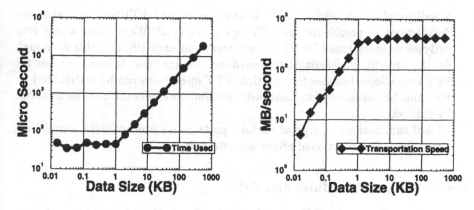

Fig. 3. Performance of global summations

1. Plural data transportations should be put together by using a work array if possible.
2. A data transportation at the most inner DO-loop in a multiple DO-loops operation should be driven as outer as possible.

Especially the latter remedy is very important. By this way, we may be able to reduce the number of the communications several orders of magnitude.

3.2 Strategy

Enhancement of vector processing speed

Prescriptions for conventional vector machines are also valid for VPP500 as follows:

1. The most inner do-loop lengths should be as large as possible. 2048 is a criterion number.
2. Successive data access is most efficient.
3. Vector operations in the most inner do-loops should be dense to a certain degree. Loop unrolling is an efficient way.
4. Compiler directive for vectorization is effective sometimes.

Although our program was already tuned fairy well for ordinary vector architecture machines, there is room to be tuned with the prescriptions mentioned here.

Minimization of frequency of network communications

1. Our program executes three dimensional FFT operations very frequently. Because latencies of the network communications are not negligible as described above and FFT operations are global, it is hard to expect to have

a well parallelized efficient FFT subroutine on the VPP500. Therefore we decided not to parallelize the FFT operation itself. We can use about fifty MBytes memory space for FFT work arrays on each PE, and this is enough for the present. Parallelization according to energy bands is convenient for FFT operations, because the multiple FFT operations can be distributed on PEs and be executed simultaneously without any network communications by this way.

2. Global summations emerge at various parts as we proceed data parallelizations. Remedies mentioned above are effective.

Partitioning of procedures and data

We mainly partition four DO variables:energy level, G-vector, indexes of ions, and FFT meshes.

1. Energy level
   The calculation of Eq.(6) is appropriate to be parallelized according to energy levels. Arrays of wave function coefficients, energy eigen values, and occupation numbers are partitioned in the same manner.

2. G-vector
   In the case of norm-conserving pseudopotentials, charge density is expanded with plane waves whose number is about 8 times of that is used to expand wave functions. Partitioning of charge density is effective if the partitioned DO loop length exceeds 2048. We decided to partition the arrays and DO loops of charge density and related values, like structure factor, exchange-correlation energy, etc. for the sake of large systems.

3. Index of ions
   Forces which are acting on atoms are partitioned.

4. FFT meshes
   We apply the generalized gradient approximation (GGA) [12] beyond the local density approximation (LDA). According to a procedure proposed by White and Bird [13], the GGA is implemented on our program. Because this method uses FFTs to evaluate the gradients of charge densities, we need to prepare some extra FFT work arrays. For large systems, we can't have these arrays as local ones. We need to partition these arrays except one FFT work array. FFT data are distributed onto PEs and gathered up in the local FFT work array when required. FFT operations themselves are not parallelized.

Concurrent Program

It is desirable to make a concurrent program which runs on other computers also, otherwise plural programs originated from a common one would evolve on various kinds of computers. This situation is inconvenient to implement new functions and to fix bugs. Fortunately, if we don't use service procedures specific to VPP FORTRAN, the program would be a concurrent one which can be compiled on other computers. However, we can't avoid to use some service procedures and need to modify algorithm for the sake of enhancement of parallelization efficiency. This kind of modification is often disadvantageous to other systems. We can resolve this problem by using C preprocessor.

3.3 Gram-Schmidt orthonormalization

The Gram-Schmidt orthonormalization is difficult to parallelize because it requires frequent network communications and grain sizes of operations partitioned on PEs changes as the procedure proceeds. We implement two kinds of parallelization:

1. Transpose transportation of wave functions.
   We can parallelize the algorithm by partitioning DO-loops and arrays of wave functions by G-vectors, when the number of plane waves used to expand wave functions is large enough as partitioned number exceeds 2048. However the other subroutines prefer the partitioning according to energy levels, we need to prepare two kinds of arrays for wave functions and to transpose these data via the network. This method is effective for calculations of large material systems and applicable for the cases that enough memory space is available.
2. Cyclic partitioning of energy levels.
   Cyclic partitioning of wave functions by energy levels make the grain sizes of operations on the PEs more uniform than the ordinary band partitioning. This method dose not require extra work arrays. Problem is that cyclic partitioning reduces the parallelization efficiencies in the other subroutines. When memory space is restricted, we use this parallelized subroutine.

4 Performance Measurement

4.1 Acceleration ratio

Fig. 4 shows the acceleration ratio for the case of diamond surface (C(001) 4×4) optimization. The calculational condition is as follows:

```
             Cut-off energy for wave functions = 36Ry
  The number of plane waves for wave functions = 35,638
            Cut-off energy for charge density = 144Ry
The number of plane waves for charge densities = 284,059
                                     FFT mesh = 100 x 72 x 72
                           The number of ions = 160
               The number of sampling points = 2
                The number of energy levels = 480/k-point
                                   Potential = Vanderbilt type
```

The CPU timings were measured for each subroutine with 8, 16, and 24 PEs. The normalized acceleration ratios derived from the measured data at the second iteration of the self-consistent calculation are plotted in the graph, because the values at the second iteration are close to the average values. The acceleration ratios of total and the three most time consuming subroutines are shown in the graph. MSDV,GRMSMD and EIGEN0 are the subroutines for the update of wave functions by the steepest descent method, the Gram-Schmidt orthonormalization, and the calculation of energy eigen values, respectively. Although the subroutine for the Gram-Schmidt orthonormalization is not parallelized efficiently compared to the other subroutines, the acceleration ratio of the total at 24PEs is good. FFT operations shares the CPU time of 17.8%, 17.0% and 16.5% for 8PEs, 16PEs and 24PEs, respectively.

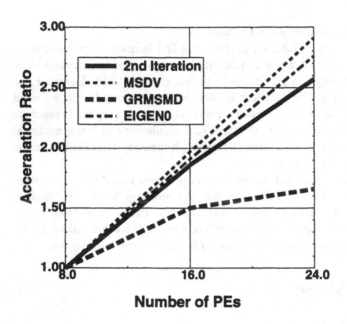

Fig. 4. Acceleration rations

We also measured timings with 8PEs for a Si surface system which contains 248 Si atoms, and found average performance of one PE exceeds 630 MFLOPS.

5 Concluding Remarks

The Car-Parrinello method fits vector parallel architecture machines like the VPP500. The reasons are

1. Partitioning of energy levels is easy. Since in most cases the number of energy levels is larger than 100, grain sizes of data partitioned by the energy levels would be uniform. Parallelization of conventional first principles programs which solve eigen value equations by Hamiltonian matrix diagonalization would be harder by far than that of the Car-Parrinello programs.

2. Each PE of the VPP500 has enough memory space of 256MB to do FFT operations locally.

We succeeded in parallelizing a Car-Parrinello program efficiently.

When we challenge to calculate much larger material systems, new problems emerge. A FFT array overflows local memory, so we will require a parallelized FFT subroutine. Other local arrays also swell so large that we have to partition and change them into global arrays. This is a time consuming job. The VPP500's latency periods of network communications are still long for some algorithms like the Gram-Schmidt orthonormalization. If the periods are shorter, many kinds of subroutine would be parallelized easier than now.

This work was partly supported by New Energy and Industrial Technology Development Organization(NEDO).

References

1. T. Uda.: Phys. Rev. Lett. to be published.
2. T. Yamasaki and T. Uda and K. Terakura.: '95 JRCAT International Symposium on Nanoscale Self Organization. p186
3. T. Miyazaki et al.: Phys. Rev. Lett.**74** (1995) 5104
4. R. Car and M. Parrinello.: Phys. Rev. Lett. **55** (1985) 2471
5. W. Kohn and L. J. Sham.: Phys. Rev. **140** (1965) A1133
6. A. R. Williams and J. Soler.: Bull. Am. Phys. Soc.**32** (1987) 562
7. G. B. Bachelet, D. R. Hamann, and M. Schlüter.: Phys. Rev. B.**26** (1982) 4199
8. D. R. Hamann.: Phys. Rev. B.**40** (1989) 2980
9. L. Kleinman and D. M. Bylander.: Phys. Rev. Lett.**48** (1982) 1425
10. N. Troullier and J. L. Martin.: Phys. Rev. B.**43** (1991) 1993
11. D. Vanderbilt.: Phys. Rev. B.**41** (1990) 7892
12. J. P. Perdew.: Physica B. **172** (1991) 1
13. J. A. White and D. M. Bird.: Phys. Rev. B.**50** (1994) 4954

Parallel Computations
with Large Atmospheric Models

Z. Zlatev[1], I. Dimov[2], K. Georgiev[3], J. Waś niewski[4]

[1] National Environmental Research Institute, Frederiksborgvej 399, P.O. Box 358, DK-4000 Roskilde, Denmark. email: luzz@sun2.dmu.dk.
[2] Institute of Informatics and Computer Technology, Bulgarian Academy of Sciences, Akad. G. Bonchev str., Bl. 25-A, 1113 Sofia. Bulgaria. email: dimov@iscbg.acad.bg
[3] Institute of Informatics and Computer Technology, Bulgarian Academy of Sciences, Akad. G. Bonchev str., Bl. 25-A, 1113 Sofia. Bulgaria. email: georgiev@iscbg.acad.bg.
[4] Danish Computing Centre for Research and Education (UNI•C), DTU, Bldg. 304, DK-2800 Lyngby, Denmark. email: jerzy.wasniewski@unidhp1.uni-c.dk

Abstract. Large atmospheric models appear often in important applications. Long-range transport of air pollutants is one of the phenomena that can be studied by using such models. These models lead, after some kind of discretization, to very huge systems of ordinary differential equations (up to order of 10**6). The use of modern parallel and/or vector machines is a necessary condition in the efforts to handle successfully big air pollution models. However, this is often far from sufficient. One should also optimize the code in order to be able to exploit fully the great potential power of the modern high-speed computers. The air pollution code developed at the National Environmental Research Institute has been optimized for several different types of high-speed computers (vector machines, parallel computers with shared memory and parallel computers with distributed memory; including in the last group the massively parallel computers).

Results obtained on two computers will be discussed in this paper. The first computer is a vector processor, a CRAY Y-MP C90A computer. The second one is a parallel computer with shared memory, a POWER CHALLENGE product of Silicon Graphics.

Key words. Air pollution models, partial differential equations, ordinary differential equations, numerical algorithms, parallel and vector computers, speed-up, efficiency.

1   Need for large air pollution models

Air pollutants emitted by different sources can be transported, by the wind, on long distances. Several physical processes (diffusion, deposition and chemical transformations) take place during the transport. Regions that are very far from the large emission sources may also be polluted. It is well-known that the atmosphere must be kept clean (or, at least, should not be polluted too much). It is also well-known that if the concentrations of some species exceed certain acceptable (or critical) levels, then they may become dangerous for plants, animals and humans.

Mathematical models are needed in the efforts to predict the optimal way of keeping the air pollution under acceptable levels. It should be emphasized here that the mathematical models are the **only** tool by the use of which one can predict the results of many different actions and, moreover, one can attempt to choose the best solution (or, at least, a solution which is close to the best one) in the efforts to reduce the air pollution in an optimal way. The mathematical models are necessarily large (the transport of air pollutants is carried out over long distances and, thus, the space domains are very large).

2   The Danish Eulerian Model

The work on the development of the Danish Eulerian Model has been initiated in 1980. First a simple transport scheme was developed (one pollutant only and without chemical reactions). The next step was the development of a simple model with two pollutants (and linearized chemical reactions). An experimental model containing ten pollutants and non-linear chemical reactions (including here photochemical reactions) was the third step. The first operational version of the Danish Eulerian Model was based on a chemical scheme with 35 pollutants. Some experiments with chemical schemes containing 56 and 168 pollutants are carried out at present. Different versions of the Danish Eulerian Model are discussed in [3] and [6].

The reliability of the operational model with 35 pollutants has been tested by comparing model results both with measurements taken over land ([4], [5]) and with measurements taken over sea ([2]). Test-problems, where the analytical solution is known, have been used to check the accuracy of the numerical algorithms ([3]).

3   Mathematical description of the model

The Danish Eulerian model is described by a system of PDE's (the number of equations in this system being equal to the number of pollutants involved in the model):

$$\frac{\partial c_s}{\partial t} = -\frac{\partial(u c_s)}{\partial x} - \frac{\partial(v c_s)}{\partial y} - \frac{\partial(w c_s)}{\partial z}$$

$$+ \frac{\partial}{\partial x}\left(K_x \frac{\partial c_s}{\partial x}\right) + \frac{\partial}{\partial y}\left(K_y \frac{\partial c_s}{\partial y}\right) + \frac{\partial}{\partial z}\left(K_z \frac{\partial c_s}{\partial z}\right)$$

$$-(\kappa_{1s} + \kappa_{2s}) c_s$$

$$+ E_s + Q_s(c_1, c_2, \ldots, c_q)$$

$$s = 1, 2, \ldots, q. \tag{1}$$

The different quantities that are involved in the mathematical model have the following meaning:

- the concentrations are denoted by c_s;

- u, v and w are wind velocities;

- K_x, K_y and K_z are diffusion coefficients;

- the emission sources in the space domain are described by the functions E_s;

- κ_{1s} and κ_{2s} are deposition coefficients;

- the chemical reactions involved in the model are described by the non-linear functions $Q_s(c_1, c_2, \ldots, c_q)$.

The non-linear functions Q_s are of the form:

$$Q_s(c_1, c_2, \ldots, c_q) = -\sum_{i=1}^{q} \alpha_{si} c_i + \sum_{i=1}^{q} \sum_{j=1}^{q} \beta_{sij} c_i c_j, \qquad s = 1, 2, \ldots, q. \qquad (2)$$

This is a special kind of non-linearity, but it is not clear how to exploit this fact during the numerical treatment of the model.

It is clear from the above description of the quantities involved in the mathematical model that all five physical processes (advection, diffusion, emission, deposition and chemical reactions) can be studied by using the above system of PDE's.

The mathematical model described by (1) must be considered together with appropriate initial and boundary conditions ([3]).

The Danish Eulerian Model has mainly been run as a two-dimensional model until now, which can formally be obtained from (1) by removing the derivatives with regard to z. The computational difficulties connected with the development of a three-dimensional version will be discussed in the following sections. Some numerical results will be presented.

4 Numerical algorithms used in the 3-D version of the Danish Eulerian Model

The mathematical model described by the system of PDE's (1) is split to five sub-models according to the physical processes involved in the model. The sub-models are: horizontal advection sub-model, horizontal diffusion sub-model, deposition sub-model, chemistry sub-model (the emission being included here) and vertical exchange sub-model. The splitting procedure leads to a series of parallel tasks. In the first four sub-models, each horizontal plane can be treated as a parallel task. In the fifth sub-model, each vertical line can be treated as a parallel task. This shows that the splitting procedure leads to a lot of natural parallelism. More details about the splitting procedure and the possibilities to run the model in parallel are given in [3] and [6].

Various numerical algorithms have been tried in the treatment of the different sub-models. The numerical algorithms currently used are listed below; the algorithms used in the numerical treatment of the model are discussed in [3].

- A pseudospectral discretization algorithm is used in the horizontal advection sub-model. Experiments with other methods (finite elements, corrected flux transport algorithms and semi-Lagrangian methods) are presently carried out.

- A semi-analytical approach based on expansions of the unknown functions in Fourier series is used in the horizontal diffusion sub-model. Some experiments with finite elements are also carried out.

- The splitting procedure leads to a deposition sub-model consisting of independent linear ODE's; these are solved exactly.

- The QSSA (the quasi-steady-state-approximation) is used when the chemical sub-model is handled numerically. Experiments with several classical time-integration algorithms are also carried out.

- Linear finite elements are used in the treatment of the vertical exchange sub-model.

5 Computational difficulties connected with 3-D version of the Danish Eulerian Model

The development of three-dimensional air pollution models leads to huge computational tasks. The computational tasks are very large even if the model is considered as a two-dimensional model. Assume that some splitting procedure has been applied and that the spatial derivatives are discretized by using an appropriate numerical algorithm (see the previous section or [3], [6]). Then the system of PDE's (1) is transformed into five systems of ODE's corresponding to the five sub-models discussed in the previous section. The five ODE systems have to be treated successively at every time-step. The number of equations in each of these ODE systems is equal to the product of the number of grid-points and the number of pollutants. In the case where the model is considered as two-dimensional, the numbers of equations in the ODE systems for different space discretizations and for different chemical schemes are given in Table 1. The sizes of the grid-squares for the three grids used in Table 1 are given in Table 2. The total size of the square space domain used at present in the Danish Eulerian Model is (4800 km × 4800 km); the space domain contains the whole of Europe together with parts of Asia, Africa and the Atlantic Ocean.

Number of pollutants	(32 × 32)	(96 × 96)	(192 × 192)
1	1024	9216	36864
2	2048	18432	73728
12	12288	110592	442368
35	35840	322560	1290240
56	57344	516096	2064384
168	172032	1548288	6193152

Table 1

Numbers of equations per ODE system in the two-dimensional versions of the
Danish Eulerian Model.

1	(32 × 32)	(150 km × 150 km)
2	(96 × 96)	(50 km × 50 km)
3	(192 × 192)	(25 km × 25 km)

Table 2

The sizes of the grid-squares for the three grids used in Table 1.

It should be mentioned here that (i) if a three-dimensional model with ten
vertical layers is used, then all figures that are given in Table 1 must be multiplied
by ten, (ii) 3456 time-steps have been used for a typical run (with meteorological
data for one month + five days to start up the model; [3]) when the (96 × 96
) grid is used and (iii) the chemical reactions lead to a very stiff (and also very
badly scaled) ODE systems; this fact causes extra computational difficulties.

6 Need for high-speed computers

The size of the ODE systems shown in Table 1 (these systems have to be treated
during several thousand time-steps) explains why the use of high-speed computers is absolutely necessary when large air pollution problems are to be handled.
It is even more important to perform carefully the programming work in order to try to exploit better the great potential power of the modern high-speed
computers. As a rule, this is not an easy task (especially on the newest parallel
computers). Finally, it should be emphasized that even when the fastest computers are available and even when the programming work is very carefully done,
it is still not possible, at present, to solve some of the biggest problems listed in
Table 1 (especially in the case where the model is treated as a three-dimensional
model). Therefore faster and bigger computers are needed in order to be able to
treat successfully big air pollution models.

Different versions of the Danish Eulerian Model have been used in runs on
several high-speed computers ([1], [3], [6]). Some results obtained by running the
three-dimensional version on a CRAY Y-MP C90A will be presented in the next

section. Preliminary results obtained by running three important modules of the model (the advection sub-model, the chemistry sub-model and a combination of the advection and the chemistry sub-models) on a POWER CHALLENGE Silicon Graphics computer will be discussed in Section 8.

7 Runs on a CRAY Y-MP C90A computer

The three-dimensional version of the Danish Eulerian Model has until now been run only on a CRAY Y-MP C90A computer. The performance achieved in the different parts of the model (corresponding to the five physical processes) is shown in Table 3. The total computing time and the overall speed of computations, measured in MFLOPS (millions of floating point operations, additions and multiplications, per second), are given in Table 4.

Physical process	Computing time	In percent
Advection	638	22.2
Diffusion	295	10.2
Deposition	138	4.8
Chemistry	1652	57.4
Vertical exchange	50	1.7
Overhead	107	3.7

Table 3
Computing times (in seconds) for the different parts of the model obtained on a CRAY Y-MP C90A computer.

Quantity measured	Value
Computing time	2880
Speed in MFLOPS	304

Table 4
The total computing time (in seconds) and the overall computational speed obtained in a run with one-month meteorological data (plus five days to start up the model) on a CRAY Y-MP C90A computer.

The performance of several subroutines on CRAY Y-MP C90A has recently been improved. The improvements made lead to an increase of the performance of the whole code by about 30%.

8   Runs on a POWER CHALLENGE product of Silicon Graphics

As mentioned above, three modules of the Danish Eulerian model:

1. the chemical sub-model,

2. the transport (advection) sub-model

3. the combination of the advection and the chemical sub-models,

have been run on a POWER CHALLENGE product of Silicon Graphics. It should be emphasized that the results produced by these modules are not directly comparable with the results presented in the previous section, where the whole three-dimensional model has been run. The modules are used to check the accuracy of the numerical methods used in the advection part and in the chemistry part as well as to check the reliability of the coupling procedure between the advection and the chemical parts of the Danish Eulerian Model (see more details in [1], [3] and [6]).

While it is difficult to find a good numerical algorithm for the chemical sub-model, it is not very difficult to achieve parallel computations with any numerical algorithm. Indeed, the computations within the chemical sub-model can schematicaly be described (at every time-step and for any numerical algorithm applied in the chemical sub-model) by the following loop.

DO I=1,NUMBER OF GRID-POINTS
   Perform all chemical reactions at the I'th grid-point.
END DO

It is clear that the number of parallel tasks is equal to the number of grid-points and that the loading balance is perfect. Therefore one should expect to obtain good speed-ups. The results shown in Table 5 show that the efficiency with regard to the speed-ups are very close to the optimal values that could be achieved. Let

– N_p be the number of the processores used,

– t_1 be the computing time obtained on one processor only,

– t_p be the computing time obtained when p processors are used,

then the speed-up and the efficiency (in percent) are defined by the following two formulae

$$Speed-up = \frac{t_1}{t_p}, \tag{3}$$

$$Efficiency - 1 = \frac{100t_1}{N_p t_p}. \tag{4}$$

It is very important to achieved parallel computations, however this is very often not sufficient (in other words even when the computations are carried out in parallel, the performance may be rather low). Therefore another measure is needed in order to achieve a full evaluation of the efficiency of the code. This second measure is the efficiency with regard to the peak performance of the architecture on which the code is run. This quantity is defined by

$$Efficiency - 2 = \frac{100 * MFLOPS}{260 * N_p}. \tag{5}$$

where MFLOPS is the computational speed achieved by the code (millions of floating point operations per second), while 260 MFLOPS is the peak performance of one POWER CHALLENGE processor.

It is seen from Table 6 that the efficiency with regard to the peak performance is rather poor. The results obtained on one processor of CRAY Y-MP C90A are:

$$t_1 = 284, \qquad MFLOPS = 422.5, \qquad Efficiency - 2 = 46.5\%. \tag{6}$$

The POWER CHALLENGE results can perhaps be improved by better tunning of the code and by trying to exploit the cash more efficiently.

The transport (advection) part can also easily be run in parallel. The computations within the transport sub-model can schematicaly be described (again at every time-step and for any numerical algorithm applied in this module) by the following loop.

> DO I=1,NUMBER OF COMPOUNDS
>> *Perform the transport for the I'th compound.*
>
> END DO

The number of parallel tasks is now equal to the number of compounds (in the present case the number of compounds is 56). This may cause problems if the number of processors is large. In the runs presented in Table 7 and Table 8, where the number of processors is up to 8, there are no problems with the above loop. The tasks are not very well balanced for the algorithm used (because there is an iterative process, and for the different compounds the number of iterations may be different). Therefore one should expect lower speed-ups (comparing with the chemical module). This effect can be seen by comparing the results given in Table 7 with those in Table 5. However, However, the speed-ups given in Table 7 are nevertheless very high (the efficiency with regard to speed-ups is higher than 80%, which is an excellent result).

The efficiency obtained with regard to the peak performance of the configurations of the POWER CHALLENGE used in the runs is greater for the transport module (comparing with the chemical sub-model); compare the results given in Table 8 with those in Table 6. Nevertheless, the results show that also here improvements are highly desirable. This becomes very clear by comparing the results given in Table 7 - Table 8 with the results obtained for this module on CRAY Y-MP C90A:

$$t_1 = 205, \qquad MFLOPS = 440.0, \qquad Efficiency - 2 = 48.8\%. \qquad (7)$$

The third module, which is very important in the checks of the accuracy of the coupling procedure (transport with chemistry), is carried out by performing, at every time-step, successively the two loops given above. Therefore one should expect the efficiency results for the third module to be between the efficiency results of the first two modules (the chemical module and the transport module). The results given in Table 9 - Table 10 confirm such a conclusion; compare these results with the corresponding results in Table 5 -Table 8.

Processor	Time	Speed-up	Efficiency-1
1	4576.1	-	100.0%
2	2390.9	1.91	95.6%
4	1179.8	3.88	96.9%
8	602.7	7.59	94.9%

Table 5
Computing times (in seconds), speed-ups and efficiency-1 results obtained during runs of the chemical module on the POWER CHALLENGE product of Silicon Graphics.

Processor	Time	MFLOPS	Efficiency-2
1	4576.1	26.2	10.1%
2	2390.9	50.2	9.7%
4	1179.8	101.7	9.8%
8	602.7	199.1	9.6%

Table 6
Computing times (in seconds), MFLOPS and efficiency-2 results obtained during runs of the chemical module on the POWER CHALLENGE product of Silicon Graphics.

Processor	Time	Speed-up	Efficiency-1
1	1963.5	-	100.0%
2	1041.3	1.89	94.6%
4	594.6	3.30	82.5%
8	305.8	6.42	80.3%

Table 7

Computing times (in seconds), speed-ups and efficiency-1 results obtained during runs of the transport module on the POWER CHALLENGE product of Silicon Graphics.

Processor	Time	MFLOPS	Efficiency-2
1	1963.5	46.0	17.7%
2	1041.3	86.6	16.7%
4	594.6	151.7	14.6%
8	305.6	295.2	14.2%

Table 8

Computing times (in seconds), MFLOPS and efficiency-2 results obtained during runs of the transport module on the POWER CHALLENGE product of Silicon Graphics.

Processor	Time	Speed-up	Efficiency-1
1	5990.3	-	100.0%
2	3040.0	1.97	98.5%
4	1632.2	3.67	91.8%
8	897.4	6.68	83.5%

Table 9

Computing times (in seconds), speed-ups and efficiency-1 results obtained during runs of the third module (chemistry + transport) on the POWER CHALLENGE product of Silicon Graphics.

Processor	Time	MFLOPS	Efficiency-2
1	5990.3	35.0	13.4%
2	3040.0	69.0	13.3%
4	1632.2	128.5	12.4%
8	897.4	233.7	11.2%

Table 10

Computing times (in seconds), MFLOPS and efficiency-2 results obtained during runs of the third module (chemistry + transport) on the POWER CHALLENGE product of Silicon Graphics.

9 Concluding remarks

It is seen that the chemistry is the most time-consuming part of the model. Therefore improvements of the chemical subroutines are most desirable (however, also the performance of the mathematical modules describing the other physical processes are to be improved). It should be emphasised that the three-dimensional model can be treated numerically at present only on a relatively coarse spatial grid, the (32×32) grid, and only for the relatively simple chemical scheme with 35 pollutants. It is necessary to improve both the numerical algorithms and the computer programs for the different modules in order to be able to run also some of the more complicated cases (refined grids and chemical schemes with more pollutants). It is crucial to select (or develop) algorithms that perform well on the new high-speed computers. Some work in these directions is carried out at present.

Acknowledgements

This research was partially supported by NMR (Nordic Council of Ministers), EMEP (European Monitoring and Evaluating Programme), NATO (North Atlantic Treaty Organization), SMP (Danish Strategic Environmental Programme) and the Danish Natural Sciences Research Council.

References

1. BROWN, J., WAŚNIEWSKI, J., ZLATEV, Z.: Running air pollution models on massively parallel machines; Parallel Computing, **21** (1995), 971-991.
2. HARRISON, R. M., ZLATEV, Z., OTTLEY, C. J.: A comparison of the predictions of an Eulerian atmospheric transport-chemistry model with measurements over the North Sea; Atmos. Environ., **28** (1994), 497-516.
3. ZLATEV, Z.: Computer treatment of large air pollution models; Kluwer Academic Publishers, Dordrecht-Boston-London, 1995.
4. ZLATEV, Z., CHRISTENSEN, J., ELIASSEN, A.: Studying high ozone concentrations by using the Danish Eulerian Model; Atmos. Environ., **27A** (1993), 845-865.
5. ZLATEV, Z., CHRISTENSEN, J., HOV, Ø.: A Eulerian air pollution model for Europe with nonlinear chemistry; J. Atmos. Chem., **15** (1992), 1-37.
6. ZLATEV, Z., DIMOV, I., GEORGIEV, K.: Studying long-range transport of air pollutants; Computational Science and Engineering, **1**, No. 3 (1994), 45-52.

Author Index

Lecture Notes in Computer Science

For information about Vols. 1–970

please contact your bookseller or Springer-Verlag

Vol. 1007: A. Bosselaers, B. Preneel (Eds.), Integrity Primitives for Secure Information Systems. VII, 239 pages. 1995.

Vol. 1008: B. Preneel (Ed.), Fast Software Encryption. Proceedings, 1994. VIII, 367 pages. 1995.

Vol. 1009: M. Broy, S. Jähnichen (Eds.), KORSO: Methods, Languages, and Tools for the Construction of Correct Software. X, 449 pages. 1995. Vol.

Vol. 1010: M. Veloso, A. Aamodt (Eds.), Case-Based Reasoning Research and Development. Proceedings, 1995. X, 576 pages. 1995. (Subseries LNAI).

Vol. 1011: T. Furuhashi (Ed.), Advances in Fuzzy Logic, Neural Networks and Genetic Algorithms. Proceedings, 1994. (Subseries LNAI).

Vol. 1012: M. Bartošek, J. Staudek, J. Wiedermann (Eds.), SOFSEM '95: Theory and Practice of Informatics. Proceedings, 1995. XI, 499 pages. 1995.

Vol. 1013: T.W. Ling, A.O. Mendelzon, L. Vieille (Eds.), Deductive and Object-Oriented Databases. Proceedings, 1995. XIV, 557 pages. 1995.

Vol. 1014: A.P. del Pobil, M.A. Serna, Spatial Representation and Motion Planning. XII, 242 pages. 1995.

Vol. 1015: B. Blumenthal, J. Gornostaev, C. Unger (Eds.), Human-Computer Interaction. Proceedings, 1995. VIII, 203 pages. 1995.

VOL. 1016: R. Cipolla, Active Visual Inference of Surface Shape. XII, 194 pages. 1995.

Vol. 1017: M. Nagl (Ed.), Graph-Theoretic Concepts in Computer Science. Proceedings, 1995. XI, 406 pages. 1995.

Vol. 1018: T.D.C. Little, R. Gusella (Eds.), Network and Operating Systems Support for Digital Audio and Video. Proceedings, 1995. XI, 357 pages. 1995.

Vol. 1019: E. Brinksma, W.R. Cleaveland, K.G. Larsen, T. Margaria, B. Steffen (Eds.), Tools and Algorithms for the Construction and Analysis of Systems. Selected Papers, 1995. VII, 291 pages. 1995.

Vol. 1020: I.D. Watson (Ed.), Progress in Case-Based Reasoning. Proceedings, 1995. VIII, 209 pages. 1995. (Subseries LNAI).

Vol. 1021: M.P. Papazoglou (Ed.), OOER '95: Object-Oriented and Entity-Relationship Modeling. Proceedings, 1995. XVII, 451 pages. 1995.

Vol. 1022: P.H. Hartel, R. Plasmeijer (Eds.), Functional Programming Languages in Education. Proceedings, 1995. X, 309 pages. 1995.

Vol. 1023: K. Kanchanasut, J.-J. Lévy (Eds.), Algorithms, Concurrency and Knowlwdge. Proceedings, 1995. X, 410 pages. 1995.

Vol. 1024: R.T. Chin, H.H.S. Ip, A.C. Naiman, T.-C. Pong (Eds.), Image Analysis Applications and Computer Graphics. Proceedings, 1995. XVI, 533 pages. 1995.

Vol. 1025: C. Boyd (Ed.), Cryptography and Coding. Proceedings, 1995. IX, 291 pages. 1995.

Vol. 1026: P.S. Thiagarajan (Ed.), Foundations of Software Technology and Theoretical Computer Science. Proceedings, 1995. XII, 515 pages. 1995.

Vol. 1027: F.J. Brandenburg (Ed.), Graph Drawing. Proceedings, 1995. XII, 526 pages. 1996.

Vol. 1028: N.R. Adam, Y. Yesha (Eds.), Electronic Commerce. X, 155 pages. 1996.

Vol. 1029: E. Dawson, J. Golić (Eds.), Cryptography: Policy and Algorithms. Proceedings, 1995. XI, 327 pages. 1996.

Vol. 1030: F. Pichler, R. Moreno-Díaz, R. Albrecht (Eds.), Computer Aided Systems Theory - EUROCAST '95. Proceedings, 1995. XII, 539 pages. 1996.

Vol.1031: M. Toussaint (Ed.), Ada in Europe. Proceedings, 1995. XI, 455 pages. 1996.

Vol. 1032: P. Godefroid, Partial-Order Methods for the Verification of Concurrent Systems. IV, 143 pages. 1996.

Vol. 1033: C.-H. Huang, P. Sadayappan, U. Banerjee, D. Gelernter, A. Nicolau, D. Padua (Eds.), Languages and Compilers for Parallel Computing. Proceedings, 1995. XIII, 597 pages. 1996.

Vol. 1034: G. Kuper, M. Wallace (Eds.), Constraint Databases and Applications. Proceedings, 1995. VII, 185 pages. 1996.

Vol. 1035: S.Z. Li, D.P. Mital, E.K. Teoh, H. Wang (Eds.), Recent Developments in Computer Vision. Proceedings, 1995. XI, 604 pages. 1996.

Vol. 1036: G. Adorni, M. Zock (Eds.), Trends in Natural Language Generation - An Artificial Intelligence Perspective. Proceedings, 1993. IX, 382 pages. 1996. (Subseries LNAI).

Vol. 1037: M. Wooldridge, J.P. Müller, M. Tambe (Eds.), Intelligent Agents II. Proceedings, 1995. XVI, 437 pages. 1996. (Subseries LNAI).

Vol. 1038: W: Van de Velde, J.W. Perram (Eds.), Agents Breaking Away. Proceedings, 1996. XIV, 232 pages. 1996. (Subseries LNAI).

Vol. 1039: D. Gollmann (Ed.), Fast Software Encryption. Proceedings, 1996. X, 219 pages. 1996.

Vol. 1040: S. Wermter, E. Riloff, G. Scheler (Eds.), Connectionist, Statistical, and Symbolic Approaches to Learning for Natural Language Processing. Proceedings, 1995. IX, 468 pages. 1996. (Subseries LNAI).

Vol. 1041: J. Dongarra, K. Madsen, J. Waśniewski (Eds.), Applied Parallel Computing. Proceedings, 1995. XII, 562 pages. 1996.

Vol. 1042: G. Weiß, S. Sen (Eds.), Adaption and Learning in Multi-Agent Systems. Proceedings, 1995. X, 238 pages. 1996. (Subseries LNAI).

Vol. 1043: F. Moller, G. Birtwistle (Eds.), Logics for Concurrency. XI, 266 pages. 1996.

Vol. 1044: B. Plattner (Ed.), Broadband Communications. Proceedings, 1996. XIV, 359 pages. 1996.

Vol. 1045: B. Butscher, E. Moeller, H. Pusch (Eds.), Interactive Distributed Multimedia Systems and Services. Proceedings, 1996. XI, 333 pages. 1996.

Vol. 1046: C. Puech, R. Reischuk (Eds.), STACS 96. Proceedings, 1996. XI, 690 pages. 1996.

Vol. 1047: E. Hajnicz, Time Structures. IX, 244 pages. 1996. (Subseries LNAI).

Vol. 1048: M. Proietti (Ed.), Logic Program Syynthesis and Transformation. Proceedings, 1995. X, 267 pages. 1996.